Black Swan
beer of W. Australia

Mont Shale
-bituminous sandstone
Tuff
Pearlite Breccia
Franciscan — Unconformed

SECOND EDITION

EVOLUTION OF THE EARTH

ROBERT H. DOTT, JR.
University of Wisconsin, Madison

ROGER L. BATTEN
American Museum of Natural History, New York
and Columbia University

Maps and Diagrams by
RANDALL D. SALE
Cartographic Laboratory
University of Wisconsin

McGRAW-HILL BOOK COMPANY

New York   St. Louis   San Francisco
Auckland   Düsseldorf   Johannesburg
Kuala Lumpur   London   Mexico   Montreal
New Delhi   Panama   Paris   São Paulo
Singapore   Sydney   Tokyo   Toronto

To our parents
and all of our other teachers.

EVOLUTION OF THE EARTH

234567890VHVH79876

Library of Congress Cataloging in Publication Data

Dott, Robert H., Jr.
    Evolution of the earth.

    1. Historical geology.   I.   Batten, Roger Lyman,
joint author.   II.   Title

QE28.3.D68   1976   551.7   75-26760
ISBN 0-07-017619-1

This book was set in Helvetica Light by Black Dot, Inc.
The editors were Robert H. Summersgill and Carol First;
the designer was Nicholas Krenitsky;
the production supervisor was Joe Campanella.
Von Hoffmann Press, Inc., was printer and binder.

CONTENTS

CONTENTS

science; teasing information owt of nature J. Rogers

PREFACE

Time present and time past
Are both perhaps present in time future,
And time future contained in time past.
T. S. Eliot,
Four Quartets

As we stated in the Preface to the first edition of *Evolution of the Earth*, we are firmly committed to the proposition that introductions to the sciences should be primarily conceptual rather than informational. Students deserve introductions that reveal the logical framework of a discipline, show relations of that discipline to the totality of human knowledge, and give them some idea of what it is like to be a participant in the discipline. At the same time, there should be depth and rigor that challenge the mind. Our experience indicates that students are stimulated greatly by constant exposure to scientific controversies, occasional spicy personal feuds, or an amusing faux pas. The student also needs to become a partner in the endless process of hypothesis testing, which is what we believe science is all about. In this book, we have tried to use these approaches.

In keeping with our strong preference for a "How do we know?" rather than "What do we know?" approach, we stress what assumptions are made by earth historians, what kinds of evidence (and tools for gathering that evidence) and what processes of reasoning and limitations of hypotheses are involved in reconstructing and interpreting the past. At the same time, we have again in this edition tried to emphasize that which is more or less unique about geology as an intellectual discipline. We feel that the tendency to neglect earth history in many introductory courses is most unfortunate, for to us it is its historical nature that sets geology apart from the more familiar nonhistorical sciences. Therefore, we have deliberately tried to draw out and to illustrate this uniqueness in the hope of stimulating a renaissance in the teaching of earth history. As readers use the book, we hope that they will keep in mind the following three maxims, which seem to us to be of transcendent importance for all educated people: (1) new concepts of time, (2) the universality of evolutionary changes, and (3) the importance throughout time of ecological interactions between life and the physical world.

Initial encounters with the sciences involve major shifts of scales, both spatial and temporal, as well as language shifts. For the average person, these commonly are very difficult. For non-science-oriented people, apathy and even antagonism may result from an early inability to grasp such shifts. In geology, the initial adjustment for the lay person undoubtedly is best made in the field where the raw data lie. Ideally, most introductory geology should be taught in the field almost exclusively. For the most part, however, this is not practical, so procedures must be used to lend synthetically some reality and to induce some student involvement in the subject. For earth history, we have had success by beginning with discussions of the more familiar geologic processes that have affected humans most directly in historic times. Gradually the discussion shifts into truly geologic frames of reference. For the more science-oriented student, we recommend also simple homework problems that deal with rates of a wide variety of geologic processes. Suggested examples are contained in a supplementary *Instructor's Manual.* We have had much success in using such problems not only to supplement the text, but also to provide departure points for discussions in small class or laboratory groups.

Evolution of the Earth developed from a second-semester course at the University of Wisconsin, Madison, taken mostly by first- and second-year college students representing a broad spectrum of science and nonscience backgrounds. The first edition was used successfully in this same course, and the modifications incorporated in this new edition reflect our own experience together with the constructive criticisms of many other teachers at a variety of different colleges having widely differing requirements. With the total length reduced and organization modified considerably, we believe that this new edition of *Evolution of the Earth* lends itself to greater flexibility of usage. The book assumes only a prior knowledge of general geology comparable to a high school earth science or first-semester college physical geology course. We have retained the classical foundations of earth and life history, but certain material, especially of a chemical and physical nature that seems necessary to us for an adequate understanding of modern geology, may prove difficult for some readers. We are confident that the skillful teacher can help students through such material by drawing out the essential highlights without intimidating them.

The historical elaboration of the development of basic principles for interpreting earth and life history invoked in the early chapters of the first edition is continued. First, it provides a kind of proxy for the reader's actually recapitulating discoveries and interpretations accomplished by past generations that make up the fabric of geologic principles. Second, it helps to make clear that the science is a human activity, and that the quest for the understanding of nature is an on-going, open-ended process in which readers themselves could participate if they should so choose. Finally, the historical approach reveals the cultural relationships of the science, which are un-

usually rich in the case of geology. Historical geology provides a great opportunity—even obligation—to present a broad integration of diverse material and to clarify the relevance of science. The first four chapters in which this approach is most prominently incorporated can be read rather quickly and understood by most students with a minimum of classroom elaboration if the instructor so chooses. Nonscience students especially like this material. Chapters 5 through 7 present some of the more complex principles and concepts basic to interpreting earth and life history. These are very important to the remainder of the book, and experience shows that they require classroom supplementations. Perhaps the most important change of organization in the second edition is the presentation of the unifying hypotheses of continental drift and global tectonics prior to—and as a background for—the detailed discussion of the actual geologic records of various parts of the earth. In the first edition, much of this material was deferred to the end of the book, where it was developed as a *consequence* of the detailed historical evidence, but global concepts have advanced so far since the appearance of the first edition, that the old organization is clearly outmoded. Large crustal displacements through time are now sufficiently well established that it seems mandatory to use them as a framework for the analysis of the local record. Therefore, Chapters 8 and 9 present a broad background leading from mountain building through continental drift and sea-floor spreading to plate tectonics. These chapters, in short, provide the *raison d'être* for the remainder of the book and so deserve special treatment.

In developing new emphases in historical geology, we have minimized the encyclopedic approach so prevalent in the past. Chapters 10 through 18 treat the actual historic record for the earth with North American data integrated much more closely with that for other continents than in the first edition. But whether a given discussion concerns North America or some other part of the world, the purpose always is to illustrate how geologists unravel and interpret the historical record more than to burden the reader with a formidable mass of factual detail and terminology. Only a handful of stratigraphic names appear in the text, and a lengthy relating of evidence and events in every corner of every continent is not presented. Instead, while a broad chronologic framework is provided throughout, chief emphasis is placed upon interpretation of physical and biological environments, reconstruction of paleogeography, and evolution. Our unifying theme throughout is that of *overall chemical evolution* of the earth! Interactions between the living and the nonliving are emphasized, as are major stratigraphic and tectonic patterns that recur both in time and space. In this way, each of the North American chapters tends to be topical as well as chronological, with only one or a few rather unique circumstances (such as evaporite deposition and organic reefs in Chapter 13) developed in each at the expense of more mundane, nonunique stratigraphic detail. Factual detail that is presented is for documentation only; students should not be expected to memorize it but rather to understand the principles and logic of the historical arguments for which the facts provide essential evidence.

The concluding chapter, as in the first edition, reminds the reader of the three fundamental maxims that follow from a study of earth history, which were stated above. Brief illustrations of each are discussed, and then the extrapolation of some implications of these maxims into humanity's future are developed briefly in order to suggest some conclusions that have relevance to all members of the human race.

It is impossible to acknowledge all of the countless individuals who have contributed directly or indirectly to the completion of this book. We especially thank L. L. Sloss and W. P. Gerould for originally encouraging us to embark on the writing of the first edition and Bradford Bayne and Robert Summersgill of McGraw-Hill for encouraging this revision. Reviews of the manuscript by R. L. Heller, Donald H. Zenger, Donald B. Moore, Hugh Rance, Gene A. Carr, William T. Gans, and Edward A. Hay have improved it materially. Carol First very efficiently supervised the editing of the new edition. Colleagues who read portions of the first edition, or in other ways made important specific suggestions, include C. J. Bowser, C. R. Bentley, R. F. Black, R. A. Bryson, C. W. Byers, C. S. Clay, L. M. Cline, C. Craddock, K. A. W. Crook, I. W. D. Dalziel, R. H. Dott, Sr., C. E. Dutton, L. R. Laudon, L. J. Maher, D. M. Mickelson, J. R. Moore, L. C. Pray, and R. Schaeffer. Some others who made unpublished material available or who provided important insights through general discussions include, in addition to the above, R. J. Adie, S. A. Born, A. V. Carozzi, D. L. Clark, T. A. Cohen, W. Compston, Grant Cottam, R. M. Gates, M. Kay, T. S. Laudon, O. Loucks, M. W. McElhinny, R. P. Meyer, A. A. Meyerhoff, N. D. Newell, N. A. Ostenso, Edna P. Plumstead, W. R. Reeder, R. Siegfried, K. O. Stanley, R. Stauffer, P. R. Vogt, and J. W. Wells. R. D. Sale, of the University of Wisconsin Cartographic Laboratory, prepared all new line drawings and modified several others, thus maintaining the prize-winning quality of the illustrations. Sources of drawings and photographs are cited in the figure captions unless they were entirely our own. Agencies that were especially helpful in providing photographs were the American Museum of Natural History, Exxon Production Research Co., Geological Survey of Canada, Geotech Marine Sciences Division of Teledyne Exploration Co., Manned Space Flight Center of the National Aeronautic and Space Agency, Marathon Oil Co., Shell Oil Corp., Smithsonian Oceanographic Sorting Center, United States Geological Survey, and U. S. National Museum. Karen E. Dott assisted in the assembly and labeling of illustrations. We are also pleased to acknowledge support of various research projects by the National Science Foundation, especially its Office of Polar Programs, and the Graduate School of the University of Wisconsin, Madison, which allowed us to visit and to study several foreign areas. Indirectly, these opportunities have enhanced greatly the final product. Lastly, we again recognize the apparent patience of our families.

And I have written books on the soul,
Proving absurd all written hitherto,
And putting us to ignorance again.
(So said Robert Browning's
skeptical philosopher Cleon.)

Robert H. Dott, Jr.
Roger L. Batten

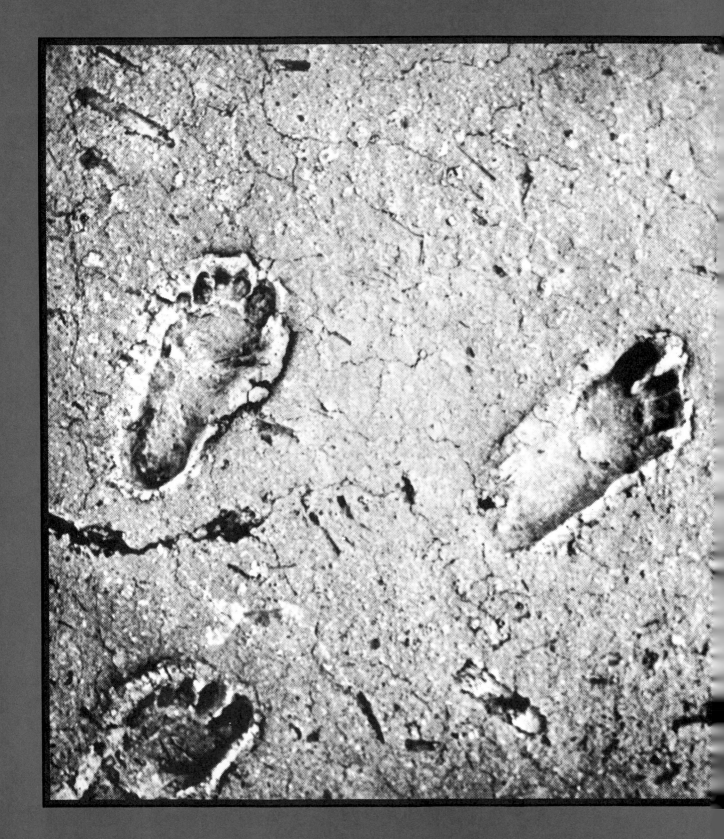

1

TERRESTRIAL CHANGE

There's nothing constant in the
universe,
All ebb and flow, and every
shape that's born
Bears in its womb the seeds of change.

Ovid,
Metamorphoses, XV (A.D. 8)

FIGURE 1.1 Fossil human
footprints in volcanic ash
near Managua, Nicaragua,
formed by people running
from an erupting volcano.
*(Photo by F. B. Richardson,
furnished by Howel
Williams.)*

It was in August of 1883 that a small island named Krakatoa, located between Java and Sumatra in Indonesia, disappeared in what was long considered the most violent explosion ever witnessed by humanity. A 300-meter-(1,000-foot-) deep hole was formed in the bottom of the sea where the island had stood. The explosion was heard 5,000 kilometers (3,000 miles) away. Volcanic ash, which was blown 80 kilometers high and fell on ships 2,000 kilometers from the island, darkened the sky for two days. Huge *tsunamis*, waves 30 meters high, crashed upon surrounding islands, drowning 35,000 people and destroying 300 towns.

Another geologic catastrophe recorded in history books was the mischief of Vesuvius volcano in Italy in A.D. 79. Earthquakes warned of eruptions, and many people fled from the Roman city of Pompeii at the foot of the mountain near present-day Naples. Those who lingered, however, were suddenly killed by poisonous volcanic gases and then immediately buried with their houses, furniture, and even pets beneath volcanic ash and mudflows.

A volcanic event of still greater historical interest occurred in the eastern Mediterranean region about 1500 B.C., but its story has only recently been worked out. Santorin, or Thera, is a crescent-shaped island between Greece and Crete (Fig. 1.2). It is a volcanic caldera, that is, a ring enclosing a submerged crater 400 meters deep—a puny remnant of a once lofty mountain. (Crater Lake in Oregon is also a caldera, but it stands well above sea level.) Many small eruptions have occurred at Thera within historic times, but old lava flows and thick volcanic ash deposits tell of a more violent earlier history. Recent sampling by oceanographers of deep-sea sediments in the eastern Mediterranean has revealed widespread buried ash layers that become thicker toward Thera, their obvious source. Carbon-14 isotope dating and archaeological remains indicate that the last very great eruptions occurred between 1450 and 1520 B.C. The material erupted was *three or four times greater* than from Krakatoa 3,400 years later.

Thera's eruptions are of special interest not only for their geologic magnitude but even more so because of their historical implications. The earliest European civilization—called Minoan—began on Crete only 125 kilometers (75 miles) south of Thera, and a well-preserved Minoan settlement has recently been excavated on Thera itself. But the Cretan civilization collapsed suddenly between 1400 and 1500 B.C. for reasons that have eluded archaeologists for generations.

Greek archaeologist Spyridon Marinatos postu-

FIGURE 1.2 Location of volcano Thera relative to historically prominent areas of the eastern Mediterranean. Shaded zone represents probable fallout pattern of volcanic ash.

lated way back in 1934 that an eruption of Thera may have ended the Minoan culture, and his hunch has recently been confirmed beyond reasonable doubt. Thera's cataclysmic eruptions would have generated *tsunamis*, waves perhaps as much as 100 meters high, which could reach northern Crete without warning in a mere 20 minutes. As around Krakatoa, such waves would do severe damage, and the Minoan capital of Knossos is known to have been virtually leveled at about this time. Other palaces, as well as temples, roads, and viaducts on the island, also were destroyed. Prevailing winds blew great volumes of ash southeastward. Most of this ended up on the bottom of the Mediterranean, but a lot must have fallen on eastern Crete (Fig. 1.2), where it would have raised havoc with cultivated fields. Poisonous gases also may have blown as far as Crete.

We have at last a logical explanation for the unusually sudden disappearance of a great culture, but that is not the whole story. Egyptian writings record the cessation of imports of Cretan cedar and of oil needed for preparing mummies during the Eighteenth Dynasty of Egypt, which lo and behold would

place it about 1500 B.C.! Those same writings tell of a period of floods and darkness when the "sun appeared in the sky like the moon." Could Thera, then, have caused the famous days of darkness of the Old Testament? And could the ebb of its *tsunamis* possibly explain the parting of the waters in the "Sea of Reeds" in northeastern Egypt to allow the Jewish Exodus? Could Thera also have inspired stories of the sinking of mythical Atlantis? Some scholars think so.

The only real certainty in this world is *change*! Upon even casual examination we can see natural changes taking place on the face of the earth today. Volcanoes are converted to shoals, or vice versa, and memories from history lessons recall other past dramatic changes. Changes in life on earth also are apparent, and presumed life crises have figured in literature, for example the biblical "Noah's Flood." People always have been deeply impressed—sometimes even rudely inconvenienced—by "catastrophic" geologic events, whereas more subtle, slower changes, even though ultimately more profound, have gone almost unnoticed. Such typically human preoccupation has strongly influenced the development of geological concepts, particularly as they bear upon the history of the earth. Therefore, it is profitable to relate a few examples of important geological processes that have acted during the past 5,000 years or so of recorded human history. Consideration of these changes in the context of a familiar span of time may provide the best basis for approaching matters of greater antiquity in our planet's diary.

CATASTROPHIC GEOLOGIC EVENTS IN HUMAN HISTORY

Another interesting example of geologic events in recorded history can be found in the famous Old Testament story of Sodom and Gomorrah in the Book of Genesis. Location of these two "cities of sin" was disputed for many years, but recent findings suggest their probable location under the south end of the Dead Sea. Genesis records that the cities were destroyed: "Brimstone and fire rained upon the cities, fire rose out of the sea, sulfurous waters poured forth so that the very rocks caught fire." It is easily seen from this graphic description how such an event, if it did occur, would make a great impression upon early Mediterranean civilizations of approximately 2000

B.C. Viewing the Jordan Valley today, one can readily understand the impact, for it is an exceedingly hot, inhospitable region of irregular topography and smelly springs.

(Modern scholars are finding the Bible to be an amazingly accurate historical document;) so let us look at what is known of the geology of this area to see if we can discover some verification and explanation of this famous catastrophic event interpreted by the Hebrews as retribution. First, we find that the Jordan Valley is a long, straight-sided depression bounded by the Jordanian plateaus and extending south to join the larger Red Sea depression (Fig. 1.3). Second, we would be impressed with evidence of present and past hot spring activity. Evidence of still greater past heat is borne out by lava flows. Straightness of the valley walls and sharp truncation of anticlinal and synclinal folds in rocks of the East and West Plateaus (Fig. 1.4), as well as the hot springs and lavas themselves, indicate the presence of great faults penetrating deep into the earth. When we discover that earthquakes are common, the whole story becomes clear. Severe tremors accompanying faulting must have resulted in slight southward tilting of the Dead Sea fault block, allowing the sea to flood southward over the cities. Hot sulfur springs probably became unusually active in the Jordan Valley as a result of the same tremors.

One of the most volcanically active regions in the world is Central America, and here, too, there have been several interesting cases of human involvement in geologic change. Perfectly preserved human footprints in fine-grained volcanic mud deposits occur near Managua, Nicaragua (Fig. 1.1). The muds obvi-

FIGURE 1.3 Gemini spacecraft photo over northern Egypt (looking southeast across Sinai Peninsula and Red Sea). Jordan Valley and Dead Sea lie just beyond left-center edge of view; heavily vegetated Nile Valley lies right of center. *(Courtesy NASA; photo no. 66-63533.)*

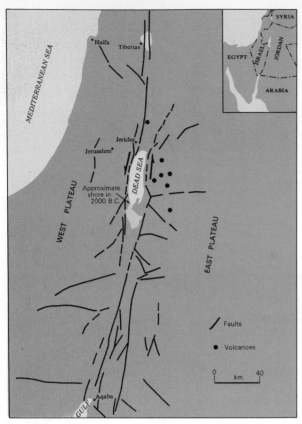

FIGURE 1.4 Faults and youthful volcanoes of Dead Sea region. Sodom and Gomorrah were located at the south end of the Dead Sea (in the shaded area) and were flooded by subsidence about 2000 B.C. *(Adapted from A. M. Quennel, 1956, 21st International Geological Congress.)*

ously were sticky when people hastened across the Managuan plains fleeing from nearby volcanic eruptions sometime between 2,000 and 5,000 years ago. Fine ash was falling continually, and buried the prints for posterity. Animals, too, lumbered over the ash; they included bison and some possible domesticated animals.

Equally dramatic in recent times have been the very devastating earthquakes such as have occurred in San Francisco in 1906 (Fig. 1.5), near Yellowstone Park in 1959, and in Alaska on Good Friday, 1964 (Fig. 1.6). Tremors are also commonly associated with volcanism, as we have already seen. A very close link has been established by observation for many years in Hawaii, where eruptions are benevolently forewarned by tremors. Though forgotten today,

the violent earthquake and *tsunamis* that destroyed ✳Lisbon and were felt all over western Europe and northwestern Africa on All Saints Day, 1755, made far (greater impact on people's minds) than any of these other ones. The timing of that catastrophe deeply impressed superstitious Europeans.

Proof that earth tremors are really the rule rather than the exception is suggested by average yearly observation of at least 200,000 measurable shocks. But it is judged that *over 1 million potentially measurable ones actually occur yearly!* Seismographs simply are too sparsely scattered to detect more than a fraction of our planet's nervous throbbing.

✳SUBTLE GEOLOGIC EVENTS IN HUMAN HISTORY

CLIMATE

Human civilizations' appeared more or less simultaneously along the lower Nile River, in the Tigris and Euphrates valleys in present Iraq, on the Indus delta of northwest India, and in northeast China. The eastern Mediterranean–Middle East region, about which we know the most, is today a harsh, arid country that seems an unlikely cradle for agricultural societies. Yet when the great Ice Age glaciers covered northern Europe, this region enjoyed a climate that was comfortably warm and more humid than now. Climatologists believe that an atmospheric high-pressure cell centered over the ice sheet pushed the moist westerly winds south of the Mediterranean Sea. Knowing this, we should not be surprised that early humans spread across the area and here developed their culture rapidly. After the great glaciers retreated, the Middle East climate harshened drastically. By 3000 to 5000 B.C., major cultural centers were restricted to the large, through-flowing river systems, such as the Nile, Tigris-Euphrates, and Indus, that were more or less independent of the increasing drought. It is from just such areas that our earliest written records have come. They tell of already highly developed agricultural systems with elaborate irrigated plantations and advanced urban centers.

Other interesting slow geologic changes have also occurred during human history. Between 1890 and 1950, careful measurements at widely scattered seaports demonstrated that sea level had been rising over the entire earth an average of 1.2 millimeters per year. This means a total rise of about 70 millimeters (about 3 inches), certainly not devastating in 60 years. But if long continued, this might be of some concern in the future, especially on an earth rapidly becoming crowded by a staggering population boom. Why should sea level be changing at all? If this could be answered, then the probable impact on coastal human habitation might be assessed better. We do not have far to look for the most obvious explanation. Over approximately the same 60-year interval, the climate was warming. For example, winters had been much less severe than in the "good old days" of great-grandfather's youth when people skated safely across a completely frozen lower Hudson River from New York City to New Jersey. As the general climate warmed from 1890 to about 1950 by from 0.5 to 5°C (depending upon latitude), glaciers in high mountains over most of the earth retreated gradually, in some cases as much as 16 kilometers (10 miles). The world's largest ice masses, those of Greenland and Antarctica, apparently retreated very slightly at their margins, too. The obvious effect of melting so much ice would be to raise sea level slightly, as has been the case.

GLACIERS AND SEA LEVEL
If we assume that melting will continue indefinitely and then gaze into the crystal ball, we can make some startling predictions. From estimates of the total vol-

FIGURE 1.5 With curious irony, this statue of famous geologist Louis Agassiz (see Chap. 18) was toppled from its mounting at Stanford University during the 1906 San Francisco earthquake. *(From Lawson, et al., 1908,* The California earthquake of 1906, *Carnegie Institution of Washington.)*

FIGURE 1.6 Hanning Bay Fault off the south-central coast of Alaska activated during the 1964 Good Friday earthquake. The fault extends along the right side of white area, which contains bleached subtidal organisms raised 4 to 5 meters during the shock. *(Courtesy George Plafker; described in* Science, *v. 148, 25 June 1965, pp. 1675–1687; copyright 1965 by American Association for the Advancement of Science.)*

ume of water still locked up in glacial ice, it is calculated that sea level would be raised through complete melting by *at least 50 meters more above its present level.* This would, of course, flood most coastal areas where many of the world's largest concentrations of civilized activity are found. In North America, Washington, D.C. would become a seaport, while New York, Miami, New Orleans, Houston, and Seattle all would become latter-day Sodoms and Gomorrahs. Memphis, now 560 kilometers from the sea, would become a major port on a new Mississippi delta, and Washington, D.C., lies at a critical elevation so that it would suffer very interesting consequences. Most of the government buildings are located on Potomac River lowlands less than 30 meters above present sea level. The Pentagon, only 11 meters at its base, would become useless. The Washington Monument, whose base has an elevation of 9 meters, would show just its top above water as a useful navigational marker.

While we are at it, we should examine the opposite effect of glaciation, namely the worldwide lowering of sea level during maximum glacial advances. Intuition alone could tell you what must have happened. The last glacial maximum occurred about 20,000 to 25,000 years ago, and from evidence of submerged ancient shoreline features, it is inferred that sea level must have been approximately 100 meters (roughly 300 feet) lower than at present.

Several geologists have carefully studied the changes of sea level during the past 20,000 years, and some have attempted to link these with historical records. They suggest that an average rise of 100 centimeters per century occurred from about 17,000 up to 6,000 years ago. About 4000 B.C., rapid rise culminated, and all subsequent rise has averaged only 12 to 15 centimeters per century. Clearly, the last phase of rapid rise may have caused flooding of early Bronze Age coastal settlements, and it is certainly interesting to find reference to a deluge in the legends of many separate, ancient cultures including Greek, Babylonian, Hindu, and Hebrew, most familiar to us in the Old Testament.

About 1925, a 3-meter-thick clay layer with marine shells was discovered beneath Ur, one of the world's most ancient cities located in the lower Euphrates valley and known to date back at least to 3000 or 4000 B.C. (Fig. 1.7). Paleolithic human artifacts underlie the clay, thus dating the oceanic incursion at between 4000 and about 8000 B.C. Similar buried clay, though thinner, was then found to underlie a large area of the Tigris-Euphrates valleys. Seemingly, proof of the ancient flood had been discovered at last! It is interesting to note the close correspondence of the age of the clay established archaeologically with the culmination of postglacial rapid rise of sea level established geologically. In 3000 B.C., Ur stood at the head of the Persian Gulf; so it is possible that rising sea level rapidly flooded the lower valley until the rivers could extend their deposits southward again by sedimentation. Since 3000 B.C. it is known that the Tigris-Euphrates delta has advanced southward nearly 175 kilometers (100 miles); today it advances 25 meters per year.

Could recent small oscillations of sea level have given rise in the Mediterranean–Persian Gulf region to the many ancient Flood legends? If the biblical story were correct, there should have been simultaneous wetting of all other great ancient coastal civilizations. But it has not been possible to establish any synchroneity of deluges reported in the many ancient traditions. Probably these events resulted from periodic local river floods, which caused devastation of the great cultural centers concentrated on low-lying deltas of large rivers. Even today the Tigris-Euphrates delta is subject to frequent floods, and annually two-thirds of Bangladesh, on the immense Ganges delta, disappears beneath several meters of water during the monsoon season. Both regions are low and swampy for hundreds of miles inland.

Even if small rises of sea level did occur, they could not have inundated the whole then-known world, as tradition would have us believe, but only low coastal areas. For the past 6,000 years, maximum vertical fluctuations of only about 3 meters above or below present sea level are indicated. On the other hand, as we have seen already, a slight warming of climate for the next 1,000 years or so might result in an additional rise of at least 50 meters. Perhaps we should look still further ahead and prepare instead for cooler times again, as our present "interglacial" stage may last only about 10,000 more years, at which time large ice sheets may begin forming and sea level falling once more.

CRUSTAL WARPING

Much of the low, northern European coastline has been flooded slowly at least since the 1600s when

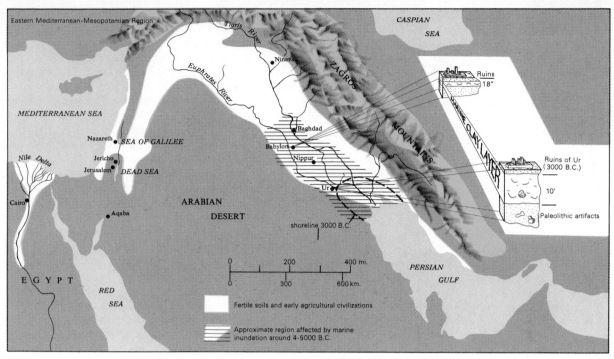

FIGURE 1.7   Eastern Mediterranean–Mesopotamian region showing the "fertile crescent," centers of early agrarian civilizations, and the known extent of the post-Paleolithic marine deposits of the Tigris-Euphrates valley.

tidal records were first made. This is to be expected from what we have just seen. In Scandinavia, on the other hand, an awareness developed about 1700 that the coastline there had *retreated,* suggesting an apparent fall of sea level of more than 1 meter in the previous century. It had been assumed that the earth's crust was fixed until, in 1765, a Finnish surveyor proved that the northern Baltic coastline was changing more in the north than in the south. He concluded that the *crust must be rising differentially rather than sea level falling.*

Amsterdam records indicate that sea level is apparently rising 20 centimeters (8 inches) per century there. But this is about twice the rate deduced for the worldwide effect of glacial retreat. How can we reconcile this discrepancy? Where the apparent rate of rise is too large, we are forced to conclude that there also has been some subsidence of the land, and, conversely, where the apparent rate of rise of sea level is too small (or nonexistent), the coastal land also must have risen as sea level has risen. Therefore, study of any coast may need to take into consideration several factors; another cause of oceanic del-

uge—namely local subsidence of the earth's crust—must be added to those already discussed.

There is no better place in the world to illustrate the interplay of land and sea level changes than northwestern Europe. A long history of careful measurements shows that, as the North Sea region has subsided to flood the low countries, most of interior central Europe as well as the northern two-thirds of Great Britain and Scandinavia have been rising at a geologically rapid rate (Fig. 1.8). Borings in the Netherlands indicate that old shoreline sediments known from their fossils to have been deposited roughly 1 million years ago during the Ice Age, now lie 2,000 meters below present sea level (i.e., beneath 2,000 meters of younger deposits). The crust of the earth here has subsided about 2,000 meters in 1 million years, or an average of about 2 millimeters per year. Measurements in central Scandinavia indicate that parts of the North Sea and Baltic coasts are now

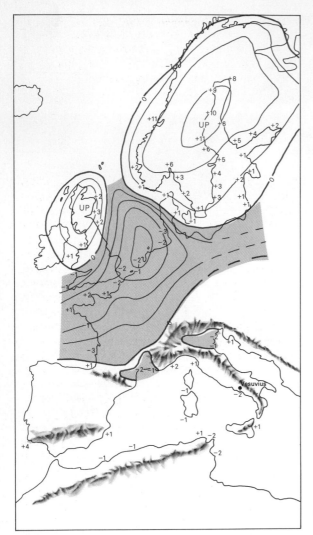

FIGURE 1.8 Relative crustal subsidence (−) and uplift (+) in northwestern Europe in the 10,000 years since the last glacial advance. Numbers indicate known rates of present coastal movements in millimeters per year. *(After R. Fairbridge,* The changing level of the sea, *1960; Copyright © by Scientific American, Inc. All rights reserved.)*

much greater, notice, than the simultaneous rate of subsidence of the nearby low countries (Fig. 1.8).

Submarine topography and dredgings indicate conclusively that much of the North Sea floor and the English Channel formed a low, swampy land connecting Britain with the Continent at the end of the last glaciation about 10,000 years ago. Glacial ice had then melted entirely from Britain and must have been shrinking rapidly from Scandinavia. Although sea level had begun rising rapidly, it did not flood this region until 6,000 or 7,000 years ago, when, you will recall, we estimate that the present general level was achieved. Easy migration of plants, animals, and early humans was possible across this area from the Continent to Britain for at least 3,000 years. As low areas were flooded, former ice cap centers of Scandinavia and Britain have risen by isostatic rebound in response to removal of the great load of ice that had weighed down the crust there. This clear cause-and-effect is duplicated for North America, where measurements prove that most of east-central Canada is also rising in response to deglaciation. This is one of our strongest confirmations of the generally plastic behavior of large parts of the earth's interior. To compensate for the rise of Scandinavia and Britain, we conclude that some subcrustal material must flow slowly under these areas, and lateral migration from beneath the North Sea may account for the subsidence there. Deep transfer of material is in accord with the principle of isostasy, which requires that a general tendency for equilibrium or balance of levels exists between different parts of the outer crust. The underlying mantle of the earth behaves plastically, buoying up crustal units according to their relative thicknesses and densities. Subsequently, we shall explore the historical implications of this fundamental condition of the earth.

HUMAN BEINGS AS GEOLOGIC AGENTS

Human beings sometimes have contributed to geologic changes, usually unwittingly. Near Los Angeles, California, people have for 30 years been encouraging encroachment of the sea. The second largest oil field in California extends beneath the coast under Long Beach harbor. This area is the only satisfactory harbor in the greater Los Angeles region; therefore, a

rising as much as 10 to 11 millimeters per year. The maximum total rise has been 250 meters (825 feet) since the retreat of glacial ice from the region about 10,000 years ago. The average apparent rate of rise, then, has been about 2.5 meters per century, an astoundingly rapid rate as we shall soon see—and

large naval base, an automobile assembly and shipping facility, and other industrial activities are concentrated along the waterfront. Over the years, extraction of petroleum and natural gas from incompletely cemented sandstones more than 1,000 meters below the surface has lowered the pressure in the pores of the strata and allowed the sediment grains to compact; subsidence has resulted. An area 8 kilometers (5 miles) in diameter with subsidence amounting to as much as 6 meters in 23 years is centered almost directly under the coastal facilities. You can imagine the consternation of irate naval and industrial landowners! To stem further subsidence, petroleum companies today replace extracted petroleum with water pumped into the sandstones at a rate equal to that of extraction.

People have acted the role of a geologic agent in this same region in other ways. First they persist in completely stripping vegetation from hillsides for new suburban housing tracts, which encourages landsliding. Fires caused by people denude the mountain slopes, further aggravating the erosion problem. Housing developments also continue to be erected across known active faults and old landslides. Human activities have had a more subtle effect upon southern California beaches. Prior to a great wartime building boom of the 1940s, rivers continually replenished the beach sands, which were slowly swept southward along the coast to Newport, south of Long Beach, there to be flushed into deep water via a submarine canyon. Metropolitan expansion brought complete usage of normal runoff waters so that rivers are completely dry except during rare winter downpours. Furthermore, through much of their lengths, the rivers now have artificial cement channels, which deprive them of considerable sand that would otherwise be added from their banks. The beaches have been starved of sand, and so winter storm waves are taking their toll of the shore and some choice homes. This calamity has necessitated building elaborate sand-drift traps and expensive haulage of sand in barges to re-feed the whole coast.

Alarming depletion of natural water supplies in the arid Southwest, and necessity for transporting water hundreds of miles at fantastic expense is another story of humans versus nature. Most ground water now being extracted in such areas fell as rain thousands of years ago, and is being withdrawn more rapidly than it is being replenished. As pressure on finite resources increases, other regions must also take notice, for many people today are living precariously on borrowed water—borrowed from the past without any adequate reinvestment program for the future's mushrooming thirsty population.

TIME AND RATES OF GEOLOGIC PROCESSES—HOW FAST IS FAST?

GEOLOGIC TIME

Having reviewed examples of different geologic processes active within the brief span of human recorded residence upon our little planet, we shall now begin to investigate the matter of geologic time. Here we step into a new frame of reference for most people. Time is to the geologist much as the immensity of space is to the astronomer. Five thousand years of recorded human history seems a long time, and if we tell you that humans first appeared between 1.5 and 2 million years ago, they will seem quite respectably antique. What would you think, though, when we tell you that *the earth is almost 3,000 times older?*

Because experience shows that realization of the magnitude of geologic time is so new to the uninitiated, we find that analogy with a more familiar measure of time is helpful. Our favorite is to compare it to the seven days of the week. If we take Monday as the first day and consider the Monday morning sun to have risen about 5 billion years ago (roughly the earth's birthday, give or take a half billion), we would find that the record of Monday through Saturday, or over three-fourths of earth history, is exceedingly obscure! For example, the first recognizable fossil animals appeared a mere 600 to 700 million years ago, or early Sunday on the last day of our analogy. We can get a truer perspective when we realize that recorded human history would begin only after 11:55 P.M. Sunday. The oldest known living things, a California bristlecone pine tree 4,600 years old and a Sequoia tree almost 4,000 years old, both sprouted "minutes" later; Christ would have been born only seconds before midnight at the close of the "week." Subsequent events—such as the industrial revolution, discovery of electricity, invention of the automobile, airplane, and television—hardly seem worth mentioning on this scale.

To illustrate geologic time still another way, let us perform some simple arithmetic to discover approxi-

EVOLUTION OF THE EARTH

TABLE 1.1 Comparison of rates of some geologic processes

Slowest				Fastest
Average denudation of continent	Cutting of Grand Canyon	Postglacial rise of sea level	Rise of Scandinavia	Advance of Tigris-Euphrates Delta
0.03 mm per year	0.7 mm per year	5 mm per year	10 mm per year	25,000 mm per year

mate rates of additional geologic changes, particularly the so-called subtle kinds of changes.

Examples of Rates

Recall that we said sea level was approximately 100 meters lower about 20,000 years ago and has risen with minor fluctuations since. The average rate of rise would then be 0.005 meter, or 5 millimeters, per year. In other words, it would take an average of 66 years to rise 1 foot. On the other hand, looking to the future and taking the view that sea level will rise another 50 meters if existing ice caps melt, and assuming the same average rate of melting and rise as above, it should take but a mere 10,000 years to produce the final flood. Table 1.1 summarizes for comparison the rates of four other common geologic processes.

Besides the question: "How fast is fast?," there is also a question: "How frequent is frequent?" To people, river floods are among the most familiar geologic catastrophes. Moreover, the flood plains of rivers are as much a part of the total watershed systems as are the main channels. This should be obvious to even the most casual observer who lives near a large river. But people long have ignored this fact and have continued to build houses, towns and factories on flood plains that inevitably are destined for dousings.

On the carefully controlled Mississippi River, severe floods have been infrequent, and so inhabitants have become complacent over many decades. In 1965 a combination of large ice jams and sudden melting of heavy snow produced a catastrophic flood. None of the oldtimers could remember such an event, nor could they recollect stories of such from their ancestors. Quickly a hue and cry went up for governmental preventive aid—totally ignoring the inevitability of floods. Flooding provides a useful example of the limitation of human time perspectives. If floods like that of 1965 occur, on the average, only once every couple of centuries, there could be no record of them in our young folklore. Similarly, an

"unprecedented" North Sea flooding of the Netherlands in 1953 was caused by meteorological circumstances so unusual that it is estimated they would occur only about once every 400 years. Though by human standards a *200-* or *400-year flood* is a rare event, such catastrophes must be considered frequent (as well as normal) on a geologic time scale.

An important question might be asked here. Namely, are measured rates of uplift and other structural changes adequate to produce great mountains over a reasonable span of geologic time, or do these important features require abnormally faster rates as might seem to be the case? Said another way, are we today witnessing a period of active mountain making in areas like California, or is this a relatively static time?

During his great round-the-world voyage in *H.M.S. Beagle*, Charles Darwin experienced several earthquakes. In 1835 following an especially severe one, he noted an uplift of the Chilean coast of about a meter, as well as nearby volcanic eruptions. He reasoned that a rise of a few feet per shock might produce, over geologic time, whole mountain ranges (Fig. 1.9). Figure 1.10 is a simple graph projecting some rates of measured uplift in several parts of California, obviously a relatively active region. This graph was prepared by an American geologist to show what changes of elevation might be expected 200,000 years in the future, a very short time geologically. The Alamitos Plains near Long Beach would be approximately 200 meters higher, and other areas shown would also gain lofty elevations. Obviously we must conclude that present, known rates of uplift can indeed produce mountains in a relatively short time.

Different Ways of Growing and Changing

In the preceding arithmetic gymnastics, we have talked only of average rates of change and for simplicity, therefore, we have assumed constant rates over long times. Figure 1.10 illustrates a graph showing this type of constant change, which plots as a

FIGURE 1.9 Discontinuity (unconformity) between tilted Pliocene strata (about 8 million years old) and flat late Pleistocene marine gravels (about 0.5 to 1.0 million years old) west of Ventura, California. After deposition of the latter, the entire coast was elevated nearly 200 meters. *(Courtesy K. O. Stanley.)*

straight line (thus termed "linear change"). However, we know that most natural processes do not proceed at constant rates. In nature "linear" change is quite the exception. For example, erosion of a newly uplifted mountain range would tend to proceed fastest early and then slow down after uplift ceased, because stream gradients would flatten. Growth in most animals slows as adulthood is approached. If you invest money, you want it to earn interest at an increasing rate through time, but, conversely, if you borrow money, you would hope to pay out interest at a decreasing rate as you pay back the principal.

We hear a great deal about the human population "explosion," and a glance at Fig. 1.11 reveals why.

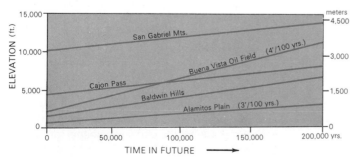

FIGURE 1.10 Present known rates of uplift in southern California projected into the future. *(Adapted from J. Gilluly, 1949; with permission of Geological Society of America.)*

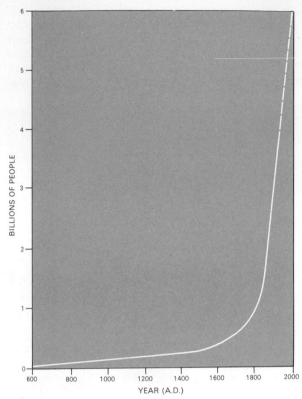

FIGURE 1.11   World human population growth curve.
(Data from United Nations estimates.)

Homo sapiens has multiplied in much the same way as you computed compound interest in elementary mathematics. In other words, population grows at an ever-increasing rate. We see that human beings in their first million years did not increase their total numbers greatly. It must have taken hundreds of thousands of years to break the first million mark. Records exist only for the past few centuries, but it is clear that the first documented doubling of population took two centuries (from 1650–1856); the second doubling a little less than one century; and today it doubles in only 40 years! The present rate is equivalent to adding the population of Britain annually. By the year 2000, people will number around 6 billion— double the present count!

Knowledge grows in much the same way, as shown in Fig. 1.11. It is estimated that more new information has appeared in the past 10 years than in the previous 100. If geologic literature alone continues to double every 10 years, by the year 2700 it could cover the earth to a depth of 10 miles. The situation already is such that recently a bibliography of bibliographies of earth science publications was published.

All the changes discussed up to this point move in one general direction (increase of size, decrease of elevation, etc.); that is, they are nonrepeating or irregular changes. Many changes in nature, however, are repeating ones. Commonly these repeat with a more or less regular time-period, as for example the fluctuations of the tides and changes of seasons. Such changes are called cyclic or rhythmic phenomena.

CATASTROPHIC VERSUS UNIFORM VIEWS OF CHANGE AND THE AGE OF THE EARTH

Ancient peoples quite understandably were impressed only by rapid and violent geologic processes. As a consequence, a view of the earth developed very early and gradually earned for itself the name *catastrophism*. By this view, which dominated human thought until 1850, most changes occurred suddenly, rapidly, and devastatingly. The most famous example of such a supposed catastrophe was, of course, the great deluge of the Old Testament. Others included volcanic eruptions or earthquakes at Pompeii, Thera, and Lisbon.

A parallel idea consistent with catastrophism was a universally accepted belief that the earth was but a few thousand years old. This conception is revealed in Shakespeare's *As You Like It*, written around 1600, where reference is made to the age of the earth being about 6,000 years. In 1654 a more definite pronouncement came from Anglican Archbishop Ussher of Ireland, who announced with great certainty that, based upon his analysis of the scriptures, the world had been created in the year 4004 B.C. on the 26th of October at nine o'clock in the morning! Some years later, another authority fixed an equally precise date for the great deluge, November 18, 2349 B.C. At that time, practically all leading intellectuals believed implicitly in a literal interpretation of the Bible, and so it is not surprising to find that a 6,000-year age for the earth was gospel for fully 200 years.

time implies catastrophe.

In the seventeenth century, the entire scriptural account of the Creation was a point of departure for science. Humans were considered early comers who lived for a time in paradise. As punishment for their sins, the deluge was sent to throw the earth's surface into chaos. Countless creatures suddenly were buried in sediments, violent waves and currents scoured out valleys and piled up rocks to form mountains, and the earth was wracked by earthquakes and volcanic eruptions. After the flood, the mountains began to be worn down and valleys to be filled.

When geology emerged as a vigorous intellectual pursuit around 1800, an alternate concept of earth change began to develop. By the new notion, much as we tried to show with some of our arithmetic above, it was held that the same common processes observed operating upon the earth were largely responsible for all past changes imprinted in the rock record. A noncatastrophic or more *uniform view* of change thus emerged, which required only a limited role for exceptionally sudden events like earthquakes, volcanic eruptions, floods, and avalanches. Today we envision the earth as dynamic, ever-changing, and evolving under the influences of many complex physical, chemical, and biological processes. These processes compete with one another constantly to produce a situation tending toward some balance among them, but it is always being modified and kept off balance by new changes. The resulting modifications are in part nonrecurring, or irregular, changes, and in part regular recurring, or cyclic, in character.

A corollary of the uniform change concept, which had great impact upon society in the 1800s, was its inescapable implication that the earth must, in fact, be much more than a few thousand years old in order to allow enough time for all observable ancient changes to have occurred in noncatastrophic ways. We begin to see here the seeds of a social controversy whose repercussions affected our entire civilization with the age of the earth and the nature of geologic changes as principal points of disagreement. The controversy was brought to a head largely by the findings of geology in the early nineteenth century. In the next chapter we shall examine this philosophical crisis as we start to develop the first principles upon which to base our interpretation of the fascinating and perplexing diary of our evolving planet.

Readings

Alaska's Good Friday Earthquake, March 27, 1964: Washington, U.S. Geological Survey, Circular 491.

Archaeological Services of Greece, 1969, International scientific congress on the volcano Thera: Athens.

Bullard, F. M., 1962, Volcanoes: Austin, Univ. of Texas Press.

Fairbridge, R. W., 1960, The changing level of the sea: Scientific American, v. 202, May 1960, pp. 70–79.

Gilluly, J., 1949, The distribution of mountain-building in geologic time: Bulletin of the Geological Society of America, v. 60, pp. 561–590.

Ojakangas, R. and Darby, D., 1976, The earth and its history: New York, McGraw-Hill. (Paperback)

2

FLOODS, FOSSILS, AND HERESIES

These shells are the greatest and most lasting monuments of antiquity, which in all probability, will far antedate all the most ancient monuments of the world, even the pyramids, obelisks, mummys, hieroglyphicks, and coins . . . Nor will there be wanting media or criteria of chronology which may give us some account even of the time when (they formed).

Robert Hooke (1703)

FIGURE 2.1 Seventeenth century illustrations of fossil ammonoids (*Cornua ammonis* or "snakestones"), prepared by Robert Hooke. *(From* Posthumous works of Robert Hooke, *1703.)*

Even up to 1800, agreement of biblical revelation with reason seemed excellent to most people. The presence of abundant fossil marine creatures in strata strewn over the land was more than adequate proof of the Flood. Before the sinful fall of humanity, the earth was described as a paradise with a mild, equable climate. Under such conditions, it was not surprising that Adam and his progeny each lived to be more than 900 years old; Noah lived to a ripe old 950 years. But after the Flood, conditions on earth were quite different and less habitable. Topographic changes presumably caused a harshening of climate, and as a result people did not live as long (Table 2.1). People now had to work much harder to win an existence. And the harsher climates bred warlike peoples. All evidence seemed fully consistent.

Until the late eighteenth century, the biblical Genesis was so perfectly rationalized with all geologic facts that only a handful of people questioned the conventional geology, and they did so at no small personal risk. Indeed, it was the adding up of ages of successive generations recorded in the Bible, some of which are shown in Table 2.1, that was the basis of estimates of the age of the earth.

WHAT IS A FOSSIL?

EARLY QUESTIONS

Among ancient peoples there were many controversies concerning geologic matters. Besides supposed catastrophic floods, probably the most important things that puzzled the ancients were fossils. What is a fossil? The word means simply "something dug up," but, as generally used, a fossil is any recognizable evidence of pre-existing life. Shells of old animals, bones, plant structures such as petrified tree trunks, impressions of plant leaves, soft worms, or jellyfish, and even tracks, burrows, and fecal material formed by animals qualify. Fossils are preserved in many different ways (see Fig. 2.2; also Appendix I).

Some of the first recorded ideas about fossils came from Greece, where shells were early observed high in the mountains. Large bones were interpreted as relics of a former race of heroic human giants, but apparent oceanic organisms hundreds of feet above sea level and miles inland posed weightier questions. Had the sea receded or had these objects "grown" in the rock much like mineral veins? Perhaps, somehow or other, fish and other animals had crawled into cracks in the rock and died, then were converted into stone by mysterious "vapors." Of course, in this view, it was supposed that fossils were much younger than

TABLE 2.1   Human life-spans before and after Noah's Flood*

Longevity before the flood		Longevity after the flood			
	Years		Years		Years
Adam	930	Noah after the Flood	350	Reuben	124
Seth	912	Shem after the Flood	502	Simeon	120
Enos	905	Arphaxad	438	Levi	137
Cainan	910	Salah	433	Judah	119
Mahalaleel	895	Eber	464	Dan	124
Jared	962	Peleg	239	Naphtali	130
Methuselah	969	Reu	239	Gad	125
Lamech	777	Serug	230	Asher	126
Noah before the Flood	600	Nahor	148	Issachar	122
Shem before the Flood	98	Terah	205	Zebulun	114
		Abraham	175	Joseph	110
		Isaac	180	Sarah	127
		Jacob	147	Kohath	133

*Reported by John Whitehurst, 1778: *An inquiry into the original state and formation of the earth.*

the rocks in which they occurred. Still another quaint idea of ancient peoples was that fossils were "figured stones" created in the rocks by "plastic forces." This, however, does not tell us very much either. Centuries later, some people thought that they grew in the rocks from seeds. It even was suggested that they grew from fish spawn caught in cracks in the rocks during the great Flood.

Four major questions about fossils were posed by early speculations: (1) Are fossils really organic remains? (2) How did they get into the rocks? (3) When did they get there? (4) How did they become "petrified"?

need for low energy existence.

FIGURE 2.2   A well-preserved ant in Pleistocene amber, an unusual mode of preservation. *(Courtesy American Museum of Natural History.)*

AN ARTIST SAW THE WAY

The most important early record of a careful interpretation of fossils was that left by the Italian genius Leonardo da Vinci, one of the most original natural philosophers of the early Renaissance. About A.D. 1500 he recognized that fossil shells in north Italy represented ancient marine life even though found in strata exposed many miles from the nearest seashore (Fig. 2.3). In opposition to the popular view that fossils had been washed in by the biblical Deluge, he argued that clams could not travel from the Adriatic Sea to Lombardy, a distance of 400 kilometers, in 40 days. Many of the shells certainly could not have been washed inland great distances because they were too fragile. He also pointed out that the associations or assemblages of different kinds of fossils found in these ancient strata were still intact and resembled living communities of organisms that he observed at the coast. Leonardo further noted that there were many distinct layers that were fossil-rich and that these were separated by completely barren, nonfossiliferous ones (Fig. 2.3). This suggested to him, by inductive reasoning from seasonal flooding of rivers, that there were many events recorded rather than a single, worldwide deluge.

NICHOLAS STENO

In the middle seventeenth century, another important man, Nicholas Steno, also made some clear observa-

" Stensen (Lutheran physician)

tions about fossils, and in contrast to Leonardo's, his writings were widely and quickly circulated beginning in 1669. His interest in geology arose largely from anatomical comparisons of a fossil shark with a modern one.

Steno, like Leonardo, felt that fossils formed with the rocks slowly, bit by bit. Steno was a careful observer and based conclusions upon those things that he actually observed in the rocks. He recognized that shell-bearing strata beneath the site of ancient Rome, which were used in constructing its buildings, must be older than the city; that is, more than 3,000 years. But Steno clung to the belief that earth history was not much longer than human history, and went badly awry in interpreting large mammal bones in central Italy as the remains of Hannibal's famous elephants; he missed their age by at least a million years!

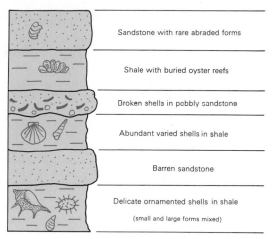

Sandstone with rare abraded forms

Shale with buried oyster reefs

Broken shells in pebbly sandstone

Abundant varied shells in shale

Barren sandstone

Delicate ornamented shells in shale
(small and large forms mixed)

FIGURE 2.3 Fossiliferous and barren strata such as those studied by Leonardo da Vinci in northern Italy.

collapsing cave theories for unconformities

STENO'S PRINCIPLES

Besides correctly interpreting fossils, Steno drew some even more important conclusions about the strata in which they occur. The result was the statement of the most basic principles for analysis of earth history. Steno showed great insight into the significance of individual strata; for example he recognized that particles would settle from a fluid in proportion to their relative weights or mass, the larger ones first and so on. Any changes in size of particles would cause development of horizontal layering or stratification, the most conspicuous single property of sedimentary rocks (Fig. 2.4). He also appreciated the importance of solution and precipitation of chemically soluble sedimentary materials. In short, he recognized the tendency for uniformity of texture and composition in individual strata, and he inferred that the characteristics of different strata reflected changes of such conditions as temperature, wind, currents, or storms.

Finally, and most significantly, sedimentary strata consisting of particles and shells could not all have existed from the beginning of time (i.e., all be the same age), but must have been deposited particle by particle and layer by layer, one on top of another. Therefore, in a sequence with many layers of stratified rocks, a given layer must be older than any overlying layers. We now call this the principle of superposition of strata, and, though it seems obvious, it is the crux of

interpretation of geologic history from the rock record. Steno was the first to state formally this and two other principles, which made him one of the first men to recognize that strata contain a decipherable chronological record of earth history (the analysis of which today falls under the heading of stratigraphy).

Steno's principles provide the ultimate bases of practically all interpretation of earth history, and so their importance can hardly be overemphasized even though they may seem obvious today. Briefly stated, they are

sedimentary rocks

1 *Principle of superposition:* in any succession of strata not severely deformed, (the oldest stratum lies at the bottom,) with successively younger ones above. [This is the basis of relative ages of all strata and their contained fossils.] *strata positioning*

2 *Principle of original horizontality:* because sedimentary particles settle from fluids under gravitational influence, stratification originally must be horizontal; steeply inclined strata, therefore, have suffered subsequent disturbance.

3 *Principle of original lateral continuity:* strata originally extended in all directions until they thinned to zero or terminated against the edges of their original area (or basin) of deposition. *Pismo Beach eg.*

The Grand Canyon illustrates all three principles (Fig. 2.5). First, the upper three-fourths of the canyon

FIGURE 2.4   Conspicuous stratification, the most characteristic feature of sedimentary rocks; interpreted fully by Nicholas Steno in the seventeenth century. *(Sandstones and thin shales of late Paleozoic Itararé Formation, Brazil; courtesy J. C. Crowell.)*

walls expose horizontal strata undisturbed for a long time. At the bottom of the canyon, however, one can see tilted strata that, by the second principle, must have been originally horizontal. They are separated from the horizontal sequence by a discontinuity or break in the rock sequence, called an unconformity. From the first principle, we see that the oldest stratum respectively in either sequence must lie at the bottom. Furthermore, we conclude that the tilted sequence as a whole is older than the flat sequence, for it underlies the latter. Finally, the third principle is illustrated by the fact that each stratum exposed on the far wall of the canyon has its clear counterpart on the nearer walls, and it requires little imagination to see that each was continuous across the present canyon area prior to cutting of the gorge by the Colorado River. Although here the original lateral continuity of strata is obvious, in most areas it is less so. Yet, everywhere that we see the eroded or broken edges of strata exposed, we know that once they were more extensive laterally than they appear to be today. This becomes of paramount importance in trying to restore ancient conditions on earth from a much-eroded and fragmentary stratified record.

UTILITY OF FOSSILS

ROBERT HOOKE'S OVERLOOKED HYPOTHESIS

About 1670, or about the same time that Steno's writings appeared, the famous British scientist Robert Hooke also argued that fossils were truly organic, and joined other challengers of the conventional notion that all had been emplaced by the single Noachian Flood. Hooke studied and meticulously illustrated fossil shells (Fig. 2.1), making extensive use of the microscope, which had recently been invented. Most significant was Hooke's suggestion that fossils might be useful for making chronologic comparison of rocks of similar age, much as old Roman coins were used to date human historical events in Europe (see quotation at the beginning of the chapter). He speculated that species had a fixed "life span," for many of the fossils he studied had no known living counterparts. This was one of the earliest hints of extinction of species, it long having been assumed that *all* life was created about 6,000 years ago, and is still living. It was nearly a century before his idea actually was tested and proven valid.

Geology had not yet arrived at a stage in which very much attention was being given to detailed study of individual strata. Until the late eighteenth century, natural scientists were preoccupied with grander problems. Combat still continued with the Deluge enthusiasts, coming to be known as *diluvialists*, who held such interesting ideas as the suggestion that the Flood had first dissolved all antediluvian matter, except fossils, and then reprecipitated the sediments in which the fossils became encased as

FIGURE 2.5 View in eastern Grand Canyon illustrating Steno's principles. A discordance (called angular unconformity) separates older tilted strata (Eocambrian) from younger (Cambrian) ones. *(Courtesy E. D. McKee.)*

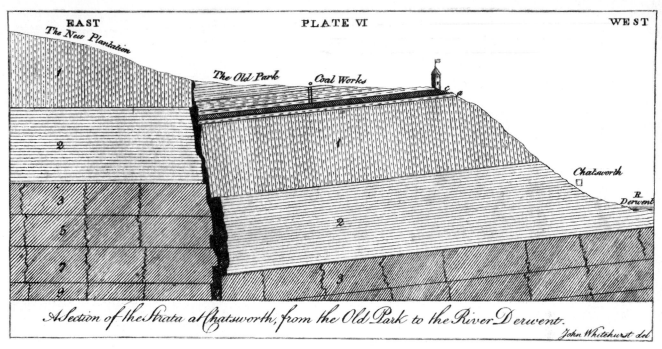

EAST   PLATE VI   WEST

The New Plantation

The Old Park   *Coal Works*

Chatsworth

R. Derwent

A Section of the Strata at Chatsworth, from the Old Park to the River Derwent.

John Whitehurst del.

they settled out of the turbulent waters. But where were the remains of men killed by the Flood? This question was of sufficient moment that a Swiss naturalist, J. Scheuchzer, in 1709 excitedly interpreted a giant salamander skeleton as the fossil remains of a man drowned by the Deluge. First things first—clearly science was not yet ready to *use* fossils! Slowly, doubts arose over the geologic role of the Flood. By about 1830, it was regarded as too brief to have altered the earth's surface significantly, though it was suggested that caves represented relicts of the prediluvial landscape now filled with sediments and bones.

✳ FOSSILS AND GEOLOGIC MAPPING

By the mid-eighteenth century, a great deal of fossil collecting had been accomplished, and the mapping of earth materials was beginning. As early as 1723, an English naturalist, John Woodward, suggested an

FIGURE 2.6   Guettard's early mineralogic map of France and England, forerunner of true geologic maps (prepared with assistance from A. L. Lavoisier). Note symbols for rock types, ores, coal; shaded oval band represents the chalk-bearing strata. (*From* Histoire de L'Academie Royal des Sciences, *Paris, 1746.*)

FIGURE 2.7   An early cross section showing the stratigraphic sequence and faulting in a coal field near Newcastle, England. The author had a quaint explanation of the tilting and faulting shown: subterranean heat caused expansion and cracking of the earth; sea water then leaked downward to be converted to steam, resulting in sudden convulsions of the crust, climaxed by the biblical Deluge. (*From J. Whitehurst, 1778,* An inquiry into the original state and formation of the earth.)

identity or correlation of certain strata on the European mainland with those in Britain on the basis of "great numbers of shells and other productions of the sea." In 1746 one of the earliest crude maps, showing mineral and fossil localities in part of France and England (Fig. 2.6), was produced by J. E. Guettard. Guettard showed that, like their modern counterparts, fossil shells commonly have others attached to them, may have holes bored through them, and may tend to be broken and worn. Although he correctly inferred their origin and noted some important localities, he did not see the full potential of fossils for identification and tracing, or mapping, of different strata.

Just before 1800, a civil engineer named William Smith was actively involved in land surveying in England. From visits to mines, he early recognized regularity to the rock succession, which already had been demonstrated locally in coal mines (Fig. 2.7).

EVOLUTION OF THE EARTH

Extensive catalogues of English fossils also had appeared about 1700. Smith studied all this information, added to it, and finally proved the enormous utility of fossils. In 1796 he wrote of the "wonderful order and regularity with which nature has disposed of these singular productions (fossils) and assigned to each its class and its peculiar Stratum." Culmination of his great effort was preparation of the first geologic map of high quality, first sheets of which were completed before 1800.

Working in a humid region where rock outcrops are rare, artificial excavations and mines were of great importance in helping Smith to formulate his ideas. As a practicing engineer, he was engaged in extensive canal excavations in southern England. The canals (begun in 1793) were intended to increase the

transport capabilities for the rapidly expanding British commercial empire in the early throes of the industrial revolution. Smith traced and mapped different strata according to their color and mineral composition as well as distinctive fossils. As his mapping extended across England (Fig. 2.8), he could recognize most of the strata over long distances by their fossils, much as one might recognize bygone eras from the distinctive styles of old coins, bottles, or cans in the layers of different metropolitan rubbish heaps.

Smith's map and its accompanying table of strata completed in 1815 was a cartographic masterpiece, for Smith used meticulous skill in locating the boundary lines or contacts between differing strata. He was equally careful in collecting, locating, and marking his fossils, a most important precaution if they were to be used for identifying particular strata. Smith's prowess as a careful observer is exemplified by his distinction of fossils still in their original place from badly abraded "alluvial" ones, the latter of which would be less reliable for tracing individual strata because they had been transported and mixed after death. The map immediately proved useful, not only for planning canals, quarries, and mines, but also as a guide to soils. Its author long since had recognized that soil develops from underlying bedrock, and, therefore, reflects some characteristics of the rock. Smith gradually was able to establish a detailed sequence of strata, which then enabled him to predict distribution of different rock bodies in new, unmapped country.

Meanwhile in the Paris region, the principle of fossil correlation and its usefulness in geologic mapping was discovered simultaneously. Young A. L. Lavoisier, a great chemist-to-be, by 1780 had become aware of some order to fossil occurrences. He recognized different assemblages of fossils associated with different rock types and reasoned that these associations reflected environmental conditions, a most advanced idea for that time. Two other Frenchmen named Georges Cuvier and Alexandre Brongniart, however, developed these ideas more fully and proved that around Paris there is a definite relationship between fossil occurrences and the succession of the sedimentary deposits in which they occur (Fig. 2.8). Like Smith, they also developed an essentially modern geologic map (published 1811) showing the distribution of various rock divisions based largely

FIGURE 2.8 Index map of Europe showing areas of the Smith and the Cuvier-Brongniart geologic maps, as well as some localities referred to in Chaps. 3 and 4.

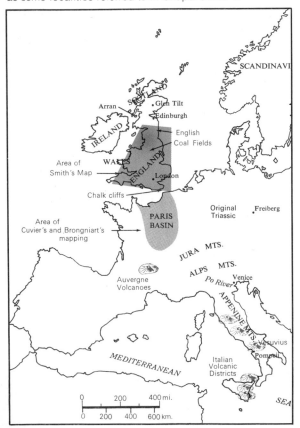

upon the similarity of fossils found in the particular strata. Cuvier said:

✳ These fossils are generally the same in corresponding beds, and present tolerably marked differences of species from one group of beds to another. It is a method of recognition which up to the present has never deceived us.

Distinction of strata by fossils is the classic example in geology of simultaneous discovery of a principle by several independent workers. In today's vernacular, the Smith-Cuvier discovery was a "scientific breakthrough." After the giant step was taken around 1810, the use of fossils was accepted quickly. The subdivision, tracing, and mapping of European strata went forth at a great pace with stimulation from the industrial revolution.

Hindsight shows that the strata might have been mapped according to the types of vineyard grapes in addition to the other features used. French wine districts show a marked relationship to bedrock types; for example the Champagne region lies upon chalk-bearing strata (Fig. 2.6).

Index Fossils and the Principle of Fossil Succession *some useful for intercontinental correlation*

The first construction of geologic maps was based largely upon the ability to compare and trace individual strata using their unique fossils to help identify them, in other words, to correlate rocks with similar fossils. Paleontologic correlation involves the most important new principles developed after those contributed by Steno. All these together form the stratigraphic basis for analysis of earth history.

From the combined work of Cuvier, Brongniart, and Smith, two important and closely related principles have come to us. The first of these is generally known as fossil succession, which forms a sort of corollary to the important principle of superposition given us by Steno, and discussed above. The principle of fossil succession states that, in a succession of strata containing fossils, obviously the lowest fossils are oldest. This is simply a direct extension of Steno's principle. The other important tenet coming from their work may be called the principle of fossil assemblages, namely that like assemblages of fossils are of like age, and therefore strata containing them are of like age (Fig. 2.9). Application of the principle in-

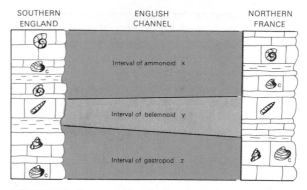

FIGURE 2.9 The use of fossils for correlation across the English Channel of strata studied by Smith in England and Cuvier and Brongniart in France. Fossils x, y, and z are useful index fossils, but clam c is not, because it ranges through the entire sequence.

volves what we now call index fossils. An index fossil is one that is particularly useful for correlation of strata; that is, it is an index to a particular stratum or group of strata.

What constitutes an index fossil, and do not all fossils qualify as such? For a fossil to be considered a stratigraphic index, it must be: (1) easily recognized (i.e., unique); (2) widespread in occurrence; and (3) restricted to a very limited thickness of strata (i.e., the organism represented was short-lived; Fig. Al.3). To have utility, all these conditions must be met. Moreover, not all fossils meet these criteria. Many are difficult to distinguish from others, some are not widespread, and many were too long-lived; they occur through great thicknesses of strata without change (for example, fossil c, Fig. 2.9). Note that, as Hooke guessed 100 years before Smith and Cuvier, an index fossil has special qualities in common with archaeological artifacts like old Roman coins, which can be used to date and correlate human historical events.

What makes correlation with index fossils actually work? Neither Smith nor Cuvier formally set down all the logic in Hooke's suggestion, which they independently rediscovered and proved, but today we can do this in retrospect. The proven utility of an index fossil requires these conditions: first, that once a species became extinct it never reappeared; and second, that no two species are identical. The method, as Cuvier observed, "up to the present has never deceived us."

It was Leonardo and Steno and a few other early workers whom we have mentioned who first pointed the way for the science by insisting upon consulting the earth itself for the answers to its riddles, unencumbered by dogmas, superstitions, and ill-founded assumptions. They were among the first to oppose blind acceptance of preachings that were in no way verifiable, that is, whose validity could not be tested by actual observation of nature. Such intellectual crusading against orthodoxy was a risky business, for it landed several men in prison and cost at least one his life.

EXPLANATIONS OF CHANGE AMONG FOSSILS

led to idea of extinction of species

Smith, a practical man with practical motives, felt little compulsion to explain why unique fossils occurred in certain strata. He simply accepted the happy fact and used it. Cuvier, the scholar, on the other hand, was very much concerned with the problem of fossil succession, and he developed an enormously influential theory to explain it. His great stature as a biologist and the harmony of his theory of the fossil record with theology earned wide acclaim.

Early in his career, Cuvier proved by comparative anatomy that some fossil and living elephants, though similar, belonged to different species. Then he discovered other fossil bones that belonged to creatures completely different from any known living forms. Cuvier had thus proved that some past inhabitants of earth had vanished completely. (He had discovered the first clear examples of known extinction of organisms.) *1811*

From their studies of strata of the Paris Basin, Cuvier and Brongniart correctly concluded that there were a number of major <u>unconformities or interruptions</u> in the sequence. As one can see from the column of strata in Fig. 2.10, there are alternations of sandstones, shales, and limestones. In addition, there are several zones of ancient soil with fragments of plants in them as well as several breaks in deposition marked by concentrations of wood, shell fragments, and small pebbles. Obviously the record of sedimentation is not a smooth, continuous one, but tells a story of marked changes and interruptions of deposition accompanied by rather abrupt changes of fossil assemblages. Cuvier's theory of fossil succession assumed that portions of assemblages represented in

successive groups of strata had become extinct before succeeding strata were deposited. He interpreted the evidence to indicate many wholesale catastrophic extinctions of assemblages of millions of organisms caused by violent oscillations of the sea. Land organisms were thought to show effects of the catastrophes even more dramatically than marine ones. Cuvier noted that fossils in successively younger strata were more like modern organisms, successive extinctions having eliminated many now-unknown species so that, with each <u>catastrophe,</u> life moved toward the ultimate "perfection" of the modern scene.

Although today Cuvier's hypothesis is regarded as inadequate, nonetheless it is important to realize that it was a reasonable conclusion based both upon local evidence and its seeming accord with popular theological assumptions. Recall from Chap. 1 that ancient peoples had regarded most major changes in the earth as produced catastrophically by unknown agents. This prejudice was woven early into the fabric of Judeo-Christian traditions, had become dogma, and thereby was perpetuated so that its grip on human thought strengthened. Against this background, and on the authority of scholars such as Cuvier, a distinct philosophy called catastrophism matured in the early nineteenth century. About 1850, Alcide D'Orbigny modified the hypothesis by postulating the creation of new species following each of 27 catastrophes. The complete theory became known as catastrophism and special creations. It said, in effect, that there was no connection in life between the index fossils of successive strata.

The alternative to Cuvier's concept of catastrophism, which was a theory of extinctions to explain fossil succession, was that fossils in younger strata are descendants from older groups of organisms by change from one species to another. Even Cuvier noted a general tendency for younger fossils to be progressively more like living organisms, but insisted it was due solely to selective extinctions. Other geologists also noted such change; for example, another Frenchman (Giraud-Soulavie) as early as 1780 subdivided strata into five "ages" based upon the gross characteristics of their fossil assemblages. Rocks of the "first age" included no forms analogous to living ones, the "second" included both extinct and living marine forms, the "third epoch" had shells of modern types only, the "fourth" was plant-bearing, and the

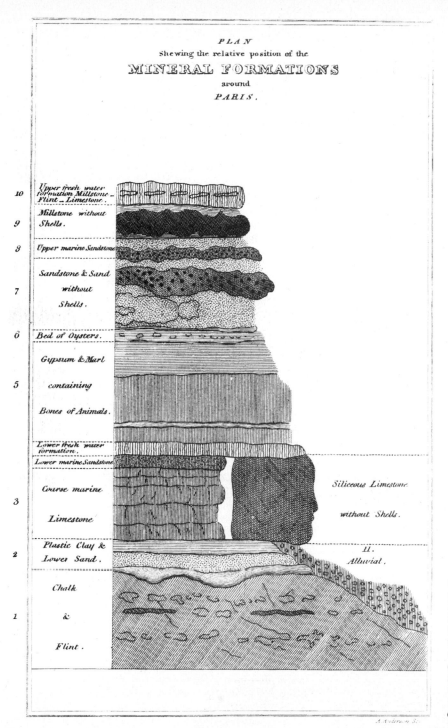

FIGURE 2.10 Stratigraphic columnar section of strata studied by Cuvier and Brongniart around Paris. It is interesting to note that in the translation of this work from the French, some English editor added what is obviously Napoleon Bonaparte's profile to the cliff for unit 3. *(From G. Cuvier, Essay on the theory of the earth, English translation, 1818.)*

"fifth" was distinguished by mammal remains. This was one of the first attempts to subdivide the stratigraphic rock record into broad divisions based upon changes in the development of life. It is a precursor to a geologic time scale, for it implies an irreversible or unidirectional trend of development—a systematic change through time—_evolution_!

Organic evolution, the alternative to Cuvier's catastrophism, was not new in Cuvier's time. The general concept had been suggested at least 100 years earlier, and two of Cuvier's own colleagues were ardent evolutionists who continually feuded with him. Apparently no urgent "need" was felt for such a theory, however, until more factual evidence had accumulated and the intellectual climate began to change near the end of the 1700s.[1] This important idea is discussed extensively in Chap. 5 after other essential background has been developed.

THE FOSSIL RECORD (incomplete)

Dendrites - crystals, not fossils

From Steno's time onward, geologists quickly recognized that fossils were the remains of once living organisms by making comparisons (where possible) with living organisms. A fossil clam shell (Fig. Al.7) is not that much different from a living clam in general appearance. By the nineteenth century, cataloging and describing of fossils had reached the point where it was possible to correlate strata with some precision, to begin to construct a picture of the complex changes in the history of life, to reconstruct past climates, and to document evolution.

[1]Feuding about evolution continued a long time after, of course. Only about 70 or 80 years ago, the acid pen of Mark Twain had this to say: "It was foreseen that man would have to have the oyster. . . . Very well, you cannot make an oyster out of whole cloth, you must make the oyster's ancestor first. This is not done in a day. You must make a vast variety of invertebrates, to start with—belemnites, trilobites, Jebusites, Amalekites, and that sort of fry, and put them to soak in a primary sea, and wait and see what will happen. Some will be a disappointment—the belemnites, the Ammonites, and such; they will be failures, they will die out and become extinct, in the course of the nineteen million years covered by the experiment, but all is not lost, for the Amalekites will fetch the homestake; they will develop gradually into encrinites, and stalactites, and blatherskites, and one thing and another as the mighty ages creep on . . . the Archaean and the Cambrian Periods . . . and at last the first grand stage in the preparation of the world for man stands completed, the oyster is done. An oyster has hardly any more reasoning power than a scientist has; and so it is reasonably certain that this one jumped to the conclusion that the nineteen million years was a preparation for _him_; but that would be just like an oyster." (From the essay _Was the World Made for Man?_)

The most common and important fossils are the *hard parts of organisms—shells, bones, teeth—for the simple reason that soft tissues usually decay and disappear rapidly following death. Hard parts may remain on the surface for some time, but the faster they become buried by sediment, the better preserved they are. For example, shells lying on the beach or in shallow water are moved back and forth by waves and tides; soon they become broken up and eventually are ground up into small particles. Indeed, many lime sands and muds in the tropics are almost completely made up of such fragments. In time, as more sediments pile up, pressure and solution cause the grains to be cemented together, forming limestone. Some shells or other skeletons of animals that lived on or in these lime sediments could become "fossils in rock" after cementation took place.

Not all fossils had hard parts when the organisms were alive. These fossils are preserved only due to unusual burials, and are important because they give us clues otherwise not available. Frequently, in volcanic ashes or in black muds (where oxygen is absent) plants, jellyfish, insects, and many other organisms may leave impressions (see Fig. Al.11) or form carbonized films after the organic molecules are reduced to the nonvolatile element carbon. Spores and pollen of plants are common in many rocks and have become very useful tools in assaying past climates and ecologic zones, as well as being used even for intercontinental correlation.

Trace fossils (indirect evidences of life) include a large variety of objects or marks made by organisms. Tracks and trails left by animals tell us about the nature of their locomotion and habits. Burrows, copro-*lites (fossilized fecal matter), and various surface or subsurface marks aid in estimating depth of water, softness of sediments before lithification, water circulation, and clues to the types of organisms living in the water and the food they ate (see Figs. 11.11 and 16.1).

*The fossil record has provided us with a vast, but incomplete, library of information to help interpret earth history. This record is being continually improved by new discoveries, but there are some built-in biases which probably can never be overcome. For example, it has been estimated that of 3,000 plants and animals living in and around modern coral reefs, fewer than 75 species are recognizable after death, and most of these have mineralized hard parts or skeletons.

Another major type of bias is the availability of fossiliferous rocks and their chance location. Hence, much more has been known about the fossil record in the Northern Hemisphere because of larger populations and more geologists. Additionally, over 75 percent of the record on continents represents land or shallow-sea environments, with only a few rare deeper water deposits. During the past several decades, deep-sea sediments have been extensively cored, and vital data have been obtained to reconstruct the history of the ocean basins. However, most of the fossils found in the cores are single-celled forms, thus presenting a bias. Further, the oldest deep-water fossils found to date are mainly Cretaceous, with possible Jurassic forms in the North Pacific; thus there is an all-too-short record of deep-water events.

Population density relative to size of a given organism is yet another form of bias. Microscopic organisms in the oceans are present in astronomical numbers, partly because of reproductive frequency and larger numbers of progeny compared to dinosaurs or elephants, which have very small populations and produce fewer offspring. Hence, the chance of finding a dinosaur is far less than that of finding a plant spore.

A very common bias relates to the random availability of rock outcrops, which may be ephemeral. Some outcrops yield vast numbers of fossils and are long lasting, but many disappear due to formation of soil, grassing over, or abandoning of quarries, which fill with ground and surface water.

SUMMARY

By now the central position of fossils in the history of geology should be clear, as well as the importance of fossils among the tools used for interpreting earth history. Cuvier even contended that we owe directly to fossils the beginnings of any real theories of the earth, for until they were studied, there was no reason to suspect distinct, successive events in the development of our planet.

Fossils are studied by geologists for three chief reasons. *First,* they are direct records of ancient life and illustrate, at least partially, the development or evolutionary change of that life through time. Put another way, they are results of organic evolution; also related to evolution is the fossil evidence of extinction of ancient life forms.

Second, fossils are important records of past environments; in fact they are among our most important tools for interpreting ancient climates and geographies. Of special importance is the proof they provide that large parts of continents have been submerged, not once but dozens of times, and certainly for more than 40 days at a time. Fossils also provided early evidence suggesting drastic past climatic changes. Hooke believed that fossils proved that England once had a tropical climate. He even suggested that the poles and equator had been differently situated so that southern England previously lay in the tropics.

Third, fossils have a great utilitarian value to geologists in the correlation of strata, as was predicted by Hooke but demonstrated by Smith, Cuvier, and Brongniart. Fossil correlation required careful collection and description of fossils and an appreciation of Steno's principle of superposition.

It is important to remember that the stratigraphic utility of fossils is quite independent of questions about "What is a fossil?" or of the controversy between catastrophism and organic evolution. The fossils could as well be nuts and bolts of distinct shapes, sizes, or colors as far as their utility alone is concerned; anything unique about any particular stratum is potentially useful for correlation! Catastrophism and evolution have to do with theoretically explaining *why* index fossils are useful, not the fact that they are.

In this chapter we have tried to illustrate something of the complex way in which geologic thought developed. One of the most important early ideas about the earth was the deluge concept, which, as we noted in Chap. 1, appears in many ancient traditions. Its ultimate origin is lost in antiquity, but it is perfectly natural that ancient Greeks would take marine fossils high in their mountains to be proof of submergence. Jews, followed by Christians, adopted this idea, renamed the event Noah's Flood, and the single-deluge origin of all fossils became entrenched for centuries. The Renaissance was marked by rejection both of Greek speculation and Medieval dogmatism in favor of the new observational or empirical method of investigation evidenced by Leonardo da Vinci, Steno, and others.

When the true origin and utility of fossils was

proven about 1800, that knowledge was immediately and widely applied, yet the philosophical showdown over catastrophism was delayed for half a century—even long after most thinkers had abandoned the single-deluge tradition. A final break with Medieval attitudes was agonizing, as shown by Cuvier's stubborn position as late as 1831:

The thread of the operations is broken, the march of nature is changed; and not one of her agents now at work would have sufficed to have effected her ancient works. (G. Cuvier, *Discourse on the Revolution of the Globe*)

Readings

Adams, F. D., 1938, The birth and development of the geological sciences: New York, Dover. (Paperback, Dover, 1954)

Clark, D. L., 1968, Fossils, paleontology and evolution: Dubuque, W. C. Brown. (Paperback)

Cowen, R., 1975, History of life: New York, McGraw-Hill. (Paperback)

Cuvier, G., 1818, Essay on the theory of the earth, with mineralogical notes by Professor Jameson and observations on the geology of North America by Samuel L. Mitchill: New York, Kirk & Mercein.

Geikie, A., 1905, The founders of geology (2d ed.): New York, Macmillan. (Paperback, Dover, 1962)

Gillispie, C. C., 1951, Genesis and geology: Cambridge, Harvard. (Paperback, Harper Torchbooks, 1959)

McAlester, A. L., 1968, The history of life: Englewood Cliffs, Prentice-Hall. (Paperback)

Rudwick, M. J. S., 1972, The meaning of fossils: episodes in the history of palaeontology: London and New York, MacDonald-Elsevier.

Schneer, C. J., ed., 1969, Toward a history of geology: Cambridge, M.I.T.

Wallace, P., 1972, Geology of wine: Montreal, 24th International Geological Congress, sec. 6, pp. 359–365.

Woodford, A. O., 1971, Catastrophism and evolution: Journal of Geological Education, v. 19, pp. 229–231.

Valentine, J. W., 1973, Evolutionary paleoecology of the marine biosphere: Englewood Cliffs, Prentice-Hall.

Von Zittel, K. A., 1901, History of geology and palaeontology to the end of the nineteenth century: London, Walter Scott.

3

"NO VESTIGE OF A BEGINNING— NO PROSPECT OF AN END"

The mind seemed to grow giddy by looking so far into the abyss of time.

John Playfair (1805, at Siccar Point)

I could get along very well if it were not for those geologists. I hear the clink of their hammers at the end of every bible verse.

John Ruskin (1851)

FIGURE 3.1   Unconformity with basal "puddingstone" or conglomerate along the River Jed, south of Edinburgh, Scotland, the second such example discovered by James Hutton in 1787. (From Theory of the earth, 1795.)

Though Columbus set out to circumnavigate the earth and prove it to be spherical, he actually did neither. Vikings had beaten him to America 500 years earlier, and it was Magellan who first completely circled the globe 30 years later. But the Greeks long before had deduced that the earth was not flat, for it cast a circular shadow on the moon during lunar eclipses, the Flat Earth Society notwithstanding. Moreover, the curvature, radius and circumference of the planet had been determined with surprising accuracy 2,000 years before Columbus!

How do important new ideas develop? Why are they sometimes quickly accepted and developed to produce "breakthroughs," while at other times are attacked or lost? Answers to these questions are, of course, complex, and relate to the whole social, economic, and intellectual history of humanity. For example, the old idea that the earth lay at the center of the universe was accepted and tenaciously held to for 14 centuries because it seemed admirably to fit the common experiences of apparent rising and setting of the sun and rotation of stars in the heavens. It also neatly fit the popular assumption that humanity, and therefore earth, lay at the center of everything.

Realization came very slowly that explanations must be based upon detailed evidence from the earth itself. Geology's awakening began with Steno's revelations about fossils and strata in 1669. Being an invention of humans, science is a special kind of social institution, which, therefore, shares the agonies of cultural change. Social, political, economic, and religious disorders of the Renaissance affected science markedly. For example, colonialism and wider trade created needs for better navigation, timekeeping, and map making, which stimulated astronomy and mathematics. Geology was affected through mining, quarrying, and clay production, especially after the industrial revolution.

In this book we shall regard *speculation* as conjecture—simply opinion or guess—for which there is little or no evidence. A *hypothesis*, however, is a logical, but tentative, explanation of verifiable phenomena based upon evidence. More than one hypothesis may appear equally capable of explaining a given set of facts; then further evidence for testing of the alternate hypotheses is needed to identify the most probable one.

As a rule, the simplest hypothesis that explains the largest number of facts is preferred. But this doctrine of simplicity must not be overworked, for there is no assurance that the simplest explanations will always be correct; many things in nature are not simple. A *theory* is distinguished here as an elegant and well-verified explanation that supersedes or encompasses a number of separate hypotheses. In science, theories finally may be elevated to the status of *laws* if rigorously tested and found to be invariable; a law, in other words, is a theory about which there is no longer any reasonable doubt. Formally speaking, there are many hypotheses in geology, few theories, and no exclusively geologic laws. The laws of physics, chemistry, and biology, however, provide the ultimate bases for understanding all geologic phenomena.

FIRST UNIFIED HYPOTHESES

THE COSMOGONISTS

The history of knowledge has been characterized by periodic formulation of hypotheses that explained most of the facts available at a given time. Science is a process of continuous refinement and testing of such explanations. Hypotheses inevitably have been colored by the temperaments, experiences, and prejudices of their advocates, which makes them all the more interesting to study.

Several Renaissance thinkers attempted to formulate general, all-inclusive hypotheses (called cosmogonies) to explain the origin and development of earth, life, and the entire universe. Though riddled with speculation, these early explanations had a long-lasting influence. It was assumed that the earth originally was hot and glowed like the sun. Subsequently, as it cooled, a primitive hard crust formed, and the water and atmosphere became segregated according to relative densities. The interior had a fiery core surrounded concentrically by other materials.

BUFFON'S BREAK WITH GENESIS

One of the most fertile eighteenth century thinkers in natural science was the Frenchman G. L. de Buffon, who published beginning in 1749 a 34-volume work entitled *Histoire Naturelle*. Accepting the idea of a molten origin, Buffon suggested that planets originated through detachment of hot portions of the sun by collision or near-collision with a great comet, an idea still held in nonscientific circles today (see Fig. 7.1, p. 92). Subsequent earth history consisted of cooling through six distinct epochs totaling 75,000 years. He was one of the first to question the literal significance of the Six Days of Creation and to postulate such a great age for the earth.

Buffon represented a major turning point in the history of geologic thought by his rejection of Scripture as a source of geologic insight and his extension of the supposed age of the earth. Even more importantly, he rejected catastrophes, believing instead that ordinary processes now operating could explain geologic phenomena.

FIRST TRUE GEOLOGIC CHRONOLOGY

Ideas commonly have arisen independently in different minds thousands of miles apart. Such was the case with the first concepts of natural subdivisions within the rocks of the earth's crust. Whereas Buffon's Epochs were largely speculative, several contemporaries working in Germany, Italy, Switzerland, and England were recognizing and naming local natural rock divisions based upon evidence observed in the field. In the English coal fields as early as 1719, a succession of strata was recognized and illustrated in detailed cross sections. Two Germans in the middle eighteenth century also discerned an order in the vertical succession of rocks. The first one distinguished stratified, nearly flat-lying, fossiliferous deposits (or *Flötzgebirge*) from more primitive, unfossiliferous, steeply tilted ones cut by dikes and ore veins. He believed that all had formed in the sea, but that *primitive* ones were deposited in disorder during the early creation while the *Flötzgebirge* formed during the Deluge. A third group of deposits postdated the Deluge and were formed by accidents of nature such as earthquakes, landslides, and volcanic eruptions. He assumed that these divisions matched the epochs of Genesis. In preparing early geologic maps and cross sections the other German observed that certain strata are characterized everywhere by land plants and coal, while others contain only marine fossils. He originated the modern concept that the strata of a given formation reflect similar formative processes and environments. Both men assumed that lithologic differences among formations reflected worldwide fluctuations of the sea.

In 1759 a respected Italian authority on mining and mineralogy named Giovanni Arduino formally distinguished in northern Italy: *Primitive Rocks* (unfossiliferous schists and veins of the high Alpine mountain core), *Secondary Mountains* (with limestone

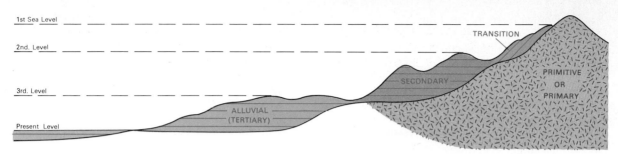

FIGURE 3.2 Early neptunian concept of deposition of strata by a receding, universal ocean. Note that, as drawn from early descriptions of the neptunian theory, positions of the strata seem to defy superposition; that is, successively younger ones do not everywhere directly overlie older. Later neptunians, such as A. G. Werner, envisioned the older strata as dipping and flattening beneath younger ones away from the mountains.

and shale containing abundant fossils), *Tertiary Mountains* (richly fossiliferous clay, sand, and limestone of the low hills), and finally the youngest *Volcanic Rocks*. The scheme became the first formal stratigraphic standard for subdivision of the rocks of the earth's upper crust. Being the simplest hypothesis, a universality of historical development of the entire earth's surface long had been assumed, and it already was recognized that fossiliferous marine strata similar to those of Italy were widespread over much of Europe. Therefore, it was natural for Arduino's chronology to be applied in Germany, Russia, and elsewhere. It was destined to have a profound influence upon the progress of geology.

NEPTUNISM *(all rocks formed in the oceans)*

A. G. WERNER

As Buffon was publishing, one of the most influential men in all of the history of geology, Abraham Gottlob Werner, was born in Germany. In 1787, Werner published a general theory of the origin and rock sequence of the crust. He became perhaps the most inspiring and persuasive geology teacher of all time. Devoted followers carried his ideas beyond Germany, thus making him the most influential figure in geology for half a century.

Most predecessors had postulated a former universal ocean, and believed that most rocks were products thereof. Mountains were assumed by many to be original irregularities of a chaotic early Creation, and the succession of strata was presumed to reflect subsequent gradual diminution of the sea and emergence of land. Because of the great emphasis upon the sea for explaining all of the crust, this "theory"

earned the apt nickname neptunism (for Neptune, Roman god of the sea) (Fig. 3.2).

Werner adopted the early neptunian concept of the earth, and amplified it through a more detailed accounting of the chronological succession of strata. But, unlike some early neptunians, Werner was not a catastrophist. He regarded the earth as much older than humanity, and did not bother correlating stratigraphic history with Scripture. Primitive rocks, such as granite and schist, were regarded as unfossiliferous and entirely of chemical origin (Table 3.1). Be-

TABLE 3.1 Standard geologic column as conceived by 1800

Alluvial rocks (also Tertiary of Arduino)	Relatively loose gravel, sand, peat and some limestones, sandstones and shale
Secondary (*Flötz*) rocks	Sandstone, limestone, gypsum, salt, coal, basalt, obsidian
Transition rocks	Hard graywacke, slate, and some limestone with first organic fossils; chiefly chemical, but with first mechanical deposits
Primitive (or Primary) rocks	Chemically formed original surface of earth (granite, schist, gneiss, serpentine, etc., with no organic fossil remains)

cause they appear in the high axes of mountain ranges, they must be the earliest rocks precipitated from the sea. As the water subsided, fossiliferous Transition rocks were deposited. These included the first mechanically formed (i.e., fragmental) impure sandstones, called graywackes, as well as some chemical limestone. The inclination of these strata (Fig. 3.2) was assumed to be due to initial deposition on the irregular Primitive crustal surface or to collapse into subterranean caverns. Lower topographically were *Flötz* (flat) rocks, later referred to as the Secondary. These comprised sandstones, limestones, gypsum, rock salt, coal, and basalt. Finally, in the lowlands and valley bottoms, the youngest or Alluvial deposits were formed as the sea receded to its present position; they are nonmarine. Arduino's ★ Tertiary and Volcanic were not recognized by Werner as separate divisions. Werner's grouping of rocks by supposed relative age soon became almost universally accepted as a standard for worldwide comparison. He had succeeded in mobilizing, at last, a conscious scientific desire to study the history of the earth.

THE HEATED BASALT CONTROVERSY

The neptunian scheme suffered from several obvious shortcomings, such as a gigantic water-disposal problem. Wernerians ignored many of these, even while proudly announcing that they rejected hypothesis and speculation and built their entire case upon irrefutable facts, a pious claim pressed with equal vigor by the opposition.

Most difficult and controversial was the origin of basalt. To Werner, volcanoes all were recent and had no great importance in the history of the earth in spite of dramatic evidence marshalled by the previous generation of anti-neptunian Italian geologists, who thought that all land was formed initially by volcanic eruptions such as they themselves had seen build new islands in the Mediterranean. Werner, instead, endorsed an old idea that volcanic eruptions, when they do occur, originate from combustion of buried coal seams; therefore volcanoes could have occurred only after deposition of the great Secondary coal layers. He contended stubbornly to his last day that basalt interstratified with sediments had been precipitated from the universal ocean; where it looked like lava, it had been fused by combustion of adjacent coals.

It now seems astounding that most geologists could seriously entertain the neptunian origin of basalt, but it is even more surprising to realize that two Frenchmen had published volcanic interpretations of famous basalts in central France (see Fig. 2.8) before Werner first published his theory (Fig. 3.3)! Together with the Italians, those men founded a different school of geologic thought that came to be called vulcanism (for Vulcan, god of fire), to which three of Werner's own students soon were converted. Subsequently, volcanic rocks were found interstratified with Secondary and Primitive deposits, proving that volcanism had occurred throughout *all* of earth history.

PLUTONISM—THE BEGINNING OF MODERN GEOLOGY

JAMES HUTTON

In the last quarter of the eighteenth century, geology finally was coming of age; the name itself came into general use then. It is a curious coincidence that the two most influential early geologists—and of opposite persuasions—lived and wrote at the same time. The great Scottish innovator, James Hutton, was a gentleman farmer and geologist. Farming generated an interest in soils and their fertility, which led to his interest in the earth. Hutton was obsessed with discovering a mechanism for maintenance of the environment for land plants and animals in the face of obvious destructive forces acting to wear away the British landscape. First announcement of his revolutionary theory was in 1785 when Werner's influence was at its peak. Eventually Hutton's theory was to provide the foundation of modern geology.

AN OLD DYNAMIC EARTH

Hutton early recognized that rocks exposed today to the vicissitudes of the atmosphere tend to decay and produce gravel and soil. He noted by analogy that many rocks contain debris derived from older rocks, which apparently had decayed similarly. Moreover, he appreciated that counterparts of ancient sedimentary rocks are still forming, chiefly in the sea. These observations led him to a cyclic view of earth change with construction of new products neatly balancing destruction of old. Such tendency toward equilibrium among dynamic natural processes today is described as a physical system in a steady state. Energy is

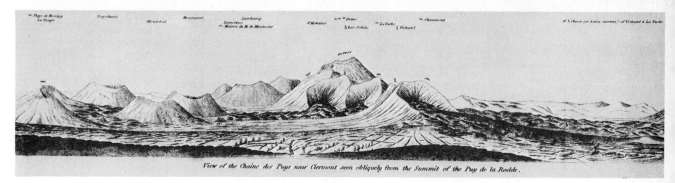

View of the Chaine des Puys near Clermont seen obliquely from the Summit of the Puy de la Rodde.

FIGURE 3.3 Auvergne volcanoes, south-central France. Note the perfect volcanic cones and hummocky-surfaced lava flows extending out from two of the craters near the center. *(After G. P. Scrope,* Considerations on volcanoes, *1825.)* puys

expended, but the system appears generally the same at any time; a beach is a modern example, for it looks essentially the same day-to-day in spite of constant pounding by the surf. A prevailing view in Hutton's day—at the beginning of the industrial revolution—was that an external intelligence had set all of nature in motion like some gigantic machine. The earth was like a great heat "engine," with volcanoes acting as safety valves, but it was also a cyclical "machine" such as a clock whose pendulum swings to and fro.

[Hutton's greatest contribution was the original full appreciation that the earth is internally dynamic and ever-changing in sharp opposition to the Wernerian view that rocks had formed on a rather static solid earth foundation. This was really the central conflict with neptunism! To Hutton, it was absurd that steeply inclined Transition and Secondary strata in mountains were originally deposited in such positions. He reasoned instead that these strata had been tilted and crumpled later by internal earth forces.

Hutton also attacked catastrophism. He recognized that valleys were cut by rivers rather than the sloshings of great oceanic floods; he objected to the belief that everything must be deluged, and found, instead, modern earth processes quite capable, given enough time, of having produced the record of the past.

Hutton saw that, if we assume that the earth is only 6,000 years old, we *must* subscribe to catastrophic earth change to explain geologic facts. The fresh condition of a 2,000-year-old Roman wall in northern England, however, convinced him that ordinary geologic processes—by human standards—act slowly. Because rocks seemed to reveal only the results of ordinary processes like those visible and acting at the present, a great deal more time must have been required for the almost imperceptible changes to have accomplished great geologic work. "What more can we require? Nothing but time," said Hutton.

"NOTHING IN THE STRICT SENSE PRIMITIVE"

Hutton's first quarrel with neptunians was over the latters' assumed uniqueness of the different chronologic rock divisions, especially the Primitive. Such was counter to Hutton's own observations as well as to his concept of an earth in dynamic equilibrium and acted upon by the same (i.e., uniform) processes through time. Presence of pebbles and sand within Scottish schists otherwise regarded as Primitive attested to the existence of still older rocks somewhere from which those fragments had been eroded. So-called Primitive rocks were not entirely of chemical origin, as the neptunians claimed; neither were they unique in lacking fossils.

BASALT (revolutionary)

Hutton was impressed early with the importance of subterranean heat, whose existence was inferred from hot springs and volcanoes. Also he interpreted clinker-like coal seams transected by basaltic dikes as having been baked. He studied a well-exposed basalt sill surrounded by sediments (Fig. 3.4), and

FIGURE 3.4   Salisbury Crags, eastern edge of Edinburgh, Scotland, a basalt sill intruded into the "Secondary" Old Red Sandstone and tilted gently eastward. This locality convinced James Hutton of the igneous origin of basalt, but later Robert Jameson was equally sure of oceanic precipitation of the same basalt. As a student, Charles Darwin was taken here by Jameson; he later commented "I do not wonder that I determined never to attend to Geology."

found clear evidence both of its hot origin and forcible intrusion after the sediments had formed.

Hutton's igneous interpretations of basalt and natural rock glass were reinforced experimentally in 1792 by a friend (Sir James Hall), who melted basalt between 800 and 1200°C, only to produce glass with rapid cooling, but basalt again with slow cooling.

GRANITE AND MINERAL VEINS

Hutton knew that mineral veins occur in areas of Primitive granites and schists. In 1764 he observed a peculiarly textured granite in northern Scotland and thought that it, as well as the complex minerals in metallic veins, could have formed only through crystallization from hot fluid material. Sea water could not hold all the different compounds simultaneously in solution, therefore veins and dikes were not simple oceanic precipitates in open cracks as the neptunians preached. Influenced by the work of a prominent

chemist friend, Hutton shrewdly inferred that great pressure deep in the earth would affect chemical reactions markedly; compounds that are gaseous at high temperature and low pressure (such as those expelled to the atmosphere from volcanoes) would not be lost as gases from hot solutions under high pressure at depth. Moreover, there could be intense heat without fire at such depths. Though his evidence leaves much to be desired, Hutton's interpretation of a hot origin for granite eventually displaced the neptunian view. Great appeal for subterranean heat to produce basalt, granite, and mineral veins earned for his theory the nickname *plutonism* (for Pluto, god of the lower, infernal world).

EVIDENCE OF UPHEAVAL OF MOUNTAINS

Hutton proposed that heat also consolidated sediments, and then caused upheaval of rocks to form mountains by thermal expansion of the crust (Fig. 3.5). Granites, veins, and basalts were assumed to have formed at such times. He deduced that dikes of granite cutting across or intruding older rocks must exist. Having predicted their ultimate discovery, he scoured Scotland for examples, and he found many such dikes transecting schists in central and western Scotland. His elation could not have been greater had they been veins of solid gold.

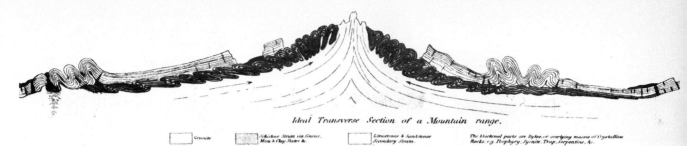

Ideal Transverse Section of a Mountain range.

Granite   Schistose Strata viz. Gneiss, Mica & Clay Slates &c.   Limestones & Sandstones Secondary Strata.   The blackened parts are Dykes, or overlying masses of Crystalline Rocks. e.g. Porphyry, Syenite, Trap, Serpentine, &c.

Hutton also discovered what he had predicted and long had sought, namely a great unconformity in the stratified rock record (Fig. 3.1). He suspected that the contact between Secondary and older rocks might reveal clear evidence of great upheaval of mountains from the ancient sea floor. In all three exposures that he discovered, pebbles of older rocks occurred just above the unconformity. In Hutton's examples, the dip of strata below the unconformity was steep, and from Steno's principle of original horizontality, one concludes that severe upheaval had tilted the rocks and formed mountains.

The unconformity, together with intrusive granite dikes, proved conclusively that mountains are neither static bumps left from a primeval earth surface nor relicts of the Deluge, but were formed by repeated dynamic convulsions of the crust (Fig. 3.6).

CHARLES LYELL'S UNIFORMITY OF NATURE

GEOLOGY'S GREATEST BOOK
Before Hutton's unifying theory and the nearly simultaneous development of principles of correlation and geologic mapping by Smith and Cuvier, modern

FIGURE 3.5 The plutonist concept initiated by Hutton of upheaval of mountains, folding of strata, and intrusion of granites, all as results of internal heat. Note the discordance between the upper (Secondary) strata and the more contorted schists. *(After G. P. Scrope,* Considerations on volcanoes, *1825.)*

geology as a unified science really did not exist. These developments finally provided the link among very diverse studies, and provided both method and purpose. Evidence of subsequent rapid growth of geology is provided by the founding of the first geological society in London (1807) and the first governmental geological survey in Great Britain (1835).

In spite of these breakthroughs around the beginning of the nineteenth century, geologic efforts were still somewhat diffused. Many people active in earth studies were not immediately aware of the impact of Hutton, Smith, and Cuvier. Moreover, Hutton was a poor writer and received bad reviews because of his anti-catastrophist preachings (Fig. 3.7). Particularly, his contention of great antiquity of the earth expressed in the title to this chapter was like a red flag

FIGURE 3.6 Diagrammatic representation of Hutton's concept of multiple upheavals, intrusion of granites, and development of unconformities to form a continuous progression of new landscapes born from the wastes of old.

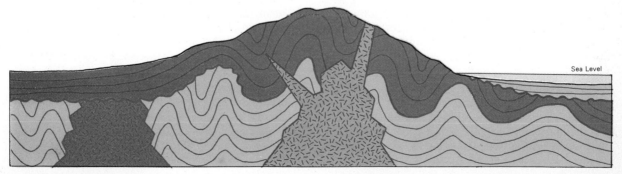

Sea Level

FIGURE 3.7 Cartoon of James Hutton, rock hammer in hand, contemplating an outcrop bristling with the faces of several antagonists. *(From John Kay's* Edinburgh portraits, *1842.)*

to a bull. In the eyes of many, no science was valid if it did not support and clarify Scriptures, which Hutton had refrained from doing, even though he was firmly convinced of some grand design or wisdom in nature.

A persuasive synthesis of mushrooming factual data with a noncatastrophic, Huttonian interpretation was supplied by an English geologist, Charles Lyell, in a great work entitled *Principles of Geology*, which was revised 11 times between 1830 and 1872. Lyell, who was born the same year that Hutton died, produced one of those rare books of almost unprecedented impact. He painstakingly illustrated the concept of uniformity of nature through time. He traveled extensively in Europe and North America and was able to show overwhelmingly that geologic processes observed today can be assumed to have operated in the past. As an antidote for Cuvier's catastrophism,

Lyell adopted a steady-state view of the earth more extreme than that of Hutton. He believed staunchly that the general intensity of processes and conditions had varied hardly at all through time. Of course some violent events like earthquakes occur, *but no more frequently in one epoch than in another*. If one continent were large, another simultaneously was small; if the climate here were cold, it was balanced there by hot conditions. By such rationalizing, he envisioned that conditions remained essentially constant through time for the earth as a whole. To Lyell, his mission was "freeing the science from Moses." To acknowledge *any* irregular variations of intensity was to leave the door dangerously ajar for catastrophism. Therefore, only uniformly repetitive or cyclic (i.e., steady-state) changes were permissible.

The uniformity doctrine actually was older than Lyell or even Hutton, for it is implicit in earlier writings of Leonardo, Buffon, and a Russian named Lomonosov. But Lyell most successfully interpreted and publicized it for society at large. His importance is underscored by the fact that early American geologists were ardent neptunian catastrophists ignorant of Hutton's theory until Lyell began publicizing the new geology, as well as by his impact upon Victorian writers. Moreover, *Principles of Geology* had a great influence upon Lyell's friend, Charles Darwin. Lyell, himself, had great difficulty in abandoning a fixed species concept, for to accept change of one species into another in an irreversible line of descent seemed to him contradictory to his unbending devotion to a uniform, cyclical earth. Evolution and a strict steady state are incompatible.

THE DOCTRINE OF UNIFORMITY TODAY
→ "the uniformity of nature's law"

The uniformity of nature, or uniformitarianism, must be examined carefully. First, we note that uniformity is an *assumption* about nature—a doctrine rather than a logically proven natural law. Hutton himself said in 1788 that "the uniformity of nature, even if not strictly true, is necessary for our clear conception of the system of nature."

Much confusion exists about the uniformity doctrine, even among scientists. The uninitiated interpret it as implying literally that the earth always has been *exactly the same*—a Lyellian influence. An old cliché that the *present is the key to the past* is misunderstood by many, too, for the present earth is unique in

FIGURE 3.8 One of many noncatastrophic, historic geological changes documented by Charles Lyell. The ancient village of Eccles in Norfolk, England, was beginning to be attacked by the North Sea around 1600. By 1839 coastal dunes had buried all but the top of the church tower (left), but by 1862 wave and wind attack had exposed the ruins again (right). (After C. Lyell, Principles of geology, 10th ed., v. I, pp. 514–515.)

terms of climate, topography, and life. In part the *past is a key to the present*, for the historical record provides a perspective against which to compare the present.

How, then, is uniformity to be regarded? Only a static earth could be completely unchanging, yet ours clearly is dynamic. Lyell allowed change, for that is what the *Principles* was all about (Fig. 3.8). His changes, however, were orderly, cyclic ones confined within narrow limits. But a Lyellian steady-state dynamic earth would defy basic laws of physics. This contradiction was recognized by the nineteenth century British physicist Lord Kelvin, who challenged strict uniformitarianism as perpetual motion—an earth machine that never ran down was a physical absurdity! Kelvin reasoned that the energy reservoir of the entire solar system must have been greater in the past and was gradually being dissipated. His position implied significant differences of intensity of past conditions, thus a noncyclic view of the earth.

Today we envision neither a violently catastrophic nor a rigidly uniform earth, but *rather an evolutionary one that has changed through an irreversible chain of cumulative historic events.*

Amid all the revolutions of the globe the economy of Nature has been uniform, and her laws are the only thing that have resisted the general movement. The rivers and the rocks, the seas, and the continents have been changed in all their parts; but the laws which describe those changes, and the rules to which they are subject, have remained invariably the same. (John Playfair, 1802)

The only assumption we make today is that physical and chemical laws are constant, which is properly called <u>actualism.</u> By inductive reasoning and analogy, the study of geologic processes acting today provides us with powerful clues to their past action, but we do *not* assume that those processes always acted with the same rates and intensities. There is confusion about what is meant by "catastrophic" processes and by a lack of appreciation of the vastness of geologic time. Geologists today routinely accept sudden, violent, and even certain unique events as perfectly consistent with contemporary earth theory. Only by substituting the term actualism for the ambiguous uniformitarianism can misconceptions be minimized.

THE SCIENTIFIC METHODOLOGY OF GEOLOGY

HISTORICAL SCIENCE

Physics and chemistry are concerned almost exclusively with phenomena controlled by presumably universal natural systems that are nonhistorical, that is, *independent of the time at which they operate.* Geology and biology, on the other hand, are historical. When a geologist focuses only upon present processes and configurations of earth materials, he is an applied physicist or chemist. But when he begins to interpret a series of past events, he becomes a unique historical scientist. While assuming all physical, chemical, and biological theory, reconstruction and explanation of history has become his chief goal. Moreover, prediction of results from known causes (deductive reasoning), so important in nonhistorical science, is replaced in geology by the interpretation of ancient causes from their historical results (inductive reasoning).

SCIENTIFIC EXPLANATIONS

a game of teasing of information out of nature

We have seen that only the objective, disciplined assault called science promises real success in understanding nature. Science consists simply of the formulation and testing of hypotheses based upon observational evidence; experiments are important where applicable, but their function is merely to simplify observation by imposing controlled conditions. Textbook treatments notwithstanding, it is rare for a scientist to make many observations without already having a tentative hypothesis in mind to test, and many brilliant breakthroughs have resulted from accidental discoveries or intuitive flashes based upon skimpy evidence. For example, Hutton had formulated his theory and deduced from it the existence of intrusive dikes and unconformities *long before he had seen either feature*.

The average person may not appreciate the differences between scientific proof and absolute proof in logic, which difference is between *probability* and *certainty*. Science deals with probabilities, that is, one hypothesis seems for the moment more correct than another. Therefore, scientific explanation is never-ending, for it produces only an approximate working model of nature as we think we understand it at a particular time. In geology our explanations are less certain than in, say, physics because of the enormous complexity of the earth and the very incomplete record we have of its history.

As a series of observations leads to many interrelated hypotheses, it becomes necessary periodically to establish broad simplifications, either unifying theoretical laws or empirical generalizations. Newton's famous laws of motion provide clear examples, and ones which can be expressed in concise, unambiguous mathematical language. Empirical strati-graphic generalizations developed by Smith and Cuvier, on the other hand, cannot be so expressed, but their significance is not lessened thereby.

MULTIPLE WORKING HYPOTHESES

Most geologic phenomena involve many interacting factors, and it is commonly impossible to restrict the number of variables under consideration or to evaluate each and every variable precisely. Moreover, the record of evidence is fragmentary. As a result, it is often impossible (at least temporarily) to discover the decisive evidence to disprove a tentative hypothesis. Because of this difficulty, an American geologist, T. C. Chamberlin, in 1890 formalized an important and widely employed procedure of inquiry called the method of multiple working hypotheses. He pointed out that if one constructs only a single explanation, he may become too strongly wooed by it, and unwittingly seek only the facts to support it or bend his hypothesis to fit new facts. Moreover, absence of an alternative explanation is no assurance that one has indeed discovered the truth. Multiple alternate hypotheses divide our affection, suggest new tests, and expose more facets of the problem. This method leads to more acute mental habits and cultivates the highest possible degree of objectivity. It is a method of analysis to be commended to all walks of life.

THE USE OF MODELS

Because of the limitations of incomplete evidence and multiplicity of controlling factors, it is helpful to devise idealized simplifications known as models. Scale models (replicas) of actual objects or processes are familiar and of obvious value, say, for studying transport of sand by water. Experimentation with scale models has inherent practical limitations, especially if very long spans of time or an excessive number of factors are involved in the process being modeled. In historical sciences, we commonly use modern phenomena as models of ancient events, in which cases a natural modern analogue (e.g., an entire river) is a proxy for an unnatural experiment. For large and complex systems we let nature do the "experiment" while we observe and measure. Advent of high-speed computers has made possible the development of mathematical models for exploring interactions in such cases; simulation of very complex geologic processes, such as the isostatic rise of a continent after deglaciation, becomes possible.

** never to be completed.

Finally, descriptive conceptual models are constructed to convey idealized interpretations of phenomena, but they are more subjective. Most of the interpretative diagrams in this book are simplifying conceptual models of geologic reality.

SUMMARY

Early Italian geologists reasoned from their experience that all land came into being through volcanism. Neptunians at the same time argued, in effect, that the earth's crust was structurally static. Mountains either were inherited vestiges of an early molten, chaotic period of the Creation, or were formed only 2,000 or 3,000 years ago by catastrophic waves and currents during the Noachian Deluge. Practically all rocks, including basalt, were seen as deposited from a receding, universal ocean. Gradually French and Italian vulcanists were persuasive in removing basalt from Werner's list of oceanic precipitates; the mere 6,000-year age of the earth was challenged by Buffon and Hutton, and the Flood became relegated to a brief, young event of little geologic importance.

Finally, at the end of the eighteenth century, modern geology began as Hutton exposed the fallacies of the neptunian arguments. As the first plutonist, he fully recognized the dynamic nature of the earth's interior. At the same time, Hutton laid the groundwork for later overthrow of catastrophism by Lyell and Darwin, through his insistence that present observable processes, given enough time, could have produced all the geologic record.

Three chief doctrines of geologic change have prevailed. Catastrophism invoked sudden, violent upheavals and floods produced by unknowable causes not now operating. Actualism assumed a uniformity of causes through time but considered that their rates, intensities, and locations have varied. Strict uniformitarianism, on the other hand, assumed that processes in the past were constant *both* in kind and intensity (or rate). While the Lyellian doctrine of uniformitarianism in its strict form is no longer acceptable, the assumption of uniformity of natural laws (actualism) is mandatory for rational historical analysis.

Actualism provides a connecting thread between the present and past and allows us to reconstruct events never witnessed by humans. Comparative or analogical reasoning both among ancient situations and between ancient and modern ones must be employed very extensively. This is a two-way process, for in some cases the past is a key to better understanding of the present. History provides a powerful test of geological hypotheses; if historical evidence demands that a certain thing occurred, then it must be possible even if difficult to conceive by existing theory. Failure of current physical or chemical theory to explain some phenomenon is hardly adequate reason to deny its possibility, though this commonly has been done. That which *has* happened *can* happen!

The most important aspect of history is that it never exactly repeats itself. Historical events involve so many complex factors that they are at least in small degree unique events, therefore the probability of duplicating exactly a known complex historical series of events at another time becomes very improbable. Individual phenomena, to be sure, have recurred many times, but sequentially related events are not repeatable *in exactly the same way*. It follows that the earth is a dynamic, evolutionary planet, not a cyclical steady-state one as Lyell thought.

Readings

Albritton, C. C., Jr., 1967, Uniformity and simplicity: Geological Society of America Special Paper 89.

Davies, G. L., 1969, The earth in decay: New York, American Elsevier.

Greene, J. C., 1959, The death of Adam: Ames, Iowa State Univ. Press, chaps. 1–3. (Paperback, Mentor Books, 1961)

Hooykaas, R., 1963, The principle of uniformity in geology, biology and theology: Leiden, Brill.

Hutton, J., 1795, Theory of the earth with proof and illustrations: London, Cadell & Davies.

Lyell, C., 1830–1872, Principles of geology (14 eds.): London, J. Murray.

Playfair, J., 1802, Illustrations of the Huttonian theory: London, Cadell & Davies.

Simpson, G. G., 1963, Historical science, *in* The fabric of geology: Geological Society of America.

Wilson, L. G., 1971, Sir Charles Lyell and the species question: American Scientist, v. 59, pp. 43–55.

Specimen No.1 Scratched
by a Glacier Thirty three
Thousand Three hundred
Thirty Three Years before
the Creation

Scratched by a cart
Wheel on Waterloo
Bridge the
day before
yesterday

Prodigious
Glacial
Scratches

Scratched by T. Sopwith

The Rectilinear Course of these
Grooves corresponds with the
motions of an IMMENSE
BODY the momentum of which
does not allow it to change its
Course upon Slight Resistances

COSTUME of   the GLACIERS

4

THE RELATIVE GEOLOGIC TIME SCALE AND MODERN STRATIGRAPHIC PRINCIPLES

If history always repeats itself, how come there's so much to learn?

(Anonymous Schoolboy)

Historians explain the past and economists predict the future. Thus, only the present is confusing.

(Anonymous)

FIGURE 4.1 Cartoon of W. E. Buckland, noted British catastrophist, who steadfastly defended the alleged role of the biblical Flood in producing many geologic features. Finally, by 1840, he reluctantly concurred (as shown) that glaciers had produced most of the features attributed to the Deluge. *(From Gordon,* Life and correspondence of William Buckland, *1894.)*

After publication of Smith's and Cuvier's first geologic maps, stratigraphic studies went forth at a rapid pace. Though the Wernerian chronology (see Table 3.1) continued to be used as a general standard of reference for supposed relative age, local names for strata became more and more numerous. Naming of distinctive local rock bodies was a natural by-product of mining and mapping by many geologists. The names developed unconsciously as a shorthand, and reflected geographic localities or peculiar rock types (Fig. 4.2). At first such names were used informally and only locally, but as studies were extended, the lateral continuity of certain distinctive strata became apparent, and it was natural that some names were extended more widely. At the same time, fossils were being collected and studied more and more, and strata with similar assemblages were *correlated* from widely separated areas even where complete physical continuity between them could not be observed because of discontinuous outcrops. Names for strata became extended still farther from the local areas where originally applied, and the geologic map of all western Europe developed.

As the more important named divisions became extended widely in Britain and northwestern Europe, relationships between different groups gradually became clearer through application of the principle of superposition. A group of Secondary strata named Juras for the Jura Mountains of France and Switzerland were found to overlie another group named Trias in central Germany and to underlie a third group in France named Cretaceous (Fig. 2.8). By application of Steno's principle of original lateral continuity, the Cretaceous name was extended from France across to England (see Fig. 2.6). Conversely, the coal-bearing strata of Britain were named Carboniferous, and correlation allowed extension of that name to the Continent.

In a rather haphazard, trial-and-error fashion, old rock divisions became subdivided into more discrete and clearly traceable, named groups of related strata called formations. The names applied to each group, once arranged in their proper vertical succession according to superposition, provided the basis of a more detailed and practical chronology that gradually displaced the older scheme.

There was a period of overlap of out-of-date concepts with the growing new chronology. For example, although the idea of a single major deluge had yielded to growing contrary geologic evidence, many people still clung to a nearly universal but brief, recent Flood. Such a flood was thought essential to account for widespread and large erratic boulders

FIGURE 4.2 Ludlow Castle, seat of power of the Tudors in Ludlow, England, was built (chiefly between A.D. 1200 and 1400) of Devonian Old Red Sandstone. Note stratification visible in many blocks.

scattered over northern Europe (Fig. 4.3); it still resembled catastrophism. The boulders and associated clay or drift (called Diluvium) were assumed to have been drifted in by icebergs. Many prominent geologists followed this belief (Fig. 4.1), and as late as 1829 the formal name Quaternary was proposed in good Wernerian tradition to include these young deposits. The occurrence of primitive Paleolithic human artifacts with such deposits led Lyell to believe that humans had existed through most of Quaternary time.

THE MODERN RELATIVE TIME SCALE

At first there was no thought of building a new systematic time scale. But there developed the need to classify and organize material into some manageable, orderly form; otherwise the study of strata would be chaos. About 1830 a conscious effort was undertaken to name formally the entire European succession. Two British geologists, Adam Sedgwick and Roderick Murchison, decided that the Transition rocks had been neglected. Therefore, they addressed themselves to the task of studying them in Wales.

In 1835 two divisions were named, the older the Cambrian (for Cambria, ancient Roman name for Wales) and the younger the Silurian (for an ancient Welsh tribe, the Silures). Sedgwick and Murchison next extended their work southward. There they encountered unfamiliar strata with marine fossils judged to be intermediate between Silurian and Carboniferous ones, thus perhaps contemporaneous with the nonmarine Old Red Sandstone that lies in a similar stratigraphic position in Wales (Fig. 4.2). A new, third division was named Devonian, and soon thereafter deposits of nonmarine (Old Red) type were found interstratified with some marine Devonian layers in South Wales, confirming correlation with the Old Red Sandstone (Fig. 4.4).

Sedgwick recognized need for a still more formal (stratigraphic classification) with different levels of subdivisions based upon more rational criteria than those of the archaic scheme. He therefore proposed the concept of very large divisions based upon the gross characteristics of fossils that were by now rather well known. Each of the large divisions would include a number of smaller subdivisions. For example he proposed the Paleozoic Era (meaning "early or old life") to comprise the Cambrian, Silurian, and Devonian divisions. The Paleozoic was to include all the divisions from the oldest recognizable fossil animals through all the divisions dominated by invertebrate animals of related types. We now know that abundant and highly organized invertebrate fossil animals first appear in Cambrian strata, though primitive ancestral forms occur in lithologically similar Eocambrian ones. The Prepaleozoic has been given many different names such as Azoic ("lacking life") and simply Precambrian, which is the most commonly

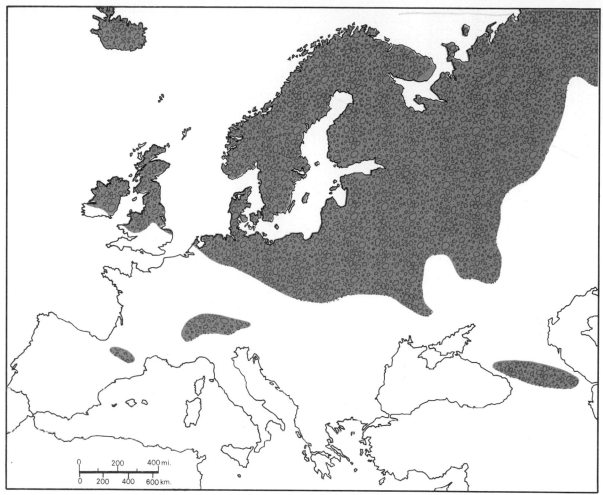

FIGURE 4.3 Distribution of the Drift or Diluvium with erratic boulders ("boulder clay") over Europe, and considered, for a time, to have been drifted in by icebergs during the Noachian Flood. After 1840 it was attributed to glacial transport. *(Adapted from L. J. Wills, Palaeogeographical atlas, 1951; by permission of Blackie and Son Ltd.)*

used term today. In formal usage we prefer Prepaleozoic, although we also use Precambrian out of long habit.

The largest time scale divisions, based solely upon fossil life, are now called eras of geologic time. The divisions of strata subdivided and named according to specific European strata and their contained fossils are called systems. Today we must distinguish between the tangible physical rock record, the systems, and abstract time. Thus the Cambrian System of rocks was formed during the Cambrian Period of time. In Table 4.1 we show the relative geologic time scale with eras subdivided into peri-

ods, and for the younger periods, into still smaller divisions called epochs.

Development of the geologic time scale was anything but an orderly affair. One of geology's many stormy feuds arose over the division and naming of strata. Sedgwick and Murchison, who were fast friends during the 1830s, later fell out bitterly over definition of the boundary between the Cambrian and

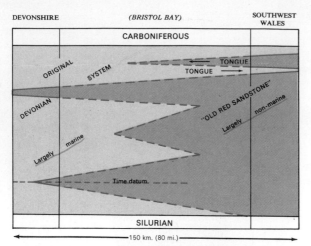

DEVONSHIRE · (BRISTOL BAY) · SOUTHWEST WALES

CARBONIFEROUS

ORIGINAL SYSTEM

TONGUE

TONGUE

DEVONIAN

"OLD RED SANDSTONE"

Largely non-marine

Largely marine

Time datum.

SILURIAN

←—— 150 km. (80 mi.) ——→

FIGURE 4.4 Relations of the nonmarine Old Red Sandstone facies of Wales to marine Devonian facies of Devonshire as inferred by Sedgwick and Murchison. Intertonguing of facies proved the Devonian age of the Old Red and also that different environments of deposition existed simultaneously.

FIGURE 4.5 Geology of Wales (left), where Sedgwick and Murchison worked and feuded over correlation of Cambrian and Silurian strata (right). The Old Red Sandstone marks the base of the Secondary.

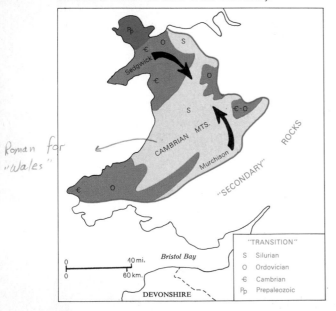

Roman for "Wales"

Sedgwick

CAMBRIAN MTS.

Murchison

"SECONDARY" ROCKS

Bristol Bay

DEVONSHIRE

0 — 10 mi.
0 — 60 km.

"TRANSITION"
S Silurian
O Ordovician
Є Cambrian
Pp Prepaleozoic

Silurian Systems. As their separate mappings in Wales drew closer together, it became apparent that each had included one particular group of rocks in "his" system. Sedgwick's topmost Cambrian overlapped Murchison's lowermost Silurian (Fig. 4.5). What to do? These otherwise proper English gentlemen ceased speaking to each other and the problem was unresolved during their lifetimes. Much later a compromise was struck by removing the rocks in question and erecting a new system called Ordovician (Table 4.1).

While feuding erupted in Wales, several epochs of the old Tertiary division, by then incorporated as a system, were defined and named (Table 4.1) on the basis of relative percentages of living species of organisms represented among Tertiary fossils in the Mediterranean region. Soon the Mesozoic Era ("middle life") and Cenozoic Era ("new life") were named. The Permian and Carboniferous Systems were named next, and, with the illogical retention of archaic Tertiary and Quaternary as systems of the Cenozoic Era, the time scale was essentially complete. Attempts have been made to replace Tertiary and Quaternary (Table 4.1), but the terms still appear.

Discussion of the time scale's development has been included to underscore that it "grew like Topsy"

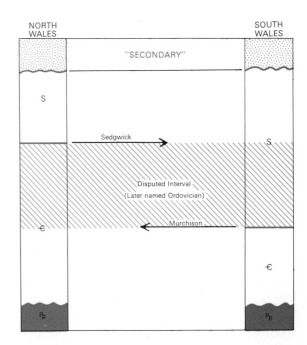

NORTH WALES · SOUTH WALES

"SECONDARY"

S

Sedgwick

S

Disputed Interval
(Later named Ordovician)

Murchison

Є

Є

Pp

TABLE 4.1   The modern relative geological time scale compared with the archaic scale

[handwritten: → Know for 2nd Test.]

Archaic scale as applied in Britain	Modern scale		
	Eras	Periods or Systems	Epochs or Series
Quaternary (1829)* Alluvium Diluvium	Cenozoic *[handwritten: 65 mil-yrs.]*	Neogene (1853)	Holocene (or Recent) (1885) (1833) Pleistocene (1839) Pliocene (1833) Miocene (1833)
Tertiary (1759)	(Recent Life) (1841)	Paleogene (1866)	Oligocene (1854) Eocene (1833) Paleocene (1874)
	Mesozoic (Middle Life) (1841) *[handwritten: ≥ 30 mil. yrs ago]*	Cretaceous (1822) *[handwritten: → creta (Latin for Chalk)]* Jurassic (1795) Triassic (1834)	
Secondary (1759) (Old Red Sandstone) Transition (of Werner) (1786)	*[handwritten: 700 mil. yrs. ago]* Paleozoic (Ancient Life) (1838)	Permian (1841) *[handwritten: "coal measures"]* Carboniferous (1882) *[handwritten: rather late in being named.]* Devonian (1837) Silurian (1835) Ordovician (1879) Cambrian (1835) "Eocambrian" (informal; discussed in Chap. 11)	North America: Pennsylvanian (1891) Mississippian (1870)
Primitive or Primary (1759)	Prepaleozoic or Precambrian (Local subdivisions are used, but their worldwide correlation is difficult; further discussion appears in Chap. 10)		

[handwritten: Europe only →] *[handwritten: → time of formal naming.]*

*Dates indicate when divisions were named.

over a wide region. Nonetheless, it has evolved into an organized, workable scheme of classification now well known even outside the walks of geology.[1] In the process of mapping and correlation of strata from their "home" or type areas, many problems were encountered, some of which are still argued heatedly today. But building of the scale illustrates the success of application of the principles of superposition, original horizontality, original lateral continuity, and of similar fossil assemblages.

Most of the systems were established in areas of dominantly marine, fossiliferous deposits, which provide admirable standards (called type sections) for comparison of many equivalent-aged strata over the world. A special problem was presented by nonmarine deposits such as the Old Red Sandstone. As shown in Fig. 4.4, the tracing of thin, datable marine layers into the nonmarine deposits was required for dating the latter.

Final proof of the inadequacy of Werner's theory of the earth also came with the new chronology. Werner's Primitive rocks were shown to include Pale-

[1]In *Life on the Mississippi*, Mark Twain wrote of an old St. Louis hotel: "The Southern was a good hotel, and we could have had a comfortable time there . . . true, the billiard-tables were of the Old Silurian Period, and the cues and balls of the Post-Pliocene; but there was refreshment in this, not discomfort; for there are rest and healing in the contemplation of antiquities."

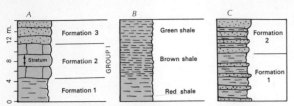

FIGURE 4.6 Designation of formations by lithology is clear in *A*. But it is not so in *B* because of subtle gradations of color, nor in *C* because of intimate interstratification of two lithologies. In the latter two cases, some arbitrary division must be chosen. In *C*, scale is important, for a 1-centimeter-thick sandstone lamina hardly has utility as a formation. (Dots represent sandstone, dashes are shale, and brick pattern is limestone.)

ozoic, Mesozoic, and even some Cenozoic ones. His Transition ended with the Silurian in Britain, Carboniferous in Germany, and the Eocene in the Alps. Clearly, lithology, structure, and metamorphism were not, after all, reliable indices of age, though this important axiom was slow to be fully recognized.

LOCAL ROCK STRATIGRAPHIC UNITS

THE FORMATION

The most basic unit of stratigraphy is the formation, first defined in Germany about 1770. The original notion of a distinctive series of strata that apparently originated through the "same formative processes" is still valid 200 years later. Formations ideally are named for type localities where they are typically displayed, and they must be distinctive in appearance in order to be easily recognizable.

Many practical problems arise in defining formations. Designation of upper and lower limits or contacts of formations is especially difficult as shown in Fig. 4.6. Obviously the definition of a formation is somewhat arbitrary and is strongly influenced by scale factors. The minimum thickness to be designated in one formation depends upon the scale of mapping as well as the character of the strata themselves. A stratum is the thinnest rock layer observable; formations include more than one stratum, and systems contain many formations.

Characteristics chosen to define a formation may include one or more of the following several types: (1)

composition of mineral grains; (2) color (reflecting some chemical property); (3) textural properties (size of grains, etc.); (4) thickness and geometry of stratification; (5) overall character of any organic remains; and (6) outcrop characteristics.

SEDIMENTARY FACIES (lateral profile)

So far nothing has been said about the mutual age or time relations of formations, or of their lateral relationships. For many years after both neptunism and catastrophism had been laid to rest, there still persisted a belief in universal patterns to the history of the entire earth's surface. It was natural to assume that obvious stratigraphic divisions being named in Europe—and destined to become the standard systems—must be universally synchronous worldwide. The finding of similar fossils in most respective systems over much of Europe seemed ample evidence. This concept of universality was also implicit in even the earliest definition of a formation. Thus a kind of layer-cake arrangement was conceived with each formation assumed to extend indefinitely laterally without change. This premise had been, of course, the cornerstone of Werner's time scale, and it was not easily given up.

It is not clear when geologists first began to suspect that distinctive, locally named formations could not extend unchanged indefinitely. The earliest clear records occur in French writings, which show a recognition that similarity of fossils in similar sedimentary rocks might reflect environmental factors even more than strict age equivalence.

In 1789 the French chemist A. Lavoisier published several diagrams and text showing the relation of different adjacent sedimentary environments and the effects upon both sediments and organisms expected from changes of relative land and sea levels (Fig. 4.7). Using comparative reasoning, he also illustrated strata around Paris that seemed to reflect such changes in the past. Lavoisier and his teacher (Rouelle) appear to have been the first to appreciate fully the importance of adjacent environments *as well as* changes through time upon developments of sedimentary deposits and their contained organisms. Shallow, nearshore (littoral) marine sediments tend to be coarser and contain organisms adapted to rough water, whereas contemporaneous offshore, deeper, and quieter (pelagic) marine sediments are finer and contain delicate bottom-dwelling organisms together with floating and swimming forms. Thus products of each environment have unique characteristics even though they accumulated contemporaneously and (grade imperceptibly into one another.) Such advanced ideas were not widely known outside France, however.

Recall that Sedgwick and Murchison, after laborious studies, determined that the Old Red Sandstone of South Wales was equivalent in age to strata originally defined as their Devonian System (Fig. 4.4). The former, with plant and fish remains, reflected environments markedly different from the clear marine condition of the latter; yet strata formed in those contrasting environments *were of the same age.*

Analogies between features such as ripple marks in the Old Red Sandstone with counterparts forming in modern sediments produced some of the earliest environmental interpretations of sedimentary rocks. Simple application of the tenet that present processes provide keys to the origins of ancient features would seem to indicate at once that many different sediment types inevitably must have formed simultaneously side by side in different environments. Yet, full impact of this conclusion was slow in developing. It did not happen until 1838 when a perfectly exposed example of lateral changes in both lithology and fossils within a single stratum was described in Switzerland.

To fully characterize strata, it became necessary to describe not only vertical but also lateral varia-tions, for which the term _sedimentary facies_ was adopted. Facies variations are lateral changes in the total aspect of strata, which may be rather subtle in nature (e.g., a gradual increase in the ratio of sand to mud). A particular three-dimensional body of sediment—a facies—grades laterally by some statistical change in its properties to another, adjacent facies. A map of modern ocean-bottom sediments or of animal communities reveals the reality of variations of products of adjacent, synchronous environments (Fig. 4.8). On land today—at the same time—we find arctic, temperate, and tropical environments of markedly different characteristics side by side.

TRANSGRESSION AND REGRESSION

classic facies change example.

An example from the historical period in the Netherlands shows some implications of facies variations through time. As indicated in Chap. 1, the North Sea coast of the Low Countries has been submerging since the last glacial advance. Whether submergence

FIGURE 4.7   A. Lavoisier's diagram of the relations of coarse littoral (*Bancs Littoraux*) and finer pelagic (*Bancs Pelagiens*) sediments to the northern French coastline. Lavoisier recognized that gravel could be moved only by waves near shore; finer sediments are carried into deeper water. He also recognized that distinctive organisms inhabited each environment. But if sea level rose, flooding the land (*la Mer montante*), both littoral and pelagic sediments would migrate landward. Conversely, if sea level fell (*la Mer descendante),* they would shift seaward. *(From A. Lavoisier,* Memoires d'Academie Royale Sciences, *1789.)*

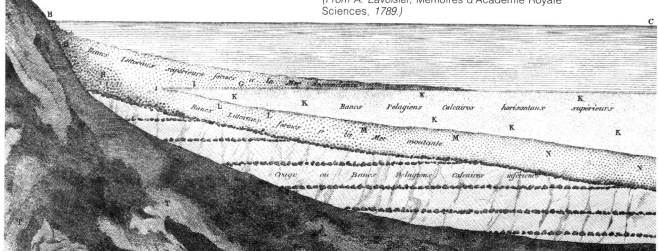

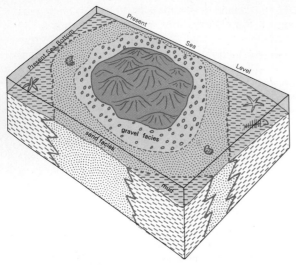

FIGURE 4.8   Sedimentary facies around a hypothetical island showing the tendency for coarser sediments to be confined to strongly agitated, near-shore environments. Bottom-dwelling organisms in the different environments differ considerably, also. The upper surface of the diagram is a map of present bottom sediment types showing lateral variations only at a moment in time; sides are *cross sections* showing vertical facies relationships through time.

was due to worldwide rise of sea level or to land subsidence (or both), the sedimentary result would be essentially the same. Bore holes have been dug in Holland for several centuries, producing an accurately dated Pleistocene and Holocene succession (Fig.

FIGURE 4.9   A historic example of transgression on the Netherlands coast, showing landward shift of facies dated by archaeological and carbon 14 evidence. Note migration of characteristic molluscan animals with the sandy facies; the fish, being a swimming form, was independent of the bottom environment and might be found in either facies.

4.9). Advance of the sea over the land, termed transgression, caused a continuous shift of environments and their sedimentary and biologic products landward. The result is a more or less continuous, nearly flat layer of sandy nearshore deposits and finer offshore ones as shown. The relation between these two sandy and muddy facies is complex, but it involves a seaward gradation *along any single datum of like age* (such as the present sea floor), from sand to mud.

If these deposits are preserved and lithified, what should future geologists designate as formations? Defined on physical, chemical, and organic characteristics, the most obvious choices would be a sandstone formation and a shale formation. But note that these two formations would *in part grade laterally into one another*; that is, they are not in simple superposition even though one tends to lie mostly above the other. Some lateral projections of sand (called tongues) penetrate into the generally shaly facies and vice versa. A sand tongue may reflect a period of violent storms, which moved sand farther offshore than normal.

It follows that neither formation is of exactly the same age everywhere along the cross section. Each is older at the seaward than at the landward end because of the progressive transgression by the sea. On the other hand, at any point *some part of one is the same age as part of the other*. Clearly, then, formations cannot be thought of as either: (1) laterally indefinite and unchanging, or (2) exactly one age throughout their entire extent. Rock divisions must be defined relatively locally and more or less independent of time. After they have been defined and mapped carefully, then it may become possible to evaluate fully their time or age relationships. However, such evaluation requires a great deal of information.

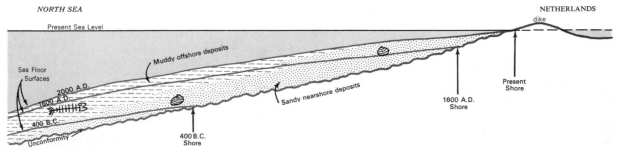

Two basic, simple types of facies patterns must be distinguished in their idealized form. The transgressive or onlap pattern is illustrated by Fig. 4.9 (and lower strata of Fig. 4.7) Transgression generally: (1) is preceded by an erosional unconformity; (2) involves shrinkage of the land; (3) produces a landward shift of sedimentary facies through time; and (4) is reflected by deposits that tend to become finer upwards at any one geographic locality. Regressive facies or offlap patterns result from a relative apparent fall of sea level (or rise of land level). In the Netherlands, a temporary retreat occurred about 3,500 years ago as the rate of rise of sea level slowed and sedimentation rates exceeded the rate of transgression. The shoreline was literally pushed seaward by excessive sedimentation. More recently, artificial diking and pumping has also led to renewed regres-

sion. If retreat of the sea from Holland were to continue indefinitely, then a facies pattern such as that of the uppermost strata in Fig. 4.7 would result. Such a regressive pattern: (1) reflects enlargement of the land; (2) may be more or less destroyed at the top by erosion during regression; (3) shows a seaward shift of sedimentary facies; and (4) shows at any one locality a tendency of the deposits to coarsen upward.

As was hinted in Chap. 1, large-scale changes of sea and land levels have been extremely important in earth history. What causes could produce such colossal changes? Worldwide sea-level changes may result from fluctuating continental glaciation or large-scale warpings of deep ocean basins. Worldwide

FIGURE 4.10 Complications of transgression and regression of a local shoreline due to interaction of widespread sea-level changes and local crustal warping.

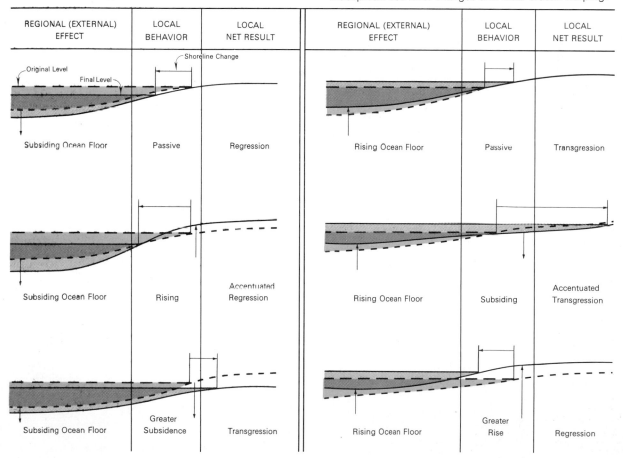

regressions during Pleistocene glacial episodes were very important to early humans, having allowed them to migrate from Siberia to America and from Europe to Britain. Local changes such as mountainous uplifts or crustal subsidence can result in local transgression or regression as well. Very rapid sedimentation also can produce local regression, although this is generally a secondary effect of climatic changes or uplift of large land areas (see Fig. 4.10).

The relief of the land will also dictate enormous differences in the magnitude of shoreline translations during transgression or regression. Low, flat land, such as the Netherlands, would be inundated or drained very widely by but a slight relative change of sea level, whereas a bold coastline such as that of California would be only locally affected (Fig. 4.11). Obviously from the fragmentary evidence preserved in ancient rocks, we cannot be very hopeful of determining the actual or absolute cause of every transgressive or regressive facies pattern; we see only an *apparent* sea-level change reflected. A great deal of continentwide or even worldwide information is required to test and reject possible alternate explanations. We shall illustrate many real examples of these important interpretations subsequently.

MODERN STRATIGRAPHIC CLASSIFICATION

ROCKS VERSUS TIME

As was indicated early in this chapter, it has become mandatory to make clear distinctions between abstract geologic time, which is continuous, and the actual rock record, which is riddled with unconformities of varying magnitude. Yet all that we are to know of the historical continuum must be gleaned from the imperfect rock record. A unit of relative geologic time has meaning only in terms of rocks formed during that time. We have distinguished the Cambrian System of rocks from the Cambrian Period of time. The rocks of this system in Wales, where they were first studied and named, provide a world standard for comparison by correlation of rocks anywhere else, which, on the basis of similar fossils, are judged to have formed during that same period of the Paleozoic Era.

A pure time division, such as the period, must include the time represented by any and all discontinuities as well as the actual preserved rocks in the corresponding rock division, the system. Figure 4.12 illustrates this difference. It follows that if elsewhere a more complete sequence of Cambrian rocks were found with fewer and smaller unconformities and more fossils, it would provide a better world standard of reference than does that of Wales. While this is true, the original European System standards are now so firmly established by long usage that such changes have not been made.

Most classifications contain different levels of subdivision. A simplification of the most generally accepted hierarchy of stratigraphic subdivisions is as follows:

Relative time divisions	Equivalent universal rock divisions
Era	———
Period	System
Epoch	Series
Age	Stage

FIGURE 4.11 Contrasting effects of sea-level fluctuations on low versus steep coastlines. Roughly half of the total continental surface today lies within a few hundred meters of sea level; thus a modest sea-level change produces a profound change in land area. Many such modifications have occurred in the past.

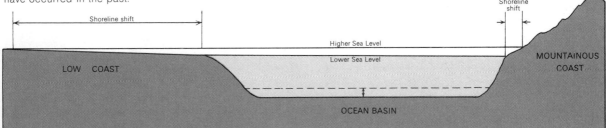

The above rock divisions are termed universal because they comprise at least continentwide if not worldwide standards as opposed to the purely local rock units such as formations. Though the rocks of Wales that define the Cambrian System have certain peculiar lithologic characteristics, all rocks considered of like age elsewhere regardless of their lithology are also referred to the Cambrian System. Comparison among these is made only in terms of age equivalence.

BIOSTRATIGRAPHIC CONCEPTS

Identity of age of strata, especially over long distances, has been established largely by application of the principle of fossil assemblages. The subdivisions of any newly studied stratigraphic sequence are correlated with the standard universal rock divisions on the basis of similar fossils. When Charles Walcott, a geologic pioneer in western America, identified thick strata in southeast California as Cambrian a century ago, he was performing a correlation with Wales based upon index fossil assemblages. Such correlations assume that similar evolutionary stages of development were reached simultaneously by particular organisms in all parts of the world. Experience shows that, at the level of precision of correlation so far achieved, this assumption is valid. The facies concept, however, makes it clear that evolution through time is not the only change operating, for environmental differences in space also must be considered as factors affecting distribution of fossils. Environmental changes are, by and large, relatively short-term affairs, and so are not very significant when considering fossil assemblages in the large universal rock divisions. However, for smaller time divisions, the vagaries of local environmental influences must be considered, for environmental changes may have been more rapid than evolutionary ones. Moreover, only certain groups of organisms—those which evolved very rapidly—can provide reliable correlations for short time intervals. Figure 4.13 shows how differing rates of evolution may compare with environmental change. Several related aspects of index fossil assemblages enter into their use for correlation. A widespread biologic datum; that is, a restricted thickness of strata characterized by a distinctive index fossil, is commonly termed a fossil zone (Fig. 4.14).

Because sedimentary environment may influence bottom-dwelling or benthonic organisms so profound-

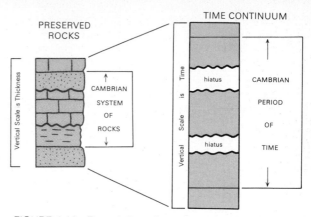

FIGURE 4.12 The relation of a preserved rock record with discontinuities and the abstract time continuum corresponding to that rock record plus its unconformities. A tangible rock record exists only for the shaded portion at the right, while the blank areas (hiatus), represent unconformity intervals.

ly, such creatures may be of little value in correlation of strata of adjacent different facies. Many fossil types are notoriously restricted to one or a few lithologies ("facies fossils"). For example, certain clam species today burrow only in beach sands, while others live only in muddy tidal flats. These would be less-than-perfect index fossils because of their narrow preference for a particular sediment type. It follows, then, that the best index fossil is one which lived more or less independent of the bottom environment where sediments form. Obviously these would be floating or swimming organisms. Fortunately, there are several such groups that also evolved rapidly and that therefore serve well for correlation even between different sedimentary facies (Fig. 4.15). Figures 4.14 and 4.16 show in practice how index fossil zones together with lithology are used for correlation and mapping.

UNCONFORMITIES, THE UNIVERSAL TIME SCALE, AND ROCK SEQUENCES

Unconformities, like rock units, can be traced and mapped to establish their physical continuity, and they also can be studied from the standpoint of age. Continuity and age are established by mapping the relationships of strata and fossils immediately above and below the unconformable contact. It may turn out that, like some formations, an unconformity surface

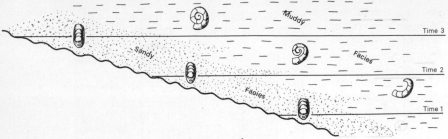

FIGURE 4.13 Significance of contrasting rates of evolution and rates of environmental shift (transgression). The brachiopod *Lingula* in the sandy facies evolved very slowly and is a poor index fossil; it has migrated with its shifting sandy environment and is unchanged biologically after millions of years. Cephalopods in the muddy facies, however, were swimming forms free of the bottom environment; they also evolved rapidly so their species are admirable index fossils for times 1, 2, and 3.

FIGURE 4.14 Correlation using three different index fossils. In practice, assemblages are more useful than a single species, but close attention must be paid to overlapping stratigraphic *ranges* of index fossils. Note also that the *range* and the *maximum* development of a single species vary slightly from place to place. This imposes a lower limit of resolution to fossil correlations beyond which it is fruitless to attempt further refinement.

varies somewhat in age from place to place. More importantly, the total time interval represented by the discontinuity may vary greatly; at one place there may be far more rock record missing than at another point along the same surface. An unconformity may even disappear laterally into a continuous, unbroken, conformable sequence of strata (Fig. 4.17). Unconformities, then, show important lateral and vertical differences equally as important as those of rock units.

Today we recognize three types of unconformi-

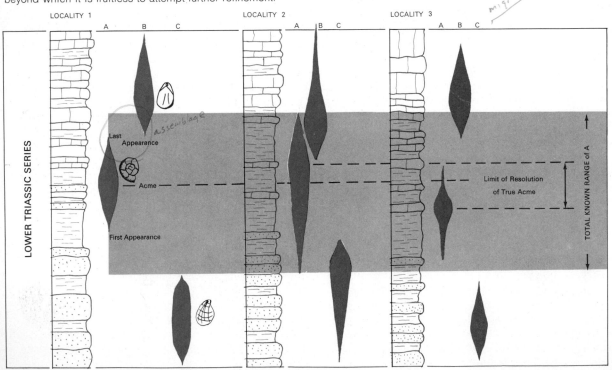

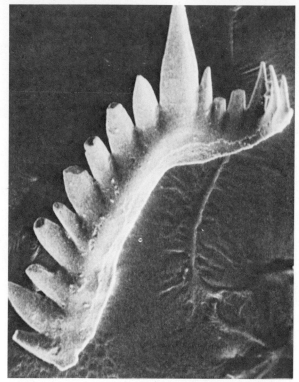

FIGURE 4.15   Lateral view of a conodont element. Conodonts are small anatomical parts of uncertain origin, but are probably supports for the digestive system of an extinct, floating organism. Evolutionary structural changes occurred rapidly, and conodonts are widespread in different rock types (they are thought to be free-swimming forms). They are ideal index fossils. This conodont element is from the Permian of Texas, x 110. *(Courtesy D. L. Clark.)*

ties: angular unconformities such as those Hutton studied, disconformities with little or no angular discordance between older and younger sets of strata, and nonconformities between younger sediments and older igneous or metamorphic rocks (Fig. 4.17). Angular unconformities and nonconformities represent profound upheaval and deep erosion, whereas disconformities represent little or no structural disturbance, only erosion or nondeposition. But disconformities may represent long time intervals.

Increasingly, geologists recognize thick, laterally extensive rock units called sequences that include many formations as packages bounded by exceptionally profound regional unconformities. No restricted time connotation (synchroneity) is attached to the unconformable sequence boundaries; in fact they are

known to be of varying age from place to place. Sequences constitute an additional type of stratigraphic division—a regional rock unit, but not a universal time division with strict age connotations. Certain unconformities are judged to be the most important ones among many. Inevitably not all geologists agree entirely on the validity of the suggested sequence boundaries, but as is reflected in the organization of the chapters on North American history, a few major unconformities serve to punctuate the stratigraphic record in a meaningful way.

APPLICATIONS OF BASIC STRATIGRAPHIC CONCEPTS

HUTTON'S SCOTLAND
Now we can illustrate and at the same time begin to interpret stratigraphy in a modern sense using Great Britain as an example. Figure 4.18 shows a cross section through Scotland, illustrating most of the important features first studied by Hutton. In both areas a great unconformity can be seen beneath the Devonian Old Red Sandstone. Rocks beneath are

FIGURE 4.16   Three-dimensional relationships of formations and index fossil zones; top surface shows how a geologic map is constructed by establishing physical continuity of formations between isolated outcrops. Age correlation by fossil zones shows that Formation 3 is synchronous everywhere, but Formations 1 and 2 vary in age due to lateral facies changes. (Compare with Figs. 4.9 and 4.13.)

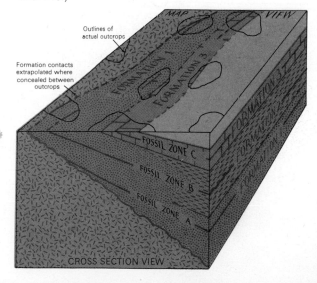

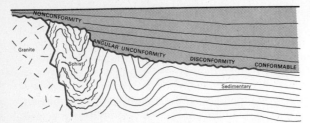

FIGURE 4.17 Variations in unconformities. Note that one type may change laterally to another.

Silurian and older, and so it is apparent that sometime in the later Silurian, and perhaps early Devonian, profound mountain building occurred in a belt running northeast through Wales and Scotland; it also affected western Scandinavia. Southeast of this old disturbed belt, as in eastern Wales, the Old Red Sandstone is more nearly conformable with the Silurian.

The fact and age of the mountain-building event is clear on the southeast (right) end of the Scottish cross section from the superpositional relations of the Old Red Sandstone over the angular unconformity. But to the north (left) of the Highland Boundary Fault, things are not so definite. Severe upheaval, metamorphism, and igneous activity occurred, but when? Did these changes happen at the same time that Hutton's unconformity developed? In-faulted blocks of Old Red Sandstone suggest that the upheaval was, at least in part, pre-Devonian. A rough clue to magnitude of events, though not to precise age, is provided

by the presence of strongly metamorphosed rocks beneath the pre-Old Red unconformity. Exposure of such rocks at an old erosion surface implies deep and rapid erosion of the crust to expose rocks that formed in a relatively high pressure and elevated temperature environment.

Law of Intrusion

RELATIVE AGE OF IGNEOUS BODIES
How can we narrow the age of all the other metamorphic rocks and granites of the Grampian Highlands where Old Red Sandstone is absent? Some important principles exist for deciphering the relative age of igneous bodies, and if such bodies in turn can be related in age to fossiliferous sediments, then the igneous rocks, too, can be related to the geologic time scale. It is self-evident, as emphasized by Hutton, that *an intrusive igneous body must be younger than all rocks intruded by it*. Commonly it is possible to establish relative age of several intrusive masses by mutually cross-cutting relations among them. Thus at the left end of Fig. 4.18, the granite is clearly younger than the metamorphic rocks, in which a few Cambrian fossils occur. The first dikes that Hutton studied were of this granite. But the granite is older than basaltic dike 1, which is in turn older than dike 2. These relations are purely relative and say nothing about how much younger each successive rock is than another. Is it to be measured in minutes, days, years, or what? It is a reasonable hypothesis that at least the granite, and possibly also the metamorphism, date from the same time as Hutton's pre-Old Red Sandstone unconformity, which developed about the end of the Silurian Period. Similarly, it is tempting to correlate dike 1 with the basalt sill and dikes

FIGURE 4.18 Cross section showing structural relationships among the rocks studied by Hutton in Scotland (vertical scale exaggerated to show details). *(Adapted from* Geological map of Great Britain, *1957.)*

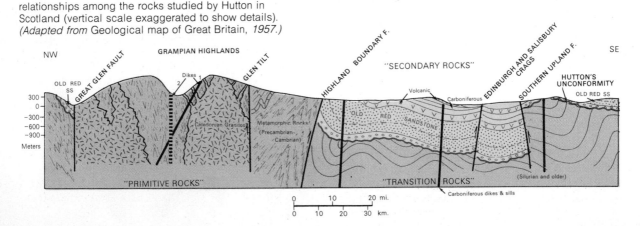

studied by Hutton at Edinburgh, which were intruded into the Old Red Sandstone (see Chap. 3). But why not correlate basalt dike 2 of the Highlands with these instead? Just how different in age are these two Highland dikes?

RELATIVE AGE FROM INCLUDED FRAGMENTS

Some new lines of evidence are needed to answer the above questions. What we require, clearly, is some definitive younger-age limit for the igneous and metamorphic rocks. If the Old Red sediments rested unconformably upon the granite, dating of granite emplacement would be simple by superposition, but this is not the case. We can seek indirect evidence for relative age in conglomerates of the Old Red itself, for coarse fragments provide direct evidence of the character of rocks exposed by erosion while they were accumulating. An important additional principle elucidated by Hutton is that *rocks represented as fragments in another rock body must be older than the containing rock.* Note that this applies to pieces (inclusions) of sediments in igneous intrusions as well as to grains or pebbles in sediments.

In our example, we find granite pebbles essentially identical to those in the Grampian Highlands in the Old Red. Therefore, our argument is tightened, and a great deal more confidence can be attached to our hypothesis. Hutton's original intrusive granites and his unconformities turn out to be closely related, and they constitute the principal evidence of one of the earth's greatest mountain-building events much as Hutton envisaged it. This event has since been named the Caledonian Orogeny (Caledonia was the Roman name for ancient Scotland; orogeny means mountain building).

RELATIVE AGE OF FAULTS

Let us return momentarily to the cross section (Fig. 4.18) and reconsider the basalt dikes. Further evidence near Edinburgh indicates that these dikes and related lavas are of Carboniferous age. It is reasonable, but by no means provable, that basaltic dike 1 in the Highlands is of like age. Dike 2 may be only slightly younger or it may be much younger. Without other evidence from beyond the confines of this cross section we can say no more, but in western Scotland there is a swarm of Cenozoic basaltic dikes, which probably are correlative with dike 2. We should also note that the faults shown can be dated as having been active after the Caledonian Orogeny and chiefly after the Devonian. There is no evidence to narrow their age further. We see then that, like igneous intrusions, *faults are younger than all rocks that they displace and older than the oldest rock that unconformably overlaps them.* Strictly speaking, however, this only dates the last fault movements; many faults have been active for very long periods, even hundreds of millions of years, and thus other indirect evidence may be required to date their first movements.

SUMMARY

In this chapter we have seen how the relative geologic time scale developed in northern Europe. Geology now had both a viable unifying theory and new methodology. Evolution of the time scale illustrates the application and refinement of the basic stratigraphic principles of Steno, Hutton, Smith, and Cuvier. Extension of mapping and of stratigraphic knowledge inevitably confronted geologists with the realization that formations must change in character laterally in response to differences of ancient sedimentary environmental influences, which imposes restrictions on correlation using index fossil assemblages. Some fossils, in fact, occur exclusively in only one facies type. Sedimentary facies patterns reflect many influences, but especially important are transgression and regression by the sea due to relative vertical changes of land and sea levels.

Today we require a more elaborate stratigraphic classification scheme that makes special allowance for distinction of rock units having tangible spatial dimensions (systems, series, etc.) from abstract time (era, period, epoch, etc.). The classification also has two levels for rock units, the universal (worldwide) standard rock units, such as the system, and the local rock units, such as the formation. Many discontinuities occur in the rock record; therefore we have an incomplete record of earth history. It is now apparent that local rock units may vary in age from place to place. Similarly, unconformities are recognized to vary in age and magnitude of time represented from place to place.

The relative geological time scale provides a standard of reference for rocks throughout the world.

But relative geologic age generally must be established by fossils and by determining superpositional relations of strata containing them. Hutton illustrated other important means of determining relative age. For igneous rocks, cross-cutting intrusive relations and the unconformable superposition of younger, datable sediments provide bases for their relative dating. Composition of conglomerate pebbles and of included fragments in igneous rocks also provide evidence for relative age. In subsequent chapters, we shall apply these principles and methods to North America.

Readings

Bennison, G. M., and Wright, A. E., 1969, The geologic history of the British Isles: New York, St. Martin's.

Berry, W. B. N., 1968, Growth of a prehistoric time scale: San Francisco, Freeman. (Paperback)

Brinkman, R., 1960, Geologic evolution of Europe: New York, Hafner. (English translation)

Carozzi, A. V., 1965, Lavoisier's fundamental contribution to stratigraphy: The Ohio Journal of Science, v. 65, pp. 72–85.

Eicher, D. L., 1968, Geologic time: Englewood Cliffs, Prentice-Hall. (Paperback)

Geikie, A., 1905, The founders of geology (2d ed.): New York, Macmillan. (Paperback, Dover, 1962)

Gignoux, M., 1955, Stratigraphic geology: San Francisco, Freeman. (Emphasis on Europe)

Harbaugh, J. W., 1968, Stratigraphy and geologic time: Dubuque, W. C. Brown. (Paperback)

Krumbein, W. C., and Sloss, L. L., 1963, Stratigraphy and sedimentation (2d ed.): San Francisco, Freeman.

Matthews, R. K., 1974, Dynamic stratigraphy: Englewood Cliffs, Prentice-Hall.

Rutten, M. G., 1969, The geology of western Europe: Amsterdam, Elsevier.

Woodford, A. O., 1965, Historical geology: San Francisco, Freeman.

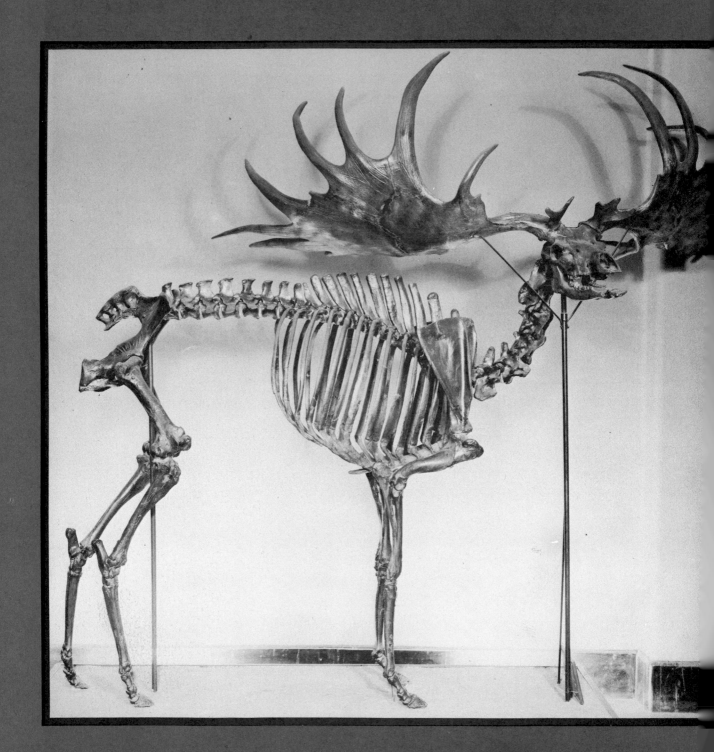

5

EVOLUTION

A STORM OF CONTROVERSY AND A TRIUMPH OF HUMAN INTELLECT

A deer with a neck that was longer by
half
Than the rest of his family's—try
not to laugh—
By stretching and stretching
became a giraffe
Which nobody can deny.

Lord Charles Neaves (1868)

(By permission of M. J. Sirks and Conway Zirkle,
The evolution of biology, © 1964, The Ronald
Press, New York.)

FIGURE 5.1   Restoration of
the Pleistocene Irish Elk.
*(Courtesy American Museum
of Natural History.)*

Since its beginnings, probably few theories have stirred so much controversy as evolution. Briefly stated, organic evolution is the cumulative change of organisms through time and is usually irreversible. By and large, the controversy over evolution is not so intense today as it was during the Victorian epoch, characterized in large part by religious conservatism and agrarian isolation. Into this austerity, the *Origin of Species* by Charles Darwin and the hammer-like logic and colorful persuasiveness of England's most popular lecturer and biologist, Thomas Huxley ("the Bishop eater"), shook the Victorians to their bootstraps.

As late as 1925 here in the United States, arguments about evolution blazed into national headlines during the famous Scopes "Monkey Trial" at Dayton, Tennessee. This trial was held to test a law prohibiting the teaching of evolution in the state. It was not until 1967 that Tennessee finally abandoned the "Monkey law," and in the fall of 1968 the U.S. Supreme Court declared all such laws unconstitutional. While legal aspects of teaching evolution have been clarified, there are still many people who are reluctant to learn about the subject because of religious precepts. In 1969 California adopted a requirement that the biblical account of the origin of plants, animals, and man be given equal treatment with evolution in public school biology courses.

Failure to separate science and technology has, in recent years, caused a reaction of some people against science as the reason behind all of society's ills. It must be realized however, in view of the trend toward urbanization and the problems inherent in a rapidly expanding population, that science is one human endeavor that can help solve today's problems.

The general concept of evolutionary change stands as one of the few really great generalizations of knowledge. Evolution is a concept that was almost inevitable, because it is a unifying principle that explains the distribution and diversity of life through its over 3-billion-year record. The theory of evolution was one of the last of such principles to develop—a full 200 years after Newton discovered the basic laws of mechanics. Yet, ideas about evolution, along with some of the questions that are fundamental to it, had been discussed as far back as classic Greek times.

SOME BACKGROUND AND A LOOK AT THE FUNDAMENTALS

Why is it that it took so long for man to realize that organisms have changed? Part of the answer lies in the fact that <u>plants and animals evolved so slowly</u> that humans could not see them change. Early thoughts on evolution were formed by explaining why there were so many different kinds of organisms. Here are some ideas that we feel might be some ingredients—thus a common denominator—in the various theories of evolution and that can provide us with a basis for asking some questions relevant to evolutionary concepts.

The appearance of a new character (or mutation) is one way we can recognize a new group on any given classification (taxonomic) level, for example, the sudden appearance of a new tooth cusp in a mammal or the enlargement of a preexisting feature such as the lengthening of a giraffe's neck or an elephant's trunk. It has been observed that not all new characters or modified characters constitute a new species, but if they are persistent and become widespread, it is with these that we can recognize differences between species and identify new species. From this oversimplification we can proceed to the "How" and "Why" questions, which serve to separate the various theories of evolution. How is the role of environment related to the origin of new species? How does inheritance affect evolutionary change? How and why do new characters come to be? And after new characters arise, how do they cause a new species to form?

We saw in Chap. 2 that the Greeks noted fossil shells far from the sea, indicating, they thought, that the waters of the earth were receding. Hence, for a long time the idea of environmental change was known. We shall come back to the all-important idea of such a change and see that it is closely linked to evolution and constitutes another basic ingredient of all such theories. The question that perhaps should have been asked is: "What effect does this change have on organisms?" But this question could not have been asked in Greek times because awareness of cause and effect relationships was not as acute as in later times. Furthermore, early people assumed that the fossil shells were simply dead modern shells. Recognition of possible extinct organisms generally did not occur in the modern sense until the 1600s. The question of the diversity of life was scarcely raised until relatively recently because of the "creationist" belief that all life was placed on earth for the benefit or detriment of man.

One of the fundamental questions leading to a theory of evolution is: "What observations will explain why there are so many kinds of organisms?" Before the 1700s, natural historians were preoccupied with other questions, such as the origin of life. All this was partly changed by the early classifiers because they brought before all a proof of the great variety of life. The giant among the natural historians who undertook encyclopedic description was Carolus Linnaeus, who published a series of works describing the fauna and flora of Sweden, and, finally, all known species of the world. By 1758, when his work was standardized, some 8,000 species of organisms had been defined. Imagine his amazement if he could realize that almost a million species now have been described and that no end is in sight!

BUFFON'S EVOLUTION

A leading intellect of eighteenth century France was Georges de Buffon. He was the first influential man to convince some of his fellow scientists and philosophers that evolution must have occurred. He defined the species for the first time, noting that it exists as a separate entity and demonstrating that no species could interbreed with another. This is a fundamental concept in biology, and has taken on an even greater importance since the development of genetics (the science of heredity). Curiously, he decided that species were stable and could not give rise to new and different species. He reasoned that since *species* are infertile to each other, it is impossible to conceive of fertile *parents* giving rise to offspring which are new species. He was the first man to understand that the environment caused variation in the offspring, but he interpreted this to be important only on the local population level.

His writings are full of contradictions, but gradually through his life he moved toward the conclusion that all forms of life must have evolved from a fewer number of forms. His one most important contribution is that all animals are closely related to the environment in which they live and that the environment somehow molds and changes the organisms. He noted that there are also changes due to inheritance of characters from both parents. Thus he had established two of the absolutely fundamental attributes of evolution; he had asked the right questions at the right time in history. He put these two concepts together, saying that modifications of organisms are due to inheritance of characters and changes caused by environment; these in turn produced a natural descent that resulted in life as we see it today. In reading his statement we are not convinced that we should classify this as a theory of evolution, for he did not explain *how* new characters could arise, or *how* the environment could cause modifications *that could be inherited*. As we have noted, no theory of evolution should be called such unless it can explain how environment and inheritance interact and how new characters arise and change. In fact, Buffon's definition of the fixity of species did much to impede the development of an idea of the origin of species. As we have noted in Chap. 3, it was sometimes hazardous to suggest theories outside the climate of the times, and Buffon allowed for the special creation of species in his *Theory of the Earth*, even though he frequently implanted the idea that the Scriptures could be wrong.

Thus the conditions for a theory of evolution acceptable to the scientific body were established by Buffon by the middle of the eighteenth century as a reasonable, albeit vague, concept to explain life as we see it. The majority of the scientific community, including Charles Darwin, were strong dissenters and put their finger on the most serious problem to evolutionists, that of *time*. How could all of life become so diversified in such a short period of time since the Creation? Experiments involving breeding of domesticated plants and animals failed to produce a new species. The geologists realized that they needed more time also, and Buffon, as we have seen, suggested that the earth was 75,000 years old. The problem of the span of time available for evolution was not settled until much later.

ERASMUS DARWIN: DID HE BEGIN IT ALL?

Buffon strongly influenced one of the important intellectuals of Great Britain, Erasmus Darwin, the grandfather of Charles. As an animal experimenter he saw that there are many changes going on during ontogeny (the life history of an individual): a caterpillar changes into a butterfly and a tadpole into a frog. He noted that in breeding, changes can be recognized from generation to generation, although the changes are very small. He realized also that there is a strong relationship between environment and heredity.

He went on to say that there are many forces which control evolutionary change. He maintained

that lust (sexual selection) and the various weapons and ornaments and protective devices (deer antlers, etc.) attendant on sexual selection were important dynamic forces of evolution. The paraphrase of this in Charles Darwin's terms could be survival of the fittest. Another point he emphasized was that species became adapted to their environment primarily by the method of gathering food. For example, the development of the elephant's trunk was to aid in tearing grass and moving it up to the mouth. The third element is security: protective coloration, wings, legs, shells, spines, etc. The results of these forces were *new characters which were acquired to fit the organism to its environment and these were passed on to its descendants*.

But again, his statement, like Buffon's, lacks our criteria for a full-blown theory of evolution: How do the characters arise? He does tell us why they might have developed and that, if the characters make organisms better fit, they are passed on to the offspring. We believe that these ideas strongly influenced the early thinking of Charles. Natural selection is laced throughout his grandfather's *Zoonomia*, and Erasmus Darwin got many intellectuals in Britain interested in evolution and provided them with the beginnings of some method of approach to the problem.[1]

LAMARCK: A STEP BACKWARD?

Lamarck's thinking represents the culmination of a long series of philosophers who attempted to explain all of the universe in a single concept. Lamarck's theory of evolution is in a sense one of the most misunderstood of all such concepts. If he were alive today and listened to the average zoologist state his theory, he would be puzzled. He summarized his philosophy in a book published in 1809. The thread going through it is: The fundamental aspect of nature is change. For example, minerals are not species of fixed composition but only stages in a continuous flux of disintegration and recombination. He believed that all groups of organisms are unreal and that species as natural entities are creations of humans. Note how different this idea of species is from Buffon's fixed species. Life, Lamarck said, is a great stream of gradual complication. He believed that this system was replenished each day by spontaneous genera-

---

[1]However, many history of science students do not believe that he did influence Charles Darwin or Lamarck.

tion from inorganic material which filled the system up from the bottom, the simplest forms today are those most recently formed. The most complicated animals are the oldest groups. Even more amazing was his belief that no animal group ever becomes extinct. If some group (or better, stage) should be destroyed locally, it will be quickly replaced by those forms coming up from below just like a hole fills up with water. One facet of his theory, which has come to be thought of as the whole concept, and in fact is the only part considered important, is called the inheritance of acquired characteristics. Lamarck perceived that the effects of the environment somehow could be inherited by offspring. Buffon and Erasmus Darwin saw this, but neither could provide a mechanism to explain it. Lamarck found one, and it is the most outstanding contribution he made, although few believed him.

THE LAMARCKIAN THEORY OF EVOLUTION AS WE KNOW IT

Lamarck believed that the environment does not directly cause animals to be different as Buffon and Erasmus Darwin believed, but changes in the environment will lead to changes in the needs of organisms. Changes in the needs will mean that the animals will be forced to change their habits. If these changes become constant, that is, if the environment continues to change, animals will acquire new habits, which will last as long as the needs that gave rise to them. New habits will give rise to new characteristics and thus a new species would arise. This was one of the few mechanisms suggested in the nineteenth century for the creation of new characters. It is a subtle concept in a way because it requires a mysterious inner force in the organism to develop a new feature, but it gets around arguments about the direct effect of the environment. For example, experiments have shown that a reduction of diet causes stunting in some animals, but such a decrease will not directly cause an animal to develop a new *genetic* line of descendants that are smaller.

One classic example Lamarck used to explain his theory is how the giraffe got its long neck. This somewhat ridiculous example was another reason why he was seldom taken seriously; other examples appear even more ludicrous. The giraffe, so theorized

Lamarck, started out as a grazing animal and, for whatever cause, there came to be a shortage of ground cover. Giraffes turned to eating leaves on trees, and as more and more turned to this source of food the supply of leaves lessened. The giraffes were forced to stretch their necks continuously to reach higher leaves. Thus, the long neck, and *voilà*, a new species. This is more or less a complete example of a theory of evolution and seems to answer all the questions we need to satisfy. The effects of a changing environment, a mechanism for developing a new character (hence a new species), and a way whereby these changes are passed on to the offspring—a far more satisfying concept than Buffon's rather ill-defined direct environmental approach. Lamarck also threw in a few "laws" to help amplify his idea. The first is built into other theories as well, namely, that everything gained by the phylogeny (the stream or history of a group of organisms) through the influence of nature is kept by heredity, provided that the modifications are common to both parents. The second idea states that the more an organ is used (a stretched neck, for example), the stronger it becomes and the larger it will be in proportion to the duration of its use.

Most students of evolution do not subscribe to Lamarck's theory today. For one important reason, there apparently is no experimental evidence to support Lamarck's ideas. Although a group of Russian geneticists believe they possess such proof, their evidence is highly controversial and now mostly discredited in the U.S.S.R. There have been some classic experiments designed to test his ideas. In the famous Weimar experiment, over 200 generations of white mice systematically had their tails removed. Yet there was no trend for a shortening of tails in any descendants, which would be expected if Lamarck's thesis regarding use and disuse were correct. It can be argued, perhaps, that mutilation experiments hardly fall into the category of a valid test.

There are some other basic questions to be asked regarding his theory. How did the giraffe neck develop in the first place? How can new habits alter parents in exactly the same way so that offspring can receive the same new habits? It is important to realize that during the time of Lamarck, most people were of the opinion that species were fixed, immutable, created and destroyed at the whim of God. Those who did think about the problem of evolving organisms objected on the very good ground, as we have seen, that there was not enough time since the creation for all species to appear; even heretics who believed with Buffon that the earth might be 75,000 years old could not envision the enormous diversity of life to be created in such a short space of time. Lamarck got around this one very neatly; he said that as organic order got more complex, evolution accelerated.

Lamarck met opposition from his former collaborator, Baron Georges Cuvier, the most influential naturalist of France. Cuvier, working in the Paris Basin, was developing his theory of catastrophism, which fitted much better with the biblical account, and seemed to account better for the diversity of life as documented by a real fossil record as proof. Where was Lamarck's proof in the fossil record? Nineteenth century paleontologists, and indeed twentieth century paleontologists as well, have not found any supporting evidence.

There is just one more point that can be raised: the characters that Lamarck chose to emphasize were functional characters; that is, some organ that was used or disused got either larger, smaller, or more efficient. What about such things as mimicry and protective coloration? Those characters are nonfunctional. How could they develop? *Pepper Moth*

CHARLES DARWIN AND NATURAL SELECTION

Lamarck produced his theory in 1809. The year 1809 is certainly an important one because it was on February 12, 1809 that two men were born who had a profound influence on the nineteenth century: Abraham Lincoln, emancipator of slaves, and Charles Darwin, whom we might call the emancipator of our primitive ideas on the development of life. Darwin attended the University of Edinburgh as a medical student, at which time he studied geology and became thoroughly grounded in all of natural history. He later became dissatisfied with the course of his life, and went to Christ College at Cambridge University as a divinity student, acquiring a great interest in botany and the collecting of insects. The single most important event that was responsible for his gaining a notion of evolution, according to him, was the fortunate trip that he made in the years 1831 to 1835 aboard the *H.M.S. Beagle*, one of the early worldwide scientific cruises. This trip represents a milestone of

scientific investigation; it was the first to study the bewildering variety and strange adaptation of plants and animals of the tropics and the Southern Hemisphere.

Darwin observed the fantastic variety of species that are found in the world, the tremendous numbers of individuals of species, and the amount of competition among populations for food. From his observation, he deduced that living organisms are enormously fertile; yet curiously, the number in a given population appears to be fairly constant. Populations are stable in size, he said, because there is a tremendous struggle for life, because of rivals or prey-predator relationships, and the inorganic environment itself. The result is that some organisms seemed to be better fitted for life than others. Notice that this is an idea that Erasmus Darwin had. Charles was profoundly influenced by this idea and also from reading Thomas Malthus' *An Essay on the Principles of Population*, which was written in 1798. Malthus said that populations increase geometrically while food increases arithmetically, and there is a continuous struggle for food. Thomas Malthus, of course, applied this to human population, saying that when the population increases have reached the saturation point for food, then famine occurs and the population will be destroyed.

In the struggle for life, not all offspring survive; in fact, there is a high rate of infant mortality, and only those that are best fit for life survive. Darwin observed that the reason that some organisms are not fit for life is that they have inherited characteristics that prevent them from surviving in their environment. It appears that nature selects the animals best fit for the environment and thus "molds" the animals. Through time, populations become stronger and more fit for their environments. He recognized, as did Lamarck, that there are environmental changes, and when the environment changes, this necessitates continual character modifications in organisms. Darwin saw that as the environment changes nature will select those individuals that are best fitted for the particular changes occurring; that *only the fittest will survive*. But this is an oversimplification of his statement, and Darwin recognized that there are many different possibilities. For example, there may be no observable environmental change, and yet evolution seems to go on; organisms are continually and *gradually* being better adapted or progressively adapted to fit an environ-

FIGURE 5.2 Darwin's finches from the Galápagos Islands; note the differences in shapes and sizes of beaks reflecting differences in feeding habits. *(By permission of Biological Sciences Curriculum Study, High school biology, BSCS Green version, 1968; Rand McNally.)*

ment. *Evolution is likely to move in the direction of greater adaptation.*

How did he come to these conclusions? To go back to the voyage of the *Beagle*, one of the stops made in their westward journey was in the Galápagos Islands. There Darwin recognized 13 species of finches. These finches were like none other in the world; they obviously were closely related to, but distinct from, those in South America. Interestingly, although finches are unable to fly great distances, their ancestors somehow had managed to get to the islands, which are 500 miles away from the Ecuadorian coast. Another interesting point is that only those species are found on the islands (other than oceanic birds), and they are living in environments unknown elsewhere to the family of finches. Darwin was greatly puzzled by this, but gradually came to the view that they had evolved from perhaps a single species that

had accidentally arrived from South America. Since there was no competition from other birds, they became adapted to bizarre habitats not otherwise occupied (see Fig. 5.2).

DARWINISM TACKLES THE GIRAFFE

Let's see how Darwin would have explained the long neck of Lamarck's giraffe. First of all, Darwin would say, there is in all populations a certain range of variability, which was inherited from ancestors. All populations show this; there is variation in coat color and pattern, length of neck, size of horns, etc. Even though there was variation in the length of neck among primitive giraffes, whether a neck was short or long may not have been important in the beginning when there was plenty of grass to eat. If their basic food should become so scarce, however, the protogiraffe would have to change its diet and there would be

Δ in environment

competition for food. Those forms having the shorter necks would have less food than those with longer necks, for the latter could eat tree leaves. Shorter-necked forms would die off because they would be less competitive. Thus there was a progressive adaptation for longer necks.

genetically

HOW GOOD IS DARWIN'S THEORY?

Is this a good theory of evolution? It satisfies all but one of our requirements for such a concept. It tells us how and why evolution occurs much more clearly and convincingly than Lamarck's ideas and even more importantly can be demonstrated experimentally (and has been so demonstrated many times over). But remember that Lamarck was the first one to come up with a plausible idea of how new characters arise. As changes occur in the environment, there is a change in the needs of organisms, resulting in a "developed" new character.

genetic randomness.

*   But Darwin had great difficulty in explaining how a new character arose. His theory works well after the new characters appear on the scene. Another flaw was his belief in a type of inheritance that came from breeding experiments, which implied that parental characters are blended in the offspring. This was a problem because the offspring should have characters that are *better* fit than the parents, not subdued by blending

The biological critics pointed out that there was not a scrap of paleontological evidence to favor his theory. In fact, it was at this time that Cuvier clearly pointed out that the record showed catastrophism to be more likely (see Chap. 2). However, Lyell, Darwin's geological mentor, did produce from Cuvier's own work, and that of others, one piece of evidence to help out. He had subdivided the Tertiary rocks of western Europe on the basis of the percentage of modern species in each epoch. This implies that species were systematically changing in time. Neither Darwin nor Lyell, however, appeared to have used this as an example of evolution. Further, continental biologists were caught up with the idea of a central plan in nature, taking their clues from recent advances in embryology. They could see no "program" in Darwin's theory (how could an eye be developed under such a random theory?). This will be an important point in our story later on.

In light of the other theories, very little of Darwin's concept appears truly unique. After all, the concept of natural selection was known well before the *Origin of Species*. The Malthusian concept of the stable population was hardly new. Why then is Darwin singled out as a genius? Primarily because he was the first to put many loose, ill-defined ideas together into a working theory capable of being demonstrated experimentally. He so overwhelmingly documented his ideas that he conquered the reader by facts. Few biologists could deny the logic of his reasoning and proofs. Many of the public did react; they were either enchanted and elated by the *Origin of Species*, or were incensed because it clearly was at odds with Genesis and was "red in tooth and claw." Natural selection by itself appears quite innocent. Mother Nature does seem to select those that are best fit for the rigors of life. But put in a Victorian context as a seeming denial of the guiding hand of the Divinity, it appeared a very perturbing idea. T. H. Huxley, a brilliant biologist and a sparkling orator, was largely responsible for placing Darwin in a convincing light before the public. He led many fascinating debates with the clergy and was known as the bishop-eater for his talent as a debator. When one bishop asked him which of his parent's ancestors was an ape, he said that he would rather have an ape for an ancestor than a bishop.

Jan/75 Sci Amer

A TEST OF NATURAL SELECTION

The theory of natural selection should be given Darwin's name, and it has stood the test of time, although modified as we shall see. It is the simplest theory that satisfies most observations (see Chap. 3). We mentioned that protective coloration could not be explained by Lamarckianism. It was from this phenomenon that one of the most interesting experimental pieces of evidence was derived to prove Darwin's theory. In the midlands of England there is a species of moth called the pepper moth. Normally it is dull yellow in color with dark specks. It commonly lives on the bark of the plane tree (our sycamore). It is a favorite food for several species of birds. In Birmingham, as the industrial revolution progressed, the buildings and woods on the downwind, east side of Birmingham became blackened with coal soot. A very shrewd scientific worker in central England observed that pepper moths on the east side of the city, while showing some variation, were primarily dark in color. *
On the west side of Birmingham, toward the prevailing west wind and away from the soot-blackened trees and buildings, the pepper moths tended to be

light-colored, reflecting the lighter, soot-free color of the bark on trees. He collected some of the light-colored pepper moths from the western part of the area and transferred them to the trees on the east side of Birmingham and reversed the procedure for the dark-colored ones. Within a very short period of time, all the light-colored moths on the east side of Birmingham had been quickly gobbled up by birds, indicating a very strong predator selection—a neat demonstration of the Darwinian theory of natural selection.

THE FOSSIL RECORD AND EVOLUTION

The world in the mid-nineteenth century had essentially no evidence whatsoever to support the theory of natural selection from the fossil record, the only place where evolution really could be proved. The Reverend Sedgwick in 1850 reviewed the fossil record and thought he saw a general progression of forms. He saw that cephalopods were dominant in the early fossil record and that these were displaced by the more advanced fish, which, in turn, were replaced by reptiles, and then finally the mammals seemed to dominate the record. Even so, Sedgwick was not convinced that this progression was accomplished by transformation, but rather by the Creator adding these important groups to the total fauna through time. Charles Lyell at the same time said that progressive development of organic life from simplest to the most complicated has but slender foundation in fact. Darwin often stated that geology assuredly does not reveal any such finely graduated organic chain because the record is too imperfect, and recognized that this is perhaps the most obvious serious objection that could be urged against the theory. It was not until 1869 that Waagen and Karpinsky provided us with the first documented evolutionary trend in the fossil record, that of the ammonoids (see Fig. 13.3). This was followed the next year by Huxley's study of the evolution of the horse, which has become a classic example of organic evolution in the fossil record.

Lyell had illustrated a good case for evolution of species but looked at his evidence in reverse. He had subdivided the marine "Tertiary" (Cenozoic) into four divisions based on the decreasing percentages of modern species. In reality species now living were gradually evolving through the Cenozoic, replacing older species, a process probably reflecting climatic changes.

THE THEORY OF RECAPITULATION AND ITS WEIRD EFFECTS

The year 1869 was an eventful one for evolution; not only did the first paleontological documentation of evolution become known, but a very important new concept appeared: the law of recapitulation or biogenetic law. It was to have a profound influence on thinking for over 50 years, principally among paleontologists. The author of this new and exciting idea was Ernst Haeckel, the leading proponent of Darwinism in Germany. A shrewd and observant embryologist working with vertebrates, he noticed that there was a direct relationship between the development of the embryo and the history of the group to which it belonged. In attempting to understand embryological development, he saw that the mammalian embryo seemed to go through stages that reflected its ancestral history. He stated that at an early stage it developed "gill slits" and a tail similar to a fish. In fact, the embryo could not be distinguished from other vertebrate embryos during similar early stages. He said that ontogeny, or the life history of the individual, recapitulates (is a short history of) phylogeny, which is the history of the race. This was counter to observations of others who said that only the young stages of the embryo resemble early stages of ancestors, later stages departing more and more from ancestral resemblance.

Using this "law" as a base, Haeckel went further and said that evolution proceeded by adding stages at the end of the individual life. This was a logical conclusion showing the relationship between the progression he saw in the embryo and the one he saw in the phylogeny. Even though Haeckel believed in natural selection, he got caught in a trap. How could new characters be added at the *end* of life and be passed on to offspring after the breeding period had ended?

Much research has been done in embryology since Haeckel's day, and we now know that there are all too many exceptions to this simple analogy, and that ontogeny does not reflect accurately the course of evolution. For example, we know that teeth devel-

"Ontogeny recapitulates phylogeny."

oped before the tongue in the evolution of the vertebrates; yet in the embryo the tongue appears first. We see now that phylogeny is the result of a series of ontogenies and not its cause (see Fig. 5.8).

STRAIGHT-LINE EVOLUTION

The next episode in our story came from paleontology. During the 1870–1880 period there was a frenzy of collecting and describing of fossils. Many complex phylogenies of the ammonoids were discovered that appeared to demonstrate Haeckel's idea. The very rich vertebrate fossil sites in the "Great American Desert" were being exploited by two colorful paleontologists—Cope and Marsh (see Chap. 16)—who tried to beat each other at describing the largest number of fossils. Thus began the golden age of paleontology in the United States. Out of this interest, intensive collecting resulted in great systematic collections of fossils. Many evolutionary sequences involving vertebrates were documented.

The picture that emerged was unexpected and formed one of the greatest puzzles of biology. Paleontologists found that groups of organisms typically began as simple, unspecialized forms, gradually becoming more specialized and usually larger (Fig. 5.3). Finally, at the end of an evolutional span, just before extinction, organisms became bizarre, frequently grotesque. This type of simple or straight-line evolution was called orthogenesis. Horned, crested, and duck-billed dinosaurs, for example, all highly specialized forms, appeared just before dinosaurs all became extinct (see Fig. 16.38). The conclusion by some of the orthogeneticists was that organisms seemed driven along the course of evolution to their goal—extinction. Antlers, for example, supposedly caused the Irish Elk (Fig. 5.1) to become extinct in the Pleistocene because the antlers became too large to be usable; organisms appeared to be trapped by their nonadaptive characters.

Today we feel that these concepts are oversimplifications and mostly descriptive. Yes, there are orthogenetic trends, but specialized forms can occur anywhere along the line, bizarre forms may not develop at all, and certainly extinction is not a goal. Even more importantly, characters do not become overspecialized, causing major groups to become extinct. The Irish Elk became extinct about 11,000 years ago probably due to a climatic change which altered the

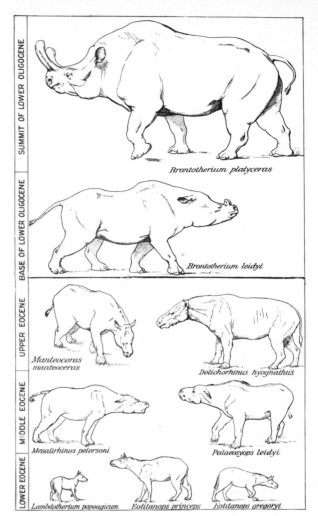

FIGURE 5.3   H. F. Osborne's evolutionary scheme of the Titanotheres of the Cenozoic, showing massive horns and large size achieved just before extinction. This is not an acceptable evolutionary trend by today's standards because we are unsure of the relationships among many of the genera. *(Courtesy American Museum of Natural History.)*

flora from a subarctic tundra to grassy open country. Other large mammals of the Northern Hemisphere also became extinct at about this time. S. J. Gould has suggested that the huge antlers were male display symbols and were not used for combat because the tines are pointed backward. Still it is important to remember that the very orderly progression of development and the tendency to become specialized near the end of a period of adaptive radiation (a period when a group may evolve rapidly and may develop a number of different lines of descent) was unexpected by Darwin or his followers.

The year 1869 was important for another historical reason. It was that year that an obscure Austrian monk by the name of Gregor Mendel published in an obscure journal the results of many years of observation and breeding of sweet peas. What Mendel had discovered was the system by which characters are transmitted from adult to offspring. He had, in short, discovered what Darwin only suspected in his later years, namely, the mechanism of heritable variation. This was one of the most important single discoveries after Darwin's work because it could have led the way to the very origin of new characters. Tragically, Mendel's work was lost and was not rediscovered for another 40 years, one of the greatest setbacks of science. Almost simultaneously in both Europe and North America, Mendel's theory of inheritance was discovered, after workers had independently arrived at the same conclusions. By 1910, when the field of genetics developed, we had learned how new characters arose and how they were inherited.

A STRIKING DISCOVERY

A striking discovery was made—new characters (or mutations) arise by chance, completely at random; thus, no purpose and no adaptation. For a period of 20 years, studies made on fruit flies demonstrated this over and over again. Here then were two opposing ideas—paleontologists believing that evolution is severely directed in straight lines leading to extinction, and the geneticists' principle that new characters arise by chance and at random. How could the complex dinosaurs be swept along on their "neat" course of evolution by waiting for a random new feature to come along? Surely even 4.5 billion years would not be enough time for its evolution. This was emphasized even more by experimental results showing that random mutations were either so small as to be insignificant (they could produce no new species) or too large so that offspring were not able to survive; besides, geneticists showed that almost all

mutations were harmful. Furthermore, the mutations arise completely *independent* of the environment with the exception of those induced by natural high-energy radiation. This fact was the telling blow to Lamarckianism, which would require the gene to mutate as a *result* of environmental forces.

THE HOME STRETCH

Such was the state of affairs until the 1930s, when a group of biologists in Europe and the United States began to see that these two ideas were not so far apart as they seemed at first. The different views were brought together to form a new synthetic theory of evolution. Essentially, Darwin's basic ideas appear to be correct. The real clue to the solution of many problems of hereditary variation came from genetics. We now know that the fundamental hereditary unit is the gene—a unit of DNA (see Chap. 7). Natural selection takes these hereditary products and preserves those that make the organism more adapted to its environment. It is also known that a gene may control the development of more than one character. Thus, the gene causing overall size increase in the Irish Elk may also have controlled the size of the antler. The mere fact that the elk existed is proof enough that it was adapted to its environment.

THE OLD QUESTION—HOW DO NEW CHARACTERS ARISE AND SPREAD?

To return to the old question of how new characters arise, we can say that they may arise by chance and at random, but how? We know that the chromosome under unknown circumstances will be arranged during cell division so that it may cross over itself or form in such a way that the chemical code for the development of features in an embryo will be altered; thus, new variability (a new character) is introduced. These mutations have been induced in the laboratory by the use of x-rays and chemical-mechanical means such as LSD. Variation was difficult for Darwin and other pregenetic evolutionists to understand in terms of how it is spread through the species. We know that half of the genes of the offspring come from each of the parents. Thus there is a "recombination" of hereditary material during each generation—one way of assuring the maintenance of variability in the population. The rate of mutation seems to be rather constant within a given group (ranging from 1 to 8 mutants per 100,000 genes in the human species). This, com-

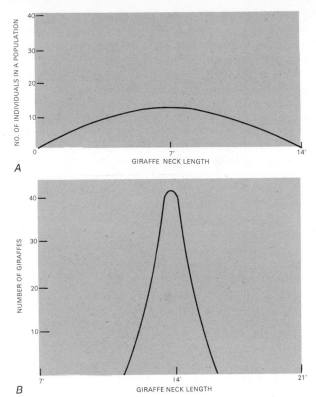

FIGURE 5.4   *A*: This population is poorly adapted for the "ideal" 7-foot height. The broad distribution of sizes indicates low selective pressure, but the population can survive rapid changes in environment. *B*. This population is very well adapted for the "ideal" 14-foot neck length. Narrow clustering around the 14-foot size reflects high selective pressure. The population would become extinct if there were a rapid change in environment, requiring, say, a 20-foot neck length, because no individuals have 20-foot neck lengths.

bined with the fact that natural selection weeds out all nonadaptive characters and preserves adaptive ones, would not seem sufficient to make any headway down the path to evolution. Yet, this is almost certainly the case.

HOW NATURAL SELECTION WORKS

Natural selection is a very creative and gradual process. We like the analogy that biologist-paleontologist G. G. Simpson used some years ago to explain it: first, the only part of a species we are interested in is the actively breeding population. Why? Because it is the only part that is involved in the

exchange of genes. This is called the gene pool. Suppose that we have a large box filled with copies of all letters of the alphabet in great abundance. Now we are allowed to draw seven letters out at a time, at random. We select seven because we want ultimately to construct the word "giraffe"—equivalent in nature to the well-adapted real thing. Obviously, it might take us a lifetime to draw out—at random—the right combination. But supposing we can throw back into the box any *g's, r's, i's, a's, e's,* and *f's*. Even further, if we draw out the seven letters *"kljnffe,"* we could paper-clip *"ffe"* together since it is the right code for getting our final adaptive result. All the time we can throw away the letters of the alphabet which do not lead toward our goal. It is obvious that we are considerably increasing the chance of getting *"g-i-r-a-f-f-e"* than when we first dipped our hand into the box (our "gene pool"). Thus even a tiny "useful" mutation will quickly be preserved in the gene pool, while a more obvious and disadvantageous mutation will be eliminated as Darwin so clearly illustrated.

The increased length of the giraffe's neck was a process involving a constantly changing gene frequency for long neck (Fig. 5.4). Now if the environmental pressure (natural selection) is acting on a population, it will eliminate the unfit and carefully preserve the fit. Evolution could thus keep up with a changing environment; understanding natural selection in this new light explains how random mutation, which is a creative process, produces characters that can be molded by the environment (Fig. 5.5).

FIGURE 5.5   A series of frequency distributions showing a succession of giraffe populations with gradually increasing neck length, keeping pace with a constant environmental change.

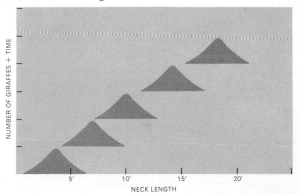

ISOLATION

New species can evolve from another cause, which may not be involved in changes—that of isolation. Supposing that we look at a wide-ranging species such as the mountain lion. It ranges, in high, semiarid regions, from southern Mexico to Canada. The extreme member populations are isolated from each other, and thus their gene arrangements could be quite different. Supposing that two adjacent populations, which have more in common, should become separated due to, say, a physical barrier such as a large river developing. If an advantageous mutation for a black coat color appears in one of them, it will spread eventually through that population *but not to the adjacent population* due to the barrier. Thus, a new, genetically isolated species would probably arise. If the mutation were a large one, and further isolation broke up the original population into smaller units with their own genetic "drift," the result might be considered a new genus.

The final chapter of the theory of evolution has not yet been written. There are many unknowns, many problems to be solved, and many questions yet to be asked. We do not know if mutations of a large nature such as the sudden appearance of a lung complex or wing results in the formation of a new order or class. Most of us feel that new classes and orders result from the accumulation of very small mutations. The reason for this puzzle is that, unexpectedly, most major groups appear suddenly in the fossil record, with few examples of transitions from ancestors, and with many groups already highly advanced when they first appear (as did the trilobites in Early Cambrian time).

THE FOSSIL RECORD

In fact, there are very few examples of continuous, gradually evolving lineages in the record such as that shown in Fig. 5.5. Eldredge and Gould believe that sudden appearances of new species can be explained by the likely possibility that new characters are more commonly introduced in small, isolated populations at the geographic periphery of the main species range. The chances of these new forms being preserved in the fossil record are small, but if successful, they could rapidly spread and become a common new species in the succeeding stratum. Hence the chances of finding a succession of species with gradually changing characters would be the exception.

It must be emphasized that there are many examples of missing links, which provide us with proof that evolution by small-step mutations (micromutations) is the best explanation for the diversity of life. For example, the oldest feathered creature, *Archaeopteryx* (Fig. 16.39) has characters of both the birds and the dinosaurs, from which they probably evolved. Many examples in the fossil record of gradual changes in a lineage through time cannot be proved, because two or three stages in the sequence may be found together at the same stratigraphic level, or an earlier stage may be found later, out of sequence. Many of these problems are due to collecting or lack of preservation of critical forms.

Figure 5.8 shows a good evolutionary series of three species of *Myalina*, a clam from the Mississippian and Pennsylvanian. *Myalina pliopetina* and *Myalina copei* are logical extensions of this evolutionary trend into the Permian; yet they are found in strata of equivalent ages in the Lower Permian. This does not mean that *Myalina pliopetina* did not serve, necessarily, as the ancestor of *Myalina copei* (it survived to live with its descendant), but we cannot prove it stratigraphically.

✷ADAPTIVE RADIATION

We shall see in Chaps. 11 to 17 that periodically in nearly all plant and animal groups, within a relatively short period of time, a large number of new species and genera will appear. In some cases, such as that of fusulinid protozoa (Fig. 5.6), there is a more or less gradual increase in numbers of genera. In other cases, the major diversification occurs suddenly (Fig. 5.7) and may gradually taper off throughout the rest of their geologic range, as in the case of the trilobites (Fig. 5.9). We are unsure of the reasons for these bursts of evolution, but the idea has been advanced that they are periods of adaptive radiation.

At the beginning of this event, a genus may move into a brand-new or "empty" ecological situation. As evolution proceeds and new species develop, the species occupy small environmental or ecological areas within the broader zone (Fig. 5.10), much like the Darwin finches did in the Galápagos Islands (Fig. 5.2). This process can be on any scale. In the case of the ammonoid cephalopods (see Fig. 16.32) in the Mesozoic, the whole subclass was involved in a radiation lasting most of the era. Whatever the cause, these bursts are very useful as stratigraphic tools

convergent / divergent evolution.

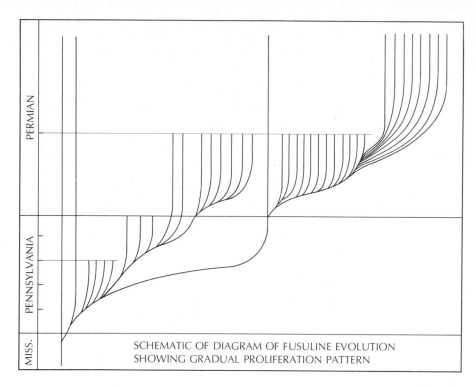

SCHEMATIC OF DIAGRAM OF FUSULINE EVOLUTION
SHOWING GRADUAL PROLIFERATION PATTERN

FIGURE 5.6 A schematic diagram of late Paleozoic fusulinid evolution, showing a relatively uniform increase in the numbers of genera.

because the species and genera tend to be short-ranged.

PHYLOGENY AND CLASSIFICATION

The course of evolution in a given group is manifested in its phylogeny, the history of the entire group. Figures 5.6 and 5.7 are examples of the phylogeny of the fusulinid protozoa and amphibians respectively. Both examples show the current interpretations of specialists regarding the derivation and subsequent history of these groups, and are subject to change as new fossils are found or as new ways of studying the characters are developed. Most specialists attempt to construct classifications of organisms to reflect phylogeny. Hence, most classifications are based on evolution.

Organisms most often are classified on the basis that "like things belong together"; implicit in this statement is the concept that they look alike because they are related. The species is the smallest formal unit, and individuals within it have more shared characters than higher, more inclusive classification groups (taxonomic units). Species are grouped into

genera, genera into families, families into orders, orders into classes, and classes into phyla. If two species share more features than other species of a genus, then we can assume that they very recently evolved from a common ancestral species. The genus

FIGURE 5.7 The invasion of land by amphibians represents an example of rapid expansion into a previously unoccupied environment. *(After A. S. Romer, Vertebrate paleontology, 1966; by permission of The University of Chicago Press.)*

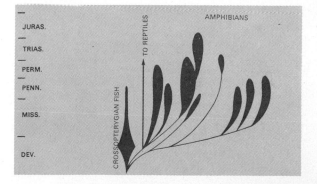

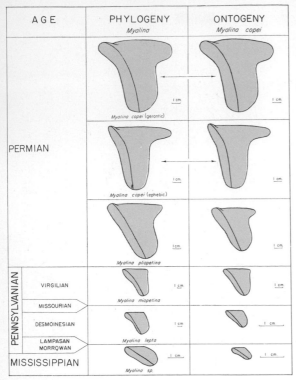

FIGURE 5.8 The evolutionary development of a series of clam species in the Upper Paleozoic. On the right is an example of recapitulation (ontogenetic column). Note that the changes in shape during the ontogeny of *Myalina copei* do not exactly reflect the actual phylogeny. *(From N. D. Newell,* Late Paleozoic pelecypods: Mytilacea. *State Geol. Surv. of Kansas, v. 10, pt. 2, 1942, with permission of the Kansas Geological Survey and N. D. Newell.)*

usually shares some, but not all, of the features of its contained species. The same is true for families, orders, etc. These features tend to be conservative, that is, established characters are less likely to be modified by selection.

On the whole this concept works well, but like most other ideas formed in the natural sciences there are annoying exceptions. One problem in comparing and analyzing relationships between organisms is that of convergence (see Fig. 5.12). In this case, unrelated organisms evolved characters quite different from their ancestors, but similar (almost never identical) to each other. This usually results from living in similar environments. For example, the ich-

thyosaurs (see Fig. 16.40), which were reptiles, are similar to sharks in outward appearance, but close inspection of the majority of characters leaves no doubt about their separate position in the classification.

The immense diversity of life in the past is intimately involved, as we have seen, with changing environments through climatic and other physical changes, the opening of "new" ecological zones (niches), or by geographic expansion or dispersal of species. For example, one important factor which controls diversity is temperature. Figure 5.11 shows that the number of species of snails decreases northward from warmer waters to cold. Warm climates support more species, but there are fewer individuals of each species. For example, in a 100- by 100-meter plot in the Amazon jungle, 60 species and 564 specimens were found. Twenty-two species were represented by a single specimen each, and only one species had as many as 33 individuals.

FIGURE 5.9 The trilobites underwent their most rapid diversification during the Cambrian, gradually losing genera through extinction.

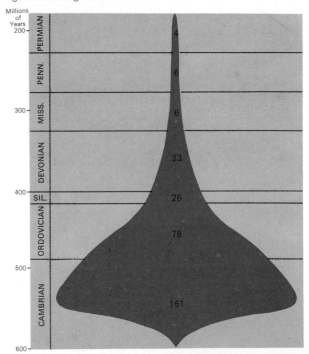

Number of trilobite genera per period

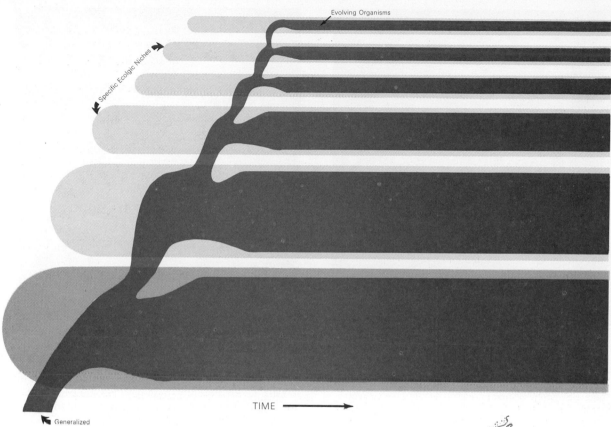

FIGURE 5.10 When a new group enters a new environment, *adaptive radiation* occurs. This process frequently involves the origin of many species resulting from "breaking down" the broader environmental zone. *(After G. G. Simpson, The major features of evolution, 1953; by permission of the Columbia University Press.)*

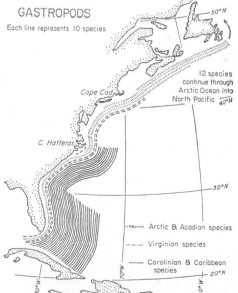

GASTROPODS

Each line represents 10 species

12 species continue through Arctic Ocean into North Pacific

Cape Cod

C. Hatteras

50°N

40°N

30°N

20°N

····· Arctic & Acadian species

- - - Virginian species

——— Carolinian & Caribbean species

FIGURE 5.11 A map showing decreasing diversity of snails toward the higher latitudes along the eastern coast of North America. *(From A. G. Fischer, Latitudinal variations in organic diversity: Evolution, v. 14, no. 1, 1960, with Permission of A. G. Fischer and the Society for the Study of Evolution.)*

FIGURE 5.12   Natural casts of two genera of snails belonging to two different families, illustrating evolutionary convergence. Even though these two are almost identical and are found at the same place, critical features identify them as members of separate families. They illustrate the difficulty sometimes encountered in proving the correct identity of index fossils. Other species of each genus do not look very much alike. *Left: Worthenia. Right: Glabrocingulum (Ananias). (Courtesy Niles Eldredge; photograph by G. R. Adlington.)*

EVOLUTION AND OUR PRESENT CULTURE

We have seen that, through Darwin and T. H. Huxley, evolution had a profound influence on nineteenth and early twentieth century life. We believe great impact was due to the religious climate of the Victorian era and to the fact that Darwinism so clearly and logically showed *for the first time* to the general public that there could be an alternative to the biblical explanation of how nature originated. The shock is over. We are in a new age, and most of us take evolution as a matter of course.

The whole conflict of evolution and religion grew out of the rejection of humans having evolved. Darwin recognized this after receiving the reaction to *Origin of Species*, and felt compelled to write the *Descent of Man*. Comparative anatomy and an extensive fossil record have so clearly demonstrated the evolution of humans that most of the thinking people of the world cannot but wonder at the complexity of our origins and development. Most perceptive theologians and biologists find no conflict between religions concerning evolution. Long ago it was realized that science is simply trying to explain the workings and not the purpose of nature.

Readings

Darwin, C., 1859, On the origin of species by means of natural selection, or the preservation of favoured races in the struggle for life: London, J. Murray.

Eldredge, N., and Gould, S. J., 1972, Punctuated equilibria: An alternative to phyletic gradualism. Models in paleobiology: J. T. M. Schopf, ed., San Francisco, Freeman Cooper, pp. 82–115.

Glass, B., ed., 1959, Forerunners of Darwin: 1745–1859: Baltimore, Johns Hopkins.

Mayr, E., 1970, Populations, species and evolution: Cambridge, Harvard.

Raup, D., and Gould, S. J., 1974, Stochastic simulation and evolution of morphology: Systematic Zoology, v. 23, pp. 305–322.

Savage, J. M., 1969, Evolution (2d ed.): New York, Holt.

Simpson, G. G., 1953, Major features of evolution: New York, Columbia.

Stebbins, G. L., 1966, Processes of organic evolution: Englewood Cliffs, Prentice-Hall.

6

THE ABSOLUTE DATING OF THE EARTH

Some drill and bore
The solid earth,
and from the strata there
Extract a register, by which we learn
That He who made it,
and revealed its date to Moses,
was mistaken in its age!

William Cowper (Late eighteenth century)

FIGURE 6.1 Humbug Mountain on the southern Oregon coast. It contains Cretaceous conglomerate and sandstone deposited following a major episode of mountain building. Diorite masses formed during that episode have yielded isotopic dates of 140 to 145 m.y., which places the mountain building in Late Jurassic time. (Photo by Henry Lowry, used by permission.)

Evolution is one of the most powerful of man's ideas, for it has revolutionized our way of thinking about natural and even social phenomena. Among great books, Darwin's *Origin of Species* probably ranks second only to the Bible in its impact on Western thought. After 1859, the basic concept of evolution, or change from one form into another through time, was also applied beyond biology. Science has come to view nature as a whole as evolutionary, thus constantly changing in a series of unique, historic events.

There are, of course, differences in the meaning of evolution in the organic and inorganic realms. While individual familiar chemical and physical processes tend to be orderly, predictable, and reproducible at any time, organic evolution seems, at first, to be random and totally unpredictable as well as nonreproducible in the strict historic sense. The mechanism of speciation through genetic mutation is essentially random, but the evolutionary *results* of natural selection are far from random, because circumstances external to a species contribute to selective pressures acting on that species. Ecologic interactions between organic and inorganic realms have been countless so that the history of one profoundly affected that of the other. Even though not wholly random, the factors involved are so numerous and their interactions so complex that the actual course of the total evolution of the earth hardly could have been predicted in advance. These interactions make up an important cornerstone of the evolutionary view of nature adopted in this book.

It appears that the entire earth is the result of one grand chemical evolution from a heterogeneous, formless, lifeless beginning to a highly organized, complex, inhabited world. The transmutation of new isotopes from radioactive parents, a subject of this chapter, clearly represents inorganic evolution. Buffon's hypothesis of epochs of cooling of the earth and of life development mentioned in Chap. 3 represents the first crude evolutionary conception of earth history. Basic to any idea of the origin of history of the planet is the question of the total age of the earth, which even Buffon greatly underestimated. In this chapter the different schemes for "absolute" dating are examined as a prelude to our study of the sequential evolution of earth as a whole. In short, we need a calendar against which we can measure time numerically rather than only in relative terms.

EARLY ATTEMPTS TO DATE THE EARTH

You will recall that, on the basis of the Scriptures, Archbishop Ussher in 1654 argued that the earth was then only 5,658 years old, having been formed in 4004 B.C. Buffon, on largely intuitive grounds, suggested in 1760 that it was 75,000 years old. Hutton, Lyell, and their followers reasoned from the implications of their uniformitarianism and the observable rates of geologic processes, that the earth must be far older than that. Finally, Charles Darwin needed as much time as possible for the evolution of the amazing array of life forms known by 1859; millions of years seemed indicated before the development of human beings.

Near the end of the nineteenth century, a British geologist estimated that all the time since the beginning of the Cambrian amounted to about 75 million years. He based his conclusion upon the maximum known thickness of strata of that age span multiplied by an assumed average rate of sedimentation derived from study of modern depositional rates. But even if he were correct in all his assumptions, the method could not yield the total age of the earth, for he did not consider Prepaleozoic rocks at all.

Use of the accumulation rate of salt in the sea as an index to the age of sea water, and at least a minimum for the earth itself, was suggested as early as 1715. However, it was not until 1899 that an Irishman named Joly actually attempted to make accurate calculations. Using the then-known average rate of delivery of salt to the sea by rivers, he found that it would have taken about 100 million years to develop the present salinity.

All these early attempts at assessing the total age of the earth in terms of absolute time involved a host of unacceptable assumptions. For example, if one were to choose an element other than sodium or chlorine to estimate the age of sea water, he would find a different apparent age. Today it is impossible to accept the simple extrapolation back in time of present rates of practically any processes. The present rates for most still are not very accurately known. It is partly because of the absurdities that arise from such simple extrapolations that we emphasized in Chap. 3 that processes had *not* been perfectly uniform as to intensity through time. Nonetheless, the early attempts to date the earth showed ingenuity and an evolution of thought in one direction, namely toward older and older estimates for our planet's birthdate. By the middle of the last century, most geologists thought in terms of a total age on the order of several hundred million years.

KELVIN'S DATING OF THE EARTH

In 1846 the great physicist Lord Kelvin launched a 50-year combat with geologists over the earth's age and the nature of its historical development. This was an outgrowth of his displeasure with strict uniformitarianism because it seemed inconsistent with physical principles. A completely uniform, unchanging earth was, to him, impossible. That temperature increases with depth in the earth was known, and was taken as proof that the earth is losing heat from its interior. Either the earth must be cooling after an initial very hot stage, or else it contains a continuing internal source of heat energy. Lyell appealed to chemical reactions in the interior to produce heat in an endless cycle allowing for a steady-state earth. Kelvin considered Lyell's untiring heat engine to be perpetual motion, and thus unacceptable. He discounted any renewing internal heat source and accepted the nearly universal assumption that the earth was originally molten. Therefore heat was being dissipated through time, and the earth could not be unchanging as long as this expenditure of energy continued. He had, in fact, endorsed an *evolutionary view* of the earth that eventually would displace Lyell's conception.

Kelvin appealed to a theory for the origin of the sun, which assumed that it formed by contraction of a gaseous cloud. Contraction would heat the cloud until it began to radiate light. Knowing the sun's mass and present energy output, the apparent time when the present level of solar radiation began could then be calculated from newtonian principles of mechanics. The initial determination indicated radiation at the present level for about 18 million years. Because the earth's origin was assumed to be linked closely to the sun, apparently the earth could not be much older than that.

From deep mines an approximate rate (or gradient) of the downward increase of temperature was known. Using this gradient (which was about three times too great,) Kelvin extrapolated the apparent rate of cooling of the earth backward to a time when the earth presumably was molten; this should be close to the total age of the earth, he reasoned. Kelvin's calculations satisfied him that the earth is between only 20 and 30 million years old. His arguments seemed flawless, for no other terrestrial or solar heat sources were known. In the eyes of most scientists Kelvin had won his campaign decisively, and left the geologists reeling. They were reluctant to accept such a modest age, but were incapable of mustering a rigorous counteroffensive until the American geologist T. C. Chamberlin in 1899 challenged Kelvin's assumption that the earth had begun as a molten body. He introduced an important new hypothesis to be discussed in the next chapter, which proposed that the planets formed by the accretion or collecting together of cold, solid chunks of matter. According to Chamberlin, then, the earth must have heated up sometime *after* its initial formation through some internal process quite independent of solar heat.

RADIOACTIVITY

The last great breakthrough, which completed the stage for modern historical geology, occurred in 1896 in the Paris laboratory of physicist Henri Becquerel. The Frenchman recognized the phenomenon of radioactivity through the exposure of photographic film tightly sealed from light but held next to a uranium-bearing mineral. Some kind of radiation invisible to the eye had exposed that film. Two hitherto unknown elements were soon discovered, radium and polonium, which form from uranium by changes in the atomic nucleus that involve hitherto unknown types of radiation. Demonstration that many elements have several nuclear species called isotopes followed quickly. Some of these species are unstable; that is, their atoms *are not permanent* as had always been assumed. The nuclear species of a single element differ in having slight variations in the number of neutrons in the atomic nucleus. Yet, isotopic species of any given element have similar chemical properties, and it is for this reason that isotopes went undetected for so long.

Isotopes with unstable nuclei undergo spontaneous change—or mutation—until a permanent, stable configuration is reached. This process of change is called radioactive decay, and it results in one or more of three types of emanations or emissions from the nucleus at *fixed average rates* for any particular isotope. It was such emanations that exposed Becquerel's film. The emanations were originally named alpha (which proved to be a helium nucleus, i.e., a helium atom stripped of electrons), beta (a free electron), and gamma rays (similar to x-rays). Radioactive

decay can be written like a chemical reaction in the following manner:

Parent isotope → stable daughter isotope + nuclear emanations + heat

Isotopes are designated by numbers representing their relative mass (or "weight") written at the top of the chemical symbol for the respective element. Emissions from the nucleus of radioactive isotopes change the identity of that isotope; thus, unstable uranium with mass number 238 (^{238}U) decays ultimately to stable lead with a mass number of 206 (^{206}Pb). Radioactive decay of uranium and some other isotopes actually produces a complex series of transformations before an ultimate stable species results (Fig. 6.2). For example, polonium and radium proved to be but intermediate daughter isotopes in the ^{238}U series.

It is important to realize that the occurrence of an individual emanation is a statistical event. That is, neither the identity of the particular atom that will decay next nor the exact time at which the event will occur can be predicted absolutely. What we can predict is what fraction of the total parent atoms present will decay in a given period of time. Observations with devices like a Geiger counter of many emanations from a particular nuclear species over an extended time period provide a *statistical average*

FIGURE 6.2 Diagrammatic portrayal of the decay history of the uranium 238 series. Half-lives for the intermediate daughter products vary from less than 1 second (polonium 214) to 1,622 years (radium 226). *(From* How old is the earth? *by P. M. Hurley. Copyright © 1959 by Educational Services, Inc. Reprinted by permission of Doubleday and Company, Inc.)*

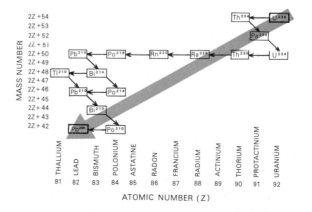

rate of activity or decay. Neither heating or cooling, nor changes in pressure, nor changes in chemical state can affect in any detectable way the average rate of spontaneous decay. Because the rate cannot be artificially changed, it is assumed that it always has been uniform for a given isotope. Decay of unstable isotopes seems to be the only change in nature that we dare consider to have a constant rate.

FIRST ISOTOPIC DATING OF MATERIALS

Further knowledge of radioactivity came with breath-taking rapidity after the turn of the present century. The British physicist E. Rutherford, after counting radium emanations for many days with a scintillometer, reasoned that total emission activity was proportional to the number of unstable parent nuclei still present. This meant that emission must decrease in some regular fashion through time, and the decay could be expressed mathematically if decay rates were known.

In 1905 a Yale chemist, Bertram Boltwood, suggested that lead was among the disintegration products of uranium. This is an ironic twist of an ancient alchemist's dream of making precious gold from lead; instead, apparently common, ugly lead was formed *from* a rare and almost equally precious element. In 1907 Boltwood, following a suggestion by Rutherford, developed the idea that a decay series, such as that of uranium, could provide a means of dating the time of crystallization of minerals that contained a radioactive element. Boltwood reasoned that in unaltered minerals

of equal ages, a constant proportion must exist between the amount of each disintegration product and the amount of parent substance . . . and . . . the proportion of each disintegration product with respect to the parent substance must be greater in those minerals which are the older.

It was assumed, of course, that all the stable daughter isotopes present had formed only from the parent isotope in a particular mineral crystal. From chemical analyses of a number of uranium-bearing minerals from localities whose relative geologic ages were known approximately, he calculated average lead-uranium ratios and discovered a very significant pattern. Without exception, the proportion of lead was progressively greater for geologically older localities (Fig. 6.3)! Boltwood felt that this systematic relationship confirmed his inference that lead was a product of uranium decay. Rutherford also had shown that helium was a product of uranium decay. Boltwood found helium gas in the same minerals that were analyzed for lead and uranium, and he inferred that indeed it was formed as another product of the decay of uranium. Attempts to date minerals by their helium-uranium ratios have met with little success because helium, being a gas, may escape readily from the mineral crystals, causing results that are "too young."

To compute the ages of minerals, Boltwood needed to know the rate of decay of uranium, which was so slow that it had not yet been measured accurately. So he estimated it indirectly from the known rate for radium, a rapidly decaying intermediate member of the uranium series (Fig. 6.2). He then calculated the ages of minerals from their lead-uranium ratios as follows:

$$\text{Age} = \frac{\text{amount of daughter isotope (Pb)}}{\text{amount of parent isotope (U)}} \times 10^{10} \text{ yrs.}$$

Boltwood's results for 10 localities on 3 continents ranged from 410 to 2,200 million years. In reality, his dates were about 20 percent too large, because the mathematics of the decay process was not fully understood until 1910, and because, unknown to him, more than one decay series—each with differing

FIGURE 6.3 Diagrammatic portrayal of decay of a radioactive isotope within a mineral crystal. Note the changing ratio of parent to daughter isotope after one and then two half-life time intervals.

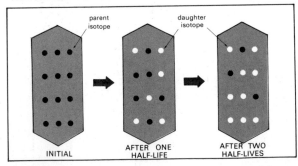

decay rates—was involved in the minerals that he studied. Even allowing for this error, the oldest apparent age listed by Boltwood was nearly 100 times greater than Kelvin's figure and 10 times the wildest guess of any geologist for the total age of the earth!

THE CONCEPT OF DECAY

To analyze many different isotopes, whose rates of decay vary widely, it is convenient to develop a generalized expression applicable to all decay series. This also should help to understand better the principles of geologic dating by the decay phenomenon. Remember that what is dated is the *time of crystallization* of a mineral; that is, the time of incorporation of an unstable isotope into that mineral (not the time of origin of that isotope in the universe, which in most cases was earlier than the origin of the earth). It is assumed that once a mineral crystallized, a closed chemical system was formed such that any daughter product now present in the mineral was formed only from the decay of the original unstable parent isotope therein (Fig. 6.3). It is something like the closed system of an hour glass, which entraps sand grains. Those grains in the top of the glass are like parent isotopes, whereas those falling to the bottom are like daughter isotopes; the ratio between them at any moment is a measure of time elapsed.

Dating is a matter of determining the ratio of parent to daughter atoms (Fig. 6.3) and then calculating a mineral's age by multiplying by the decay rate of the parent. More atoms decay early in the history of a mineral, and emanations then decrease in a regular way.[1] Figure 6.4 shows a decay-curve graph with

---

[1] $Age = \dfrac{\text{amount of daughter isotope}}{\text{amount of parent isotope}} \times \dfrac{1}{\lambda}$

Being exponential, decay can be conveniently expressed in logarithmic form such that:

$\lambda t = \ln \dfrac{N^o}{N^t}$

where λ = decay constant
   N^o = number atoms of parent isotope at time zero (crystallization)
   N^t = number atoms of parent after time t of decay
   $\ln$ = natural logarithm (base 2.78)

In other words, the time of decay is proportional to the natural logarithm of the ratio of N^o to N^t. It is convenient to have some fixed unit of decay time that would express concisely the decay characteristic of a particular isotope. Such a unit is the half-life ($t_{1/2}$), i.e., the

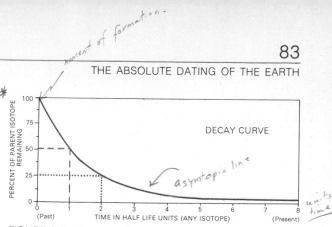

FIGURE 6.4 Simple arithmetic plot of an isotopic decay curve. After one half-life has elapsed, 50 percent of the original parent isotope remains; after two half-lives, half of that, or 25 percent, and so on.

time expressed in half-life units. The half-life, which is the time required for one-half of a given amount of any particular unstable nuclear species to decay, is a more convenient expression of rate for purposes of illustration than is the decay constant (λ). Half-life can be better envisioned by considering an analogy. Imagine a glass full of water to represent all the parent isotopes in some mineral crystal. After one half-life elapses, you drink half a glass of water. Then after one more half-life passes, and finding your thirst still unquenched, you drink half of what remains. After a third half-life elapses, you drink half of the remainder, and so on. The decrease of water in the glass is exactly like the decrease of parent isotopes as decay proceeds in a mineral. By using half-life as a time-scale unit, our curve applies to the decay of any isotope regardless of the fact that half-lives vary enormously, as shown in Table 6.1. Undoubtedly some short half-life isotopes contained in the early earth have so completely decayed that we cannot detect them; yet their daughter products must remain. For example, two isotopes of the gas xenon are thought to be decay products respectively of iodine 129 and plutonium 244, which do not occur naturally on earth (although they have been produced by nuclear explosions). Their half-lives are so short that, if they were present in the early earth as is supposed,

---

time required for one-half of an original amount of a nuclear species to decay. At one half-life, the N^o to N^t ratio equals 2, therefore:

$\lambda t_{1/2} = \ln 2$

$t_{1/2} = \dfrac{\ln 2}{\lambda} = \dfrac{0.693}{\lambda}$

EVOLUTION OF THE EARTH

TABLE 6.1 Principal decay series used for mineral and total-rock dating

Parent isotope	Half-life	Ultimate stable product	Effective age range
Rubidium 87	47 billion years (b.y.)	Strontium 87	>100 million years (m.y.)
Thorium 232	13.9 b.y.	Lead 208	>200 m.y.
Uranium 238	4.5 b.y.	Lead 206	>100 m.y.
Potassium 40	1.3 b.y.	Argon 40	>100,000 years
Uranium 235	0.71 b.y.	Lead 207	>100 m.y.
Carbon 14	5,710 (±30) years	Nitrogen 14	0–60,000 years

after only a few hundreds of millions of years, they had diminished to unmeasurably small amounts.

Weathering or other chemical modifications of minerals, particularly heating during subsequent metamorphism, may disturb the original parent-daughter ratio. This obscures the true age of crystallization of the mineral, generally producing an apparent age that is "too young." In interpreting the geologic significance of isotopic dates, one must be alert for such resetting of isotopic clocks, as we shall see in later discussions.

MODERN ISOTOPIC DATING

THE DECAY SERIES USED

During the middle twentieth century, "absolute" or isotopic dating of minerals increased tremendously. By the mid-1920s uranium-lead dating had progressed to the point that several dates had appeared that still agree with modern results, and it could be said that the end of the Cambrian Period was about twice as old as the Permian. In 1934 the first isotopic time scale was published, and geology textbooks were beginning to recognize isotopic dating. Discovery about 1940 of the rubidium 87–strontium 87 (^{87}Rb–^{87}Sr) series and the proof in 1948 that radiogenic potassium 40 decays to the inert gas argon 40 (^{40}K–^{40}Ar) initiated the use of other isotopes as geologic clocks in addition to those of well-known uranium and thorium. The most important decay series in current use are shown in Table 6.1. The differences in rates of decay impose certain limitations on the relative usefulness of each series for minerals of different relative ages. Carbon 14, for example, decays so rapidly that after 70,000 years there is not enough of the parent left to be accurately measured with present analytical techniques. Therefore, this isotope is useful only for late Pleistocene and Holo-

cene times; it is especially useful in archaeology, climatology, and studies of the movement of ground water and ocean circulation.

The long half-life isotopes are of no use for dating minerals less than about 100 million years, because no perceptible change will have occurred in such a short time. Long half-life isotopes, however, are useful through most of the geologic time scale (Table 6.2). Uranium, thorium, and rubidium are especially applicable to Prepaleozoic rocks (i.e., over 700 million years). The potassium-argon series is slightly less useful for very old rocks. This is not because of its decay rate, but because the ultimate stable daughter product argon, being a gas, is lost from crystals more easily than other daughter products. Argon leakage is a particularly serious problem if a mineral has been reheated through burial or metamorphism to temperatures exceeding 150°C after initial crystallization. In short, potassium-bearing minerals can have their isotopic clocks reset more readily than certain other minerals. Dating metamorphic events, however, may be just as important as dating the original crystallization of a rock. In fact there is growing awareness that isotopic dates, strictly speaking, are dates of cooling of minerals down to some critical temperature at which daughter isotopes become fixed in mineral structures. This date appears in many cases to be geologically identical with the date of initial crystallization of the minerals, but there are important exceptions. Such discordant dates generally are detected by studying at least two isotope series that have responded differently during a rock's history.

DISCORDANT DATES

Daughter isotopes are usually trapped within the crystal structures of minerals in which they originated. But metamorphic recrystallization of a rock tends to purge minerals of most daughter products formed

TABLE 6.2   The isotopic time scale*

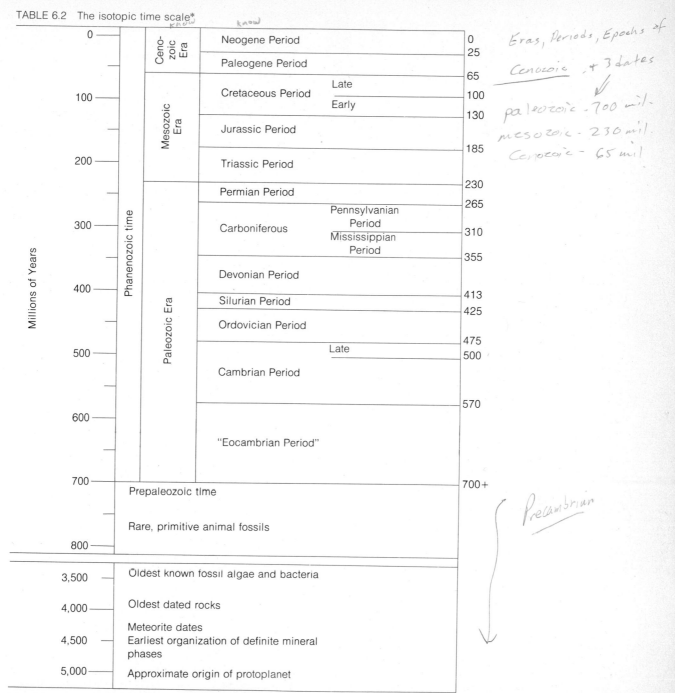

Millions of Years				
0 —	Phanenozoic time	Ceno-zoic Era	Neogene Period	0
				25
			Paleogene Period	
				65
100 —		Mesozoic Era	Cretaceous Period — Late	100
			Cretaceous Period — Early	130
			Jurassic Period	
				185
200 —			Triassic Period	
		Paleozoic Era	Permian Period	230
				265
300 —			Carboniferous — Pennsylvanian Period	
			Carboniferous — Mississippian Period	310
				355
400 —			Devonian Period	
			Silurian Period	413
				425
			Ordovician Period	
				475
500 —			Cambrian Period — Late	500
600 —				570
			"Eocambrian Period"	
700 —				700+
	Prepaleozoic time			
	Rare, primitive animal fossils			
800 —				
3,500 —	Oldest known fossil algae and bacteria			
4,000 —	Oldest dated rocks			
4,500 —	Meteorite dates / Earliest organization of definite mineral phases			
5,000 —	Approximate origin of protoplanet			

*Modified from J. L. Kulp, 1961, *Science,* v. 133, and W. B. Harland, 1971, *The Phanerozoic time scale:* London, Geological Society Publication No. 5.

FIGURE 6.5 (1) The effect of metamorphism in redistributing daughter isotopes (circles) in a rock that crystallized originally 1,000 million years ago; (2) after 500 million years, some parent isotopes (dots) in a feldspar crystal had decayed; (3) metamorphism 480 million years ago drove the daughter atoms out of the crystal, but they were retained in the surrounding rock; (4) today dating of the feldspar would reveal the metamorphic event, while a whole-rock date would reveal the original crystallization 1,000 million years ago, assuming that the rock has remained a closed chemical system with respect to the daughter isotope.

therein, because daughter isotopes are chemically dissimilar from their parent species, and so cannot form true chemical bonds in the mineral (Fig. 6.5). Helium and argon may diffuse completely out of the original rock. Other daughter products, however, will become incorporated in new minerals, while remaining parent isotopes begin a new decay history in purged crystals. The latter grains will yield isotopic dates of the metamorphic, clock-resetting event, whereas if none of the premetamorphic daughter products were lost from the rock as a whole during its recrystallization, the parent-daughter ratio for the *total rock* still will indicate the time of original formation of the rock (Fig. 6.5). In refined dating today, several isotopic dates are determined for the total rock and for several different minerals; cross checks with different isotope series also are desirable. In this manner, the maximum possible can be learned about a rock's isotopic history.

CARBON 14—A SPECIAL CASE

The value of carbon 14 for dating was discovered by W. F. Libby in 1947. Unlike most other isotopes used for dating, the only unstable isotope of carbon is produced in the upper atmosphere. High-energy cosmic particles shatter the nuclei of oxygen and nitrogen atoms there, releasing neutrons, which in turn produce ^{14}C when they collide with other nitrogen atoms. Unstable ^{14}C decays eventually to stable nitrogen 14. Most of the ^{14}C produced is oxidized to carbon dioxide (CO_2), some of which enters the

hydrosphere. Production of ^{14}C in the atmosphere is assumed to be at a nearly constant rate; therefore ^{14}C abundance should have long since reached a steady state. That is, a constant balance should exist between its production and its decay such that the ratio of ^{14}C to stable carbon is assumed to be constant.

From either the air or water, organisms acquire small amounts of ^{14}C along with the stable isotopes. Any dead organic matter, such as wood, charcoal, some bone, human artifacts such as cloth, and also shells and young limestones are all potentially datable. The number of ^{14}C and ^{14}N atoms present is so small that direct measurement is impossible. Instead, dating is done by use of a radiation counter to compare the radiation activity (i.e., the level of beta emissions) of the specimen with that of a standard modern sample whose activity is known very accurately. The number of emissions or "counts" in a given time period is proportional to the number of atoms of ^{14}C present. The older the specimen, the less will be its activity level. At just over 5,700 years, the half-life of ^{14}C is rather short. Together with the limit of sensitivity of the counting device, this imposes an upper limit of measurable age of about 70,000 years. Since the industrial revolution, combustion and atmospheric nuclear testing have altered artificially the ^{14}C content of the total carbon reservoir, which has created problems for maintaining reliable modern standard samples. Several sources of loss or addition of ^{14}C to specimens and fluctuations of past atmospheric ^{14}C abundance also impose limitations on the method.

RATIOS AMONG DAUGHTER ISOTOPES

Dating by comparison of ratios of only daughter products is being used increasingly. Comparison of different daughter lead isotopes, such as ^{206}Pb and ^{207}Pb (as in Table 6.3), provides some of the most accurate rock dates possible for the entire uranium series. Because these daughter isotopes result from series with different decay rates, proportions of the lead isotopes change through time. Their changing ratio provides a special isotopic clock.

An ingenious method has been developed for dating Pleistocene deep-sea clays using the ratio of protactinium 231 (half-life 34,300 years) to thorium 230 (half-life 80,000 years), both intermediate daughter products of uranium decay; thorium 230 and uranium 234 also are used. All these isotopes are

TABLE 6.3 Concordant isotopic ages from three different decay series for uraninite from a Prepaleozoic pegmatite dike, Black Hills, South Dakota*

Isotope ratios used	Age
$^{206}Pb{:}^{238}U$	1,580 m.y.
$^{207}Pb{:}^{235}U$	1,600 m.y.
$^{207}Pb{:}^{206}Pb$	1,630 m.y.

*After Wetherill et al. 1956, *Geochimica et Cosmochimica Acta*, v. 9.

taken from sea water by clay minerals, and their ratios in sediments become a function only of the *time since absorption from the water*. This method allows dating of sediments as much as 175,000 years old, or nearly three times the range of carbon 14 dating. The range of 200,000 years to 1 million years ago has been the least accessible, although refined K-Ar techniques now allow the dating of favorable material considerably younger than 1 million years.

THE ACCURACY OF ISOTOPIC DATES

Several factors limit the accuracy of "absolute" isotopic dates. First, the statistical nature of the decay process itself means that, even under the best of conditions, only average decay rates are determinable. This introduces a small uncertainty, though for very old minerals it is negligible. Some decay rates are better known than others; for example, that of ^{238}U is known to ±1 percent, ^{235}U and ^{87}Rb to ±2 percent, and ^{40}K to ±3 to 5 percent. A third source of uncertainty originates in the laboratory analyses of the isotope ratios. The mass spectrograph provides by far the most sensitive analyses of isotope abundance, with from ±0.2 to 2.0 percent accuracy possible.

Because of the various contributing analytical limitations, isotopic dates generally are reported with an uncertainty figure expressed, such as 100 ± 5 million years. This means that analytical limitations do not allow one to say more than that the age lies between 95 and 105 million years. Under ideal conditions of rock freshness, analytical care, and relatively old material (with accurately measurable amounts of daughter material present), the analytical error for isotopic dates can be as low as ±2 percent. For a mineral 10 million years old, that would be only ±0.2 million years; for one 100 million years old, a mere ±2.0 million years; for one 1,000 million years old, ±20 million years. It is important to realize that some ✳ uncertainty exists for *every* isotopic date, but in an

introductory book such as this, the uncertainty factor is of little consequence, and so it will be omitted from most dates quoted.

THE ISOTOPIC TIME SCALE

Most of the datable minerals containing radioactive elements originate in igneous rocks; therefore the bulk of isotopic dating must be confined to such rocks. Detrital minerals containing unstable nuclear species occur in sediments, to be sure, but to date them is not to date the time of sedimentation but rather the original time of crystallization of the mineral in a parent igneous (or metamorphic) rock or mineral vein prior to erosion, transport, and deposition in a sediment.

Because the relative geologic time scale is based upon fossiliferous sedimentary rocks, obviously it is urgent somehow to establish absolute dates for the sediments. Direct isotopic dates of sedimentary rocks are possible for only a few minerals that contain unstable species *and that crystallize in the environment of deposition*. Such a mineral is glauconite, a green mica-like silicate mineral containing potassium ($FeKSi_2O_6$). It apparently forms only in marine environments, where deposition is slow. The potassium in glauconite consists in part of ^{40}K; thus, the mineral can be dated by the K–Ar method. Potassium-bearing feldspars form in certain sediments soon after deposition, and these, too, are potentially datable. Unfortunately neither glauconite nor the nondetrital feldspars are very common, and they give rather unreliable results. The uranium method has been used for black shale containing minute amounts of uranium-bearing minerals associated with carbon compounds. The uranium compounds formed in the sediments, and so they provide dates of sedimentation. Carbon 14 can be used in certain very young sediments

Because most dating must be done on igneous or metamorphic rocks, it is necessary to relate such ages indirectly to the relative geologic time scale, and thus to establish an "absolute" age scale (Table 6.2). Here the basic principles of stratigraphy come into play again. A volcanic formation interstratified with fossiliferous sediments and containing datable minerals clearly is of the same geologic age as those sediments, providing an ideal point on the isotopic time scale. An intrusive igneous body, however, must

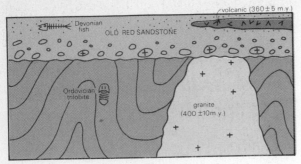

FIGURE 6.6 Idealized diagram of the Scottish rocks studied by Hutton, showing how key dates are established for an isotopic ("absolute") time scale. The granite clearly is post-Ordovician and pre-Devonian, thus is of Silurian age. A date for it therefore establishes that the Silurian Period is about 400 million years old. The date for a lava flow within the Old Red Sandstone establishes that the Devonian Period is about 360 million years old.

be younger than all rocks through which it cuts. If the latter are fossiliferous strata, we have an older geologic age limit for the intrusion. If, perchance, younger fossiliferous sedimentary rocks rest unconformably upon the intrusion, or at least contain pebbles of it, then an upper geologic age limit can also be established (Fig. 6.6). If the age difference between the older and younger fossiliferous sediments is not great, the intrusion is closely dated in the relative geologic scale, and its isotopic date provides another precise control point on the absolute time scale. The scale is constantly being refined, and techniques for analysis of smaller and smaller amounts of isotopes are being developed continually.

Isotopic dating has value for the field geologist in the dating and correlation of local rock bodies where fossil evidence is poor or lacking. This is simply reversing the process used for originally establishing points on the isotopic scale. Because of the many problems that enter into the whole dating process, this sort of use has received little application as yet. Commonly the different rocks in a limited area under study are so close in age that very great sensitivity is required to discriminate them isotopically.

MEMORABLE DATES

At last we can inquire intelligently of our planet's true birthdate. The chemical composition of the solar system must be relevant to its age as well as to its mode of origin. It is thought that most of the natural chemical elements in the universe evolved through thermonuclear reactions in stars; additional ones also developed outside the stars, as in the earth, through radioactive decay of unstable isotopes. The elements in our part of the universe apparently evolved from the lightest and simplest (hydrogen, 75 percent; and helium, 24 percent by weight) to the heaviest and rarest ones (such as uranium and lead). Presence still in the earth of moderate half-life isotopes such as ^{40}K ($t_{1/2}$ = 1.3 b.y.) places an upper limit on the formation of the earth. If the earth were more than 6 or 7 billion years old, essentially all those would have decayed.

No individual isotopic date represents the time of origin of a parent isotope, for most of them originated before the earth was formed; the majority of elements in the solar system probably are about 7 to 10 billion years old. The dates instead indicate times of incorporation of a given isotope into a mineral crystal. Isotopes had been decaying ever since their extraterrestrial origin, but only after incorporation into the closed system of a crystal (or rock) could the decay products be trapped with the parent isotope to provide a measurable isotopic ratio useful for dating (Fig. 6.3). Uranium, lead, and rubidium dating have revealed that igneous rocks as old as 3 to 3.5 billion years occur in South Africa and northwestern Russia near Finland, and over 3.7 to 3.8 billion years in southwestern Greenland and in Minnesota. Thus some of the outer crust of continents is around 3.5 billion years old. Some of the moon's crust appears to be of about the same age according to isotopic dating of the Apollo-mission specimens (3.5 to 4.2 billion years).

The oldest rock dates still do not represent the total age of the earth, for it must have taken the crust a while to form, as we shall see in the next chapter. An ingenious way to estimate the elusive total age was suggested through work with lead isotopes in the 1930s. It was found that proportions of lead isotopes differ greatly in lead minerals, such as galena (PbS), of different geologic ages. One isotope, Pb 204, is not known to be forming today by any radioactive decay process now operating on earth. Significantly, it is more abundant in old lead minerals, whereas geologically younger minerals contain progressively more of the radiogenic lead isotopes produced by uranium and thorium decay (Table 6.1). In other words, as decay has created radiogenic lead, some of that lead has been freed from its parent uranium-bearing minerals by melting or weathering, and has become recombined with other leads in later-formed lead minerals. Such dilution of the original nonradiogenic lead has increased through geologic time, apparently in direct proportion to the continuous generation of new radiogenic lead isotopes through decay.

The systematic change in the ratios among the lead species in younger and younger minerals provides a special isotopic clock because each species is formed at a different rate. Today there is about one part of new radiogenic lead to two parts of original "common" lead in the crust. Meteorites contain leads but no uranium or thorium parents; therefore the proportions of different lead species therein are assumed to have remained fixed since the meteorites formed. And lead ratios in meteorites are taken as a standard representing the approximate primary or primordial lead ratios in the early earth. By comparing ratios of the four lead isotopes in varying-aged minerals, we can extrapolate back to ratios identical to the primordial leads of meteorites. Such extrapolation indicates that the primordial ratio existed on earth about 4.7 to 4.8 billion years ago. This date apparently represents the time when elements in the earth first became arranged into definite chemical compounds, and it is taken as the beginning of truly geologic time. It must be close to the planet's birthdate.

How much earlier the very first assembly of the earth began is more difficult to fix. Astronomers believe, from theories of stars and galaxies, that the Milky Way galaxy is about 10 billion years old, and that our sun formed about 6 billion years ago. Meteorites can be dated by the U-Pb, K-Ar, or Rb-Sr methods, and they yield results averaging between 4.3 and 4.5 billion years, nearly identical with that indicated by lead ratios for the initial formation of earth minerals. Moon rocks also yield dates at least as great as 3.8 billion years. Assuming that meteorites, as well as the planets and moons, always have been as closely linked to the sun as they are today, they can hardly be older than that star. Therefore, it appears that the earth as a solid sphere is not more than 5.0 billion years old, and that the outer crust of continental areas started forming at least 3.7 billion years ago (Table 6.2).

The first bona fide records of fossil life come from rocks that are about 3.0 to 3.5 billion years old, which contain structures and chemical compounds formed by bacteria and possibly primitive algae. The first recognizable animal fossils are a mere 1 billion years old or so, but the first abundant animal record began only 600 million years ago; man is a mere 1.5 to 2.0 million years old. Therefore, for one-fourth of earth history there is no recognized record of any life form, and only the last quarter, often referred to as Phanerozoic time, has an abundant record, particularly of index fossils useful for geologic dating and correlation of strata. Inevitably, then, we know the most about the latest, brief part of our planet's history. If we were to proportion this book strictly according to the absolute time scale, 16 chapters should be given over to the Prepaleozoic, 1.81 to the Paleozoic, 0.6 to the Mesozoic, and only 0.25 to the Cenozoic. However, we actually do almost the reverse because of the relative detail of our knowledge, which is the reverse of age.

SUMMARY

In this chapter we have seen that estimates of the age of the earth have been increasing constantly up to the present-day figure of 4.5 to 5.0 billion years. The maximum figure does not seem likely to increase very much more, however. Early dating methods, such as those based upon increasing saltiness of the seas or total deposition of stratified sediments, were ingenious, but suffered from erroneous assumptions and inaccurate knowledge of true rates of the processes involved. We can never assume constant rates of most processes over long intervals of geologic time. So far, only radioactive decay shows apparent uniformity of rates, and even this has been questioned.

Lord Kelvin's calculation of age from the earth's internal heat gradient was equally ingenious and infinitely more persuasive because of the author's great stature and the compelling logic of his mathematics. But the most brilliant argument is no better than its weakest assumption! Kelvin's argument was doomed by the incorrect notion that all the earth's internal heat was residual from a hot origin. Radioactivity was discovered near the end of his life, and so he had no way of knowing that the earth contains its own internal heat-generating mechanism. Isotopic dating has exonerated Hutton and Lyell by showing

EVOLUTION OF THE EARTH

the earth to be very old indeed, in fact many times older than even their greatest estimates.

Today from isotopic dating we can also assess actual lengths of the geologic periods. From the results of extensive analyses, the dates of most period boundaries are now fairly well established as summarized in Table 6.2. Average length of the periods was 30 to 40 million years. The Cambrian apparently was longest (100 to 125 million years), and the Silurian (only 10 to 12 million years) was the shortest. Isotopic dating also allows rough comparisons of rates of change, such as mountain building, denudation, and sea-level changes. Throughout the book, we shall examine many events in terms of absolute, as well as relative, geologic age and rates.

Truly, isotopic dating of minerals has revolutionized geology in the twentieth century! Besides offering absolute dating, radioactivity also provides an internal source of earth heat, the chief form of terrestrial energy. Radioactive heat generation was about five times greater early in history than now because of the presence then of many more atoms of unstable isotopes, and it may have been sufficient to melt part of the interior for a time.

The discovery of radioactivity provided a dramatic example of inorganic evolution—the creation or mutation of new elemental species from older ones. Modern thermonuclear theories of stars and of the origins of the elements provide a compelling evolutionary conception for the entire solar system, as well as for the earth, which is explored more fully in the next chapter.

Readings

Burchfield, J. D., 1975, Lord Kelvin and the age of the earth: New York, Science History.

Dalrymple, G. B., and Lanphere, M. A., 1969, Potassium-argon dating: San Francisco, Freeman.

Faul, H., 1966, Ages of rocks, planets, and stars: New York, McGraw-Hill. (Paperback)

Harland, W. B., 1964, The phanerozoic time scale: London, Geological Society, and 1971, Publication No. 5.

Harper, C. T., 1973, Geochronology: Radiometric dating of rocks and minerals: Benchmark papers in geology: Stroudsburg, Dowden, Hutchinson, and Ross.

Hurley, P. M., 1959, How old is the earth?: New York, Doubleday-Anchor. (Paperback S 5)

Joly, J., 1909, Radioactivity and geology: London, Constable.

7

EARLY PHYSICAL AND CHEMICAL EVOLUTION OF THE EARTH

The ways in which men came into the knowledge of things celestial appears to me almost as marvelous as the nature of these things themselves.

Johannes Kepler

There is something fascinating about science. One gets such wholesale returns of conjecture out of such a trifling investment of fact.

Mark Twain,
Life on the Mississippi

FIGURE 7.1 Artistic rendition of Buffon's hypothesis of planetary origin by passage of a comet near the sun. *(From de Buffon,* Histoire naturelle, *1749.)*

Early explanations of planetary births had in common an assumption of a hot origin followed by cooling; they also assumed the sun to be considerably older than the planets. But we no longer accept the storybook pictures of the earliest earth as a red ball of boiling lava shrouded in ominous steam clouds. G. L. de Buffon popularized the long-held hot origin in the eighteenth century with a vivid hypothesis that the gravity fields of passing comets had torn away hot masses from the sun, which then cooled to form planets (Fig. 7.1).

Buffon's hypothesis was abandoned largely because space is so vast in relation to the nearly insignificant size of individual celestial bodies that collisions or near-collisions between comets and stars are judged to be extremely unlikely. Spatially, the planets in our solar system can be likened to a few lonely squirrels running around over the entire earth. After Buffon, several different suggestions were made also calling for condensation of the planets from a hot, gaseous cloud (nebula) surrounding the sun. But these hypotheses all had major defects. During our century, more complex hypotheses have been proposed to overcome difficulties of the earlier ones. While each improves upon others in certain respects, no fully satisfactory explanation exists even yet.

The average density of rocks exposed at the earth's surface is about 2.7 times that of water (i.e., 2.7 grams per cubic centimeter), with the very densest ones being slightly over 3.0 times. The densities of these rocks were easily measured long ago, and in the late eighteenth century a clever experiment showed that the overall bulk density of the entire earth is about 5.5 times that of water. Obviously there must be very dense material in the earth's interior. Although early cosmogonists postulated differential settling of matter according to density from an assumed chaotic early stage, resulting in a concentric arrangement of solids, liquids, and gases, determination of bulk density was the first scientific clue that the earth's interior is somehow zoned. Now we know from classical mechanics that a planet with a gravity field must tend to develop density differentiation.

To explain the planet's origin as well as the mechanism and history of differentiation of solid earth, ocean, and atmosphere is the greatest challenge for geologists. Because it provides a unifying focus for all branches of geology and geophysics, the present chapter is devoted to this important though speculative phase of earth history. First we summarize opinions on the origin of the planets, then we shall review the evidence available for interpreting the present interior makeup of the earth, and finally we shall examine hypotheses for the origin of the crust, oceans and atmosphere.

PROBABLE ORIGIN OF THE EARTH

A COLD BEGINNING

After 1900, a hypothetical origin of planets through aggregation of cold clouds of dust and gases became preferred. According to this scheme, much of the internal heating of the earth is presumed to have occurred *subsequent* to its aggregation through gravitational contraction and radioactive decay within the mass. The energy produced by such heating apparently has caused the continual disturbances seen in earthquakes, volcanoes, formation of mountains, and the like.

The first important suggestion of a cold origin of the planets was the planetesimal hypothesis of Chamberlin and Moulton formulated early in the twentieth century. (Chamberlin was a prominent American geologist; Moulton an astronomer.) Their hypothesis called upon close passage of another star and our sun. Pull of the intruder's gravity field presumably extracted solar gaseous material that condensed in space to form planetesimals. Then countless numbers of these small, cold, meteorite-like particles aggregated to form planets. As the planets grew, their strengthening gravity fields attracted still more particles.

THE SOLAR NEBULA

In the mid-twentieth century, the planetesimal concept was modified by two astronomers, von Weizäcker and Kuiper. They, for the first time, tried to explain the simultaneous origin of the entire solar system in a unified nebular hypothesis. They considered both the differences existing among the planets as well as the recent discoveries of nuclear physics. It is important to realize that our sun is a rather ordinary star and that space contains countless ones like it. Many others probably have planets circling them, too. Some astronomers even postulate 100 million planets in the universe capable of supporting life. In trying to explain the origin of our solar system, then, we cannot assume that it is unique.

The solar system is a small part of the Milky Way galaxy composed of an estimated 100 billion stars. The solar system rotates through the galaxy at 200

kilometers (120 miles) per second, and we are now in about the same relative position as in the late Paleozoic Era. Astronomers estimate that the Milky Way galaxy is about 10 to 15 billion years old. From the regularities of rotational direction and the revolution of planets in practically the same plane around the sun, as well as from thermonuclear theory for stars, von Weizäcker and Kuiper reasoned that about 5 or 6 billion years ago a giant, disk-shaped cloud of gases and "dust" was spinning in our part of the galaxy. It probably consisted chiefly of hydrogen and helium with lesser amounts of oxygen, neon, carbon dioxide, methane, and ammonia. Rotation was induced by motion of the galaxy as a whole. Gravity would concentrate more mass at the center of the disk, and the resulting compression would raise the temperature there to perhaps several million degrees. Thermonuclear reactions would then begin, providing the sun's heat energy as most of the heavier elements were manufactured thereby.

Spin of the embryonic solar system would inevitably induce turbulence in the envelope of gas and dust—the solar nebula—surrounding the growing and heating sun. In the earlier version of the nebular hypothesis, planetesimal-like bodies were thought to have condensed from the nebula first. Then, in regularly spaced eddies in the spinning nebula, cold planetesimals, dust, and gases tended to concentrate and collide to form so-called protoplanets. Moons may have formed in like manner as sisters in secondary eddies associated with the planetary ones, or they may have been captured later as slaves to their respective host planets. Isotopic dates for moon rocks suggest that our moon, wherever it formed, is the same age as the planet. As the gravity fields strengthened, the protoplanets would enlarge by sweeping up still more material from the dust cloud. After perhaps several million years, the protoplanets were essentially complete, but were hundreds of times larger in diameter than now (Fig. 7.2). Although their masses were much greater, their densities would have been very small, for they consisted at first primarily of hydrogen and helium with very minor amounts of the rare, heavy elements. As their gravity fields strengthened, the protoplanets would have contracted and become more dense. Apparently they also experienced a separation or differentiation of most heavier elements toward their centers. It was long assumed that the composition of all protoplanets initially was about the same, but because hydrogen and helium are very light, they could readily escape the modest gravity fields of the smaller planets. Accordingly, the four inner planets, like Earth (densities 5.1 to 5.5), presumably lost far more of their lighter elements than did the larger, outer ones (densities 0.71 to 2.47).

More recently it has been argued that the protoplanets accreted directly from the hot solar nebula and had initially different compositions. Sophisticated chemical theory provides an accurate idea of the probable temperatures and order of condensation of different compounds from a hot nebula, to form either planetesimals or perhaps the planets directly. (The nebula was heated by its own gravitational contraction, remember.) The most refractory elements, calcium, aluminum, and titanium, would condense into chemical compounds in the range 1500 to 1300°C. Elements such as iron, nickel, cobalt, magnesium, and silicon would condense and form compounds in the range 1300 to 1000°C, while most remaining elements would condense between 1000 and 300°C; water would condense below 100°C. Today the relative iron content of the planets decreases with distance from the sun, suggesting that within the solar nebula there was a temperature gradient such that more iron condensed in the hotter neighborhood of the sun. Because it is relatively depleted in iron, the moon might have begun condensing a little later than the earth. Further modification of compositions presumably resulted from a relative depletion of the light gases from the inner solar system by the sun's radiation (the "solar wind") literally blowing them to the outer part of the system, where they were swept up by the large planets.

Some geochemists believe that there is too much Fe and Ni left in the earth's mantle for one-stage accretion and differentiation of the protoplanet (Fig. 7.2) to be an adequate hypothesis. They propose instead a two-stage accretion by which about 80 percent of the earth first accumulated and the core differentiated as described above. Then the outer 20 percent was accreted to form the upper mantle rich in the more volatile elements, but also containing considerable iron.

Some moons, and possibly even early planets, probably collided and broke up to form meteors, which careen through space and occasionally fall into planetary or lunar gravity fields. The asteroids may represent orbiting debris from a disrupted plan-

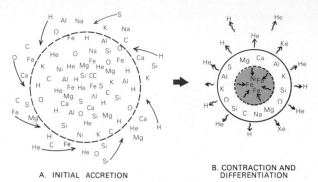

FIGURE 7.2 Hypothetical stages in the evolution of the early earth from a large, low-density, homogeneously heterogeneous *protoplanet* to a smaller, denser, internally differentiated planet.

et, and many meteors may come from there. Dating of moon rocks suggests extreme meteor activity 3 to 4 billion years ago with a great reduction since.

Either variation of the nebular hypothesis explains simultaneously the formation of sun and planets and accounts for regular spacing and concentric orbits, as well as for the direction of rotation and revolution of most planets. Coincidence of isotopic dates both for meteorites and the initial mineralogic organization of the earth and moon also is explained. In short, because it accounts for many seemingly diverse phenomena and provides simultaneous origin of sun and planets *as part of one great evolutionary process*, we believe some variation of this hypothesis to be the best available working model of the solar system. But further analysis of this fascinating subject belongs to astronomy. In the remainder of this chapter we shall explore the probable events following the protoplanet stage.

DISTRIBUTION OF THE ELEMENTS

Before proceeding farther, we must compare the overall chemical composition of the earth with that of the solar system as a whole. Only about one-quarter of the one-hundred-odd known elements are at all common, over 95 percent of the universe being comprised of only the two lightest elements, hydrogen and helium. In a general way, the heavier elements tend to be progressively less abundant. With the development of nuclear physics in the 1930s, it was concluded that most of the elements evolved successively from hydrogen through a series of complex thermonuclear reactions in stars. Only very large celestial bodies develop sufficient internal gravitational pressure (and temperature) to trigger such reactions and form stars, which is why the puny planets are just planets.

We must emphasize that practically all the atoms of various elements making up our earth were formed long before the protoplanets. A relatively few, however, have evolved within the earth. These are the daughter products of radioactive decay of initially incorporated unstable isotopes. Only 20 elements can be called major constituents of the earth, and some of these are really not very conspicuous. To the 15 elements listed in Table 7.1 must be added hydrogen (H), carbon (C), nitrogen (N), neon (Ne), and argon (Ar). Helium (He), though second only to hydrogen in the universe, is almost totally missing

from earth by loss to space. But why is there *any* hydrogen, which is still lighter? Because much was trapped by combining with oxygen (O) to make relatively heavy water molecules (H_2O). Clearly the protoearth underwent important changes of bulk composition as it shrank in size and became much more dense under the influence of its own gravity field (Fig. 7.2).

THE MOON AND OTHER PLANETS

Recent studies of other planetary bodies (Fig. 7.3) show that they differ significantly in composition from the earth, but, like the earth, all seem to have undergone chemical differentiation. The low (maria) areas of the moon have rocks most like terrestrial basalt, but they are much richer in titanium, zirconium, yttrium, and chromium, whereas they are relatively depleted in sodium (Na), potassium (K), and rubidium (Rb).

FIGURE 7.3 Footprints on the moon. Mount Hadley rises 4,400 meters above the dust-covered plain; note inclined planar structures on the mountainside. *(Apollo 15 surface photo; NASA Photo AS15-87-11849. Courtesy Paul D. Lowman and National Aeronautics and Space Administration.)*

TABLE 7.1 Comparison of estimated abundances of principal elements in the earth and in meteorites*

Major elements	Earth's crust, %	Bulk earth, %	Meteorites, av. %
Iron (Fe)	5.6	35.0	29.0
Oxygen (O)	45.0	28.0	32.0
Magnesium (Mg)	2.0	17.0	12.0
Silicon (Si)	28.0	13.0	16.0
Sulfur (S)	0.03	2.7	2.1
Nickel (Ni)	0.007	2.7	1.6
Calcium (Ca)	4.2	0.61	1.3
Aluminum (Al)	8.2	0.44	1.4
Cobalt (Co)	0.002	0.20	0.12
Sodium (Na)	2.4	0.14	0.60
Manganese (Mn)	0.09	0.09	0.21
Potassium (K)	2.1	0.07	0.15
Titanium (Ti)	0.57	0.04	0.13
Phosphorus (P)	0.10	0.03	0.11
Chromium (Cr)	0.01	0.01	0.34

*Adapted from B. Mason, 1966, *Principles of geochemistry* (3d ed.): New York, Wiley.

Their densities are 3.1 to 3.5. The lunar highlands contain less-dense rocks rich in plagioclase feldspar (anorthositic gabbro). These lunar "continents" are rich in aluminum like the earth's continents, but are depleted in sodium, potassium, silicon (Si), and rubidium. Apparently the moon's core is poor in iron and nickel, but its gabbroic crust is six to eight times thicker than the earth's crust. Judging from heat-flow measurements, the moon is two to three times richer in isotopes of uranium (U) and thorium (Th). Dates of around 3.8 to 4.2 billion years for its crust suggests very early gravitational differentiation accompanied by great igneous activity and almost complete loss of volatile elements. Subsequently the moon has been nearly static except for bombardment by meteorites and solar radiation. Space-probe data suggest that Mars and Venus have crusts more like the earth, and strongly support differentiation of these bodies as well. Their atmospheres differ considerably, however. Venus has an atmospheric pressure 90 times that of Earth's, and its atmosphere is composed of 97 percent carbon dioxide (CO_2) and <0.1 percent hydrogen (H_2); water and free oxygen (O_2) are essentially nonexistent. Mars has a much less dense atmosphere, with nitrogen (N_2) and CO_2 dominant, and water vapor and oxygen only a fraction as abundant as in Earth's atmosphere. Why such differences?

Perhaps after all, each planet had a unique initial composition.

PHYSICAL NATURE OF THE EARTH'S INTERIOR

ANALOGIES DRAWN FROM METEORITES

A major clue about the earth's interior came from, of all things, meteorites. In 1873 an American geologist named J. D. Dana studied the mineral composition of various types of meteorites and suggested that the interior of the earth might be made up of similar materials. If so, it would account nicely for the bulk density of the earth, for meteorites are much denser than crustal rocks. The assumed kinship of meteorites at least with the inner planets is strengthened by very similar ratios of certain elements in both the earth and meteorites (Table 7.1).

There are two major types of meteorites, metallic iron-nickel ones (25 percent) and nonmetallic or stony ones (75 percent). While their combined, overall composition approaches that of the total earth, it shows at least five times as much iron and only three-fourths as much oxygen and silicon as are found in crustal rocks (Table 7.1). Why should there be different kinds of meteorites, and why are they so different from the accessible earth's crust? According to the protoplanet hypothesis, all embryonic planets underwent gravitational collapse simultaneously, and presumably density differentiation was universal among them; recent studies of the moon, Venus, and Mars confirm it. If one or more bodies disintegrated later for unknown reasons, fragments both of very dense core material (presumably metallic) and less dense outer material (stony) would have been cast into space as meteorites.

The prevalent interpretation of meteorite origin contains the possibility that metallic meteorites do represent, as Dana suggested a century ago, materials like the earth's interior. The similarity of isotopic dates for initial organization of both as noted in Chap. 6 strengthens the comparison.

SEISMOLOGICAL EVIDENCE OF INTERNAL STRUCTURE

Below the deepest mines and wells, we have compelling, though indirect, geophysical evidence of inter-

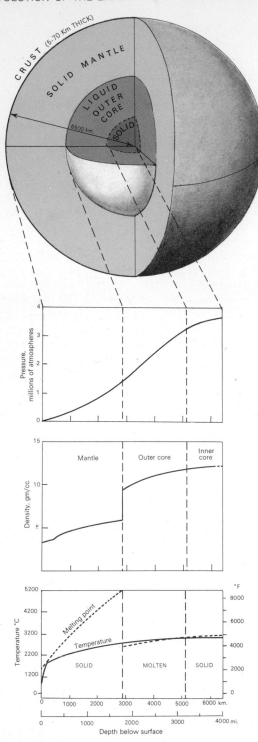

nal density zonation. Results from seismology, especially, led to recognition of the zones in the earth's interior shown in Fig. 7.4 and in Table 7.2.

A near-surface discontinuity in earthquake shock-wave transmission is named the Mohorovicic discontinuity or Moho; the velocity of wave propagation accelerates below it. The Moho generally has been taken as the base of the earth's crust. It lies an average of 5 kilometers below the sea floors and 33 kilometers below continental surfaces, but is as deep as 70 kilometers under some portions of continents (Fig. 7.5). Rock materials below the Moho must have higher rigidity and slightly greater density than the crust to satisfy geophysical measurements. But a zone of relatively low seismic velocity and low rigidity—probably due to partial melting—is present within the upper mantle between 60 and 200 kilometers deep. Apparently it is of great importance in major structural adjustments. A second great seismic discontinuity at about 2,900 kilometers below the surface separates the solid mantle from the liquid outer core (Fig. 7.4).

EVIDENCE ABOUT THE CORE FROM THE MAGNETIC FIELD

The magnetic field must be related to overall earth physics, and it has had a history which, as we shall see later, has a profound bearing upon the interpretation of the history of the crust. About A.D. 1600 the field was recognized as similar to one surrounding a gigantic two-poled bar magnet aligned with the earth's axis. Precise observations indicated that the field was not completely fixed; the poles were found to migrate in a crudely circular path spanning approximately 20 degrees of longitude in five centuries. Also, it has been shown that the field has reversed its polarity many times. During the last 5 million years, it has reversed 10 times, but the period of reversals has varied considerably, as we shall see in Chap. 9.

The cause of the magnetic field has not been proven, though interesting hypotheses have been proposed. Several features must be explained: (1) the

FIGURE 7.4   The earth's interior. *(Adapted from A. N. Strahler, 1960,* Physical geography, *with permission of John Wiley and Sons; and A. N. Strahler, 1963,* The earth sciences, *with permission of Harper & Row, Publishers.)*

TABLE 7.2 The major zones of the earth*

	Thickness or radius, km	Volume, 10^{27} cm³	Volume, %	Mean density, g/cm³	Mass, 10^{27} g	Mass, %
Atmosphere	——	——	——	——	0.000005	0.00009
Hydrosphere	3.80 (av.)	0.00137	0.1	1.03	0.00141	0.024
Crust	30	0.015	1.4	2.8	0.043	0.7
Mantle	2870	0.892	82.3	4.5	4.056	67.8
Core	3471	0.175	16.2	10.7	1.876	31.5
Total earth	6371	1.083	100.0	5.52	5.976	100.0

*After B. Mason, 1966, *Principles of geochemistry* (3d ed.): New York, Wiley.

field has two poles located near the geographic poles; (2) it shows irregular variations both in position and polarity; (3) these variations bear no relation to the crust, and so must have their origin deep in the earth.

The most widely accepted view is that internal electric currents produce a field much like that formed around a wire transmitting a current. The assumed electrical currents require some driving mechanism to maintain them, however. An iron-nickel-rich core (similar to metallic meteorites) would be a good electrical conductor, and a fluid outer part

of such a core would allow mechanical motion of electrical charges. Physicist W. M. Elsasser suggest-

FIGURE 7.5 Comparison of available factual observations bearing upon chemical and physical properties of the crust and upper mantle with interpretations of the character of inaccessible regions. This shows a long-held conception of crust and upper mantle. It now appears that the M discontinuity was overemphasized; instead, the upper mantle now seems linked to the crust, and the "low-velocity zone" marks a greater physical discontinuity. (d = density; V_p = velocity in kilometers per second of compressional seismic waves—V_p for "low seismic velocity zone" of mantle is from 7.7 to 7.8 km/sec.)

THE FACTS

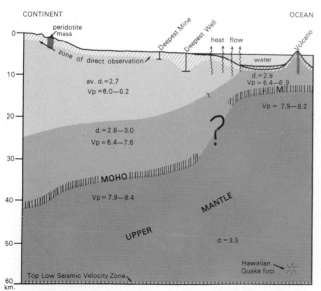

THE INTERPRETATION

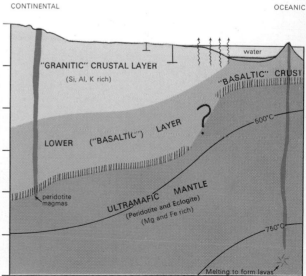

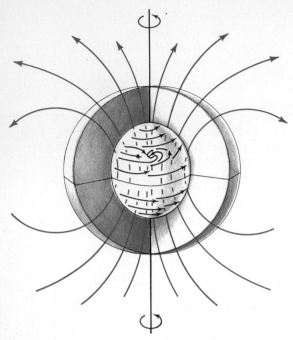

FIGURE 7.6 Cutaway diagram illustrating Elsasser's dynamo theory of the earth's magnetic field. Differential rotation of solid mantle (faster) and fluid core (slower) presumably induces eastward flow of electrically charged, ferromagnetic material in the outer core. Flow of electrical charges induces a strong field (large curved arrows), which is maintained by the earth's rotation. Large anomalies may be caused by complex eddies in the outer core. *Unipolar Dynamo Theory*

ed in 1939 that interaction of motion and electrical currents in the outer core could generate and sustain the magnetic field (Fig. 7.6). Near coincidence of the field and spin-axis orientations is consistent with this explanation, and the relatively short-term irregular movements of the field and its poles all can be explained by variable eddy motions in the core. Elsasser assumed that the outer core is an electrical conductor and is in motion due to heat convection therein. Recently it has been suggested that wobble (precession) of the earth's rotational axis, coupled with the Coriolis effect (see p. 211), may drive the dynamo instead. In any case, it is well known that if a conductor moves within a magnetic field, an electric current is generated within it. An analogy can be made with a conventional electric generator. The earth's core is the conductor moving within an elec-

tromagnetic earth field; it is as though the current being generated were put back through the electromagnets to maintain the field.

According to Elsasser's dynamo hypothesis, the field results directly from core motions, and rotation of the earth affects both orientation and strength of the field. Core motions, which otherwise might be random in direction, are preferentially oriented by the earth's spin, producing a strong field whose axis on the average over time closely parallels the spin axis. The reversals of polarity are not fully understood, but may result from instabilities of motion within the fluid outer core. Apparently the reversals are completed suddenly in a geologic sense.

The field shields life from damaging cosmic radiation. That the field is old is suggested by presence of a record of shallow marine life on earth for over 3,000 million years and of land life for at least 400 million years. Measurements of remanent magnetism in very old rocks indicate approximately constant intensity of the field for at least 2,700 million years. There is even a hypothesis of accelerated genetic mutations on the one hand and of mass extinctions of life on the other by radiation during temporary weakening accompanying pole reversals, but estimates of the magnitude of increased radiation and poor correlations between more ancient reversals and extinction-evolution rates tend to cast some doubt upon this hypothesis. At the least, more compelling evidence is needed.

CHEMICAL COMPOSITION OF THE EARTH

THE DEEP INTERIOR

Besides analogies with meteorite compositions and evidence from the magnetic field, there is other evidence that can be brought to bear upon the important question of composition of the earth's interior. First we assume that the interior consists chiefly of relatively common elements (see Table 7.1) arranged in mineral phases that would yield observed seismic velocities and densities at inferred temperatures and pressures. Observation of heat flow from the interior indicates that radioactive isotopes cannot now be very concentrated below the crust, or the earth would be much hotter. Finally, analogies long have been drawn from the densities and seismic properties of known surface-rock materials and meteorites in order to satisfy conditions at depth.

From the sum total of all evidence, it was reasoned years ago that the bulk composition of the crust beneath oceans is close to that of basalt (density 2.9), continental crust is close to that of granitic rocks (density 2.7), and the upper mantle seems to be most closely comparable with the unusual ultramafic rocks (density about 3.3), which occur locally within structurally complex mountain belts. Ultramafic rocks such as peridotite consist almost exclusively of dark, so-called mafic minerals (ma-, magnesium, and -fic, ferric or iron), such as the pyroxenes, olivine, and garnet, and lack feldspar and quartz, so are relatively poor in potassium and silica. They are similar to the nonmetallic (stony) meteorites.

An ultramafic mantle of density 3.3 to 4.5 still would not satisfy the overall earth density of 5.5. Therefore, the core must be much denser because it makes up only 16 percent of the earth's volume; it must be 10 to 12 times as dense as water (Table 7.2)! Iron-nickel meteorites are the only natural materials known that approach this figure, and so partly by a process of elimination, composition of the core was judged to be much like that of such meteorites. Only the seismic properties of the core can be measured, and they give information primarily about rigidity. It has been shown by comparing observed density and seismic characteristics of the core with experimental results for many chemical elements that the transition metals satisfy all known conditions. Of these, only iron and nickel are abundant enough in nature to form the major part of the core. But if the core were pure iron-nickel, it would be more dense and would have greater seismic velocities than are observed. Therefore, some lighter element must be present to form an iron-nickel alloy. About 10 percent sulfur would best satisfy geophysical observations, and would also explain why the rest of the earth seems to be greatly depleted in sulfur.

DIRECT PETROLOGIC EVIDENCE ABOUT CRUST AND UPPER MANTLE COMPOSITIONS

The accumulation of circumstantial evidence relative to the character of the earth's interior reads like a detective story, but there is one last kind of more direct evidence that strengthens the case. Besides the ultramafic rocks, samples of the lower crust and upper mantle are provided by many volcanic eruptions and by some very rare intrusive igneous rocks (Fig. 7.5).

In oceanic areas such as the Hawaiian Islands, earthquakes accompany volcanic eruptions. Apparently some energy released by quakes is converted to heat that melts rock locally to form magma. Resulting lava should provide a sample of the chemical composition of the focal area of the quake 50 to 60 kilometers below the surface provided there has been no significant chemical change en route upward. Average Hawaiian lava is olivine-rich basalt, and so it is inferred that the parent upper mantle material has a similar *chemical composition* (Fig. 7.5). Because of great pressure, however, the parent material can be assumed to have neither the same *mineral composition* nor the same density as surface basalt flows. Surface materials reflect the low-pressure environment of the shallow crust; they contain minerals such as feldspar, which has a relatively open arrangement of atoms. In the mantle, identical elements must be arranged in more dense mineral phases.

In some volcanic areas, deep-focus earthquakes occur from 300 to about 700 kilometers below the surface—well below even the thickest crust. This suggests that the upper mantle is solid and, with respect to sudden shocks, shows a moderate degree of rigidity. That is, when strained beyond a critical point it cannot recover, and so fails by sudden rupturing. Either sudden release of pressure or increase of temperature due to friction resulting from rupture causes local melting. Most materials decrease in density upon melting, and so the resulting magma tends to rise in the earth's gravity field. Compositional variations among resulting extruded lavas suggest either origins in different mantle-depth zones of varying composition or partial melting of a different fraction of mantle rocks.

Dense ultramafic rocks seen at the surface (mostly peridotite) represent unusual samples of the upper mantle squeezed up into the crust in zones of intense shearing (Fig. 7.5). Rare blocks included in some basalts and in very unusual diamond-bearing intrusions are also actual samples of the mantle. These blocks are made of an ultramafic rock even rarer at the surface than peridotite, which is called eclogite (composed of garnet and a pyroxene called jadeite). Overall chemical composition of eclogite is the same as familiar olivine basalt, but it consists of different, more dense minerals, reflecting the higher pressure mantle environment. As early as 1930, laboratory experiments at very high pressure indicated the importance of physical phase changes, that is, changes from minerals of low density to other minerals of high density (or vice versa) with no change of overall chemical composition. Thus the same elements are arranged in quite different minerals according to pressure and temperature. Laboratory studies in recent years have provided great insight

FIGURE 7.7   Estimates of changing heat production from major radioactive isotopes through geologic time. Note that most heat results from decay of ^{235}U and ^{40}K. *(Adapted from B. Mason, 1966, Principles of geochemistry (3d ed.); by permission of John Wiley & Sons, Inc.)*

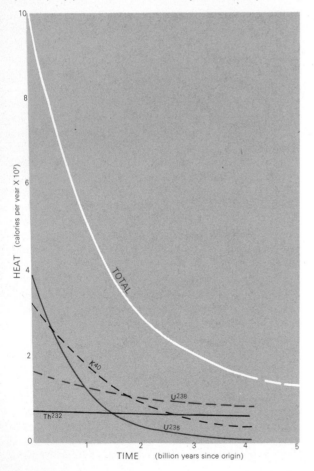

into the different phases that probably occur in different zones of the deep interior. If upper mantle material melts and rises, low-density basalt will crystallize at the surface (except in mountain belts), but if mantle material is squeezed up or is carried as blocks into the crust, either dense peridotite or eclogite results. Therefore we conclude that at least the upper mantle is composed of mixtures of peridotite and eclogite. Several seismic velocity discontinuities occur in the upper 800 kilometers of the mantle, indicating some lack of uniformity. The deeper mantle seems to be more homogenous, however, and must be composed of still more dense, magnesium-rich mineral phases, which are totally foreign to our experience at the surface. It is now felt that the Moho discontinuity at the base of the crust is not nearly as fundamental a physical boundary as the low seismic velocity zone between 60 and 200 kilometers deep (at least beneath continents, the Moho is probably only a basalt-eclogite phase change). In the next chapter we shall see why the low-velocity zone now seems so important.

THE DAWN OF EARTH HISTORY

CHEMICAL AND THERMAL EVOLUTION
Let us now explore the implications of our model of the earth's interior in terms of its probable thermal history and overall chemical evolution. An originally molten planet certainly would differentiate readily according to relative weight of the elements as it began cooling, but how could a relatively cold aggregation of planetesimals do so? Initial heating by gravitational contraction of the protoplanet would raise the temperature at the center to about 1000°C, and the early earth must have had about five times as much radioactive heat production as now (Fig. 7.7), which would have raised the temperature another 2000°C or so. A third source of heat would be large meteorite impacts, for which the moon provides abundant evidence prior to 3 billion years ago.

Assuming an original bulk composition for the earth somewhat like that of chondritic stony meteorites (Table 7.1), American geophysicist Francis Birch postulated that, within its first 500 million years or so, the protoplanet's deep interior heated up to or near the melting point of iron and nickel. These two elements presumably separated and sank to form the core between 5.0 and 4.7 billion years ago, the latter time being fixed by the earliest isotopic dates we have for the organization of distinct mineral phases both in the earth and in meteorites. Lighter silicate minerals formed the mantle. Separation would have been greatly facilitated by convection of heat within the early hot interior. Persistence of a liquid outer core beneath a solid mantle may seem puzzling at first, but magnesium- and iron-rich silicate minerals typical of the mantle have relatively higher melting points than iron at any pressure and temperature combination. Present temperature in the outer core is estimated to be near 3000°C, that is, above the melting point of core material but below that of mantle silicate minerals at that depth (Fig. 7.4). Recent estimates of the temperature necessary to melt pure iron and nickel in the early earth seem impossibly high, but the presence of sulfur with iron would lower the melting temperature by almost 1000°C. This makes the major differentiation of the earth possible within a more plausible temperature range.

Is the earth as a whole today heating up, cooling off, or in a thermal steady state? Volcanoes, hot springs, and high temperatures in deep mines and wells provide proof that heat is still flowing from the interior, and although much of the original heat-producing radioactive decay has ceased (Fig. 7.7), insulation by the crust and mantle impedes heat dissipation so much that the interior may not yet have cooled very much. In any event, the interior is still evolving. Consequences of this chemical and thermal evolution are important, for the heat energy causes structural disturbances suffered by the earth. We shall see that mountain building, the most conspicuous of such disturbances, has gone on throughout earth history. Eventually structural turmoil must decline as the thermal energy reservoir is exhausted, but we cannot judge where in this energy dissipation history the earth is today.

ORIGIN OF THE CRUST

We now turn our attention to the outer crust of the earth, which is of chief concern for the remainder of the book. Origin of the crust must fit into the overall chemical segregation of the earth by further separation of light mineral phases from original mantle material. It was suggested in 1928 by N. L. Bowen that the crust of continents, whose average properties approach that of granite, originated by chemical

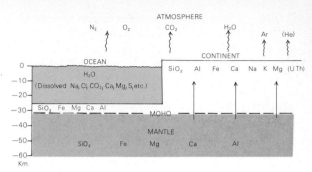

FIGURE 7.8 Present distribution of major chemical elements and also uranium, thorium, helium, and argon, which reveals the elemental differentiation of the atmosphere, sea water, and crust from the earth's mantle. For each division, elements are listed in approximately decreasing abundance from left to right. (Sea water depth is greatly exaggerated to accommodate lettering.)

fractionation from a mantle with properties closer to olivine-rich basalt or stony meteorites. It is theoretically possible by fractional crystallization and concentration of relatively lighter elements from a basaltic magma to distill off a granitic residue representing 7 or 8 percent of the original magma. The entire granitic crust totals volumetrically only about 0.01 percent of the present mantle! According to this concept, then, only a relatively small, light-element fraction of a "basaltic" mantle need have been extracted to make the continents, which suggests that a great deal more fractionation is still possible.

As the mantle differentiated, relatively light silicon, oxygen, aluminum, potassium, sodium, calcium, carbon, nitrogen, hydrogen, and helium, and lesser amounts of other elements, rose to the surface to form the primitive crust, sea water, and atmosphere (Fig. 7.8). The most familiar rock-forming minerals were produced in the proportions shown in Table 7.3. Atomic weight was not the only factor to determine their ultimate residence, however, because atomic size and electrical charge are more important for heavy elements such as uranium, thorium, and the rare earth elements. They are too large to fit into closely packed crystal structures of dense silicate minerals found in ultramafic rocks, but are comparable in size to potassium or calcium. It is not surprising, therefore, to find them preferentially concentrated in the more open structures of the minerals found in crustal rocks. Everything that we have learned about the distribution of elements in the earth, meteorites,

TABLE 7.3 Proportions of commonest rock-forming minerals in the crust*

Alkali feldspar ($KAlSi_3O_8$ and $NaAlSi_3O_8$)	31.0%	Total
Plagioclase feldspar ($NaAlSi_3O_8 \leftrightarrow CaAl_2Si_2O_8$)	29.2	feldspar 60.2%
Quartz (SiO_2)	12.4	
Pyroxenes (Ca [Mg, Fe] Si_2O_6)	12.0	
Iron and titanium oxides (magnetite, hematite, ilmenite)	4.1	
Biotite mica (complex K, Mg, Fe, Al, Ti hydroxyfluo silicate)	3.8	
Olivine ([Fe, Mg]$_2$ SiO_4)	2.6	
Muscovite (complex K, Al hydroxy-fluo silicate)	1.4	
Other minerals	3.5	
	100.0	

*After L. H. Ahrens, 1965, *Distribution of the elements in our planet:* New York, McGraw-Hill.

the moon, and other planets points to a wholesale chemical differentiation of lighter crust from a heavier interior.

If the entire earth had as much radioactive material as the continents, the globe would be entirely molten. Even though not exactly abundant in granites, the 0.0006 percent uranium there is 10 times as much as in basalt and 1,000 times more than in ultramafic rocks! Isotopic dating of continental rocks suggests that continental crust, for whatever reasons, did not become stable until about 3.6 to 3.7 billion years ago, or about 1 billion years after separation of the core and mantle. On the other hand, it is unlikely that the entire surface was ever molten; probably crust formed, was assimilated and re-formed repeatedly. With roughly five times present radioactivity, apparently there was ample heat to keep the crust unstable through almost continuous volcanic activity. But for at least the past 3.5 billion years, large portions of continental crust have persisted.

ORIGIN AND EVOLUTION OF THE ATMOSPHERE AND OF SEA WATER

THE ATMOSPHERE
Origin of the atmosphere and oceans also should fit into the great chemical differentiation of the earth. Even the origin of life seemingly was linked with differentiation, for it required certain unique chemical characteristics of the early atmosphere and oceans. Once formed, life influenced the further development of both. The atmosphere has undergone important changes through time, and is evolutionary like the solid earth. To investigate the possible nature of its development, we compare our present atmospheric composition with that of a large, outer planet, Jupiter, which probably retained a much greater share of its original gases, and with meteorites, which are regarded as resembling the early protoplanets in composition. From Table 7.4 it is clear that these compositions are all markedly different, and so no simple explanation of the earth's atmosphere is immediately forthcoming.

Two major hypotheses exist to explain the evolution of the present atmosphere from a primeval one, and there is no reason why both mechanisms could not have contributed. There is one common feature, namely the assumption that considerable free hydrogen and helium have been lost to space because of their very small atomic masses. What hydrogen remained was mostly oxidized to form sea water. The greatest single problem facing any hypothesis of atmospheric development is to explain the abundance of free oxygen, which is missing from both Jupiter and the meteorites (Table 7.4).

PHOTOCHEMICAL DISSOCIATION HYPOTHESIS
The first hypothesis assumes an early atmosphere like that of Jupiter today, with methane, ammonia, and some water vapor. The atmosphere would be devoid of any free oxygen and, therefore, of free ozone (O_3), a special form of oxygen gas composed of three instead of two oxygen atoms. Ozone filters out most of the lethal, short-wavelength ultraviolet radiation from the sun, making the lands habitable (Fig. 7.9). An ozone layer occurs at the upper part of the oxygen-molecule-rich lower atmosphere (about 15 to 30 kilometers). It is manufactured in the atmosphere by high-energy ultraviolet radiation from the sun. Oxygen molecules (O_2) are broken, and the stray oxygen atoms then combine with other molecules to form ozone (O_3). Ozone is unstable, however, and so the third atom soon goes astray to combine with other strays to form O_2 again. The process is a steady-state one; that is, ozone is constantly forming at the same rate it breaks down, thus maintaining a nearly constant amount. High-energy ultraviolet (uv) radiation can trigger other photochemical or light-induced reactions. Such reactions are known to occur in the

TABLE 7.4 Comparison of present atmosphere with the gases of other planetary bodies and volcanoes
Listed in approximate decreasing order of abundance

Earth's present atmosphere			Jupiter's atmosphere	Meteorites, av.	Volcanoes, av.	Geysers and fumaroles
Major:						
Nitrogen	(78%)	Stable	Methane	Carbon dioxide	Water vapor (73%)	Water vapor (99%)
Oxygen	(21%)	Unstable (reacts with Fe and C)	Ammonia	Carbon monoxide	Carbon dioxide (12%)	Hydrogen
Argon	(0.9%)	Stable	Hydrogen	Hydrogen	Sulfur dioxide	Methane
Carbon dioxide	(0.03%)	Unstable (reacts with silicates)	Helium	Nitrogen	Nitrogen	Hydrochloric vapor
Water vapor	(Variable)	Unstable	Neon	Sulfur dioxide	Sulfur trioxide	Hydrofluoric vapor
Minor:	(Traces only)					
				Methane*	Carbon monoxide	Carbon dioxide
Ozone		Unstable		Nitrous oxide*	Hydrogen	Hydrogen sulfide
Neon		Stable				
Helium		Stable		Carbon disulfide*	Argon	Ammonia
Krypton		Stable		Benzene*	Chlorine	Argon
Xenon		Stable		Toluene*		Nitrogen
Hydrogen		Unstable (reacts with oxygen to form water)		Naphthalene*		Carbon monoxide
Methane		Unstable		Anthracene*		

*Only present in nonmetallic carbonaceous chondrites.

upper atmosphere today and presumably were more common throughout the atmosphere before ozone accumulated. It is estimated the ozone would begin to form when the oxygen content reached about 10 percent of its present abundance. Changes resulting from such reactions in the primitive atmosphere might have been as follows:

1 Dissociation of primeval water vapor into hydrogen and oxygen with most hydrogen escaping to space:

$$2H_2O + uv \text{ light energy} \rightarrow 2H_2 \uparrow + O_2$$

2 Newly freed oxygen reacted with methane to form carbon dioxide and more water vapor:

$$CH_4 + 2O_2 \rightarrow CO_2 + 2H_2O$$

3 Oxygen also would react with ammonia to form nitrogen and water:

$$4NH_3 + 3O_2 \rightarrow 2N_2 + 6H_2O$$

4 After all CH_4 and NH_3 were converted to CO_2 and

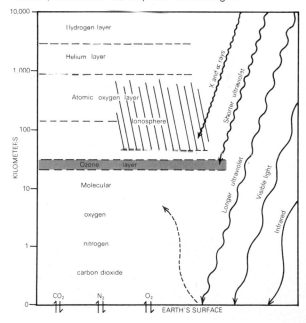

FIGURE 7.9 Profile of the atmosphere, showing density and elemental stratification, types of radiation, radiation shields, and surface-atmosphere interchanges.

N_2, *then* free O_2 could accumulate as further dissociation of water vapor occurred.

In this manner, the present nitrogen–carbon dioxide–oxygen atmosphere might have formed.

OUTGASSING HYPOTHESIS

The second explanation, which has been urged by geologists since the American W. W. Rubey proposed the idea in 1951, invokes an origin of most atmospheric gases from the interior by gaseous transfer to the surface through igneous processes. It had been suggested way back in 1910 that the traces of atmospheric helium had originated by radioactive decay of uranium, and in 1937 it was suggested that atmospheric argon, which is surprisingly abundant compared with the other rare gases, was derived similarly from decay of potassium 40 in the earth. Both the present abundance of argon 40 and the decay rate of potassium 40 are in good agreement with this hypothesis.

Volcanoes and hot springs are known to expel steam, carbon dioxide, nitrogen, and carbon monoxide (Table 7.4). By assuming outgassing (or degassing) to be a normal part of overall density differentiation, we could explain all the atmospheric nitrogen, carbon dioxide, helium, argon, and water vapor. In addition, the overwhelming preponderance of steam in the expelled gases provides a ready source of sea water.

SOURCE OF OXYGEN

Both hypotheses ultimately resort to photosynthesis by plants as the major source of free oxygen. Primitive plants had appeared by 3.0 billion years ago, but undeniable evidence of a first strongly oxidizing atmosphere occurs in rocks formed only 1 to 1.5 billion years ago. Therefore, plant proliferation and accumulation of much free oxygen may have taken as long as 2 billion years. Once free oxygen was present, ozone could form and provide an ultraviolet shield for further life development. There is a delicate interrelation between organisms and the atmosphere; ozone provides a shield for all land life, and yet apparently organisms themselves produced the oxygen from which it forms.

PROBABLE EVOLUTION OF THE ATMOSPHERE

A plausible atmospheric evolution is summarized in Fig. 7.10. If the atmosphere is indeed evolutionary, then we must inquire if it is still changing, and if so, at what rate. Although organisms generally are remarkably adaptable, it is difficult to conceive of drastic changes of atmospheric or oceanic composition since the appearance of highly organized fossil marine animal forms nearly 1.0 billion years ago. Complex interactions between life and its total ecologic environment provide a strong argument for relative stability of the atmosphere and oceans for at least the past billion years or so. Moreover, sedimentary rocks of late Prepaleozoic age are not different from those of younger ages, suggesting that the chemistry of the seas and atmosphere has not changed drastically since.

Important exchanges occur between life forms and the atmosphere. Certain bacteria probably released much of the nitrogen to the atmosphere, for today they are important in overall cycling of nitrogen between earth and atmosphere; some release it, while others fix it in nitrogenous compounds. Plant photosynthesis in general requires carbon dioxide and releases oxygen, whereas animal respiration consumes oxygen and releases carbon dioxide. Carbon is temporarily removed and stored in coal, and carbon dioxide is similarly stored in large volumes of carbonate rocks (limestone and dolomite). It is estimated that over 600 times as much CO_2 is so stored as there is now in the atmosphere, hydrosphere, and biosphere combined. Oxygen is used in large quantities in the oxidation of minerals at the crust surface, a process of major importance for at least the past 1.5 to 2 billion years.

Humans have produced recent changes in the atmosphere, especially through additions of fuel combustion products. Pollution eventually could cause a general climatic change by altering the transparency of the atmosphere to incoming solar radiation and to outgoing radiation from the earth's surface. Once started, such changes may be difficult to reverse; therefore, atmospheric pollution poses serious threats to life. Humans have the capability to upset within a few centuries an ecologic equilibrium between life and earth that has existed for at least 2 billion years!

ORIGIN OF SEA WATER

Sea water is not difficult to explain, because our hypotheses for the atmosphere also provide abundant water. The origin of the oceans becomes largely a

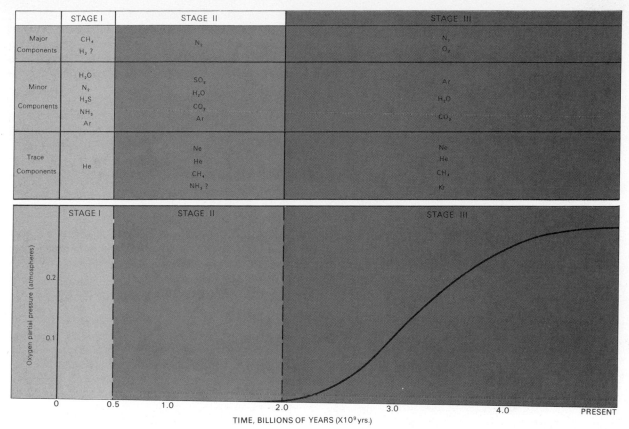

	STAGE I	STAGE II	STAGE III
Major Components	CH_4 H_2 ?	N_2	N_2 O_2
Minor Components	H_2O N_2 H_2S NH_3 Ar	SO_2 H_2O CO_2 Ar	Ar H_2O CO_2
Trace Components	He	Ne He CH_4 NH_3 ?	Ne He CH_4 Kr

FIGURE 7.10 Possible evolution of composition of the earth's atmosphere (upper) and probable rate of accumulation of free oxygen (lower). *(Adapted from H. D. Holland, 1964, in* The origin of the atmosphere and oceans; *by permission of John Wiley & Sons, Inc.)*

question of the beginning of and the rate of accumulation of water and of dissolved salts.

Following the outgassing hypothesis, rate of accumulation of sea water would be tied directly to atmospheric production and, therefore, to chemical fractionation of the solid earth. Rubey reasons that the volume of sea water has grown in direct proportion to the increase in volume of continental crust through time. Did atmosphere and seas (and crust) accumulate slowly at a more or less uniform rate, or did they accumulate early and rapidly? A rough index might be obtained from the rate of release of helium and argon 40 to the atmosphere, assuming the outgassing hypothesis to be correct. If water were released at a comparable rate, it would suggest that the oceans accumulated over a long period of time, but more rapidly early in history when greater abundance of radioactive elements produced about five times as much heat as now. Thus a growth curve for the

atmosphere and sea water probably would look like the white curve of Fig. 7.7 *turned upside down*. Furthermore, as we shall see subsequently, there is isotopic evidence that most of the present area of granitic (continental) crust already existed at least 2 billion years ago. Therefore, most of the outgassing of the atmosphere and sea water also should have been completed by then.

The oceans undergo exchanges of chemical materials with the solid earth, atmosphere, and life. Gases dissolve in sea water in some proportion to their abundance in the atmosphere, helping to stabilize the composition of both. Some of the salts dissolved in sea water presumably resulted directly from outgassing, but chemical weathering of rocks provides large volumes of salts, too. Some salts are reprecipitated promptly in marine sediments so that proportions of dissolved material in sea water are very different from those in streams. Important subtractions include vast layers of ancient rock salt and of carbonate rocks, but these represent only temporary "bank deposits" from which withdrawals can be made later by erosion. In spite of such changes, the overall chemical content of the seas seems to have been rather constant. Today there is a kind of chemical steady state between oceans, atmosphere, rock weathering, and life.

In this section we have considered the origin of sea water, but origin of the oceans is a twofold problem. We have not yet explained the ocean basins. In the next chapter we shall assess that problem and suggest alternate hypotheses proposed to explain the distribution of both oceanic and continental crust.

THE ORIGIN OF LIFE

EARLY CONCERNS

Ever since humans began to think, we have wondered where we came from and how all living things came to be as they are. Before the theory of evolution was developed, people thought in terms of multiple origins. They observed that higher animals mated to produce offspring, though the exact process of conception eluded understanding until the invention of the microscope. One important early concept was that, because the female produced the offspring, only she was involved with heredity. With the development of the microscope, the sperm was discovered, which led to the curious idea that the sperm contained a complete miniature adult, and the female provided only the environment for growth of the fetus. A more vexing problem was that of small living things that seemed to spring to life from inorganic matter in a putrefying environment. Experiments by Pasteur on putrefaction, along with the use of the microscope, demonstrated that even the lowest forms of life, like larger ones, produce offspring by transferring a portion of their cell nuclei.

Thus prior to the microscope, people confused reproduction with the actual origin of life. During the Middle Ages, many scholars were preoccupied with

the idea of spontaneous generation, believing that putrefaction somehow produced a miraculous metamorphosis of nonliving to living matter. Father Athanasius Kircher (1602–80), professor of science in the College of Rome, for example, divided animals into two groups, one that reproduced sexually and another that formed continuously by spontaneous generation. "It is obviously pointless to give these latter forms a place in the already encumbered ark," he wrote in discussing the passenger list for Noah's voyage. Even Buffon believed that organic molecules, released by putrefaction, came together to form simple organisms. It was Pasteur who proved, in the early 1860s, once and for all, by his famous sealed-flasks experiment, that there is no spontaneous generation. He showed that putrefaction, the result of airborne microbes, could not occur under sterile conditions, and so no life is created in this manner.

CONDITIONS FOR LIFE

Until the middle of this century, there was very little information to provide a clear picture of the origin of life. With the work by Rubey and others on the origin and nature of the early atmosphere and the discovery of the nature of DNA (deoxyribonucleic acid), we now have a basis for speculating rationally on the beginnings of life. The notion that life originated as a natural—perhaps inevitable—consequence of chemical conditions on the nonliving early earth first was developed in the 1920s. The earth contains abundant carbon (with remarkable combining properties), hydrogen, oxygen, and nitrogen, the major elements present in all organisms. Moreover, it is well endowed with the universal solvent, water, which, by virtue of the earth's critical temperature relation to the sun, exists largely in the liquid state. No other planet in our solar system now has all of these conditions. Perhaps Pangloss, in Voltaire's famous story *Candide*, was correct that this is, after all, the "best of all possible worlds." The modern biochemical theory calls for arrangement of the basic elements of life into more and more complex molecules until, finally, true organisms developed.

We have seen that current views on the nature of the early (anaerobic) atmosphere favor one poor in free oxygen but rich in water vapor and compounds containing hydrogen, carbon, and nitrogen. Whichever hypothesis proves correct for the subsequent changes to produce our present nitrogen-carbon dioxide-oxygen atmosphere (aerobic), or how quickly this occurred, the change was intimately associated with the origin of life.

All life shares one common feature: the presence in the nucleus of each cell of the complex molecule DNA, which, because of its composition and structure, serves as a code-carrier with the complete information for the specific entity of which it is a part. Importantly, DNA-borne information serves as a chemical template to initiate biochemical reactions that will allow the living entity, whether a diatom or an elephant, to replicate every single part of itself during growth. Much is now known about the construction of this remarkable molecule of life and its function. For example, it is estimated that the human body contains a total of about a yard of DNA in all its cell nuclei. The constituents of the DNA molecule can be arranged in 4×10^{109} different ways, which accounts for all the many characteristics of different life forms.

DNA itself, while it is the information carrier which specifies the making of proteins, does not directly take part in the actual formation of the replication. This is accomplished by the messenger RNA (ribonucleic acid), which carries the genetic information to the protein formers. The formation of proteins requires that raw material obtained from outside the cell be converted into simple nutrients such as glucose, and from these building blocks amino acids are formed. Energy for this system is induced by ATP (adenosinetriphosphate), which, like DNA and RNA, is found in all living cells.

The key to the origin of DNA, and thus life itself, is in the natural synthesis of amino acids, which are the materials from which proteins are made. To oversimplify, it is a matter of linking more and more chains of molecules (polymerizing) to construct the complex DNA double helix. The historical sequence could have been the origin of amino acids, then the chemical linkage of amino acids into larger molecules (proteinoids), and next the formation of proteins and of nucleic acids from nucleotides (four different nucleotides compose DNA and RNA). At the protein stage, it was necessary to develop a chemically stable environment to assure the next stage leading to DNA, whose replicating function marks the level we call life. But DNA is not self-sustaining; therefore development of an enclosed cell was essential to ensure the development of life. It should be kept in mind that it was a far greater leap from small amino

acid molecules to the first cell capable of self-replication than from that first cell to the organism that is now reading this page! Much experimental evidence is available to help us construct hypotheses to explain this sequence. For a full discussion of the biochemical details, see L. E. Orgel, 1973.

THE PRODUCTION OF AMINO ACIDS

The prime factors necessary for the production of amino acids are an elevated temperature, a high energy source, and the appropriate chemical elements. An experiment was performed in 1953 by Miller and Urey at the University of Chicago that demonstrated one way in which amino acids could have been produced. An electrical discharge was introduced into a chamber containing methane (CH_4), ammonia (NH_3), hydrogen, and water, all of which were heated to 100°C. After a week, four amino acids common to protein were synthesized. Since 1953 a series of experiments have been conducted using different energy sources, including ultraviolet light and other high-energy radiation, applied to very high concentrations of basic gases or chemicals in a hot water environment. The most recent experiments suggest that if methane and ammonia were dominant in the early atmosphere, the action of solar radiation could lead to the formation of the poisonous gas hydrogen cyanide (CHN), urea, and amino acids. In fact, laboratory syntheses have been made under plausible primitive earth conditions for nearly all complex organic molecules essential to DNA, basic enzyme systems, and to photosynthesis. We are unsure exactly what the environment was that allowed the first amino acids to form, but we feel that perhaps they could have developed in pools of water near volcanoes where elevated temperature would favor synthesis.

SYNTHESIS OF LARGER MOLECULES AND CELLS

However amino acids originated, nearly all theorists believe that for a time they formed continuously and concentrated in water. Some feel that this occurred in open seas, while others feel it more likely that concentration occurred in pools of hot water. Dr. Sidney Fox of the University of Florida feels that synthesis of the larger molecules could occur only if amino acids were greatly concentrated and then dried. Fox splashed dried amino acids with water, and formed simple proteins or proteinoids. While relatively small amino acid molecules and proteinoids probably formed rapidly, formation of larger molecules such as enzymes would be very improbable events under most circumstances. What was required was some additional factor to increase the probability of catalyzing these far more complex molecules. Such conditions could include hot acidic water, lightning discharge, or ultraviolet radiation.

Of even greater importance was some mechanism for transferral of information. Only when this became possible, could replication and selection begin. One of the most intriguing suggestions is that clay minerals or other crystalline materials may have acted as template surfaces on which complex, large organic molecules were synthesized. The crystal structures of clays tend to have minor imperfections that might have influenced the polymerization of organic molecules adsorbed to clay surfaces in aqueous solutions. These irregularities among different clay particles might constitute a kind of "information" that could influence the synthesis of slightly differing organic molecules, providing a kind of primitive genetic system. Some of these variable molecules would be more stable and would tend to have greater survival value.

Sidney Fox and the Russian biochemist A. I. Oparin have shown how the next step to cells might have occurred. In the experiment in which Fox created proteinoids, amino acids in a hot, dry state were splashed with water. Instant polymerization took place to form proteinoids bound together in a double-walled, spherical, cell-like envelope (termed microspheres). These have been seen to divide into smaller spheres, not unlike the process of cell division; they could be called precells. Structures very similar to these have been found in banded iron deposits in western Australia (about 2.7 billion years old). Fox's experiments point the direction toward possible formation of even more complex proteins within their own protective spheres! The next step would be synthesis of DNA.

Up to this point, it was essential to have an anaerobic environment, for if free oxygen were present, the organic molecules would be destroyed quickly by oxidation. With the formation of a cell membrane, protection could be afforded for the appearance of anaerobic bacteria (although it is likely that they were preceded by a virus-like intermediary).

Anaerobic single-celled plants probably also appeared very early. When the free oxygen-carbon dioxide-nitrogen atmosphere developed, no more amino acids could be produced outside a living organic system, and probably none are forming today. A most important danger confronting early life was the presence of ultraviolet radiation in lethal quantities. However, organisms would have been afforded protection in a moderate depth of water or by surviving in the shade of rocks.

The primitive anaerobic plants produced carbon dioxide as a metabolic by-product. Some plants such as sulfur bacteria may have been the precursors of plants which began synthesizing chlorophyll and produced oxygen. The blue-green algae were contributing free oxygen, but this oxygen was probably locked up by other elements such as iron during the early Prepaleozoic. An oxygen-rich atmosphere did not form until later.

THE DEVELOPMENT OF PHOTOSYNTHESIS

From the chemical analysis of rocks and other data, we know that an oxidizing atmosphere developed by about 2 billion years ago. The change from the original atmosphere was gradual and was accompanied by the formation of an ozone layer high above the earth (although at first it was probably close to the surface) which reduced the amount of lethal ultraviolet radiation reaching the earth. This dominance of free oxygen was almost certainly produced by plants.

The anaerobic plants living during the time of the primitive atmosphere were heterotrophs—that is, they could not manufacture their own food, but rather consumed the inorganically formed amino acids and other organic molecules existing in their environment. Only when photosynthesis developed were plants able to manufacture their own food from inorganic elements (becoming autotrophs) which later (including today) provided the basic food eaten by all heterotrophs—including humans.

This became possible through the development of photosynthesis, which provides an energy source for biological utilization of solar energy to manufacture food (i.e., stored chemical energy as organic molecules). It is the chief way that inorganic matter can be converted to food. The basic equation is

$$CO_2 + H_2O + light \xrightarrow{\text{chlorophyll}} (CH_2O) + O_2$$

The synthesis of the complex chemical chlorophyll[1] is the key to the process of photosynthesis.

The exact manner in which this changeover from anaerobic to aerobic metabolism took place is not known, and it may have taken millions of years for chlorophyll to be formed and free oxygen put into the atmosphere by plants. From fossil evidence, it appears that by the time the oldest known (South African) blue-green filamentous algae appeared nearly 3.4 billion years ago, chlorophyll was present, in part substantiated by the presence of some chemicals which are usually found in aerobic organisms. These chemicals generally result from the alteration of an alcohol portion of chlorophyll during decomposition. If this is true, the chlorophyll-producing plant fossils are in rocks that are among the oldest known.

However chlorophyll did develop, it is found in plant cells in minute bodies called chloroplasts. Interestingly, chloroplasts in some organisms may be living fossils, because microbiologists recently have found that they are able to live independently and thus may be symbionts (separate organisms living within another organism for mutual benefit). The new plants were not, at first, the main contributors of free oxygen to the atmosphere, but as they became more common, they gradually took over as chief suppliers of free oxygen.

In spite of the lithologic and paleontologic evidence of an oxidizing atmosphere about 2 billion years ago, there is no evidence that animals had evolved by that time. In fact, it took over 1 billion more years for them to appear. Animals probably evolved in plant communities as single-celled mobile forms. The first animals in the fossil record were already diverse and possessed complex features, thus indicating a profound and unfortunate gap in the record.

SUMMARY

The discovery of radioactivity provided a dramatic example of inorganic evolution—the creation of new elemental species from older ones. Modern thermonuclear theories of stars and of the origins of the elements, together with the protoplanet hypothesis, provide a compelling, unifying evolutionary conception for the entire solar system. Although the differences among the planetary bodies defy any simple explanation, nonetheless it is generally believed that the planets and their moons formed as some part of the overall evolution of the solar system.

There is much geophysical and geological evidence of concentric arrangement of material within the earth. We have seen impressive evidence that, starting as a homogeneous, accreted cold protoplanet about 5 billion years ago, the earth contracted and heated internally—primarily by radioactivity. Density differentiation produced a very dense core and intermediate-density mantle (by about 4.5 b.y. ago), less dense crust (beginning no later than 3.7 b.y. ago), and least dense sea water and atmosphere (approaching their present character at least 2 b.y. ago). It appears that the earth still is undergoing chemical evolution; energy continues to be dissipated by volcanoes, earthquakes, hot springs, and uplift and subsidence of the crust.

The crust probably formed by repeated selective fusion and rise of lighter elements from the mantle. We now know that the mantle is not homogeneous; important contrasts beneath continents and ocean basins probably relate to separation of continental material.

Biochemists believe that on the early, chemically evolving earth, given carbon, nitrogen, and oxygen and hydrogen combined as liquid water, and sources of heat and other energy, the development of the simpler organic compounds, such as the amino acids, was virtually inevitable. Synthesis of more complex molecules, such as the nucleic acids and enzymes, must have required very special conditions. Once DNA and a protective cell membrane developed, true life was formed and natural selection could begin. We see that already an impressive interaction was occurring between the living and nonliving realms during the first three-quarters of the earth's evolution. Apparently the origin of life was possible only in an oxygen-free environment, but the later development of respirative organisms required the oxygen generated by earliest creatures. Animals are not only indebted to plants for their food, but also for the oxygen that they breathe.

Readings

Ahrens, L. H., 1965, Distribution of the elements in our planet: New York, McGraw-Hill. (Paperback)

---

[1]Chlorophyll is composed of porphyrine-based cyclic compounds and has the generalized formula $C_{55}H_{72}N_4Mg$.

Bernal, J. D., 1967, The origin of life: London, Weidenfeld & Nicolson.

Birch, F., 1965, Speculations on the earth's thermal history: Bulletin of the Geological Society of America, v. 76, pp. 133–154.

Brancazio, P. J., and Cameron, A. G. W., eds., 1964, The origin and evolution of the atmospheres and oceans: New York, Wiley.

Cloud, P. E., Jr., 1968, Atmospheric and hydrospheric evolution on the primitive earth: Science, v. 160, pp. 729–736.

Elsasser, W. M., 1958, The earth as a dynamo: Scientific American, May, pp. 1–6.

Gaskell, T. F., ed., 1967, The earth's mantle: New York, Academic.

Jacobs, J. A., 1963, The earth's core and geomagnetism: Oxford, Pergamon.

———, Russell, R. D., and Wilson, J. T., 1973, Physics and geology (2d ed.): New York, McGraw-Hill.

Kopal, Z., 1973, The solar system: Oxford, Oxford Univ. Press.

Lowman, P. D., Jr., 1972, The geological evolution of the moon: Journal of Geology, v. 80, pp. 125–166.

McLaughlin, D. B., 1965, The origin of the earth, in Kay, M., and Colbert, E. H., Stratigraphy and life history: New York, Wiley, chap. 27.

Mason, B., 1966, Principles of geochemistry (3d ed.): New York, Wiley.

Orgel, L. E., 1973, The origins of life: molecules and natural selection: New York, Wiley.

Press, F., and Siever, R., 1974, Earth: San Francisco, Freeman.

Ringwood, A. E, 1975, Composition and petrology of the earth's mantle: New York, McGraw-Hill.

Robertson, E. C., ed., 1972, The nature of the solid earth: dedicated to Francis Birch: New York, McGraw-Hill.

Schopf, J. W., Haugh, B. N., Molnar, E., and Satterthwait, D. E., 1973, On the development of metaphytes and metazoans: Journal of Paleontology, v. 47, pp. 1–9. (Contains bibliography of recent papers on Prepaleozoic life)

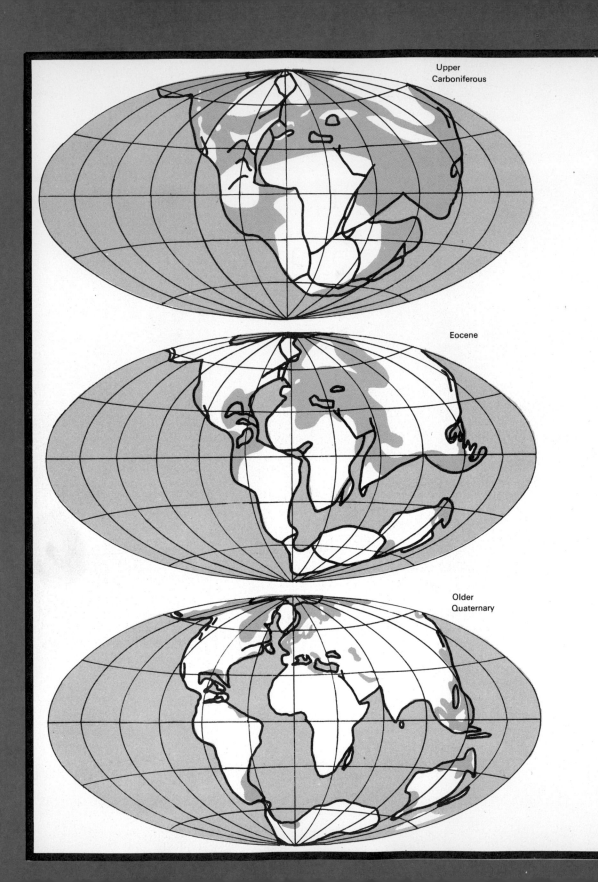

Upper
Carboniferous

Eocene

Older
Quaternary

8

MOUNTAIN BUILDING AND DRIFTING CONTINENTS

What matters it how far we go?
His scaly friend replied,
There is another shore, you know,
Upon the other side.

The Mock Turtle,
Alice in Wonderland

FIGURE 8.1 Famous reconstruction of the continents by Alfred Wegener, showing three inferred stages of continental drift. Brown represents sea areas at the respective times. *(Adapted from Wegener, 1929, Die Enstehung der kontinente und Ozeane; by permission of F. Vieweg and Sons.)*

Evidence for an overall chemical differentiation of the earth suggests that the crust might once have been uniform over the entire earth, even though it is not so today. As we have seen, relatively low-density, outer continental crustal material apparently represents the ultimate chemical distillate for the solid earth. Apparently it formed by a sweating out of some fraction of the earth's light elements (see Fig. 7.8). But was the rate of continental accumulation constant? Once formed, have continents remained fixed in position, or have they been displaced laterally? How the crust has changed through time is of the greatest interest in geology. Therefore, in this chapter we shall formulate working hypotheses to explain it. Through the remainder of the book, we shall test them against known historical and structural evidence.

There are two fundamentally different groups of hypotheses. One group postulates an early, uniform, complete continental-type crust, with the present discontinuous distribution of continents having arisen through one of at least three different mechanisms. The second group postulates an original, uniform oceanic-type crust, with later growth of continental crust at the expense of the oceanic through time (see Fig. 8.2). One variation of the latter calls for growth of a few huge supercontinents followed by fragmentation and separation. In this chapter, then, we shall examine two separate issues: *first*, was continental or oceanic crust the more primitive, and *second*, regardless of how they were formed, have continents remained fixed in their positions? Also we shall review the role of isostasy in maintaining continents in the face of erosion, the relations of geosynclines to mobile belts, and the hypothesis of continental drift and the possible relation of drift to mountain building.

AN EARLY, COMPLETE CONTINENTAL CRUST?

The continental assimilation hypothesis (Fig. 8.2 1A) assumes that the ocean basins are areas of "oceanization" of large portions of an original, all-encompassing continental crust by addition from below of elements such as magnesium, iron, and calcium, or perhaps just the intrusion of countless basaltic dikes. The resulting more dense mass then subsided to form the basins. The chief objection to this hypothesis is that it runs counter to the general chemical and density differentiation by which it appears that the least dense crust has formed *from* more dense material, not the reverse (see Fig. 7.8). Nonetheless, certain geologists favor some variation of this

EVOLUTION OF THE EARTH

multiple working hypotheses

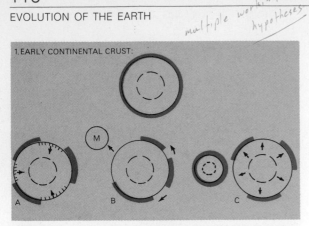

1. EARLY CONTINENTAL CRUST:

M

A   B   C

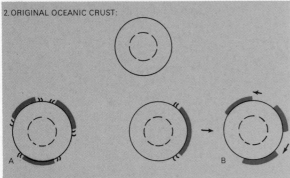

2. ORIGINAL OCEANIC CRUST:

A   B

FIGURE 8.2   Five alternate hypotheses of crustal origin, showing crustal thickness greatly exaggerated. In 1, an early complete continental crust has been modified in one of several ways, while in 2, an original complete oceanic crust has been reduced in volume as continental crust formed and changed; see text for further discussion. *(Suggested by a diagram in J. H. F. Umbgrove, 1947, The pulse of the earth, 2d ed.; by permission of Martinus Nijhoff Publisher.)*

→ Rouche limit criticism

explanation. One of the oldest hypotheses is that of George Darwin (son of Charles Darwin), who, a century ago, postulated the extraction of the moon to form the Pacific Ocean basin (Fig. 8.2-1B). He suggested that the earth's interior was liquid and that originally there was a uniform continental crust. The sun's gravitational pull presumably set up tidal oscillations in the crust. Bulging of the crust increased until a large chunk of crust and upper mantle was thrown out and launched into orbit around the earth. The mass reshaped into a sphere with a core of mantle-like material and a thick crust of continental-like material, yielding a bulk lunar density of 3.3 to 3.4. Meanwhile,

the earth's remaining crust became fragmented and redistributed by profound stresses resulting from this "catastrophic" event, but leaving the Pacific Ocean basin as a scar. Darwin's idea was abandoned by most geologists when the solid nature of most of the earth's interior was demonstrated by seismology.

An alternative explanation of the crust, which seems even more incredible, is expansion of the earth (Fig. 8.2-1C), first seriously suggested about 1925. This idea derived from the simple observation that present continental crust could cover a sphere with about half the surface area of the present earth. But a large expansion is difficult to imagine, for no apparent mechanism is now known that could produce it, and we should expect all mineral grains to show profound physical cracking due to such expansion, although they do not. Radioactive heating could cause expansion only on the order of 100 kilometers of radius. Weakening of the earth's gravity field, which has been suggested but not demonstrated, could cause expansion.

AN ORIGINAL, COMPLETE OCEANIC CRUST?   *more likely.*

The lateral growth or accretion of continents at the expense of original oceanic crust by gradual fractionation and concentration of potassium, aluminum, silicon, oxygen, sodium, and other elements is an old hypothesis with much to recommend it, as we shall see (Fig. 8.2-2A). It fits overall chemical differentiation of the earth, but unlike the idea of an early, complete continental crust, the process of crustal fractionation would be extended over much of geologic time. Traditionally, advocates of the accretion hypothesis have assumed that continents accumulated in complex patterns where they are now located, but subsequent rearrangement by displacements is not incompatible with accretion (Fig. 8.2-2B).

PERMANENCY OF POSITIONS?

It is natural (and uniformitarian) to assume that continents have always been located in the same places with respect to each other and to intervening ocean basins, regardless of how their relative sizes may have changed. Long ago, workers observed the curious parallelism of outlines of the opposing coasts of South America and Africa, suggestive of separated pieces of a jigsaw puzzle. About 1750, the Frenchman Buffon argued, on the basis of similarity of fossil land animals and plants on either side of the Atlantic, that North America and Europe had once been joined. In 1855 the first map fitting South America against Africa was published. Then in the early part of our century, the concept of large-scale rearrangement of continents or continental drift began to receive considerable attention, especially through the writings of a German meteorologist, Alfred Wegener (Fig. 8.1). Today, some form of continental separation is widely acclaimed by most geologists, though still loudly disclaimed by some. A recent modification of continental displacement is the suggestion by geophysicists of spreading sea floors. Presumably, oceanic crust moves laterally outward from ocean ridges, where new crust forms; continents may be rafted along as if on conveyor belts. According to this new view of crustal structure, the earth's surface today is subdivided into several immense plates that are actively spreading apart along oceanic ridges, while being telescoped together either along deep oceanic trenches or in mountain ranges like the Himalayas. The boundaries and patterns of movements of such plates presumably have changed through time in response to profound subcrustal changes. This revolutionary theory is outlined fully in the next chapter after some essential background has been developed.

STRUCTURE OF THE CRUST

WHAT ARE CONTINENTS?

As we pointed out in Chap. 7, the earth is a huge energy system, and earth history records the expenditure or dissipation of energy. Understanding of the evolution of the earth's crust is a combined problem of chemical and dynamic changes. We have examined the chemical aspect of the problem in some detail. Now we shall review the large-scale structural features of the earth's crust as they exist today, for we must know the product of evolution before we can ask intelligently how it evolved. Geologically speaking, continents extend beyond their present coast lines to depths between 200 and 2,000 meters below sea level. We have shown already that sea level is transitory; frequently it has risen to cause shallow flooding

of almost the entire present land area, while, conversely, it has also fallen well below its present level, most recently during Pleistocene glaciation. Sea level is an elusive datum; therefore, a more fundamental geologic boundary is needed. Most of the earth's surface lies at two levels. One-third averages about 300 meters above sea level (continents) and the other two-thirds (the ocean basins) nearly 5,000 meters below. The principal topographic break between the two lies at the shelf edge (−200 meters). For the illustrations in this book, we have shown the edge of the present continents as the outer edge of the continental shelves because geophysical studies show that the thickness (and presumably also the composition) of the crust is essentially "continental" at least to the shelf edge.

HOW DO CONTINENTS PERSIST?

Why should there be two prominent topographic levels with almost negligible intermediate ones? The principle of isostasy first proposed 100 years ago seems to provide the answer. Based upon gravity measurements, it was reasoned that high mountains might be buoyed up at depth more or less hydrostatically (according to Archimedes' principle) to maintain their height. Mountains were variously explained as (1) of less dense material than average for continents, (2) of equal density but thicker than average, or (3) a combination of these (which is presently favored). Seemingly, first-order topographic differences between continents and oceans also can be explained by isostasy. From seismic evidence, it is clear that continental crust (30 to 40 kilometers thick) is from six to eight times thicker than oceanic (5 kilometers); see Fig. 7.5. Recall also that continental crust has a slightly lower average density (about 2.7) than oceanic crust (2.8). Consequently, if isostasy holds, then a thicker and slightly less dense continental crust *should* stand higher than a thinner, more dense oceanic one.

Dramatic confirmation of isostasy comes from such things as a measured subsidence of the crust of nearly 200 millimeters (8 inches) under the load of impounded water after the building of Hoover Dam in 1935, and from the evidence of depression of the crust by huge continental ice sheets during Pleistocene time followed by uplift or rebound after melting (see Fig. 1.8).

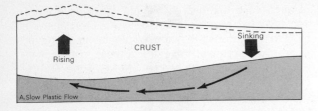

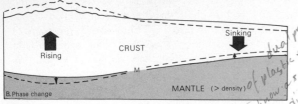

FIGURE 8.3 Two different possible mechanisms of isostatic uplift and subsidence of the crust are as follows. *A*: Lateral plastic flowage of material in the mantle, which is more plausible deep in the mantle. *B*: Physical phase changes at the Mohorovicic discontinuity from more to less dense material, or vice versa (change to lesser density thickens and raises crust while change to greater density thins and depresses crust to maintain isostatic equilibrium).

THE MECHANISM OF ISOSTASY

How can large segments of the earth's crust rise and sink to maintain isostatic equilibrium? Either the mantle must be plastic enough to flow from or to different areas in response to changes in crust level above, or the crust and upper mantle must be physically transformed to more- or less-dense phases, thus changing their mass-volume relations (Fig. 8.3), or both. In a mountainous area, erosion tends to thin the crust from above, and isostasy dictates that, to maintain equilibrium, such crust should rise, just as melting of glaciers caused the crust to rise. Eventually, erosion and isostatic uplift would reach a steady state beyond which, barring new structural disturbances of the crust, no further change will occur. When this is achieved, mountainous terrain will have been reduced to a plain. If, on the other hand, the lower crust or uppermost mantle then were to become more dense (say, due to cooling), such mass change would cause the overlying crust to sink. There is growing evidence to suggest that this may occur at the Mohorovicic discontinuity beneath continents (Fig. 8.3).

Can the mantle material flow laterally? The deep earthquake foci extending into the mantle suggest appreciable rigidity for the upper 700 kilometers, but many of the physical properties of materials in the earth are strange to us here at the surface. Many solids, including most rocks, show different types of strain, depending upon pressure, temperature, fluids present, and rate of application of stress. A sudden shock may cause brittle rupture, as has occurred along faults, but a very slowly applied stress (particularly at great pressure and temperature) may produce *in the same material* a kind of plastic flow without any rupturing. An example of such dual properties is provided by familiar silly putty. At room temperatures it has elastic properties such that it can be bounced like a rubber ball. It will break in a brittle fashion when suddenly stretched; yet if deformed slowly, it flows plastically like taffy. Most rock materials both in the deep crust and upper mantle appear to show similar properties. Below 700 kilometers, however, no earthquake foci are recorded; pressure and temperature must be high enough for entirely plastic behavior. Because plastic flow is so dependent upon rate of stress, it is apparent that over the vastness of geologic time, very large-scale deep flow is entirely possible. Based upon microscopic studies of old mantle rocks now at the surface and of samples deformed at high pressure in the laboratory, the flow or creep apparently occurs by minute slips both within mineral crystals and between grains, accompanied by some recrystallization (Fig. 8.4). Especially important for flow is the zone of relatively low seismic velocity, a less-rigid or "weak" zone (of lower viscosity) in the upper mantle between roughly 60 and 200 kilometers—well below the crust (Fig. 7.5). We might expect any major lateral movements in the outer earth to be concentrated there.

In summary, structural disturbances of the crust, and isostatic adjustments to mass changes in the crust and upper mantle, profoundly affect relative land and sea level. The stratigraphic record is the indirect result of isostatic responses coupled with the influences of mountain building, climate, and biologic processes.

PRESENT STRUCTURAL FEATURES

Besides the two most obvious topographic entities, the continents and the deep-ocean floors (abyssal plains), important second-order structural divisions of

FIGURE 8.4 Deformation by creep revealed microscopically in an ultramafic mantle rock (left) and in ultramafic rock deformed experimentally at 1100°C (right). In both cases, large olivine and pyroxene grains have been flattened and broken down into smaller grains by creep. (Microphotographs taken through polarizing microscope; each area is 2 mm long. *After Kirby and Raleigh, 1973*, Tectonophysics, v. 19, p. 179.)

the crust are also prominent. Average structural behavior of a large area of the crust *through a long time interval* is conveniently termed its tectonic character (tectonic derives from a root meaning "to build"). The relatively most stable tectonic subdivision of continents is the craton, which has subdued topography and lies fairly near sea level (Table 8.1). Tectonic disturbance of this stable division is primarily by broad, gentle warping (termed epeirogenesis).

Most abyssal plains, formerly thought to be as stable as cratons, are divided by oceanic rises or ridges (Fig. 8.5). Rises are long, high prominences that are volcanically active, show greater-than-average heat flow from the interior, and suffer many shallow-focus earthquakes (0 to 100 kilometers deep). Their summits are very irregular and protrude locally to form midoceanic volcanic islands such as Iceland. Composition of the lavas on the rises is largely that of typical oceanic basalt. The study of ocean rises is recent, and has produced some revolutionary insights into global tectonics. As noted above, new oceanic crust appears to be added in the rises, causing lateral sea-floor spreading and impingement against continents; continents may even be displaced in this manner. But more of this later.

In many ways, the most interesting and baffling of all major tectonic features are the long, narrow, and typically arc-shaped zones of greatest structural instability referred to as mobile belts (Fig. 8.5). Such belts today are the loci of volcanoes, granitic batholiths, regional metamorphism, and frequent earthquakes (including *all* deep-focus ones from −300 to −700 kilometers). The trenches show less-than-average heat flow. Severe structural disturbances

include folding and all types of faulting, but especially distinctive is low-angle overthrusting. Characteristic mobile-belt disturbances collectively are called orogenesis ("mountain building"); thus such zones also are called orogenic belts. Mobile belts and ocean ridges are contrasted in Table 8.2.

We can classify three species of mobile or orogenic belts based upon *present* spatial relations to cratons and ocean basins (Table 8.1). The western North American Cordilleran and South American Andean mountain complex is a marginal belt; the Himalayan Ranges make up an intercratonic belt; whereas many of the arcuate volcanic island and deep-sea

TABLE 8.1   Major structural elements of the crust

Relatively stable:	Continental cratons
	Oceanic abyssal plains
Relatively unstable:	Oceanic rises or ridges
	Mobile belts (as now seen):
	1. *Marginal* (between continental cratons and ocean basins)
	2. *Intercratonic* (between neighboring cratons)
	3. *Intraoceanic* (most island arc-trench systems; some trenches extend to −12,000 meters)

MAJOR WORLD TECTONIC FEATURES

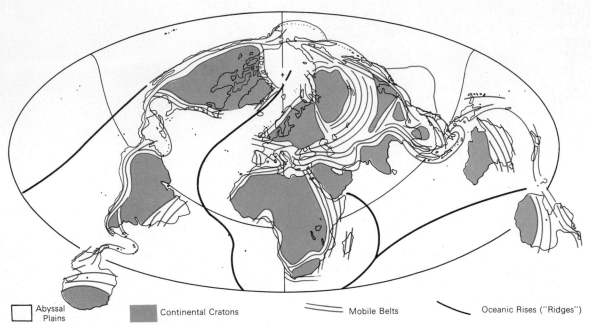

| | Abyssal Plains | | Continental Cratons | | Mobile Belts | | Oceanic Rises ("Ridges") |

FIGURE 8.5 Major tectonic features of the earth's crust for the past 1 billion years. Note the asymmetric distribution of continents and the great complexity of mobile belts with respect to continental cratons. Note also the parallelism of the opposing Atlantic coastlines. *(Base map, Nordic Projection, through courtesy John Bartholomew and Sons, Ltd., 1950,* The advanced atlas of modern geography *(2d ed.), McGraw-Hill.)*

trench systems (as in the western Pacific) are intraoceanic mobile belts with many of the same structural features as are found in continental mobile belts. We must be careful, however, not to assume (uniformitarianly) that the ancient belts were *always* situated exactly as they are now. If continents have in fact moved together and apart, then we may find that what is now an intercratonic belt began as a marginal one, and what is now marginal, may at some time have been intercratonic.

Figure 8.5 shows that the patterns of mobile belts and cratons are very complex. They are, in fact, so asymmetrically distributed and irregular in relation to each other as to almost defy rational explanation. Clearly, whatever causes mobile belts does not al-

ways discriminate neatly between crust types; yet, one naturally is inclined to seek some general simplifying hypothesis for all. As we shall see in Chap. 10, presence of granitic batholiths and extreme regional metamorphism in old mobile belts led, just before the present century, to beliefs that it was in such belts that new continental (i.e. "granitic") material was generated. Therefore, conventional wisdom has long held that the marginal mobile belts are sites of lateral growth or accretion of new continental crust outward from old nuclei—the cratons. But the role of nonmarginal belts (intercratonic and intraoceanic) is not at all clear at first glance.

The distribution of modern earthquakes and volcanoes indicates that mobile belts and ocean rises must play the major roles in the dissipation of the great energy generated in our planet. We have taken pains in preceding chapters to emphasize that the earth is structurally dynamic, and we shall concentrate henceforth upon seeking an explanation of the relations of these major active tectonic elements to overall crustal evolution.

TABLE 8.2   Contrasts between ocean ridges and mobile belts (including modern volcanic or magmatic arcs)

	Ocean ridges	Mobile belts
Seismicity	Shallow only (0 to −70 km)	Shallow and deep (to −700 km)
Volcanism	Chiefly basaltic	Chiefly andesitic
Plutonism	Gabbro or basalt dikes	Granitic batholiths
Structure	Extensional (rifted)	Compressional (complex folds and thrust faults)
Metamorphism	Minor local thermal effects	Major high-grade regional meta-morphism (gneiss, schist, etc.)
Heat flow*	High (2.0–3.5 × 10^{-6} cal/cm²-sec)	Low in trenches (0.99 × 10^{-6} cal/cm²-sec)

*Heat flow is measured in calories of heat per square centimeter of surface area per second of time. Average flow for continental cratons is 1.4 × 10^{-6} (or 0.0000014) cal/cm²-sec, while that for abyssal plains is 1.3 × 10^{-6} cal/cm²-sec.

THE GEOSYNCLINAL CONCEPT

Beginning in 1857, a great American geologist, James Hall of New York, laid the groundwork for one of geology's great early generalizations. For many years, geologists in both Europe and America had been speculating about the crumpled nature of strata seen in mountain ranges (Fig. 8.6). In his presidential address to the American Association for the Advancement of Science (which originated as a geological society), Hall noted that, in the Appalachian Mountains—now classed as a mobile belt—the strata of the Paleozoic Systems are not only more deformed but are also 10 times thicker than are their counterparts in the Mississippi Valley region farther west—today classed as part of the craton. He suggested a cause-and-effect such that the greater load of sediments in the Appalachian belt simply depressed the crust as deposition occurred (Fig. 8.7). Finally, the crust could bend down no farther, and so it failed and the strata crumpled.

In 1873 another leading American geologist, J. D. Dana, coined the term geosynclinal (large or "earth syncline") for the old zone of thick strata. On isostatic grounds, he rejected Hall's simple explanation of the great subsidence by sediment-loading alone. A "geosynclinal" was inferred to be a *result*—not a cause—of the fundamental structural instability of mountain belts. Dana argued that buckling of the crust beneath mountain belts was the primary cause, first, of subsidence and geosynclinal sedimentation, and ultimately, of mountain building (Fig. 8.7).

European geologists quickly adopted the basic geosynclinal concept of thick sediments as the ancestors of mountains. But, unlike the Americans, who believed that sedimentation always kept the subsiding geosyncline full of shallow-marine and nonmarine deposits, the Europeans believed that geosynclines began as deep furrows on the sea floor and that much of the thick sediment contained therein was of deep-water origin (Fig. 8.7). By 1930 the Europeans had gone farther in suggesting a close similarity between the sediments of geosynclines and those of deep-marine trenches, particularly those of then Dutch-held Indonesia. This began an important series of proposed links between ancient mobile belts and the modern island arc-trench systems. From these early ideas developed one of the greatest unifying concepts in all of geology, which brought for the first time a recognition that both the stratigraphy and structure of mountain belts are closely related.

GEOSYNCLINE, MOBILE BELT, AND OROGENESIS

For our purposes, we shall define a geosyncline as a great elongate belt of relatively thick strata, reflecting greater subsidence of a portion of the earth's crust relative to stable cratons. As definitions go, this is passable, but it also is essential to recognize that the Hall-Dana conception of a geosyncline was linked closely with mountain belts and orogenic structures. That is, ancient geosynclines characteristically show tightly compressed folds, various types of faults, including thrusts, and most contain granitic batholiths

FIGURE 8.6 One of the first accurate cross sections of a mountain range showing folds in strata in the Appalachian chain. *(From Rogers and Rogers, in* Transactions of American Geologists and Naturalists *for 1840–1842, pp. 474–531.)*

and regionally metamorphosed rocks. Thus geosynclines have developed within structurally mobile belts. But is a geosyncline simply the same thing as a mobile belt? Emphatically not, for mobile belt is a structural term summarizing a fundamental tectonic behavior of parts of the crust. Geosyncline is primarily a stratigraphic concept summarizing thickness of the strata lying within long, structurally mobile zones as contrasted with those deposited in more stable (cratonic or abyssal) regions. The problem is that there are mobile belts, such as the modern intraoceanic island arc-trench systems, with many of the same structural qualities as ancient mountain belts, *but with very little sedimentation.* A geosyncline, then, is any

FIGURE 8.7 Contrasting explanations of geosynclines by the Americans Hall and Dana and by early European geologists.

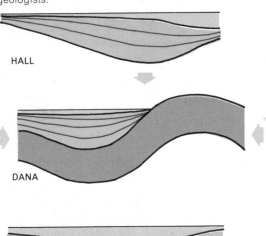

HALL

DANA

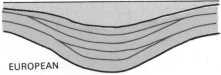

EUROPEAN

mobile belt that received a great accumulation of strata; as we think of geosynclines, they simply represent sediment-filled mobile belts.[1]

Several factors seem required to produce a geosyncline. We feel, like Dana, that great structural mobility is the most fundamental condition in such belts. Formation of a geosyncline appears to depend, *first*, upon a degree of structural mobility to depress a long zone of crust, and *second*, upon availability of a large source of sediments to fill subsiding troughs within a mobile belt. Modern intraoceanic belts (Fig. 8.5) qualify in the first but fail in the second requirement; they are *potential* geosynclines only—in other words, unfilled mobile belts.

As the study of mountain belts progressed, it became apparent in the early twentieth century that the development of mobile belts involved all facets of geology. Besides early recognition that granitic batholiths were formed in mobile belts, it was also realized that volcanic rocks make up a sizable proportion of the stratified geosynclinal sequences. Therefore volcanism somehow must play an important role in the evolution of mobile belts, which further linked modern volcanic arcs with ancient mobile belts as the Europeans had suggested in the 1930s. A "geosynclinal cycle" was developed by 1940, which claimed that crustal evolution has proceeded by (1) crustal downbuckling and geosynclinal accumulation of sediments and volcanic material, (2) orogenic upheaval accompanied by metamorphism and formation of batholiths (which provide new continental crust), and (3) erosion that finally produces isostatic equilibrium and stability. This "cycle" was much oversimplified, as we shall see. We now believe, for example, that mobile belts *destroy* at least as much oceanic crust as they may generate of new granitic crust.

[1]Unfortunately, geosyncline has acquired almost as many meanings as there have been geologists using the term. One is reminded of Humpty Dumpty's pronouncement in Lewis Carroll's *Through the Looking Glass*: "When I use a word, it means just what I choose it to mean—neither more nor less."

S.E.

A SHRIVELING EARTH (AND CRUSTAL SHORTENING)

The observation in mountain belts, at least by 1800, of tightly folded strata and the recognition of great overthrust faults in southern Germany in 1826 naturally cried out for explanation. The plutonists (e.g., James Hutton) had explained the folding as wrinkling due to a simple uplift of mountain ranges by the rise of molten granite into them; the superficial strata either were pushed or they slid to the sides (see Fig. 3.5). Beginning about 1825, the crumpling was envisioned as the result of vise-like compression within the mountain belts (Fig. 8.8), which implied a lateral shortening of the crust.

Late in the nineteenth century, a relatively simple and appealing explanation of geosynclines and mountains was offered. From the almost universal assumption that the earth was cooling, it followed that the globe should be contracting. But a cold, rigid crustal rind must yield by bending and rupture as the interior shrinks beneath it. The wrinkled skin of a drying apple provided a favorite analogy. Where else should adjustments be concentrated but at boundaries between continents and oceans? The Appalachian and western (Cordilleran) mountain belts lie precisely at such boundaries (i.e., are marginal mobile belts). Shrinkage of the earth would cause compression of the cold crust, which seemed amply evidenced in the folded structures of mountains. From folds it was estimated that the crust had been shortened a few hundred kilometers in each mountain belt. First, downbending at the continent margin occurred

as oceanic crust pushed against the continent, and then thick geosynclinal strata accumulated there. Further compressive stress—like a closing vise—exceeded the strength of the rocks, causing rupture and intense, accordion-like squeezing and uplift of the geosynclinal strata to form mountains. *Voilà!* Here was a unified concept of geosyncline *and* mountain formation. Unfortunately, it was wrong in at least one basic assumption. There is no evidence that the earth *as a whole* is contracting; its advocates knew nothing yet of radioactive heating. Moreover, as we have seen (Fig. 8.5), most of the world's mobile belts are not as neatly marginal to the continental cratons as they are in North America and as was implied in the contraction hypothesis.

CONTINENTAL DRIFT AS A CAUSE OF MOUNTAIN BUILDING—FIRST OUTRAGEOUS HYPOTHESIS

The general idea of large-scale continental displacements began to be discussed seriously in 1908, when F. B. Taylor, an American geologist, suggested that drifting of the continents had caused wrinkling of the crust to produce all the great Cenozoic mountain

FIGURE 8.8 Deformational structures produced in a laboratory scale model by compression over a 2-week period; note thrust fault at right. In scaling, it was necessary to use shoemaker's stitching wax to represent the strength of layered rocks properly. *(After W. H. Bucher, 1956, Bulletin of the Geological Society of America, v. 67, p. 1302; used by permission; photo courtesy Exxon Production Research Company.)*

systems (Fig. 8.9). Presumably the leading edges of the moving continents depressed the oceanic crust ahead to form troughs in which thick sediments could accumulate to form geosynclines. Further movement then compressed and upheaved the geosynclines to produce the present lofty mountains rimming the Pacific. Upheaval of the Alpine-Himalayan chain carried the process a step farther by culminating in the direct collision of continents. For a mechanism to drive his continents, Taylor speculated that a catastrophic tidal action was induced in the solid earth by capture of the moon during Cretaceous time. Although his explanation is totally unacceptable, Taylor had given birth to an important new theory—albeit incomplete—linking mountain building and continental displacements.

The first detailed reconstruction of the continents was attempted almost immediately (1911) by another American, H. H. Baker. He postulated an original, single landmass or supercontinent that suddenly split to form the present Arctic and Atlantic Oceans at the end of Miocene time (Fig. 8.10). Baker was at least as

imaginative as Taylor in devising a cause. He speculated that variations in their orbits brought Earth and Venus close enough together to produce such severe tidal distortion that a large portion of original continental crust was torn from the Pacific to form the moon (à la George Darwin in 1882; see p. 116), and the remaining continental crust ruptured and slipped toward the Pacific void. There is, however, no record in Cenozoic strata of a catastrophe of such magnitude.

Others soon began to drift, the most important of whom was a German meteorologist, Alfred Wegener. Wegener drew upon evidence from geology, geophysics, biology, and climatology in developing the first complete and influential statement of the drift theory in 1912. He believed that drifting apart of the continents occurred over a long period during the Mesozoic and early Cenozoic Eras (Fig. 8.1). He became the first to attempt reconstruction of the former supercontinent by fitting edges of continental shelves rather than present coastlines, a more geologically realistic approach. Besides the parallelism of continental margins, Wegener appealed chiefly to apparent paleoclimatic indicators in late Paleozoic rocks, including glacial, desert, and tropical rainforest deposits, to reconstruct Permian climatic zones (Fig. 8.11). A South African geologist, A. L. Du Toit,

FIGURE 8.9 F. B. Taylor's and A. Holmes' conceptions of formation of Cenozoic mountains by continental drifting. Note parallelism of mid-Atlantic submarine ridge to opposing continent margins. *(Adapted from Taylor, 1910, Bulletin of the Geological Society of America; Holmes, 1929, Mineralogical Magazine.)*

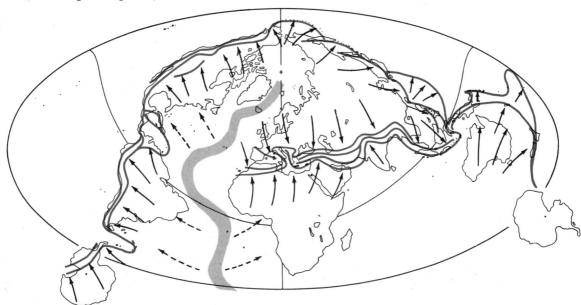

FIGURE 8.10   H. B. Baker's *Replacement Globe* reconstruction of continents prior to supposed separation. Note rotations and squeezing together of islands and peninsulas to improve fit. *(After Baker, 1912; with permission of Michigan Academy of Sciences.)*

completely consumed by drift fever, began in 1921 to strengthen the concept with a more detailed assessment of supporting geologic and paleontologic evidence (especially from the Southern Hemisphere) than Wegener could offer. Using additional evidence, Du Toit also developed more refined pre-drift reconstructions. Near-identity of many fossils, especially of land plants and animals, on the now-widely-separated continents lent support to drift. How could land plants with large seeds incapable of wind transport have been dispersed across wide oceans? And how could heavy reptiles have swum across? Even the earthworm has been used to support drift. It is said that the modern worms of Madagascar are more like those of India than of nearby Africa, and those of eastern North America supposedly are more like those of Europe than those of the American Pacific coast.

FIGURE 8.11   Wegener's reconstruction of the continents and Carboniferous paleoclimatic zones inferred from peculiar rock assemblages. Based upon paleoclimatic data, paleoequator and pole positions then were deduced which are remarkably similar to positions indicated by paleomagnetic data four decades later. An oceanic north pole would produce warm northern climate. Because map was drawn on present coordinates, paleolatitudes seem distorted. *(After Köppen and Wegener, 1924,* Die klimate der Geologischen Vorzeit, *by permission of G. Borntraeger Co.)*

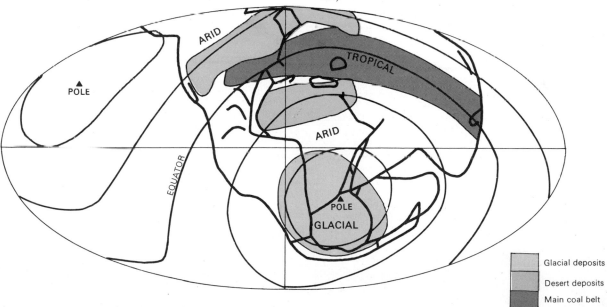

The first phase of the drift theory culminated in 1937 with publication of Du Toit's *Our Wandering Continents*, in which he scorched the conservative orthodox antidrifters (mostly American) with an attack upon their "groundless" worship of the dogma of fixed continents and ocean basins. He ridiculed, for example, the compromise of inventing countless narrow land or island bridges between continents to explain the undeniable similarities of many fossil and modern land organisms on opposite sides of the Atlantic. How could his opponents "build bridges as easy as a chef makes pancakes" (to quote Darwin), yet deny the drifting of continents? But more of Du Toit later.

THERMAL CONVECTION —PANACEA FOR MOUNTAIN BUILDING AND DRIFT?

Wegener believed that, in spite of apparent high viscosity of subcrustal materials, small forces acting over very long periods of time could cause that material to yield and allow crustal blocks to flow slowly through the upper mantle. Though it did (and still does) sound incredible, we have learned to be cautious about shouting "impossible." As noted above, modern knowledge of material behavior proves that slow plastic flow and creep do indeed occur. The question then becomes: "What force might have caused the drifting?" Wegener considered centrifugal effects of the earth's spin, tidal effects in the solid earth, and wobble of the axis as possible

FIGURE 8.12 The convection current mechanism for drifting continents and forming new ocean basins as conceived by Arthur Holmes in 1928. *Upper:* Convection begins to stretch an overlying continental block. *Lower:* That block has been ruptured and new convection moves its two fragments apart as if carried on conveyor belts. *(From A. Holmes, 1965, Physical geology (rev. ed.), Ronald Press copyright by Thos. Nelson and Sons, Ltd. 1944; used with permission.)*

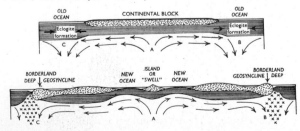

contributors, but all these were rejected by physicists as inadequate.

The possibility of thermal convection in the earth was suggested as early as 1839, and in 1881 it was proposed as a possible mechanism for localizing volcanoes and also for producing mountains by dragging and wrinkling the crust laterally. In 1928, a brilliant Scottish geologist, Arthur Holmes, who very early became interested in the implications of radioactivity, presented the first serious suggestion of thermal convection in the mantle as a possible driving mechanism both for mountain building and for continental drift (Fig. 8.12). The basic idea is simple. If more heat is generated in one portion of the deep mantle due to irregular distributions of radioactive isotopes, there would tend to be a very slow plastic flow of the hotter material upward. As it rose, it would cool and flow laterally—presumably beneath the crust—and finally when cooler than the average mantle below, it would sink. Thus a slow convective overturn of the mantle was proposed like the convective circulation of warm air in a room heated by a radiator. Holmes then assumed that where the convective flow was upward (hot), it might raise and even rupture the crust; where it was lateral, it might drag the crust along conveyor-belt-fashion (see Fig. 8.12); where mantle motion became downward (cool), the light continental crust, being isostatically buoyant, would refuse to sink, and so would buckle up to form mountains either against another continent or against oceanic crust.

Gravity measurements by Dutch geophysicist Vening Meinesz in the Indonesian region around 1930 soon suggested that the crust beneath the deep-sea trench of that modern intraoceanic mobile belt was being forcibly buckled down by some great compressive force (there is the vise again). Although since reinterpreted, this apparent huge crustal downbuckle (then named tectogene) under a modern mobile belt spurred interest in thermal convection cells as a possible mechanism for the downbuckling—first to form a geosyncline and eventually to upheave a mountain system. In the late thirties, American David Griggs performed ingenious scale-model experiments, which seemed to show that Holmes's convection could indeed produce mountains (Fig. 8.13). From then to the present day—although not a proven fact even yet—mantle convection in some form has been the favorite mechanism both for mountain making and—if one were a drifter—for rearranging continents.

MORE RECENT DOGMAS

After a hot controversy over drift during the 1930s, the matter cooled in the 1940s, with geologists around the world tending to throw their lot with either the pros or the cons and getting on to other tasks. Former president of the American Philosophical Society, W. B. Scott, expressed the prevalent American view in the 1920s by describing the drift theory as "utter, damned rot!" It was un-American to be "soft" on drift even as late as 1960. Reasons for strong opposition were twofold: *First*, the idea of such dramatic shiftings of continents, and seemingly only once in history, seemed counter to the uniformitarian doctrine. Certainly continents had changed, but it seemed more uniform for them to do it in place. Yet if they had to move, then they "should have" moved uniformly (i.e., continuously), or at least several times; a *single* great jump smacked of catastrophism! *Second*, no theoretical mechanism was known that could drive the large but very thin continents—slabs 200 times longer than they are thick—over the earth's surface. Geophysicists asked how could weak continental rocks be pushed through or over equally weak oceanic ones? It would be like trying to push huge sheets of tissue paper. Zealots like Du Toit, however, kept the faith and "accepted the inescapable deduction from the wealth of geological evidence available to the unfettered mind." They were not overwhelmed by the protestations of the physicists, because they remembered all too well the error of Lord Kelvin's mathematical arguments about the age of the earth (see Chap. 6). Here is an important illustration of divergent attitudes. The field observer's seat-of-the-pants feeling is that if "something *did* happen, it *can* happen," whereas the theoretician tends to disbelieve anything as dramatic as drift unless he can conceive of a sound mechanism. As in any other game, each side wins some and loses some.

PALEOMAGNETISM *Early* *Fred Vine*

Rebirth of interest in drift had to await new approaches, which came—ironically—largely through physics

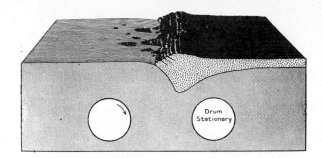

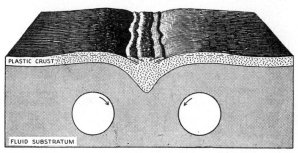

FIGURE 8.13 Laboratory scale models of effects of possible thermal convection currents in the mantle. Asymmetric downbuckling and thrust faulting seemingly like the Appalachian belt was produced by rotating only one drum in the viscous "mantle" (substratum), and bilaterally symmetrical buckling and thrusting was produced with two rotating drums or convection cells. If this model is correct, note that the model "thrust faults" actually reflect cratons or oceanic crust *underthrusting* the mobile belt rocks. *(From D. Griggs, 1939,* American Journal of Science, *v. 237, pp. 611–650; by permission.)*

in the 1950s and 1960s. The study of rock magnetism gave the first new impetus. Physicists had begun measuring the magnetic properties of rocks about 1850 and had discovered that many young lavas showed magnetization parallel to that of the earth's present field. About 1900, studies of ancient bricks and pottery showed that magnetization acquired during firing had been retained for 2,000 years. Contrasts of inclinations in lavas, dikes, and adjacent baked sediments with inclinations in the unbaked older sediments showed that igneous rocks tend to acquire and retain magnetic orientations parallel to the earth's field at the time of their cooling (Fig. 8.14). Further studies showed that the magnetization is stable indefinitely unless the rock becomes heated to about 600°C, at which temperature the magnetization is lost. Consistency of rock magnetism is demonstrated by

EVOLUTION OF THE EARTH

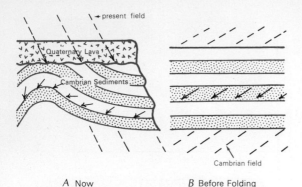

A Now   B Before Folding

FIGURE 8.14 *A:* Orientation of remanent magnetism with respect to the present magnetic field in two different-aged rocks. *B:* Restoration of the ancient magnetic field orientation from Cambrian rocks after correcting for folding.

identical orientations in several different rock types of the same age in a local area.

In the late 1920s a French physicist, Mercanton, suggested that because orientation of the magnetic field now bears a close relation to the earth's rotational axis, it might be possible to test the theory of continental drift with the magnetic characteristics of certain ancient rocks. This fertile suggestion was not followed up until after World War II, when rock paleomagnetic data began to be gathered on a large scale, initially by a group of British physicists.

Several minerals (chiefly certain oxides of iron) have magnetic properties that make them become oriented in a magnetic field. As such minerals crystallize in a cooling magma, they tend to be oriented parallel with the earth's magnetic field. If such minerals are deposited gently in sediments, they will tend to assume a similar orientation. Thus, rocks, as well as pottery, containing these minerals should today retain some evidence in their remanent magnetism of the orientation of the magnetic field when and where they were formed unless later heated. If perchance, the field shifted its position through time relative to the present position of such rocks on the earth's surface (or vice versa), then we might find a record of such shifts "fossilized" in the rocks. We could, in fact, study the history of changes of relative orientation (and also of changes of polarity) of the magnetic field, and at least one branch of geophysics would gain a historical dimension that it had lacked before.

Paleomagnetic studies based upon the above early findings have been conducted at an ever-increasing pace since 1950. The data show undeniable evidence of great changes in the magnetic field relative to present geography, but the results are still difficult to interpret. If we assume that the field was always two-poled as it is today and that it owes its orientation to the spin of the earth (see Fig. 7.6), *then the two magnetic poles should always have been more or less coincident with the geographic poles of the earth's spin axis.* If these assumptions are correct, then determination of ancient magnetic pole positions from carefully studied rock specimens should also reveal the approximate geographic paleopole and paleoequator positions. Figure 8.15 shows the orientation of magnetization in a modern rock in relation to the earth's field (left) and in an ancient rock (right). The right diagram illustrates how different the *relative* orientation of the ancient field must have been based upon the paleomagnetic evidence. Both the inclination angle (with the horizontal) and declination angle (or direction seen in a horizontal plane relative to a present north-south direction; Fig. 8.16) of the remanent magnetism for a rock are determined in the laboratory using a very sensitive magnetometer. Corrections of orientation must be made if the rocks have been tilted after they formed (Fig. 8.14). A steep inclination angle indicates formation of the rock at high latitude (Fig. 8.15), whereas a very low angle indicates a location near the equator. Declination angles for several widely separated samples *of the same age* also help to indicate the paleopole position, as shown in Fig. 8.16. Although paleomagnetism does not provide a longitudinal location, what a powerful paleogeographic tool this offers for determining the latitudinal positions of an area through time!

At first, paleomagnetic data were gathered chiefly from Europe and North America. As older and older rocks from both continents were studied, it appeared that the *relative* positions of these two continents with respect to the magnetic field, and therefore presumably also the rotational axis, had changed markedly. But it was not a random change, for the apparent North Pole seems to have been in the mid-Pacific Ocean 1 billion years ago and has migrated or wandered across Siberia to its present position, which it reached in mid-Cenozoic time (about 25 million years ago). By 1955 it was concluded that somehow the earth's rotational axis had shifted with respect to the crust so that continents indeed had occupied different latitudes through time. After 300 years of speculation, polar wandering now had acquired credence.

Although at first it was assumed that the magnetic field had shifted relative to a fixed crust, comparison of *apparent* pole migration or wandering paths for two or more continents revealed unexpected discrepancies. They did not agree, and only movements of continents could make them do so. Assuming that the field always had been a simple two-poled one, as today, then no hypothesis of polar wandering alone seemed adequate, for *the apparent ancient pole positions should have been the same for all continents at any given time in the past if only the poles (and magnetic field) were moving relative to the crust.* At first, sampling was so sparse that little confidence could be placed in the results, and some authorities still believe that there is too much scatter to constitute proof of continental displacements. But geophysicists concluded that, if the magnetic field had always been of the same sort as today (see Fig. 7.6), then they *must* revive some hypothesis of continental displacements in order to explain their data, no matter how painful that might be. *Seemingly the continents had to have moved relative to each other.* This was a curious irony, for recall that previously the physicists had most vehemently protested against drift because no adequate physical mechanism seemed available. But now the tables were turned, as expressed eloquently by American geophysicist Walter Munk at the height of renewed interest in polar wandering.[2]

In this controversy between physicists and geologists, the physicists, it would seem, have come out second best! They gave decisive reasons why polar wandering could not be true when it was weakly supported by paleoclimatic evidence, and now that rather convincing paleomagnetic (i.e., physical) evidence has been discovered, they find equally decisive reasons why it could not have been otherwise.

Paleomagnetic studies not only provided a compelling test of the drift theory, as envisioned by their originator, but once displacements seemed proven, paleomagnetism also furnished an independent, non-

[2]*Nature,* 1956, v. 177, pp. 553–554.

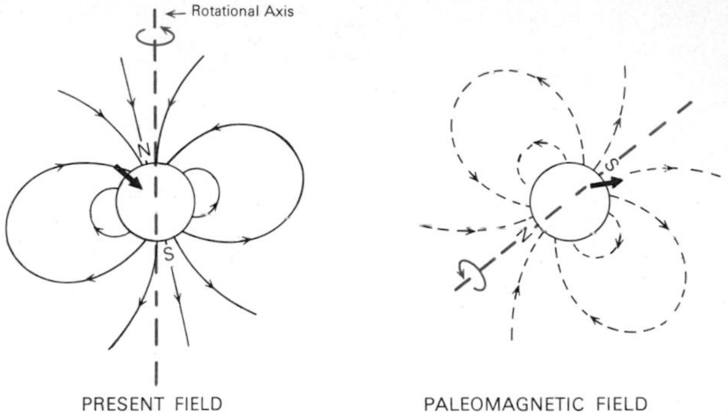

FIGURE 8.15 Relationship of earth's magnetic field to the orientation of observed remanent magnetism in rocks (heavy arrow), showing the basis for restoration of the paleomagnetic field and rotational axis *relative to* an observation station (compare Fig. 7.6). At right, field has to be shifted in a relative sense to conform to paleomagnetic data for an ancient rock.

FIGURE 8.16 The method of restoration of continental positions of the past using paleomagnetic data. The declination angle (between a "fossil magnet" shown by arrow and a north line) points to a former magnetic pole position. Continent positions must be adjusted until ancient pole positions for two continents coincide as shown in upper diagram.

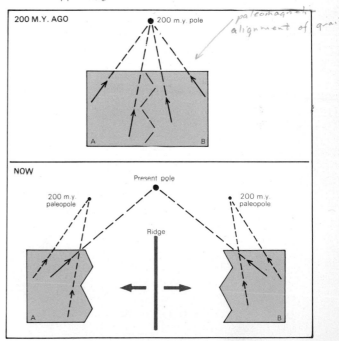

decoupling of the earth's crust.
- partial decoupling / European + American data.

geologic tool for reconstruction of the predrift continents and for helping to date drift history (Fig. 8.16). Most resulting reconstructions show impressive consistency with independent geologic evidence. For example the paleomagnetic reconstructions of the continents are almost identical with those old ones of Wegener and Du Toit (Fig. 8.1), which are based upon quite different evidence.

EXPANSION OF THE EARTH—THE MOST OUTRAGEOUS HYPOTHESES

Expansion of the earth first was suggested in the nineteenth century to explain parallelism of Atlantic shorelines. In 1933 it was revived, and reconstructions were made on small globes to achieve a hypothetical uniform covering of continental crust (Fig. 8.17). An expansion of more than 2,000 kilometers of radius was required! Although this figure is staggering, expansion does have some intuitive appeal in that it allows for an originally uniform continental crustal covering, which would imply a more complete

FIGURE 8.17 Hypothesis of an expanding earth originally uniformly covered by continental crust. In this conception, continental crust is considered primitive, while oceanic crust formed in scars produced by radial expansion. Sea water would have to have been released from the interior at a rate equal to expansion in order to maintain full ocean basins. *(Adapted from Hilgenberg, 1933,* Von Wachsenden Erdball; *Carey, 1959; drawn by D. E. Owen.)*

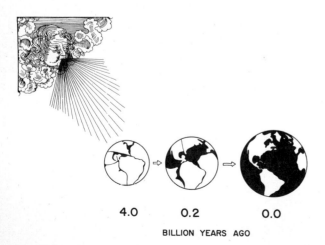

4.0 0.2 0.0

BILLION YEARS AGO

early chemical differentiation of the earth than does the present complex configuration.

Large-scale earth expansion seemed so outrageous and speculative that it was not taken seriously until about 1955, when the revival of interest in continental displacements began. It was suggested at that time that continental outlines, disjunct mobile belts, similar stratigraphic and paleontologic sequences, oceanic ridges, and perhaps even paleomagnetic data all might be explained as well by radial expansion of the earth as by lateral continental displacement. Even the new data favoring sea-floor spreading seem compatible with this different mechanism for expanding the oceans. Paleomagnetic pole restorations, however, cannot be reconciled as well with expansion as with lateral displacements of the crust. Calculations from paleomagnetic data suggest no more than a few hundred kilometers of expansion—if any. If expansion *has* occurred, note that sea water would have to have been added at a rate just sufficient to keep the enlarging ocean basins filled, for the stratigraphic record shows shallow marine deposits on continents throughout most of geologic time.

What possible mechanism could cause expansion of the earth? The first thought is heat from radioactive decay. Most materials expand when heated, but calculations show that the maximum change of earth radius by this mechanism would be on the order only of tens of kilometers.

Based upon the expanding universe theory, it was suggested in 1937 that the gravitational constant (*G*) may be inversely related to the age of the universe, and that the earth might expand as a result of decreasing *G*. Physicists estimate that 100-kilometer radial expansion might be possible, but as much as 1,000 kilometers is impossible. The order of permissible expansion would allow for approximately a 5 or 6 percent increase of area, which could just about account for all known late Cenozoic rifting, but falls far short of explaining complete displacement of continents. An additional physical consequence of decaying *G* and expansion of planetary bodies is that the sun would have to have burned its mass so quickly that it should already be in the red giant stage, which it is not. Also, heat flow from the interior per unit of surface area of an early, small earth would have made the surface temperature too great for life, yet we have a clear fossil record extending back to within 1 billion years of the estimated origin of the earth.

SUMMARY

In this chapter we have outlined a number of hypotheses to explain large-scale earth structure. The crust probably formed by the selective rise of lighter elements from the mantle. The two fundamental types of crust have an irregular distribution today, which long has puzzled geologists. A basic question is whether oceanic or continental crust is the more primitive. Hypotheses assume either that an early continental crust covered the entire earth or that continental crust has developed slowly at the expense of a primitive oceanic crust. Regardless of the original nature of continental crust, its known existence for 3 billion years seems to require, in light of isostasy, rejuvenation of continental material from below in order to maintain topographic continents in the face of denudation. Although continental crust has suffered repeated structural disturbances and erosion, average elevation of the continental plateau has remained close to sea level, at least for the past 2 to 2.5 billion years. The generation of granitic and andesitic rocks today in the active volcanic arcs, and the presence of huge granitic batholiths in the deeply eroded ancient mountain zones seem to indicate that the mobile belts have played a major role in the generation of continental-type crust through most of earth history.

Together with the mobile belts, which include the modern volcanic arc-trench systems, ocean ridges are the most tectonically active features. Between these more mobile zones lie the much larger stable divisions, continental cratons and submerged abyssal plains. Geosynclines represent extremely thick sedimentary and volcanic rocks that accumulated within subsiding portions of mobile belts. Early workers believed that a geosyncline was required somehow to produce mountains, but it is now evident that geosynclines are the results of mobile-belt processes, not their cause. The subsidence that allows thick sediment accumulations may be caused at least in part by any of the following: (1) sediment loading of the crust, (2) compression of the crust causing it to buckle downward, (3) removal of large volumes of magma from below to form lavas, or (4) change of the lower crust to a more dense phase resulting in isostatic sinking. Of these, structural downbuckling seems to be the principal cause of geosynclinal subsidence.

Half a century ago the earth was assumed to be contracting as it cooled from a hot origin. Contraction presumably buckled the crust at the edges of continents, geosynclines were formed, and ultimately these were upheaved into mountains. Because the earth apparently has not cooled in any such simple way, this hypothesis must be rejected. Conversely, it has been claimed that the earth may be expanding, but the possible amount is trivial. It is also claimed that the continents are not fixed in position, but have been displaced at least once. Under certain conditions, the outer earth behaves in a brittle fashion to stress, as evidenced by faulting; yet, under other conditions (particularly high pressure and temperature and very slowly applied stress), the lower crust and upper mantle behave plastically. Importance of geologic time is apparent, for a stress acting slowly over immense time can produce effects that would be impossible if the same stress acted briefly. In this way it now seems possible to explain the drifting of continents as the passive upper portions of much thicker plates. Heat convection in the earth's mantle is the most plausible cause of the motions of those plates. Where two are compressed together, mobile belts are formed.

After continental drift was proposed as a mountain-building mechanism in the early twentieth century, the idea lost enthusiasm until paleontologic and stratigraphic evidence favoring drift were supplemented dramatically in the 1950s with the discovery of paleomagnetic evidence. In addition to supporting large displacements of continents, paleomagnetism also provides evidence of paleolatitude that is of great general value in interpreting paleogeography and paleoclimatology.

Readings

Birch, F., 1968, On the possibility of large changes in the earth's volume: Physics of the Earth and Planetary Interiors, v. 1, pp. 141–147.

Blackett, P. M. S., Bullard, E., and Runcorn, S. K., eds., 1965, A symposium on continental drift: Royal Society of London.

Carey, S. W., convener, 1959, Continental drift: a symposium: Hobart, Univ. of Tasmania.

EVOLUTION OF THE EARTH

Dana, J. D., 1873, On some results of the earth's contraction from cooling, including a discussion of the origin of mountains, and the nature of the earth's interior: American Journal of Science, ser. 3, v. 5, pp. 423–443; v. 6, pp. 6–14; 104–115; 161–172.

Du Toit, A. L., 1937, Our wandering continents: Edinburgh, Oliver & Boyd.

Gondwanaland revisited: new evidence for continental drift, 1968: Proceedings of the American Philosophical Society, v. 112, pp. 307–353.

Hall, J., 1859, The natural history of New York, Part 6—Paleontology, J.3: Albany; Van Benthuysen, Introduction, pp. 1–85.

Holmes, A., 1965, Physical geology (rev. ed): New York, Ronald.

Marvin, U. B., 1973, Continental drift: Washington, Smithsonian Press.

McElhinny, M. W., 1973, Palaeomagnetism and plate tectonics: Cambridge, Cambridge.

Oakeshott, G. B., 1975, Volcanoes and earthquakes: geologic violence: New York, McGraw-Hill. (Paperback)

Phinney, R. A., ed., 1968, The history of the earth's crust: Princeton, Princeton.

Poole, W. H., ed., 1966, Continental margins and island arcs, An international upper mantle project report: Ottawa, Geological Survey of Canada Paper 66–15.

Scientific American, 1972, Continents adrift: reprints from Scientific American, San Francisco, Freeman.

Wegener, A., 1966, The origin of continents and oceans: New York, Dover. (Paperback, S 1708, English translation of 4th ed., 1929)

THE DEEP-SEA FLOOR AND PLATE TECTONICS

They [plates] can't curl down; they
must curl up
To form a kind of dish
To stop the oceans spilling out
And losing all the fish

. . . .

Yet others seem to hit or slide
Performing curious functions,
And where they can't make up their
minds
You there have triple junctions.

B. C. and G. C. P. King (1971)[1]

FIGURE 9.1   Deep-sea drilling vessel
Glomar Challenger owned by Global
Marine Inc. and operated by Scripps
Institution of Oceanography for the
National Science Foundation. The
ship weighs 10,400 tons, is 400 feet
long, and the derrick stands 194 feet
above the waterline. It carries over
24,000 feet of 5-inch drill pipe. Drill
bits can be changed and re-entry can
then be made into a small borehole 3
miles beneath the ship. *(Photo
courtesy Scripps Institution of
Oceanography.)*

If continental displacements are valid, then according to the drift theory, the Atlantic, Indian, and most of the Arctic Oceans did not exist prior to Mesozoic time. Clearly, historical and structural evidence from these ocean basins should shed much light upon the whole question of crustal evolution. Indeed, the deep-sea floors provide the most definitive tests of the entire concept of continental displacements.

The deep seas were first probed in 1872 when the British *Challenger* Expedition was launched. Discovery of the mid-Atlantic submarine ridge (Fig. 8.9) about 1910 seemed to support drift, because the ridge neatly parallels the opposing coastlines; perhaps it was a scar away from which the continents moved. But the greatest period of geologic investigation has been since World War II. As with rock magnetism, geophysics led the assault upon the deep seas that has produced a revolution that not only supports continental drift but goes far beyond it to provide a unifying new theory of global tectonics. Most surprising was the realization that oceanic crust, rather than being the oldest on earth, actually dates back only into the Mesozoic. Most older ocean floor seemingly has been thrust down into the mantle beneath ocean trenches as American geologist Bailey Willis prophesied way back in 1907 just as a hunch. Small amounts of old sea floor also seem to be preserved as shreds within many old mountain belts as Europeans thought 80 years ago (and noted in Chap. 8).

In the 1960s, the sea floor-spreading hypothesis suggested that new oceanic crust is constantly added through igneous activity at ocean ridges, which are sites of extension or pulling apart. Meanwhile, the sea floor is thrust beneath the trench-arc systems, which are zones of compression as if caught within a vise. Earthquake zones define about a dozen large plates covering the surface of the earth. The different relative motions among these lithosphere plates apparently can explain all the major tectonic features of the earth's surface, as well as the drifting of continents. The all-encompassing theory of plate tectonics has produced the greatest revolution in the earth sciences since Darwin and Lyell. The plate-tectonic revolution began just before we published our first edition of *Evolution of the Earth.* The revolution has continued at such a rapid pace that it has necessitated a complete reorganization of this second edition. Plate theory has also unified the earth sciences to a degree never before possible, and it has lent an excitement to the field that we hope will be apparent as you read ahead. It appears that the evolution will continue for some time yet.

GENERAL NATURE OF THE SEA FLOOR

Until recent years, most geologists assumed that the crust beneath ocean basins was very old, topographically featureless, structurally tranquil, and essentially permanent as to position. All these assumptions appear to be incorrect, for nowhere in deep oceanic sediments have fossils older than Late Jurassic or Early Cretaceous as yet been definitely established. To our utter amazement the continents turn out to be much the older! Moreover, the total thickness of sediments on the deep-sea floor is small. Even modest assumed rates of deposition suggest that much of the present deep sea may have received significant sediment for only the past 100 to 200 million years. In the northwestern Pacific Ocean, 200 meters of unconsolidated sediments underlie a zone of known Lower Cretaceous fossils, and from seismic data, it appears that older, more consolidated sediments may underlie them. It seems possible that there sedimentation may be recorded back to Triassic or possibly late Paleozoic time. Unfortunately, a final definitive answer on the maximum ages of the different deep basins must await more fossil evidence from deep drilling through the entire sediment column (Fig. 9.1).

FIGURE 9.2 Typical seismic reflection profile across North Atlantic abyssal plain showing irregular oceanic basement surface and smoothing by sedimentation. Interpretive section appears below. Note prominent A reflecting zone, which is of Cretaceous age in at least one area east of the Bahama Islands. Considerable thickness of underlying sediments suggests Atlantic basin began forming before Cretaceous time, continental drift evidence suggests Triassic. *(After J. Ewing, et al.,* Science, *v. 154, December 2, 1966, pp. 1126–1132; copyright 1966 by the American Association for the Advancement of Science.)*

Continental drifters obviously would find it embarrassing if Cambrian fossils turned up in widely scattered drill holes. But *presently available evidence* at least is consistent with a relatively youthful (Mesozoic) origin for the present ocean basins.

Precision profiles established by reflection of low-frequency sound waves from the sea floor and buried layers beneath have shown that the ocean floors are anything but smooth. Broad oceanic ridges or rises, deep trenches, escarpments, and countless submerged seamounts characterize it instead. Indeed, the pristine surface of the oceanic crust is more rugged than most continental areas, and sedimentation has served to smooth the topography in some areas by burying original irregularities (Fig. 9.2).

RATES OF EROSION AND SEDIMENTATION

North America is now being denuded at a rate that could level it in a mere 10 million years, or to put it another way, 10 North Americas could have been eroded since Middle Cretaceous time 100 million years ago. If all present continents were reduced to present sea level, and their refuse were spread uniformly over the abyssal plains, a layer of sediments about 300 meters thick would result. The observed average total thickness of deep-sea sediments is only about 600 meters, or an amount equal to the erosion of the present-sized continents *only twice during the past 200 million years* for which time there is a known record in the deep-ocean basins. From this discrepancy, as well as from other evidence, it is clear that, in the past, the rate of erosion and/or the volume of land above sea level has been much less *on the average* than now.

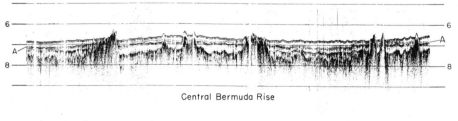

Central Bermuda Rise

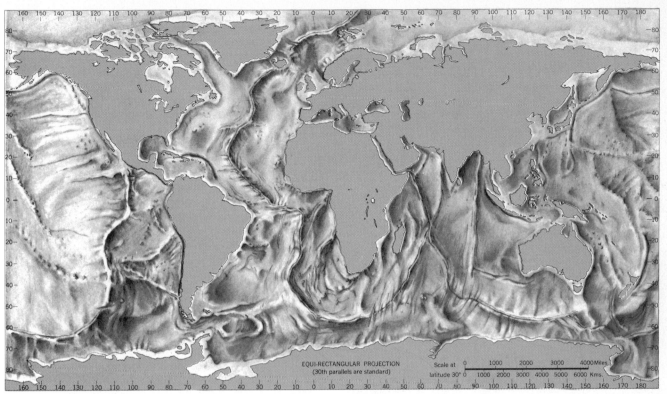

FIGURE 9.3 World's ocean-floor topography, emphasizing roughness of much of the sea floor; note ocean ridges, linear escarpments, and deep trenches. Being less masked by sediments and vegetation than continents, this topography has great structural significance.

The apparent youthfulness of the entire present deep-sea floors came as a great shock, for you will recall that in Chap. 7 we argued that basaltic oceanic crust should be a chemically more primitive type than granitic continental crust. Now we discover that not only do continents apparently contain the only record older than 200 million years, but also that there is a shortage of deep-sea sediments compared to what would be expected through long erosion of continents. Apparently the puzzle can be answered only by losses through removal both of deep-sea sediments and oceanic crust through time.

OCEANIC RISES OR RIDGES

The most striking features, especially of the Atlantic and Indian Ocean floors, are submarine rises or ridges, mentioned before (Fig. 9.3). Their symmetrical positions invite speculation that they are scars of a predrift configuration of the crust. But if so, drift was more complex than envisioned by Taylor in 1908, when it appeared that the Americas simply drifted west from the mid-Atlantic ridge while Africa and Europe moved east. Discovery of additional ridges nullified this simple interpretation, for the symmetry of ridges *surrounding* some continents makes it impossible to have moved the same continents simultaneously the same distance away from two or more ridges (note Africa in Fig. 9.3).

The bulk of the broad ridge features is made up of basaltic lavas, as evidenced by islands along their crests, by samples dredged from their submerged portions, and from underwater photographs (Fig. 9.4). Oceanic ridges have several important characteristics that set them aside as unique, major structural features of the crust (see Table 8.2). Besides volcanism, they display great shallow seismicity beneath their axes (unlike trenches, which also have deeper focus earthquakes). Ridges are also characterized by

FIGURE 9.4 Submarine ellipsoidal ("pillow") lavas on the flank of South Pacific–Antarctic oceanic ridge at a depth of 2,800 meters (longitude 145°E, latitude 56°S). (Compare Figs. 10.4 and 12.2.) *(Official NSF photo, USNS Eltanin, Cruise 15; courtesy Smithsonian Oceanographic Sorting Center.)*

greater-than-average flow of heat through the crust along their axes (Table 8.2). Ridges clearly are zones of release of much subcrustal thermal energy, but *in response to different stress conditions than prevail beneath island arcs.*

Another peculiarity shared by most ridges is a narrow depression that extends along their axes for thousands of kilometers. In Iceland, the largest exposed ridge area on earth, prominent graben structures are conspicuous across the center of the island from north to south (Fig. 9.5), and 30 active volcanoes occur along this zone. There is a close parallel of scale and morphology with African rift structures that suggests similar origins, and the northern Indian Ocean ridge even passes into the continental African rifts. Prevalence of topographic rifts along many ridge crests, together with seismicity and volcanism, all suggest—by analogy with structures of land—that ridges are zones of extension along which the crust is being torn open. Ridge patterns may be complicated by branching; a point where three branches join is called a triple junction.

The East Pacific Rise is unique in being asymmetrically located in its ocean basin and in lacking a prominent crestal depression. It extends northward to or beneath the west edge of the North American continent. Significantly, the crust and mantle there are geophysically anomalous and display evidence of unusually widespread extensional structures in the western United States, suggesting that the ridge has somehow disturbed the edge of the continent there much as the Indian Ocean ridge affected eastern Africa.

SEA-FLOOR SPREADING—ANOTHER OUTRAGEOUS HYPOTHESIS

It was submarine extension or ocean spread constantly in progress . . . Asia was therefore formed not by overthrusting, but by underthrusting. (Bailey Willis, 1907)

Recall that, in 1928, Scottish geologist Arthur Holmes presented a hypothesis of convection in the mantle as a cause of mountain building (Fig. 8.12), and in the same year, Alfred Wegener also accepted convection as one possible mechanism for continental drift. Since that time, convection has been the most widely endorsed mechanism to explain large-scale tectonic features. In 1962, Princeton University geologist H. H. Hess proposed a bold new hypothesis that two opposing thermal convection cells rising beneath ocean ridges produce tension in the crust (thus rifting) and also cause the abnormal heat flow observed there. As rifting occurs, earthquakes are generated beneath ridges, and new crustal material is erupted volcanically at ridge axes. Hess envisioned that, finally, the slow, convective flow laterally away from ridge axes carries older oceanic crust along as if on a conveyor belt, causing the spreading of sea floors through time. Original as Hess's idea was, it appears from the above quotation that he had been partially anticipated 55 years earlier.

The dramatic hypothesis of sea-floor spreading postulates a youthful origin of the Atlantic and Indian Ocean basins (and their crusts) through disruption of former continental areas by rifting over rising convective cells followed by progressive separation of the dismembered continental fragments as juvenile oceanic crust is generated between (Fig. 9.6). Presumably, eastern Africa and Arabia are today experiencing the beginning of a new phase of such disruption in response to an assumed shift of mantle convection

Incipient oceans.

FIGURE 9.5 Aerial view of western part of Thingvellier graben, southern Iceland, marking surface exposure of part of crestal rift of mid-Atlantic ridge; scarp at right is about 40 meters high (view looking south). Displacements of fences prove extension of Iceland in historic times accompanied by volcanism and earthquakes. *(Photo by Sigurdur Thorarinsson.)*

patterns, which brought rising cells beneath that region. It is significant that the Red Sea–Gulf of Aden chasm lies at the landward end of the western Indian Ocean (Carlsburg) ridge. Spreading along the extension of this ridge tore open the Aden–Red Sea–Suez–

FIGURE 9.6 The hypothesis of sea-floor spreading and its combined explanation of oceanic ridges (with rifts, high heat flow, volcanism, and earthquakes), island and guyot patterns, continental displacement, and active continental margins, here all explained by thermal convection in the mantle (arrows). Note that layering seems to demand drift also of upper mantle beneath continents; the zone of major movement may lie in the *seismic low-velocity zone* about 100 to 200 kilometers below the surface. P refers to different lithosphere plates. *(Adapted from Hess, 1962; Dietz, 1963; Orowan, 1964, Science, v. 146, p. 1003.)*

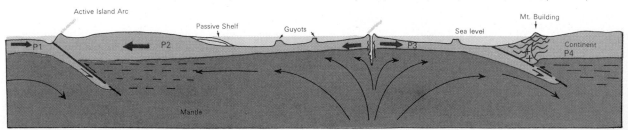

FIGURE 9.7   Gulf of Aden and southern Red Sea from a Gemini spacecraft. Note how well Africa (below) could fit against Arabia (above), suggesting that Africa was torn away to form the new ocean basin between and inaugurate a new phase of continental drifting. *(Courtesy NASA; photo No. 66-54536.)*

Aqaba chasm in the crust (Fig. 9.7). African rifts during the past 25 million years seem to typify the initial breakup of huge supercontinents—the earliest stage in continental drift.

If correct, sea-floor spreading offers, at one and the same time, an explanation of the origin of at least some of the oceanic crust and also a possible means of displacing continents. It has the merit of explaining the seemingly ambiguous parallelism of most ridges with two continental margins symmetrically posi-

tioned equidistant from them, a problem noted above. And very importantly, it helps to explain why much of the ocean floor is so young. Thus it is truly a unifying simplification of knowledge, which accounts for its immediate appeal among earth scientists.

AGE DISTRIBUTION OF OCEANIC ISLANDS AND SEAMOUNTS

About the same time that Hess first stated his concept of sea-floor spreading, Canadian geophysicist J. Tuzo Wilson had postulated that oceanic islands, which are practically all volcanic in origin, tend to be symmetrically distributed as to relative age outward from submarine ridges. Youngest islands tend to be at or near the axes of ridges, while those more distant seemed to be progressively older. Moreover, still farther removed from ridge axes are submerged volcanic mountains (seamounts) at progressively deeper levels (Fig. 9.6). Many of these were islands in the past, as evidenced by dredgings of shallow-marine fossils from their tops. Obviously the fate of oceanic islands, once they have ceased to be volcanically active, is to be more or less eroded by surf and to gradually subside to abyssal depths. Sea-floor spreading offers a consistent explanation of Wilson's conception of island age distributions. Most islands, he said, formed at ridge axes and have existed as such as long as they remained volcanically active (generally about 20 to 30 m.y.). As spreading of the crust progresses, they move laterally, conveyor-belt fashion, become inactive, and then proceed to sink as spreading carries them farther down the ridge flanks.

CORAL ATOLLS AND
FLAT-TOPPED SEAMOUNTS

In the nineteenth century, low rings of coconut-tree-studded islands surrounding shallow lagoons in the tropics (between 30° north and 30° south latitude) invited explanations by many naturalists. The several hypotheses advanced to explain these islands generated much controversy but Charles Darwin in 1842 struck upon the correct interpretation. He concluded that these atolls, which are nothing but rings of low islands composed of debris washed up by waves from coral reefs just offshore, are but one stage in an evolution of typical Pacific and Indian Ocean islands, which involves gradual subsidence of the sea floor beneath them.

Darwin noted that some tropical islands consisted of a prominent volcano surrounded by a fringing reef at its margin with little or no water between. Others had barrier reefs with narrow doughnut-shaped lagoons between the reef and the central volcano. Others had a very small volcano and much wider lagoon, and still others, the true atolls, had a ring-shaped series of islands with a broad, central lagoon and no volcano at all. He said that all atolls began as fringing reefs surrounding active volcanoes and that in the course of time, they gradually sank isostatically as the volcano became extinct. Because coral growth is most active on the outer edges of reefs, the reefs expanded outward as well as upward. More important, the coral growth had to keep pace with the subsiding volcano. If it sank too rapidly

beneath the critical lighted (photic) zone, the reef would be killed. Eventually the now-extinct volcano would disappear beneath the sea, and the atoll would have advanced to its mature state surrounding an open lagoon.

Darwin recognized that the only way his theory could be tested was by drilling into an atoll to see what lay beneath. It was not until 1952, at Eniwetok Atoll in the course of preparations for United States-Pacific nuclear testing, that drilling operations penetrated submerged volcanoes. Most of what lay between the surface and the volcanic basement consisted of reef limestone and reef debris. Furthermore, the Pacific reefs were found to be relatively old—going back in some cases at least to early Cenozoic time (Table 9.1). This was even more forcefully brought to the fore by the great depth at which submerged volcanic foundations were encountered (nearly 1,000 meters). Only the Darwinian subsidence theory for atolls can account for this great depth. In the Bahama Islands, drilling has penetrated a nearly continuous reef sequence dating back at least to Early Cretaceous time (Table 9.1); a volcanic foundation was penetrated at 6,000 meters. Many Pacific reefs have been growing continuously since at least early Cenozoic time. Some West Indies reefs began in the Cretaceous, but have had an interrupted history during Cenozoic time. The Red Sea reefs span only late

TABLE 9.1 Sea-floor subsidence indicated by known depths of reef carbonate rock columns*

Location	Depth drilled or dredged, meters	Oldest age encountered
Pacific		
Bikini	725	Oligocene or Late Eocene
Eniwetok	350	Miocene
Funafuti	330	Pliocene
Mid-Pacific mountains	1,900	Mid-Cretaceous
Philippine Sea		
Kita-daito-Jima	400	Early Miocene
Western Atlantic		
Bermuda	125	Early Oligocene
Bahama Islands	4,400	Cretaceous

*After K. O. Emery et al., 1954, U.S. Geological Survey Professional Paper 260-A; E. L. Hamilton, 1956, Geological Society of America Memoir 67; and N. D. Newell, 1955, Geological Society of America Special Paper 62.

EVOLUTION OF THE EARTH

Cenozoic time. Some modern reefs apparently began only during the Pleistocene, especially in the West Indies.

In the western and northern Pacific basin, there are many submerged, flat-topped volcanic sea-mounts. These peculiar features, called guyots (Fig. 9.6), were discovered during World War II by extensive deep-sea soundings. H. H. Hess interpreted them as oceanic volcanic mountains whose tops had been planed off by wave erosion before they subsided to depths of hundreds of thousands of meters. Proof of his interpretation came at the end of the war with revelation that dead, submerged reefs, and probably former atolls, existed on the tops of some mid-Pacific guyots west of Hawaii now at depths of more than 2,000 meters! The reefs contain Lower Cretaceous organisms that originally lived at sea level. Obviously here the corals could not keep pace with crustal subsidence. According to sea-floor spreading, all the western Pacific guyots are on crust that is more than 30 million years old.

MAJOR OCEANIC FRACTURE ZONES AND TRANSFORM FAULTS

Exploration in the northeastern Pacific basin following World War II revealed a group of long, east-west-trending submarine escarpments explicable only as great zones of faulting (Fig. 9.3). Such fracture zones, as they are called, have been found elsewhere, too, and seem to be characteristic of sea floors. In general they are perpendicular to ocean ridges. In the mid-Atlantic, for example, the ridge crest has been offset along such zones. The Indian Ocean floor displays the most amazing array of escarpments, suggesting wholesale fragmentation of the crust along largely north-south fractures that look like railroad tracks along which India might have traveled northward (Fig. 9.3).

With important exceptions like the great San Andreas fault of California, such long, nearly straight fractures are rare on continents. Indeed, most of the Pacific fracture zones terminate at the continental margins; thus they reflect structural processes restricted to oceanic crust. That the dominant movement along these zones is lateral or strike-slip is indicated by offsets of linear magnetic anomalies, offsets of the ocean-ridge crests, and from studies of motions during earthquakes that occur along them. From the apparent offsets of the mid-Atlantic ridge near the equator, it appeared at first that the motion there was all left-lateral, for the north side of each fault seems to be displaced left (westward). Cumulatively, such movement might account for the great mid-latitude bend of the ridge. But some of the faults have sharp terminations at the ridge crest, and quite a different interpretation of movement became possible in light of the sea-floor-spreading hypothesis. J. Tuzo Wilson, in 1965, conceived of a mechanism that produces what he termed transform faults. By this scheme, the transverse faults actually have formed by movement *opposite* to that implied by mere apparent offset of ridge crests (Fig. 9.8); lateral motion is transformed at ridge crests by spreading. If spreading from a young ridge crest became irregular as to rate, a rupture might result, and adjacent portions of the ridge thenceforth would spread differentially, with a transform fault formed between those segments. If spreading takes place continuously all along the ridge, then movement between adjacent segments actually is right-lateral; because of spreading, apparent offsets of ridge crests are deceptive. Careful studies of seismic records for earthquakes in the mid-Atlantic region confirm that the chief motions associated with tremors in the shear zones are right-lateral transform in nature. Apparently most great oceanic fracture zones are transform faults related to

FIGURE 9.8 Comparison of transform faults with ordinary transcurrent or strike-slip faults. Wide brown lines represent ridge axes; A and B represent two different times for each case. Note that because spreading proceeds as transforms move, fault motion suggested by offset of ridge crest is opposite of actual motion. In the lower case, spreading has ceased before faulting; therefore offset of the dead ridge does provide an accurate idea of displacement. *(Modified from Wilson, 1965, Science, v. 150, pp. 482–485.)*

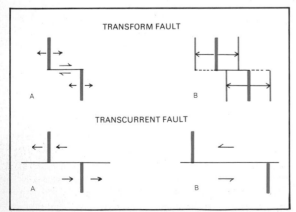

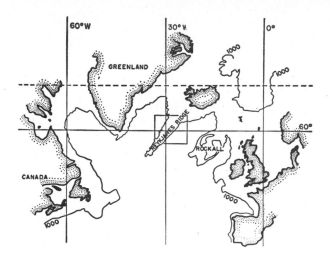

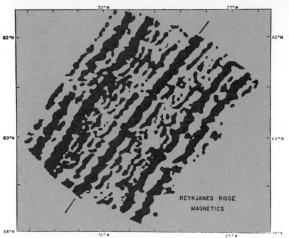

FIGURE 9.9 "Zebra-stripe" magnetic anomalies symmetrical around axis line of the mid-Atlantic ridge southwest of Iceland (in line with graben of Fig. 9.5). Symmetrical relation of stripes to ridge axes here and elsewhere led to the hypothesis that such anomalies reflect magnetic-field-polarity-reversal events during a history of sea-floor spreading. *(From Heirtzler et al., 1966, by permission of* Deep-Sea Research.*)*

sea-floor spreading. Besides connecting ocean-ridge segments, transforms may also connect ridge and volcanic-arc segments, or two arc segments.

MAGNETIC ANOMALIES AND SEA-FLOOR SPREADING

Among many new kinds of information about the sea floors was the discovery in 1961 of linear magnetic anomalies apparently unique to the abyssal oceanic crust, for the patterns terminate near the continental shelves. The anomalies first were mapped in detail in the northeastern Pacific Ocean and then across the mid-Atlantic ridge south of Iceland (Fig. 9.9). Surveys elsewhere have now provided magnetic profiles across much of the remainder of the world ocean-ridge system, and all show a consistent pattern of narrow, alternating anomalies. Near Iceland, the stripes showed two striking features: *first*, they parallel closely the ridge axis, and *second*, they show a remarkable bilateral symmetry such that those on one side of the axis tend to mirror those on the other (Fig. 9.9). The same symmetry is found elsewhere (Fig. 9.10).

What do the oceanic magnetic anomalies mean? They were mapped by ships or planes towing magnetometers along repeated traverses that crisscross the ridges. Resulting data record variations in total intensity of the magnetic field across the sea floor. At first it

might seem that the stripe patterns represent alternating zones of extreme contrasts of magnetic susceptibility (and remanent magnetism) in the oceanic crust. But the magnitude of the anomalies would require improbable alternations of deep blocks only about 20 kilometers wide of nonmagnetic and very strongly magnetic materials. Necessary geometry and susceptibility contrasts are so unlikely as to boggle the mind. Alternatively, the anomalies could represent polarity reversals in successive bands of rock materials of uniform magnetic susceptibility; thus the black bands in Figs. 9.9 and 9.10 are thought to represent normal polarity relative to the present earth field, and the white ones reversed polarity. This explanation is preferred because independent evidence of polarity changes of the earth's field have been known for many years.

As early as 1905, it was found that polarity in magnetically susceptible minerals in some late Cenozoic lavas was opposed to that of the present main field. In the 1920s and 1930s, examples of reversals were documented in Cenozoic lavas from several continents. The last reversal, which produced the present field configuration, occurred in Quaternary

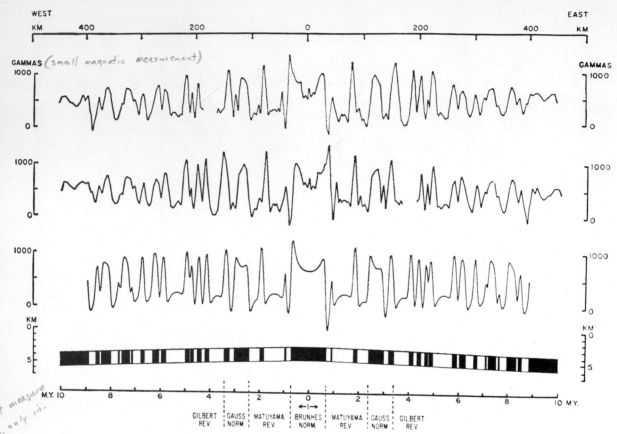

WEST / EAST

KM 400 200 0 200 400 KM

GAMMAS *(small magnetic measurement)* / GAMMAS

(handwritten, left margin) * doesn't measure reversals, only in-tensity

GILBERT REV. | GAUSS NORM. | MATUYAMA REV. | BRUNHES NORM | MATUYAMA REV. | GAUSS NORM. | GILBERT REV.

FIGURE 9.10 Ocean crust magnetic anomalies plotted as profiles of measured magnetic intensity along straight-line crossings of the South Pacific–Antarctic oceanic ridge. Note striking symmetry of anomalies on either side of the broad ridge-axis anomaly (upper curve is same as middle, but reversed to underscore the symmetry). Note also sharp contrasts of magnetic intensity peaks, which represent alternating "zebra-stripe" patterns of the anomaly maps (e.g., Fig. 9.9). Known field-polarity-reversal episodes are shown at bottom in black and white bar. *(From Pitman and Heirtzler,* Science, *v. 154, December 2, 1966, pp. 1166–1171; copyright 1966 by the American Association for the Advancement of Science.)*

(Pleistocene) time nearly 1 million years ago (Fig. 9.11). In thick sequences with many Cenozoic lavas, a number of successive reversals have been documented, and the flows now can be dated isotopically, thus providing a geomagnetic-reversal time scale. Similarly, polarity studies have been made of deep-sea sediments from submarine cores, and the results compare closely with those for lavas (Fig. 9.11). As magnetically susceptible sedimentary minerals settled to the sea floor, they assumed a remanent magnetism imposed by the magnetic field; thus, as in lavas, they preserve clues to past polarity.

Reversals of polarity of the earth's field have been important phenomena, with universal effects on minerals forming anywhere in the world in various types of rocks over at least the past 400 or 500 million years and, we assume, much longer. The period between reversals has varied enormously. During late Cenozoic time, for which data are by far the best, reversals have been occurring with a frequency of approximately 1 million years, although refined measurements show some periods as short as 10,000 years. Relatively short polarity episodes also characterized earlier Cenozoic time (Fig. 9.12), but late Mesozoic and late Paleozoic times had much less frequent reversals; early Mesozoic and middle Paleo-zoic times, conversely, were characterized by more reversals (Fig. 9.12). The actual reversing process is not understood. Apparently the poles migrate errati-cally, and the strength of the field varies considerably while reversing is underway; the entire change is thought to be geologically sudden, spanning only a few thousand years.

Discovery of an apparently universal polarity-reversal history for the earth's magnetic field, which is acceptable within the Elsasser dynamo theory for the field (see Fig. 7.6), represented a major recent break-through in geologic knowledge. It is suggested that symmetrical oceanic magnetic anomalies represent fossilized polarity episodes associated with the re-versal history. If true, they would greatly strengthen the sea-floor-spreading hypothesis. The anomalies appear to record each polarity-reversal event as the sea floor was spreading; as new crust formed in the central ridge rifts and cooled to the temperature at which magnetism becomes fixed (Curie point), it acquired and retained the polarity of the field at that time. Distance of each anomaly pair from a ridge axis is, therefore, proportional to its age (i.e., the time since it lay at the axis). It is important to stress that the linking of the anomalies with known magnetic polarity reversals is a *hypothesis*, but it is greatly strength-

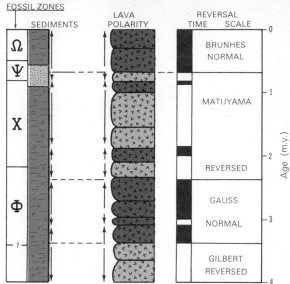

FIGURE 9.11 Diagrammatic representation of magnetic polarity reversals observed in late Cenozoic lavas in many parts of the world (which can be dated by isotopic methods) and in deep-marine sediments dated by fossils. Fossil zones, designated by Greek letters, tend to correlate closely with reversal episodes; therefore either parameter provides a tool for submarine stratigraphy. Polarity studies have led to a formal magnetic-reversal time scale, only a part of which is shown. *(Adapted from Cox et al., 1963, Nature, v. 198, p. 1049; Opdyke et al., Science, 1966, v. 154, p. 349.)*

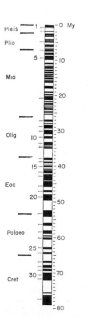

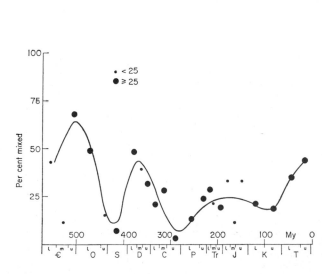

FIGURE 9.12 *Left:* Magnetic-reversal time scale well established for past 80 million years. *(From Heirtzler et al, 1968, Jour. Geophys. Res., v 73, p. 2123; copyrighted by Am. Geophysical Union). Right:* Approximate relative frequency of reversals for last 600 million years; greater percent mixed means more reversals. *(From McElhinny, 1971, Science, v. 172, p. 157–159; copyright 1971 by Amer. Assoc. for Advancement of Science).*

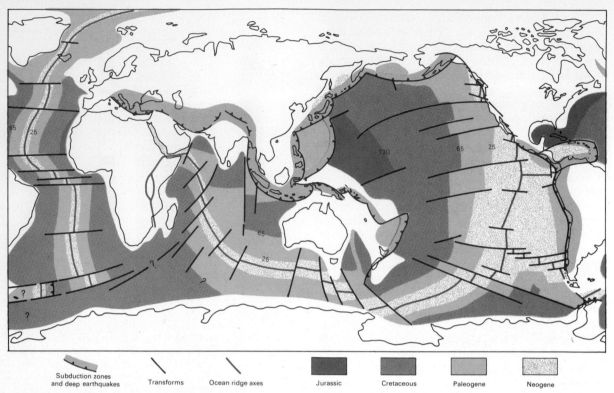

Subduction zones and deep earthquakes | Transforms | Ocean ridge axes | Jurassic | Cretaceous | Paleogene | Neogene

FIGURE 9.13 Ages of the oceanic crust based upon analysis of magnetic anomalies and deep-sea-drilling results (numbers are ages in m.y.) (compare Fig. 9.3). Also shown are active volcanic arc-trench systems (subduction zones) with associated deep earthquakes, active ocean spreading ridges, and major transform faults. *(Adapted from Karig, 1971, Journal of Geophysical Research, and Pitman and others, 1974, Geological Society of America Map of age of ocean basins.)*

ened by the fact that deep drill holes have encountered fossiliferous latest Cenozoic sediments resting upon oceanic basement on the central ridges, but progressively older ones outward toward the continents. Spreading is also supported by a two-times-greater heat flow from ridge crests as compared to deep-sea trenches, which is consistent with the convection mechanism (Fig. 8.12).

SPREADING HISTORY

The oceanic magnetic anomalies have been likened by F. J. Vine to a tape recorder attached to a sea-floor conveyor belt. With D. H. Matthews in 1963, Vine

proposed the relation of the anomalies to polarity reversals and spreading. These British workers showed that the anomalies can be correlated between different ridge systems with amazing confidence. Vine infers that a worldwide reversal history has been "taped" with great fidelity, and by comparing anomalies outward from ridge axes (Fig. 9.10) against the known polarity-epoch time scale established on land (Fig. 9.12), both average rates and an estimate of duration of spreading can be determined for different ridges. Resulting apparent rates of spreading vary from about 1 centimeter per year south of Iceland to more than 9 centimeters per year in the south Pacific, with the average being 2 to 3 centimeters per year. Knowing the reversal history from lavas on land, we can extrapolate out from ridge-crest regions, which represent only the past 4 or 5 million years, and estimate the total duration of spreading recorded in the anomalies across a given ocean basin (Fig. 9.13). Such analysis provides an estimate of the minimum age of the ocean basin; for examples, spreading from the mid-Atlantic ridge apparently began between 150 and 200 million years ago, from the northwestern Indian Ocean ridge between 80 and 100 million years ago, and Australia and Antarctica did not separate until 60 million years ago. Note that these figures match closely most independent estimates of inception of postulated continental separations. They also suggest that little if any of the present oceanic crust can be as old as the Paleozoic Era (i.e., more than 225 m.y.)! Much of the western Pacific crust is Mesozoic, but most of the rest of the present oceanic crust is Cenozoic, which is confirmed by fossil evidence available so far from the deep seas (Fig. 9.13). (Some geologists still believe, however, that, as more and deeper drilling of the sea floor proceeds, some older fossils will eventually be found).

Extrapolations of the kind just mentioned are approximations only. In reality, spreading apparently varies considerably along even the same ridge. Authorities also believe that the ridge-axis spreading sites themselves shift laterally as spreading occurs; sometimes they jumped suddenly. Changes in magnetic anomaly patterns and slight discontinuities in sediment thickness and age distribution on several ridges suggest a possible discontinuity in spreading rate about 10 million years ago (early Pliocene). On the East Pacific Rise, only Pleistocene sediments have been found on the crest, whereas Pliocene and Miocene deposits occur on the flanks. A late Mesozo-

ic period of active spreading apparently was followed by a mid-Cenozoic slackening, in turn superseded in Miocene time by the present spreading cycle.

PLATE TECTONICS

We are presented with a seemingly simple contrast of crustal strain in the major tectonic features of the crust. As noted before, ocean ridges and the volcanic arc-trench systems are overwhelmingly the most seismically active zones today (Fig. 9.13). In the middle 1960s it was argued that earthquakes define boundaries of huge, rigid plates on the outer earth. The structure of island arcs and ancient mobile belts has long suggested great zones of dominant compression, and now the oceanic ridges (together with continental rift zones) suggest extension. Such motions are confirmed by earthquake motions (Fig. 9.14). At present the broad plates between these zones are relatively stable.

An American student of ocean basins, R. S. Dietz, was an early advocate of sea-floor spreading and was one of the first to speculate, in 1963, about an all-encompassing tectonic theory based upon spreading. He suggested that the push of a spreading sea floor against a continental margin could—at least in some cases—bend down the continental margin, crumple the sediments there, and eventually upheave

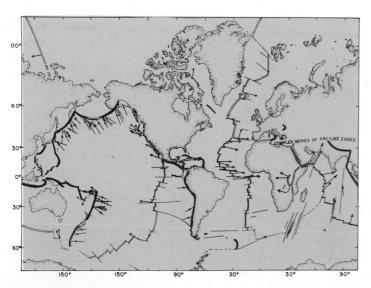

FIGURE 9.14 Slip vectors from earthquake studies. Arrows show relative horizontal motion of crustal blocks. Oceanic ridges are denoted by double lines; arc-trench systems by heavy single lines. Note that motions in general appear to be *away from ridges* (extension), but *toward arcs* (compression). *(After Isacks, Oliver, and Sykes, 1968; by permission of* Journal of Geophysical Research.*)*

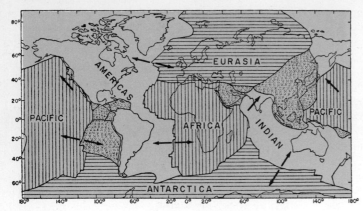

FIGURE 9.15 Eight major lithosphere plates of today (arrows indicate inferred directions of movement of plates). The plates include the crust and from 20 to 40 kilometers of the upper mantle. Their margins are delineated by the earth's major zones of earthquake and volcanic activity; note that some plates include both oceanic and continental portions that are physically coupled together. *(Adapted from J. T. Wilson, 1968, Proceedings of the American Philosophical Society, v. 112, pp. 309–320; by permission of The American Philosophical Society.)*

them; metamorphism and partial melting of the deeper sediments also were postulated, thus producing all the characteristic features of mountain belts. Dietz envisioned that new material was plastered onto continents from the ocean basins to produce gradual enlargement. The continents also could be passively swept along by spreading in conveyor-belt fashion; thus continental drift became a *by-product* of sea-floor spreading. Other workers soon argued that some continental margins (such as the Atlantic margin of the Americas today) are firmly linked or coupled to their adjacent spreading oceanic crust and so move quite passively with it and are not presently being deformed (Fig. 9.6). Such margins are *passive trailing edges* of continents. Only along the *active leading edges* of plates is active mountain building in progress (for example, the Pacific margin of South and Central America). Also, however, oceanic crust may be crushed against and under *other* oceanic crust, as appears to be the case in the western Pacific volcanic island arc-trench systems (Fig. 9.14).

In 1968 a group of seismologists at Columbia University drew together sea-floor spreading and continental drift into a broad, unifying theory called plate tectonics. Based largely upon new seismic data from all over the world, they argued that the seismic zones (Fig. 9.13) define five or six large lithosphere plates comprising the earth's outer 200 kilometers or so (Fig. 9.15). Each plate seems to behave as a rigid unit—moving away from a neighbor at an ocean ridge—a divergent plate boundary—and toward another plate along a volcanic arc and trench—a convergent boundary (Fig. 9.13). Different plates also

may touch at triple junctions or at transform faults. Not only do earth motions during earthquakes and the folds and thrust faults seen in island arcs and young mountains suggest convergence, but in the 1930s it was found that the deeper quakes (> 100 kilometers) associated with the arcs define a zone dipping 30 to 45° beneath an arc (Fig. 9.16). A recently detected speedup of seismic waves beneath this inclined seismic zone indicates the presence there of a huge slab of more rigid, apparently colder rock material extending downward into the mantle to a maximum depth of 700 kilometers. An idea from the 1930s of a huge inclined zone of shearing as much as 100 kilometers thick beneath volcanic arcs was thus confirmed, and, even more dramatic, the thrusting—called subduction—of a slab of oceanic material approximately 100 kilometers thick beneath the arc seemed inescapable. New high-resolution seismic profiles actually reveal the fault beneath trenches (Fig. 9.17). Thus, divergence of oceanic plates at ridges is indeed accompanied by subduction of the front of the plate at its convergent margin. The subducted cold material is gradually heated and probably is very slowly assimilated into the mantle below 700 km. Within the upper side of the downgoing slab, friction produces the frequent earthquakes observed beneath arcs, and local melting either of adjacent mantle material or of the downgoing plate itself seems to give rise to the magmas erupted from volcanoes of the arc (Fig. 9.16). But these magmas are not basaltic as on the ridges. Rather, fractional melting deeper in the mantle beneath arcs apparently produces magma richer in K, Na, and SiO_2 than does melting under ridges. These andesitic arc magmas are "continental" in average composition so that the accretion or addition of new continental rock material really does occur in these young mobile belts.

Seismology has also shown that the moving plates involve much more than the crust. The lithosphere plates are on the average about 100 kilometers thick, thus including the crust as a thin superficial rind as well as a much thicker portion of the upper mantle down to the low-seismic-velocity zone at about 60 kilometers (or asthenosphere of Fig. 9.18). That zone, recall, because of the decrease of seismic velocity within it, must be of lower strength than the

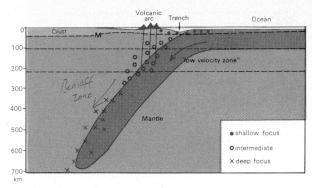

FIGURE 9.16 Relations of depths of earthquake foci to a typical island arc and trench and to *seismic low-velocity zone* of the upper mantle (only mobile belts suffer *deep-focus* quakes). Most lavas in arcs seem to originate within the low-velocity zone as a result of earthquakes. Inclined nature of the *focal zone*, together with refined studies of seismic records, indicate that lithosphere is being shoved downward beneath trench and arc along a *subduction zone. (Adapted from Dickinson and Hatherton, Science, v. 157, 18 August 1967, pp. 801–803; copyright 1967 by the American Association for the Advancement of Science.)*

FIGURE 9.17 Seismic profile across the Java trench, Indonesia, showing inclined subduction zone extending from below "trench" at left to 11 kilometers below sea level at right edge. Below this zone is subducted oceanic basement; above are complexly thrust-faulted and contorted Cenozoic strata. Cross section above seismic profile shows its location near center of section. *(Courtesy R. Beck, Royal Dutch Shell, The Hague; previously published in* Australian Petroleum Exploration Association Journal, *1972, v. 12, pp. 7–28.)*

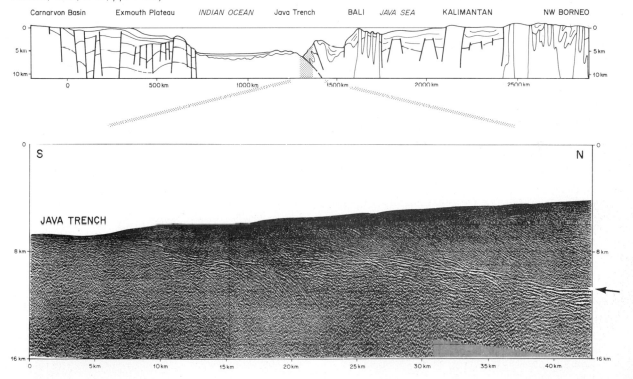

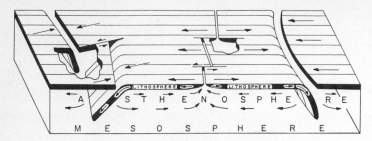

FIGURE 9.18 Diagrammatic summary of sea-floor spreading at central ocean ridges (accompanied by transform faulting) and bending of more dense *lithosphere plates* (crust plus upper mantle) at marginal arcs along subduction zones down into an inferred more plastic, less dense zone of the mantle (*asthenosphere*). This model seems best to fit all recent seismic and other data. *(After Isacks, Oliver, and Sykes, 1968; by permission of* Journal of Geophysical Research.*)*

plate above or the deeper mantle below (mesosphere of Fig. 9.18). We must now think not of the crust as the fundamental unit of the outer earth, but rather of much thicker lithosphere plates all in motion relative to one another.

WHAT DRIVES THEM?

The driving of the plates is attributed primarily to convection—our old panacea—within or below the low-seismic-velocity zone. An illustrative analogy for plate motions is provided by lava lakes in Hawaiian craters (Fig. 9.19). Dense, chilled lava crusts form on the surface and tend to sink, allowing less dense, still-molten convecting lava to rise as ridges. Lava in these ridges becomes chilled, is pushed aside, and in turn sinks slowly into the underlying molten pool to be reassimilated. Geophysical studies of the present stress across the interiors of the relatively rigid lithosphere plates indicates that they are experiencing compression within. The large but thin plates composed of rather weak rocks are like huge sheets of tissue paper, and so it is impossible that they can be either simply *pushed* outward from ridges or *pulled* downward from beneath trenches (although gravity certainly must help once a plate starts down). This must mean that they are being *dragged* passively along piggyback fashion as by convection within the underlying mantle. Such dragging would produce modest compression within the main parts of the overlying plates.

From isostasy, we can readily see that a plate with a very thin, oceanic crustal rind would more readily sink along a break than would a plate with a thick, light continental rind, which would make up about one-third of the total plate thickness. The subduction of oceanic beneath continental material can be readily understood, as can the subduction of one plate with a *thin* oceanic rind beneath another of the same type as is happening in the western Pacific. It is argued, however, that if two plates with thick continental rinds impinge, an awesome collision will result, but neither can be subducted significantly beneath the other because of their great isostatic buoyancy. (There are a few cases, however, of at least local subduction of continental material.)

Collisions between plate margins, especially between two continental masses, are then of great importance. Along any old mobile belt that is *presently* intercratonic, it appears that two continents collided (e.g., the Himalayas). In these cases, intensely compressed and metamorphosed ultramafic rocks and basalts commonly define a suture zone where two plates collided. Such zones are all that is left at the surface of a former oceanic zone that separated the two continents before collision. *Apparently these narrow, mangled zones in continents are the only relics we can expect of any sea floor older than about 200 million years!* We find, then, that convergent plate boundaries are destructive and that staggering areas of the earth's surface of the past have completely disappeared. It seems inescapable that several hundred billion cubic kilometers of lithosphere have been consumed in the past 3 to 4 billion years.

HOT MANTLE PLUMES

Cratonic rift structures apparently represent the initial stage of continental drifting, as we have suggested. The Red Sea–Aden rift system is especially revealing because the Carlsburg spreading ridge extends from a triple junction with the Indian Ocean ridge directly into the Gulf of Aden. A recent hypothesis of continental breakup involves postulated hot mantle plumes rising beneath the continent. Such plumes are thought to be reflected at the surface by midplate volcanic centers 100 to 200 kilometers in diameter that are associated with uplifted areas. They are not associated with magmatic arcs and may or may not be associated with ocean ridges. About 200 possible late Cenozoic plumes have been identified for the earth as a whole, with Africa being especially well endowed. The Hawaiian Islands are the best-known probable example in an oceanic plate.

The suggested evolution of rifting and continental drifting by rising hot mantle plumes is as follows. First, the uplift of a flat, brittle plate results in the formation of three (or more) radiating cracks. Then, as extension of the cracks continues, two arms of the trio tend to open up to form a new ocean basin. The third arm, which is nearly perpendicular to the opening seaway, becomes a sediment-filled graben (called aulacogen by Russian geologists). It gradually becomes stable and is buried. The southward-trending Ethiopian rift seems to be such a blind arm of the Red Sea–Aden spreading arms. While the Ethiopian–East African rifts are tectonically active, they certainly have not opened up as has the latter system. The mantle plume idea is a most ingenious one and will deserve further attention, but at the time of this writing, theory was advancing faster than the accumulation of new facts in support of it.

GLUING TOGETHER BROKEN PLATES

Because sea-floor spreading and the all-encompassing plate theory were derived from studies of

FIGURE 9.19 Spreading center defined by molten lava (light line from top to bottom) in Mauna Ulu lava crater, Hawaii, during a 1971 eruption. As fresh molten lava rises by convection, slightly older chilled crust is shoved aside to sink at some other (compressive) point in the crater. Note the transform fault near middle of view and the differential rate of spreading revealed by the different widths of recently chilled crust on either side of the spreading line. Front edge of view is about 50 meters wide. (Courtesy W. A. Duffield, U.S. Geological Survey Hawaiian Volcano Observatory; see Duffield, 1972, Journal of Geophysical Research, v. 77, pp. 2543–2555.)

EVOLUTION OF THE EARTH

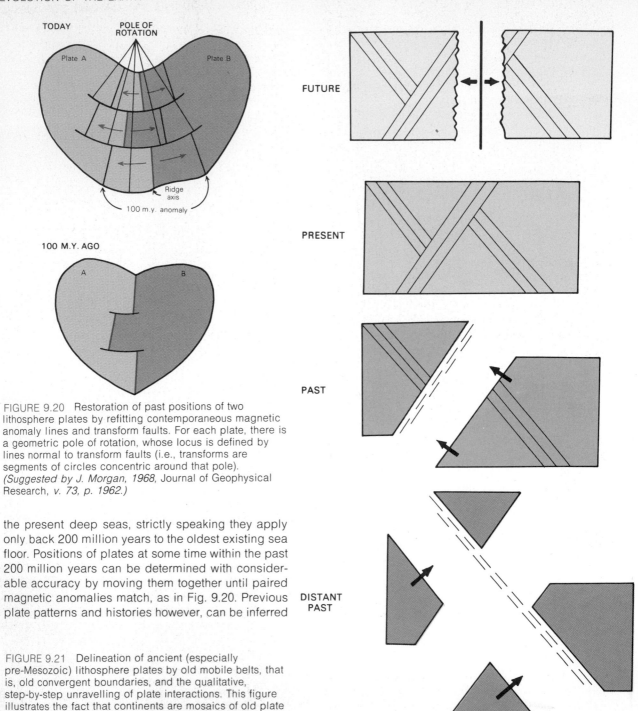

FIGURE 9.20   Restoration of past positions of two lithosphere plates by refitting contemporaneous magnetic anomaly lines and transform faults. For each plate, there is a geometric pole of rotation, whose locus is defined by lines normal to transform faults (i.e., transforms are segments of circles concentric around that pole). *(Suggested by J. Morgan, 1968, Journal of Geophysical Research, v. 73, p. 1962.)*

the present deep seas, strictly speaking they apply only back 200 million years to the oldest existing sea floor. Positions of plates at some time within the past 200 million years can be determined with considerable accuracy by moving them together until paired magnetic anomalies match, as in Fig. 9.20. Previous plate patterns and histories however, can be inferred

FIGURE 9.21   Delineation of ancient (especially pre-Mesozoic) lithosphere plates by old mobile belts, that is, old convergent boundaries, and the qualitative, step-by-step unravelling of plate interactions. This figure illustrates the fact that continents are mosaics of old plate fragments laced by ancient convergent (mobile) zones of varying ages.

only from continents, and of course the evidence becomes much more scant as we proceed back through history. Plate theory tells us that all ancient mountain or mobile (orogenic) belts represent former convergent plate boundaries, because it is believed that the impingement of plates is the mechanism of mountain building. Therefore, we can use old mountain belts to define old convergent boundaries (Fig. 9.21), and with the help of rock magnetism and every available geologic tool for reconstructing ancient geography, we may try to restore ancient plate relationships.

The present plate pattern has lasted about 200 million years so far, but significant changes of direction and rates of motion have occurred during that time. It is now generally assumed that the basic lithosphere plate mechanism has operated for at least the last half of earth history (2.5 b.y.), but plate boundaries and patterns of motions have changed drastically several times. Six or seven episodes of extreme mountain building on a nearly worldwide scale probably record wholesale reorganizations of plates. If so, then the average duration of a certain plate pattern seems to be between roughly 300 and 500 million years. Also we see that continents are mosaics assembled from pieces of plates of different ages broken apart and glued together in a crazy-quilt fashion. Small wonder that it has taken geologists 200 years to devise a truly comprehensive theory of global tectonics.

SUMMARY

Oceanographic research since World War II culminated in the 1960s with a twentieth century revolution in the earth sciences equal in importance to the seventeenth century newtonian revolution in physics and astronomy and the nineteenth century Darwinian revolution in biology. The plate-tectonics theory was accepted with almost unprecedented rapidity in a mere 5-year period. Only the continental glacial theory proposed in the 1840s was accepted with anything approaching such rapidity. Why should any theory be adopted so promptly and widely? Apparently it was because so much diverse evidence had been accumulating over a long period that the time was ripe for a general explanation.

The greatest surprises from postwar oceano-graphic work were the proof that the deep-sea floor is at most only about 200 million years old and that it is structurally very active. We now know that only the continents contain any pre-Mesozoic historic record, whereas it was long assumed that the deep seas would yield an unbroken sedimentary and paleontologic record dating back to near the beginning of the earth. Deep-sea topography, sediment distributions, and magnetic anomaly patterns first provided evidence that the sea floors are spreading along the ocean ridges. Then in the late 1960s, seismic evidence proved not only that the sea floor is moving away from the ridges, as postulated for spreading, but also that it is moving toward the trenches. Finally, seismology also provided evidence that cold lithosphere slabs are being subducted beneath the trench-arc systems.

The popularity and excitement of plate tectonics lies in the fact that it explains so much in a relatively simple, unifying manner. It allows one for the first time to envision relations of very different features scattered widely over the globe as parts of one, grand dynamic scheme on a chemically evolving planet.

No generalization is worth a damn—including this one. (Oliver Wendell Holmes)

Readings

American Geophysical Union, 1972, Plate tectonics: collected papers from the Bulletin of the American Geophysical Union.

Burk, C., and Drake, C., 1974, The geology of continental margins: Heidelberg, Springer-Verlag.

Bird, J., and Goodman, K., 1976, Global tectonics: New York, McGraw-Hill. (Paperback)

Burke, K., and Dewey, J. F., 1973, Plume-generated triple junctions: key indicators in applying plate tectonics to old rocks: Journal of Geology, v. 81, pp. 406–433.

Dewey, J. F., and Bird, J. M., 1970, Mountain belts and the new global tectonics: Journal of Geophysical Research, v. 75, pp. 2625–2647.

Dietz, R. S., 1963, Collapsing continental rises: Journal of Geology, v. 71, pp. 314–333.

Hallam, A., 1973, A revolution in the earth sciences: Oxford, Oxford.

Heezen, B. C., 1960, The rift in the ocean floor: Scientific American, v. 203, pp. 98–110.

———— 1972, The face of the deep: Oxford, Oxford. (Nearly 600 underwater photos)

Heirtzler, J. R., 1968, Sea floor spreading: Scientific American, v. 219, December, pp. 60–70.

————, 1975, Where the earth turns inside out: Project Famous—man's first voyages down to the Mid-Atlantic ridge: National Geographic, v. 147, pp. 586–615. (Excellent sea floor photographs)

Hess, H. H., 1962, History of ocean basins, *in* Petrologic studies: a volume in honor of A. F. Buddington: Geological Society of America.

Irvine, T. N., ed., 1966, The world rift system, an international upper mantle project report: Ottawa, Geological Survey of Canada Paper 66–14.

Isacks, B., Oliver, J., and Sykes, L. R., 1968, Seismology and the new global tectonics: Journal of Geophysical Research, v. 73, pp. 5855–5899.

Menard, H. W., 1964, Marine geology of the Pacific: New York, McGraw-Hill.

———— 1969, The deep-ocean floor: Scientific American, v. 221, pp. 127–137.

Mitchell, A. H., and Reading, H. G., 1969, Continental margins, geosynclines, and ocean floor spreading: Journal of Geology, v. 77, pp. 629–646.

Orowan, E., 1969, The origin of the ocean ridges: Scientific American, v. 221, November, pp. 103–119.

Seyfert, C. K., and Sirkin, L. A., 1973, Earth history and plate tectonics: an introduction to earth history: New York, Harper & Row.

Sullivan, W., 1974, Continents in motion: New York, McGraw-Hill.

Takeuchi, H., Uyeda, S., and Kanarmori, H., 1967, Debate about the earth; approach to geophysics through analysis of continental drift: San Francisco, Freeman, Cooper.

Vine, F. J., 1966, Spreading of the ocean floor: new evidence: Science, v. 154, pp. 1405–1415.

———— and Matthews, D. H., 1963, Magnetic anomalies over ocean ridges: Nature, v. 199, p. 947.

Wilson, J. T., 1965, A new class of faults and their bearing on continental drift: Nature, v. 207, p. 343.

Wyllie, P. J., 1971, The dynamic earth: New York, Wiley.

10

PREPALEOZOIC HISTORY

AN INTRODUCTION TO THE ORIGIN OF CONTINENTAL CRUST AND THE EARLY HISTORY OF LIFE

Mente et Malleo
By thought and dint of hammering
Is the good work done whereof I sing,
And a jollier crowd you'll rarely find,
Than the men who chip at earth's old rind,
And often wear a patched behind,
By thought and dint of hammering.

Andrew C. Lawson,
formerly of
the Geological Survey of Canada
and University of California

FIGURE 10.1 Early working conditions on Canadian Shield. Geologic party in large bark canoe on Lake Mistassini, Quebec, 1885. (*Courtesy Geological Survey of Canada.*)

Prepaleozoic time, or the Precambrian as it is commonly termed, included about 80 percent of total earth history, that is, from nearly 5 billion years to about 700 or 800 million years ago. Yet for the first 1.0 to 1.5 billion years or so, there is no decipherable geologic record, and it now seems unlikely that much will be found. Apparently the crust had not developed sufficiently to become permanent until about 3.7 billion years ago. In Chap. 7 we outlined the probable earliest history and some of the consequences of such a history. Beginning in this chapter, we shall concentrate attention upon North America as a tentative model of continental development. First, we shall examine in some detail the preserved geologic record for the interval from about 3.5 billion to 0.7 billion years ago, i.e., the Prepaleozoic. Although we find it logically more sound to refer to the first three-quarters of earth history as Prepaleozoic time, you should realize that to most geologists it is simply "Precambrian."

The economic importance of Prepaleozoic rocks is almost inestimable. Thousands of years ago Indians mined copper in Michigan and traded it widely over the continent. Since the industrial revolution, their inheritors have derived untold more wealth from these old rocks. The major source of iron ore is from peculiarly banded "iron formations," which are unique to the Prepaleozoic. Other major metallic sources of gold, silver, copper, nickel, chromium, and uranium, to list but a few, also are enclosed in these ancient rocks. Perhaps half of the world's metallic mineral resources comes from Prepaleozoic deposits. This seemingly unlimited reservoir is, however, finite, and our increasing demands upon it are cause for alarm. Therefore search for more goes on.

At the outset, we must acknowledge that the Prepaleozoic record is far more obscure than that for subsequent time. To be sure, many Prepaleozoic rocks are severely deformed, metamorphosed, and deeply eroded (Fig. 10.2), but others are almost as youthful appearing as Cenozoic ones. Overwhelmingly the most important single characteristic of the older record is its lack of index fossils. As we saw in Chap. 2, the first use of index fossils for correlation and mapping was the major breakthrough that allowed the construction of a valid geologic time scale. But in Prepaleozoic rocks, so far we have been denied this important tool.

Considering the handicaps, it is remarkable what the pioneers of Prepaleozoic geology accomplished in unravelling a chronology that had to be based solely upon physical stratigraphic criteria of relative age. It was chiefly those pioneers who perfected the

FIGURE 10.2 Banded high-grade metamorphic rocks exposed by glaciation along Sondre Stromfjord, southwestern Greenland. Some of the oldest dated rocks in the world (3.7 b.y.) occur near here. It is a question of long standing whether or not banding in such rocks represents relict stratification of original sediments. For many years, it was assumed that most Prepaleozoic rocks were of this sort and very old. In reality much of the Prepaleozoic record is far less metamorphosed.

use of such criteria to a high order. More recently, isotopic dating has revolutionized the study of the Prepaleozoic even more than of younger eras. In general, it has confirmed much of the basic chronology established purely by field geologic methods.

DEVELOPMENT OF A PREPALEOZOIC CHRONOLOGY

SEDGWICK IN WALES

You may recall from Chap. 4 that Adam Sedgwick first suggested that the eras, largest divisions of the geologic time scale, be named for relative differences of life development. He also recognized clearly the relations of older, unfossiliferous rocks to fossiliferous early Paleozoic ones in Wales. One of the first names proposed for Prepaleozoic time was Azoic, meaning "without life"; subsequently, Eozoic and Archeozoic ("ancient life"), and Cryptozoic were suggested when presumed fossils were found. The most amusing term was Agnotozoic, proposed by a Wisconsin geologist who doubted that alleged fossils were truly organic.

Sedgwick recognized that Prepaleozoic rocks in Wales are somewhat more deformed and metamorphosed than overlying Paleozoic ones, from which they are separated by an unconformity. In many areas of the world, however, this is not so, and the two are virtually identical in appearance except for near-absence of fossils in the older. Because upper Prepaleozoic strata are not different from those of the lower Paleozoic, designation of a boundary is somewhat arbitrary where no unconformity is present, as is the case in many of the earth's ancient mobile belts. In general the lowest stratigraphic appearance of Cambrian index fossils has defined that boundary, but no human being was around to paint a stripe on the rocks for us; so, in continuous or conformable sequences of strata, what assurance is there that Cambrian index fossils appear at a position synchronous with their lowest position in Sedgwick's Welsh sequence? What if environmental or ecologic factors were unfavorable for the organisms in certain areas at the precise moment when the curtain raised on Act 1, the Cambrian Period, in Wales? Moreover, in strongly deformed or metamorphosed strata, precise location of this boundary is almost hopeless. We may hope that some day a criterion other than fossils will be found for defining this time boundary; even today it is selected by isotopic dating in some areas.

THE CANADIAN SHIELD

Though Prepaleozoic rocks were first recognized formally in Britain, the important development of a Prepaleozoic chronology occurred largely in the Great Lakes region of America. Later it was worked out in Scandinavia as well.

The Prepaleozoic rocks on all continents are most widely exposed in the stable continental cratons. The region of more or less uninterrupted Prepaleozoic exposures in North America occupies primarily the eastern two-thirds of Canada, the United States margins of Lake Superior, and most of Greenland (Fig. 10.3). It is called the Canadian Shield. Younger strata also covered most shields in the past; therefore, shields are largely accidents of erosion wherein the stripping off of later deposits has exposed the Prepaleozoic basement (Fig. 10.3). The tectonic term craton is more useful because it defines the overall relative structural stability of a large portion of the earth's crust through a long time interval regardless of what aged rocks are exposed there today.

THE GREAT LAKES REGION

The first pioneer to probe the geologic secrets of the Canadian Shield was Sir William Logan, who in 1842 established the Geological Survey of Canada. Working conditions demanded heroic efforts, for much of the country was barely penetrable. In bush areas, transport was largely by canoe or on foot (Fig. 10.1); occasionally, boats had to be fashioned on the spot. Logan's successor as Director of the Survey even

FIGURE 10.3 Shields and other major exposures of Prepaleozoic rocks of the world; Prepaleozoic rocks extend beneath all white areas of the continents. (*Bartholomew's Nordic Projection; with permission of John Bartholomew, Edinburgh.*)

SHIELDS AND OTHER MAJOR AREAS OF EXPOSED PREPALEOZOIC ROCKS

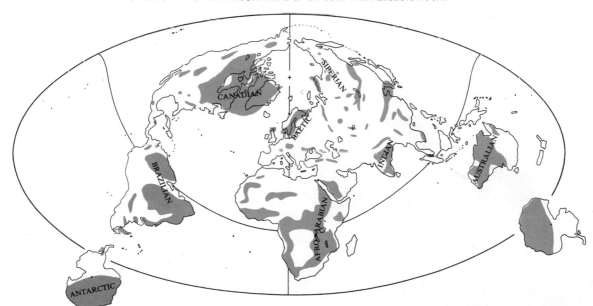

suffered the heartbreaking indignity of having a faithful, but hungry, horse eat an entire field notebook at the end of a summer season.

Logan felt that granitic gneisses near Ottawa were among the oldest rocks of the continent, although they actually cannot be dated satisfactorily by field evidence alone. North of Lake Huron, Logan and his successors found and named several divisions of sedimentary, metamorphic, and igneous rocks. The oldest rocks there stand in vertical positions, making it difficult to determine their original superpositional sequence. A number of sedimentary, volcanic, and structural features allow distinction of original top and bottom in such sequences, and shield geologists perfected their application. North of Lake Huron the oldest known rocks so revealed seemingly were metamorphosed volcanic ones rather than granites (Fig. 10.4).

In 1882 a young man named Andrew C. Lawson (author of the poem at the beginning of this chapter) joined the Geological Survey of Canada, and established a chronology for the Ontario-Minnesota border area. He proved conclusively that the oldest recognizable rocks are metamorphosed andesitic and basaltic volcanic ones (named Keewatin; Fig. 10.5). The most primitive crust was not granitic as had been assumed since pre-Wernerian times!

As mapping progressed in Minnesota, younger rock divisions were recognized and named (Fig. 10.5). The work was stimulated by discovery of iron ore near Marquette, Michigan about 1850, and the start of iron mining in northeastern Minnesota in 1884. The Minnesota ore-bearing rock (Fig. 10.6) occurs within a sequence of metamorphosed lavas called "greenstone." Outcrops are very discontinuous in the swampy, forested country around Lake Superior; therefore, tracing of peculiar rock types, which serve as marker datums, facilitated mapping. Much of the iron ore is strongly magnetic and first was located because of the erratic behavior of compasses; for the same reason, iron-bearing strata were easily traced beneath concealed areas.

Strata containing iron ores south of Lake Superior soon were correlated with those of Minnesota, although in the former area they have been much more severely disturbed. Several features were important in determining superposition in the rocks; among these were graded bedding (Fig. 10.7), cross stratification (Fig. 10.8), as well as pillow structure in volcanic rocks (Fig. 10.4). Studies in this region were carried out largely by an army of United States Geological Survey workers under the guidance of C. R. Van Hise, later to become president of the University of Wisconsin. Copper, which has been mined in Michigan

FIGURE 10.4 Ellipsoidal or "pillow" structures in Early Prepaleozoic greenstones 15 kilometers west of Marquette, Michigan. These metamorphosed lavas now are vertical, but pimple-like protrusions of the left sides of several ellipses indicate the original bottom direction. Spherical or cylindrical lava masses congeal quickly, especially when quenched by water. While cooling, ellipsoids are formed by flattening, but protrusions are squeezed downward only (compare by rotating page).

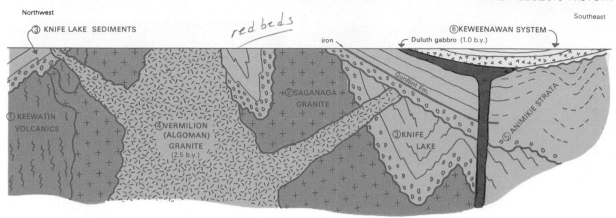

Northwest

③ KNIFE LAKE SEDIMENTS

red beds

iron

⑥ KEWEENAWAN SYSTEM

Duluth gabbro (1.0 b.y.)

Southeast

① KEEWATIN VOLCANICS

② SAGANAGA GRANITE

④ VERMILION (ALGOMAN) GRANITE (2.5 b.y.)

Gunflint Fm.

③ KNIFE LAKE

⑤ ANIMIKIE STRATA

SECTION—WESTERN ONTARIO—MINNESOTA

FIGURE 10.5 Cross section for the north shore of Lake Superior to northern Michigan. Stratigraphic relationships studied by A. C. Lawson and early United States geologists are shown, as well as isotopic dates. Prepaleozoic formations have been richly endowed with memorable—if unpronounceable—Indian names.

longer than iron, was an additional incentive for their work. It occurs in a thick succession of gently northward-dipping basalts and red-colored clastic sediments in the Keweenaw Peninsula of Michigan. A similar succession occurs on the north shore of Superior, where it dips south (Fig. 10.5), thus forming a broad syncline beneath the lake (Fig. 10.9). Simple structure and distinctive rock types made correlation of these Keweenawan rocks across the lake simple. Where structure is much more complex, however, correlation and historical interpretation is far more difficult.

GREAT LAKES CORRELATIONS

From stratigraphy alone, four major divisions of sedimentary and volcanic rocks were recognized. They were known to be separated by unconformities that reflected at least four major orogenies during which extensive granites were formed (Fig. 10.10). The orogenies punctuate the rock record in such a way as

FIGURE 10.6 Contorted Early Prepaleozoic Soudan Iron Formation near Armstrong Lake, 15 kilometers east of Soudan, Minnesota. Note the characteristic prominent alternating bands of chert (light) and magnetite (dark). It was called the Soudan Iron Formation for the Soudan Mine, which was named (though misspelled) for the Sudan in Africa, the hottest place that some wag miner could think of after his first frigid Minnesota winter. (Courtesy of Carl E. Dutton.)

EVOLUTION OF THE EARTH

FIGURE 10.7   Graded bedding in an Early Prepaleozoic graywacke composed of volcanic detritus, Lake Vermilion Formation, Tower, Minnesota. The grains settled from a fluid in order of decreasing mass, and the gradation of size indicates the direction of the original top of the strata, where they are now vertical or overturned. (*Courtesy R. Ojakangas, University of Minnesota, Duluth.*)

to provide several natural divisions (Table 10.1). But there were some thorny problems of correlation that could be resolved only through isotopic dating.

Isotopic dating has helped immensely to anchor age relations all around the Great Lakes. The numbers in Table 10.1 indicate the most important dates, and make possible a workable standard Prepaleozoic chronology for North America.

CORRELATION BEYOND THE GREAT LAKES

Relative degree of deformation and of metamorphism commonly have been used to argue that certain rocks were very old or very young. Great caution is required in exercising such criteria because many misinterpretations are on record. To wit, recall the neptunian error in regarding all metamorphosed rocks as Primitive when, in fact, some of them turned out to be Mesozoic and even Cenozoic. The assumption that severe disturbance is necessarily related to great

FIGURE 10.8   Medium-amplitude cross stratification in the Baraboo Quartzite of southwestern Wisconsin (probably about 1.5 b.y. old). This structure typifies sands transported by vigorous water or wind currents; it represents preserved internal laminae of migrating dunes. Cross strata tend to be sharply truncated at their tops (compare Fig. 10.17). (Pencil at top shows scale.)

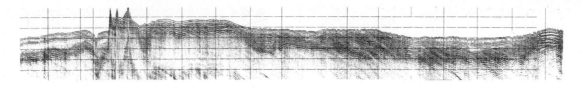

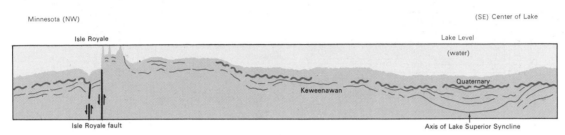

Minnesota (NW)

(SE) Center of Lake

Isle Royale

Lake Level

(water)

Quaternary

Keweenawan

Isle Royale fault

Axis of Lake Superior Syncline

FIGURE 10.9   Subbottom seismic profile across Lake Superior. Such a profile is an electronic recording of sound pulses reflected from various buried modern-sediment and ancient-rock layers. Low-frequency sound is generated in the water by an electric spark or explosives; it is reflected and recorded much as are radar and sonar pulses. Note fault that displaced Keweenawan rocks at left and clear reversal of dip at the axis of the Lake Superior syncline at right. (*Courtesy Richard Wold.*)

FIGURE 10.10   Giants Range Granite (light) intruded into Early Prepaleozoic Knife Lake sediments (dark), north of the Mesabi Iron Range, Minnesota. The cross-cutting dikes clearly date the granite as *relatively* younger than the sediments (isotopic dating indicates that this granite is about 2.5 b.y. old). (*Courtesy Carl E. Dutton.*)

TABLE 10.1 Standard chronology of Prepaleozoic events for the Canadian Shield

Great Lakes states				Canadian Classification	
Paleozoic		Eocambrian System	0.6	Hadrynian Era	
			0.7		
Prepaleozoic or Precambrian	Proterozoic — Late era	"Grenville" Orogeny		Helikian Era	Grenville Orogeny
		Keweenawan System	1.3		Keweenawan System
		Unnamed orogeny			Elsonian Orogeny
		Unnamed system	1.8		Unnamed
	Middle era	Penokean Orogeny		Aphebian Era	Hudsonian Orogeny
		Animikean and Huronian Systems	2.4		Animikean and Huronian Systems
	Archeozoic — Early era	Algoman Orogeny		Archaean Eon	Kenoran Orogeny
		Timiskamian System	3.0		(Diverse local sequences)
		Saganagan Orogeny			
		Keewatian	3.5		

antiquity is too simple, for geographic or structural location may be more important. Cenozoic rocks in the Swiss Alps are far more deformed than are the late Prepaleozoic ones around Lake Superior. And Logan's "old" gneisses near Ottawa proved to be younger than the mildly folded Late Prepaleozoic strata on the shores of Lake Superior. More significant than age is whether a given group of rocks was deposited within a mobile belt or a craton.

Before the advent of isotopic dating, it became popular to think of the Prepaleozoic rocks as falling broadly into two groups, an older, intensely metamorphosed one (Archeozoic) and a younger, less-disturbed one (Proterozoic). Far beyond the Great Lakes region, rocks were identified with one or the other of these solely on the basis of metamorphism and structure. Many errors were committed that could be corrected only through isotopic dating, which provides the only valid basis for long-distance time correlations of Prepaleozoic rocks.

With the increase in isotopic dating, it has become clear that the continent contains several distinct isotopic date provinces (Fig. 10.11), that is, large

FIGURE 10.11 Isotopic-date provinces of North America, showing regions within which respective ranges of plutonic and metamorphic dates cluster on a statistical basis. Gross structural trend or grain within each province also is shown. Several provinces overlap considerably; others are markedly discordant. Dates designate major periods of orogenesis.

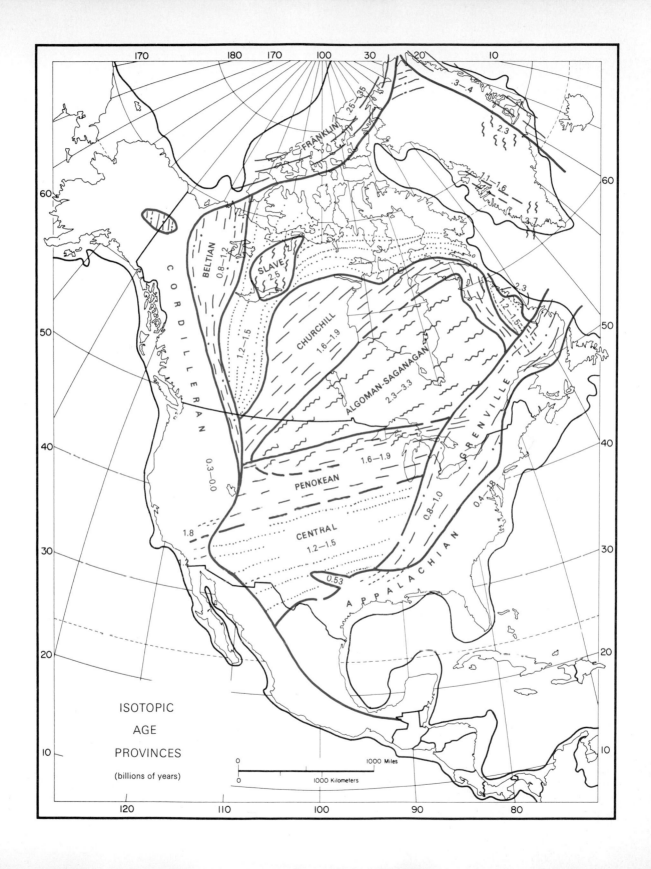

ISOTOPIC
AGE
PROVINCES
(billions of years)

regions within which a certain range of dates predominates. Such provinces are by no means simple, and they typically yield a considerable total range of ages, but statistically each has a discrete cluster of dates falling within about a 0.3 billion-year interval. It is important to remember that most of the dates upon which Fig. 10.11 is based are dates of orogenic events; thus each province is a mosaic of related mobile belts. The rocks found in these old belts are composed of volcanic materials, sediments, and large masses of granitic and high-grade metamorphic rocks, all of which attest to long-continued mobility. Many of the rocks have been metamorphosed more than once, thus their "isotopic clocks" may have been reset several times. Dates obtained from such rocks generally record only the last readjustment of isotopes during an episode of heating. Rarely the rocks retain two or more discordant dates, which reveal a complex history of multiple orogenies.

Most of the oldest isotopic dates (2.5 to 3.8 b.y.) tend to occur in the center of the continent from western Ontario southwest to Montana and Wyoming. But at least one other nucleus of similar antiquity also occurs in the far northwest corner of the shield. Surrounding each nucleus are more or less concentric mobile belt provinces of younger age.

As isotopic dating of mobile belts has supplied more data, it has become clear that the dates from all continents tend to cluster in a roughly similar way. The groups seem to represent crudely synchronous major granite-forming and metamorphic episodes. Such peaks may eventually provide a basis for a worldwide standard Prepaleozoic time scale. Different mobile belts existed for about 0.8 to 1.2 billion years, during which several periods of thick (geosynclinal) sedimentation, deformation, metamorphism, and granite formation occurred. The individual tectonic events, with periods of from 0.2 to 0.4 billion years, have been blended together to make up each of the Prepaleozoic age provinces.

INTERPRETATIONS OF CRUSTAL DEVELOPMENT FROM IGNEOUS AND METAMORPHIC ROCKS

THE IMPORTANCE OF GRANITE

Recall from Chap. 8 that, in light of the long erosional history of continents and of isostasy, some rejuvena-

tion or addition of continental crust through time seems inescapable. Partly from these considerations, but also from the igneous rock types of the early Prepaleozoic, together with the arrangements of old mobile belts, there has grown the theory of continental accretion. Geologists in the early twentieth century realized that rocks rich in silica, aluminum, and potassium, notably granites, seemed to be unique to continents. But field evidence showed that granite-rich continental crust was not original, and so it must have increased in volume through time. Apparently the original crust was thin and composed largely of basalt. By later partial melting and redistribution of elements through weathering, erosion, and igneous activity, it was postulated that some of the original crust was converted somehow to granite to form embryonic continents. A century ago it was suggested that much granite formed at depth either by selective melting and recrystallization or by differential replacement of certain layers within geosynclinal sedimentary-volcanic sequences to produce new granitic material. Through time, more and more granitization of volcanic and sedimentary rocks presumably has occurred, and the continents have grown thicker and larger as a result. Granitic continental material, being of slightly lower average density than the remainder of the crust or mantle, stands topographically higher than the ocean basins in accord with isostasy. As accretion proceeded, basaltic lavas presumably became less important, at least in the interior regions of the growing continents.

The concept of granitization clearly has developed as a handmaiden of continental accretion (or perhaps vice versa). In fact it is presumed by many to have been the very mechanism of accretion through ultrametamorphism of sedimentary and volcanic materials to gneissic and granitic crystalline rocks (Fig. 10.2). This is not, however, to deny completely the much older Huttonian concept of intrusive granitic plutons. The "granitizationist" envisions wholesale, in-place recrystallization of stratified rocks deep in the crust, but some of the material formed was fluid or plastic enough to be squeezed up into higher crustal levels as relatively small intrusions.

There is a troublesome kind of perpetual motion implied by continental accretion through granitization. In its extreme form, granitization preached that *all* granitic rocks originated by conversion from other crustal rocks and none from truly new or "juvenile" (subcrustal) magmas. Yet if continental crust has increased appreciably in volume, it is clear that either juvenile igneous material must have been added from below, or else oceanic crust has somehow been converted to continental material. Otherwise we are forced to envision only a recycling of the same material by sedimentation, granitization, re-erosion, and redeposition as new sediments again, then reconversion to granite, and so on. No matter how often repeated, there could be no appreciable increase in continental volume in such a closed system without magically pulling rabbits out of hats.

The volcanic rocks that were erupted in mobile belts, especially andesites, seem to provide an answer. Apparently they formed from juvenile magma from the mantle and as suggested in Chap. 9 so provided new raw materials for possible later granitization. More about the origin of granite appears in Chap. 16.

SIGNIFICANCE OF ISOTOPIC DATE PATTERNS

The patterns of isotopic date provinces appear to have an important bearing upon the hypothesis of continental accretion. The more or less concentric, younger-outward zones suggest that the continents may have accreted from two (or more) old nuclei by the successive formation and ultimate consolidation of mobile belts around them (Fig. 10.11). First, the nuclei seem to have been knitted together into a single craton, which then grew still bigger through addition of newer mobile belts. It is thought that after long histories of orogenies, new granitic material was formed in each successive mobile belt, thus stabilizing and transforming the behavior of each. Deep erosion accompanying isostatic uplift reduced the mountains to low plains, exposing the bowels of the old mobile belts that we see today. Shields, then, are mosaics of complexly interlaced ancient mobile belts now stabilized within cratons and deeply eroded. It is estimated that the original depth of formation of some of the great granitic batholiths was as much as 20 to 30 kilometers.

The hypothesis of accretion presents an extremely appealing and relatively simple generalization that seems to unify many different types of geologic data and to offer a logical explanation of crustal development through lateral areal enlargement of the continents. The whole mobile-belt concept, igneous and metamorphic petrology, and overall geochemical differentiation of the earth all seem to be explained in a related manner. But as study progresses, situations generally appear to be more complex than at first supposed.

CRITIQUE OF THE PATTERNS

We have no really satisfactory knowledge of the nature of the original crust of any continent. Apparently volcanic rocks were overwhelmingly the most important until about 3.0 billion years ago. Sedimentary and metamorphic rocks can be derived only from still older rocks; therefore, it is inescapable that the very earliest crust must have been composed of some kind of igneous rock. But it is difficult to speak dogmatically of lateral accretion of a crust whose ultimate ancestry is unknown. Also, we cannot rule out the possibility of some losses of crust at the edges of continents, the margins of which are largely submerged and inaccessible to us today. Early isotopic dates gave accretion a great new boost. But interpretation from them that *wholly new* continental crust has been generated in the concentric mobile belts is tenuous. All that the concentric patterns really tell us is that preserved isotopic evidence shows that orogenies occurred in a crudely concentric fashion around multiple nuclei. Perhaps the area of continental crust did not change appreciably, but was simply worked over repeatedly by successive mobile belt developments in a crudely concentric way. This alternate working hypothesis explains the same data (Fig. 10.12).

Now examine the patterns of Figs. 10.11 and 10.13, which are necessarily simplified. Note that some orogenic zones of very different isotopic age are almost wholly *superimposed upon one another* rather than being side-by-side. The Algoman Orogeny around Lake Superior seems to have been so imposed upon the much older Saganagan mobile belt (Fig. 10.5). In other cases, particularly in Labrador, younger belts cut off older ones at high angles rather than being neatly concentric. Ghosts of reworked Middle Prepaleozoic trends can be discerned by careful field work within the younger Grenville belt, where the rocks have been almost completely reconstituted (Fig. 10.11).

In most cases where an older basement can be clearly discerned, the *mobile belts developed on a pre-existing granitic crust* and do not seem to repre-

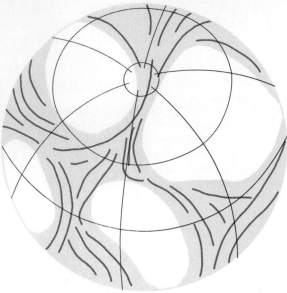

A. 2.5 b.y. ago

B. 1.8 b.y. ago

FIGURE 10.12 *A:* Hypothetical early accumulation of most continental crust (by 2.5 b.y. ago). *B:* This is followed by remobilization or tectonic reworking in successive generations of mobile belts with varying orientations. Note that this mechanism is an alternative to the old view of slow, continuous accretion of continents.

sent large additions of completely new continental material. To a great extent, then, either these Prepaleozoic mobile belts only reworked pre-existing continental crust, or else evidence of original oceanic crustal basement beneath them has been obliterated.

Returning to the argument about the persistence of continents through the past 2.5 or 3.0 billion years developed in Chap. 8, some form of crustal renewal through history seems mandatory to counterbalance denudational processes—if not, continents should disappear. It is estimated that all the sediments produced throughout geologic time represent the erosion of approximately the equivalent of the present total volume of continental crust (5000×10^6 km³). Seemingly we have to account for about one complete "turnover" of continental crust through time. Possibly accretionists have erred in thinking of continental growth almost solely in terms of area. It could be accomplished as well by slow additions at the base of the crust of continental material by differentiation (or phase transformations) of the mantle more or less in balance with the overall rate of denudation at

the top. In such manner, continents could be maintained isostatically at or above sea level and not require appreciable lateral or areal increase at all. Meanwhile, mobile belts simply may have laced and tectonically reworked old crust complexly through time with little respect for continental and oceanic boundaries. This is suggested strongly by the patterns of Figs. 10.11 and 10.13.

INTERPRETATION OF CRUSTAL DEVELOPMENT FROM PREPALEOZOIC SEDIMENTS

BIAS OF THE RECORD

Further insight into crustal history can be gleaned from Prepaleozoic sediments. The Lake Superior region contains very thick, deformed sedimentary and volcanic accumulations, which, for the most part, represent geosynclinal fillings of ancient mobile belts. Deep downfolding in the mobile belts preserved sediments from subsequent erosion. Conversely, easily eroded cratonic sequences are very poorly represented in the older Prepaleozoic record. To further confound us, many Prepaleozoic mobile belts suffered so much metamorphism that little can be said about their original sediments. Nonetheless, in the Great Lakes region it is clear that at least by

FIGURE 10.13 Prepaleozoic and earliest Paleozoic isotopic-date provinces of basement rocks of continents (middle Paleozoic and younger mobile belts shown blank emphasizing their general discordance with older mobile-belt patterns). Note complexity of discordances between provinces. Quality of data varies enormously, being least complete in South America and Asia. (*From many sources, e.g., Cahen and Snelling, 1966, The geochronology of equatorial Africa; Compston and Arriens, 1968, Canadian Journal of Earth Sciences; Gastil, 1960, 21st International Geological Congress: Holmes, 1965; Hseih, 1962; Hurley et al., 1967, Science, v. 157, op. 54–542; Jenks, 1956, International Geology Review.*)

Provinces (billion years)

Early Paleozoic
0.4–0.7

Prepaleozoic:
Late 0.8–1.2

Middle { 1.3–1.5
{ 1.6–2.0

Early > 2.0

(Maximum Reported 3.5)

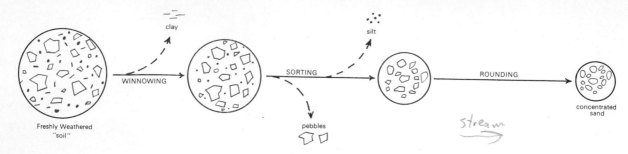

TEXTURAL EVOLUTION OF SAND

FIGURE 10.14 Idealized development of textural maturity of sand through extended abrasion and separation or sorting of different-sized grains.

Middle Prepaleozoic time, there were stable cratonic and mobile belt areas. In the following sections, we shall illustrate some important principles of interpretation of sediments, but *keep in mind that the same principles apply equally to younger rocks.*

TERRIGENOUS VERSUS NONTERRIGENOUS CLASTIC SEDIMENTS

The composition of any clastic or fragmental sedimentary rock reflects, more than anything else, the sources from which it was derived. Of course, climatic conditions may modify composition through weathering, as may chemical changes after deposition, but for the present we shall ignore these complications. It is necessary to distinguish terrigenous clastic sediments chiefly of silicate minerals such as quartz,

FIGURE 10.15 Comparison of relative sorting of sands by different processes (as measured by the statistic *standard deviation* in units of a special *phi* size scale). Several processes, e.g., surf and wind, produce markedly overlapping sorting characteristics. Additional criteria are necessary to distinguish completely the origin of most ancient sandstones; nonetheless, sorting readily eliminates some processes from consideration.

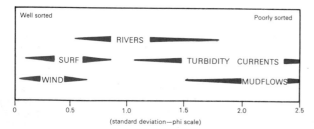

derived from erosion of older rocks in land areas, from nonterrigenous material formed within an aqueous depositional environment. The latter includes chemically precipitated sediments, such as the evaporites (salt and gypsum), and carbonate rocks composed of fossil skeletal debris or precipitated calcareous particles. In this and the next chapter we shall consider primarily the terrigenous fragmental sediments; the nonterrigenous types are discussed in Chap. 13.

TEXTURAL MATURITY

Clastic textures reflect primarily the rates and intensities of physical sedimentary processes. Maximum size reflects the power of transporting agents such as running water or wind. Wind normally moves only sand and silt, whereas moving water can carry gravel as well. Mudflows and glaciers carry immense blocks for long distances because of their greater density and viscosity. In a general way, size tends to decrease with time and distance of transport. The reason is twofold, including a tendency for diminution of carrying power or competence with distance for most agents of transport, and also reduction of particle size through continual abrasion. Degree of rounding of sharp corners of fragments also is related to intensity and time of abrasion as well as to toughness of the materials themselves. The range of sizes in a given clastic sediment, generally described as size sorting, reflects primarily the total time of transport and constancy of physical energy of transportive agents. A sediment subjected to long and constant agitation (e.g., beach sand) tends to be well sorted because there is maximum opportunity for the early dropping out of particles of large mass and for the removal or winnowing away of fine materials. The latter are

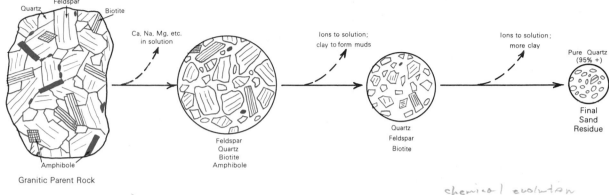

COMPOSITIONAL EVOLUTION OF SAND

chemical evolution

FIGURE 10.16   Typical changes through time of the mineral composition of sand derived from erosion of a granitic source. Less stable minerals are broken down both physically and chemically.

deposited ultimately in less physically agitated environments. In this way, fractionation of different-sized materials occurs, with a possible final result being deposition of gravel near the source, well-sorted sand in another place, and well-sorted particles of fine silt and clay in still a third place. Generally, clastic sediments become finer as they are moved farther from their source.

From the above considerations, we can formulate a very useful generalization about the ideal textural evolution of terrigenous clastic sediments that will help us to interpret the history of any specific sample. Obviously with greater and greater abrasion and winnowing by currents or waves, size of particles will be reduced, sorting of sizes will improve, and rounding will increase as shown in Fig. 10.14. This idealized clastic evolution can be summarized as the degree of textural maturity. Usually ill-sorted, dark sandstones are loosely termed graywackes (from an old German mining term, "wacken" for waste or barren); the sand grains in a graywacke may be of any composition. Bouldery and cobbly muds, as in glacial till or mudflows, represent the extreme of poor sorting. Such textural immaturity may result from very rapid deposition of sediments or from transport by an unusually viscous medium such as mud or ice, with much greater resistance to flow than water or air. In such media, winnowing of particles according to relative mass is ineffective (Fig. 10.15).

COMPOSITIONAL MATURITY

The mineral composition of a clastic sediment also will change as its particles are subjected to repeated physical crushing and chemical destruction of the less stable minerals. Rock fragments tend to be ground down rapidly to their separate mineral grains, and dark (mafic) minerals, such as pyroxenes and amphiboles, suffer rapid chemical breakdown. This leaves a dominantly sand-sized (0.125 to 2.0 millimeters) ultimate residue of the more resistant material: quartz, chert, feldspar, and some mica, as well as very rare, but durable, accessory grains of minerals with high specific gravity (e.g., zircon, garnet, and magnetite). Of these, quartz is overwhelmingly the most abundant. Feldspar, though more abundant in parent igneous rocks, is of intermediate durability and so runs second place to quartz in sediments. The others, though more durable than feldspar, are simply far less abundant in source materials. Chert is the most durable material that originates in sedimentary environments and is common in many conglomerates and sandstones.[1]

An ideal evolution of compositional maturity exists (Fig. 10.16) with progressive loss of the less

[1]Scientific revolutions resulting from invention of the telescope in the sixteenth century and compound microscope in the seventeenth century are well known. Of equal impact to geology was the invention in 1829 by a Scot named William Nicol of a device made from crystals of calcite for polarizing light. Nicol found that if minerals and rocks are cut and ground down to very thin, translucent slices, their compositions and textures can be studied microscopically if controlled polarized light is passed through the slices (see Figs. 8.4 and 10.19).

stable and concentration of more stable constituents as the vicissitudes of physical and chemical degradation take their toll. Obviously the ultimate product of such an evolution is the concentration of a residue of very pure quartz- or chert-bearing sand and gravel. But there also is a close interrelation between destruction of the less stable coarse particles and the increase of fine residue products of that destruction. For example, the most abundant sedimentary rock is shale, which is composed largely of fine clay particles derived from weathering of feldspars, the most abundant minerals in the crust. Within shales there also is a spectrum of compositional maturity such that the clay species present reflect relative thoroughness of chemical decay of parent minerals.

Broad grouping into clans of sands and gravels according to composition is very useful as a simplification of knowledge. The clans recognized in this book are the quartz-chert (or siliceous) clan, the feldspathic clan, and the rock-fragment (or lithic) clan. These categories provide convenient summary nicknames, but their limits are statistical in nature. As with any classification imposed by man upon nature, there is a degree of unavoidable arbitrariness in defining categories.

STRATIFICATION

The nature of stratification also provides clues about processes of deposition. More than a century ago it was noted by the British genius H. Clifton Sorby that ripple marks and inclined or cross stratification in sand or fine gravel indicate moderately strong current action that rolled and bounced particles along by traction. At certain velocities, turbulence develops in such a way that a rippled or corrugated sand surface is more stable than a perfectly smooth one. Because internally the inclined laminae reflect the lee faces of ripples or dune forms, they dip in the down-current direction (Fig. 10.17). Tractional currents are driven by gravity, as in rivers, or by the wind and tides, as in lakes and shallow seas. Very fine, parallel-laminated sediments, on the other hand, commonly suggest vertical settling of particles that had been carried up in suspension in the transporting fluid and not in contact with the bottom. This implies a minimum of direct current agitation of the bottom, therefore a relatively still environment. We may speak of cross stratification as representing a turbulent condition (i.e., agitated more or less constantly), and of delicate, parallel lamination in fine muds as indicating a

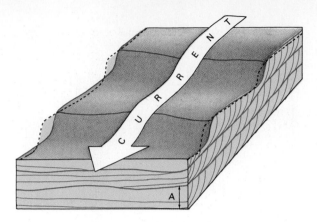

FIGURE 10.17 Origin of cross-stratification and ripple marks by migration of dune forms produced by vigorous bottom (tractional) currents. Grains roll and bounce over the dune crests, coming to rest on the lee faces. Successive laminae, inclined in a down-current direction, form as the lee face migrates; each lamina is a buried fossil lee face. Cross stratification not only reveals the original top and bottom of deformed strata but also ancient current directions. (A indicates amplitude of cross stratification.)

nonagitated condition. In many sediments we actually find evidence of alternations from dominantly tractional transport by waves or currents to dominantly suspended transport.

Graded bedding (Fig. 10.7) represents episodic introduction of abnormally coarse material into a usually still environment characterized by delicately laminated fine materials. Such spasmodic deposition occurs principally where infrequent density currents flow beneath less dense water and carry coarser debris to normally tranquil areas. Turbidity currents are the most important of such agents. They derive their driving energy from the presence of fine sediments thrown into suspension by earthquake shocks or by severe storm activity. Sediment is kept in suspension by turbulence as the slightly more dense, muddy water mass flows downslope beneath less dense, clear water (Fig. 10.18). Such a current maintains its kinetic energy and continues to flow until the turbid water encounters a topographic depression or until it becomes diluted by mixing with clear water. In Lake Mead, where natural turbidity currents have been studied most, muddy Colorado River flood water (density 1.003) sinks beneath clear lake water (density 1.001) and flows 100 kilometers along the bottom to Hoover Dam, where the currents are halted. Very fine sand and silt with graded bedding is

FIGURE 10.18 Experimental turbidity current of more dense, muddy, turbulent water flowing from right to left down a sloping laboratory flume approximately 50 centimeters deep beneath less dense clear water. Such currents are driven solely by gravity because of the density contrast between water masses. (*Courtesy G. V. Middleton and R. G. Walker, McMaster University, Ontario.*)

deposited from them along the lake floor. Such currents were not anticipated when Hoover Dam was built, and they are filling the reservoir faster than expected.

About 1930 a British geologist, E. B. Bailey, noted that conspicuously cross-stratified sediments tend to occur apart from those with much graded stratification. He interpreted this dichotomy to reflect fundamentally different sedimentary environments and depositional processes. He also extended the interpretation of the environmental differences to a still more fundamental tectonic distinction. Cross stratification was deemed restricted to shallow, agitated water characteristic of seas over relatively stable cratonic regions, whereas graded bedding was considered confined to the most structurally mobile regions where presumed deep, usually less agitated water prevailed, but into which coarser material occasionally was dumped. As a first approximation, this is a useful generalization, but the two important types of stratification really only reflect contrasting degrees of agitation and modes of transportation of particles.

EARLY PREPALEOZOIC SEDIMENTS

Applying the concept of compositional and textural maturity to the Early Prepaleozoic clastic sediments (probably about 3.5 b.y.), we find that those associat-

ed with the Keewatin volcanic rocks are very immature. Prevalence of pillow structures in the lavas suggests largely subaqueous eruptions so that lands must have been very small—probably only local volcanic islands. A true continent apparently did not yet exist. The preserved Keewatin sediments are conglomerates and sandstones composed chiefly of volcanic rock fragments and feldspar, in general, poorly sorted and rounded (Fig. 10.7). The sediments are both compositionally and texturally very immature. Unusually large percentages of chromium and nickel in South African sediments older than 3.0 billion years suggest large, exposed source areas of ultramafic rocks in the early crust. Timiskaming sediments about 2.8 to 3.0 billion years old are volumetrically far greater than Keewatin ones. Volcanic material and feldspar still were prominent, but quartz had become important among the sand grains for the first time. The quartz requires presence of granitic or rhyolitic source rocks, or both. Moreover, presence of granitic pebbles in conglomerates prove that some granitic or continental crust was present and being eroded by this time.

Early Prepaleozoic sediments of the Great Lakes region are poorly sorted and are rounded; they show considerable graded bedding, but are scantily endowed with cross stratification (Fig. 10.7). Their constituents were eroded and deposited rather rapidly in a structurally mobile belt after only modest weathering and transport. Vigorous, long-continued current or wave action is not suggested by their texture and stratification. In the words of E. B. Bailey, such

FIGURE 10.19 Photographs taken through the polarizing microscope of two contrasting Middle Prepaleozoic sandstones. *Left:* Graywacke sandstone composed of a mixture of quartz, feldspar, and diverse rock fragment grains surrounded by a considerable *matrix* of dark clay; note poor sorting and angularity of grains (Gowganda Formation, White Lake, Ontario). *Right:* Pure quartz sandstone composed almost solely of well-rounded quartz grains (note lack of dark matrix material); this is both compositionally and texturally mature (Ajibik Quartzite, near Marquette, Michigan). (Larger grains in each are about 1.0 millimeter in diameter.)

sediments look "poured in or dumped" with a minimum of abrasion and fractionation. The most outstanding quality of such clastic rocks is their dark or "dirty" appearance, resulting primarily from an abundance of fine, dark clay and other material as a matrix among the sand and gravel fragments (Fig. 10.19).

RELATION OF TECTONICS TO THE SEDIMENTS

Following eruption of Keewatin lavas, the major Saganagan Orogeny [2.7 to 3 b.y. (?) ago; Table 10.1] apparently produced important high lands. From these were eroded immature materials to be deposited rapidly in adjacent low or subsiding areas. Although we assume that the sediments were deposited in the sea, there is no absolute proof, because fossil environmental indicators are unknown in such ancient strata. The Great Lakes rocks were upheaved in turn in the Algoman orogeny (2.4 to 2.6 b.y.).

By Middle Prepaleozoic time, conditions in North America had changed enormously. The great extent

of Algoman granites (circa 2.5 b.y.) suggest that much continental crust had formed. Subsequently, a new (Penokean) mobile belt formed along the Lake Superior–northern Lake Huron trend and apparently extended northeast through western Labrador.

MIDDLE PREPALEOZOIC SEDIMENTS

Middle Prepaleozoic sediments show some striking contrasts with most of the preserved older ones. Ill-sorted graywackes still formed, particularly in Michigan and Wisconsin, but light-colored, well-sorted pure quartz sandstones also are volumetrically very abundant (Fig. 10.19). The latter have much less fine matrix material, and, though most of those in the Penokean belt have been recrystallized to metamorphic quartzites, there is still evidence that they were texturally as well as compositionally mature. Middle Prepaleozoic sandstones thus are of two principal types, differing chiefly in texture: (1) well-sorted quartz sandstones and (2) quartz-rich graywackes. Feldspar is only sparingly present in the graywackes, and other constituents, such as volcanic rock fragments present in older graywackes, are notably sparse in these. The pure quartzites almost invariably show cross strata and ripple marks (Fig. 10.8). Conglomerates, where associated, generally contain well-rounded, durable pebbles of quartz or chert. Interstratified with the quartzites are thick sequences of mudstone or slate with zones of dark graywackes and prominent, banded iron formations consisting of alternating iron and chert bands (Fig. 10.6). Associated with some pure quartzites are limestones, many of which contain wavy laminated structures presumed to have been formed by marine, bottom-dwelling primitive (filamentous) algae (Fig. 10.20).

What conditions could produce the first appearance of significant limestones associated with mature sandstones? A low (or very distant) stable land as a source of quartz and deposition in shallow, strongly agitated water are indicated. The wide extent of distinctive formations suggests, by analogy with younger strata, a marine origin. Algae fixed on the sea bottom cannot grow in water deeper than about 100 meters due to absorption by water of the sunlight needed for photosynthesis. Finally, other evidence of strong agitation over wide areas also argues for a very shallow sea susceptible to regular stirring by winds. Statistical analysis of measurements of cross stratification and ripple-mark orientations in Middle Pre-

FIGURE 10.20 Large stromatolite structure from Middle Prepaleozoic carbonate strata in the Northwest Territories, Canada. (*Photograph courtesy Paul Hoffman, Geological Survey of Canada.*)

paleozoic quartzites suggests persistent currents flowing dominantly from the north or northwest toward the south and southeast over most of the northern Great Lakes region. This sequence was deposited on a continental shelf.

Abundance of quartz sand indicates long and profound weathering of very large volumes of granitic or rhyolitic rocks. The volume of pure quartz sandstone provides a rough minimum index to the volume of material that must have been eroded to provide this concentrate. As an example, to produce the total volume of quartz residue represented by one of several prominent quartzite bearing formations north of Lake Huron would have required the complete weathering and erosion of at least 10,000 cubic kilometers (2,500 cubic miles) of granitic rock that contained a volumetric average of 25 percent quartz. To account for *all* the incalculable quartz sand in the exposed part of the Penokean mobile belt, this figure must be multiplied at least a hundredfold. The staggering volume of Middle Prepaleozoic quartz sand found both in well-sorted sandstones (now quartzites) and in ill-sorted ones (graywackes) implies several cycles of weathering, erosion, and concentration within a few hundred million years immediately following the Algoman orogeny. In each successive cycle, there would have been removal of more and more unstable mineral grains and a gradual distillation of quartz. It is inescapable that the Middle Prepaleozoic continent already was large and contained a great deal of granitic rock. Much of it was relatively stable, and it had been deeply eroded from about 2.5 to 2.0 billion years ago. The sedimentary evidence, therefore, supports the concept of rapid early growth of continental crust.

All Middle Prepaleozoic strata in the Great Lakes area become thicker from north to south; volcanic rocks are present south of the lakes, and the degree of deformation and metamorphism increases markedly

in the same direction (Fig. 10.21). Thus there is ample evidence of the existence of an east-west trending Middle Prepaleozoic (Penokean) mobile belt, which developed after deposition of the earlier shelf sequence. A stable craton still lay directly north of that belt, on which relatively thin and less-disturbed sandstones, shales, some limestone, and iron sediments were deposited. Clastic debris was derived almost wholly from the weathering and eroding of that craton. The mobile belt subsided profoundly, and, therefore, could receive very thick accumulations of sediments. It may have been the site of thrusting of a great plate of crust and mantle from the south (e.g., Fig. 9.18) beneath the craton to the north.

LATE PREPALEOZOIC ROCKS

The Penokean belt was upheaved, in its turn, into a vast mountain range, and, like the older mountains, was reduced by erosion. Late Prepaleozoic strata deposited over the old mountains are nearly flat-lying,

FIGURE 10.21 Simplified cross sections summarizing the history of the Penokean mobile belt. Most of the volcanic rocks were erupted onto the sea floor and so did not form significant volcanic islands.

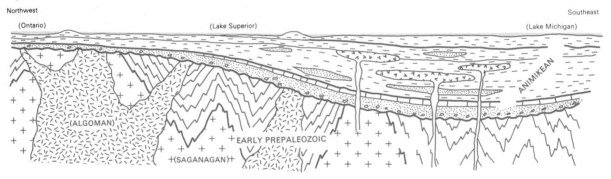

A. During Animikean Deposition

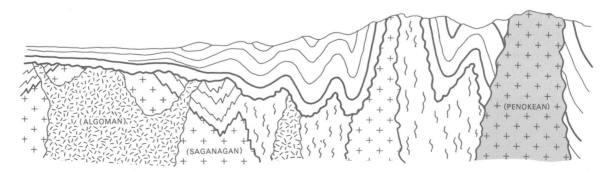

B. Following Penokean Orogeny

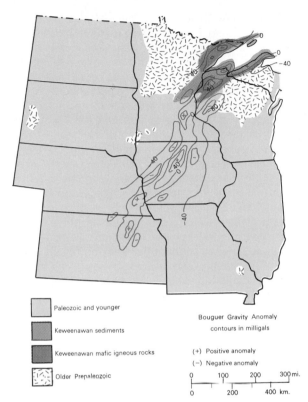

FIGURE 10.22 The mid-continent gravity anomaly, one of the largest such features known. Large positive anomalies are associated with Keweenawan basalt and gabbro outcrops at Lake Superior. Extension of similar dense rocks southward beneath Paleozoic strata is indicated by extension of similar anomalies. Keweenawan lavas probably covered a still larger region originally. (*Adapted from E. Thiel, 1956,* Bulletin of the Geological Society of America, *v. 67, pp. 1079–1100; by permission of* Geological Society of America.)

and would seem to represent stabilization of the Lake Superior region at last. But sedimentation gave way to extensive outpourings of the great Keweenawan basaltic lavas accompanied by the intrusion of vast diabase sills and the Duluth gabbro mass (1.15 b.y. ago; Fig. 10.5). Following this unusual igneous episode, sedimentation resumed, producing widespread bright red-colored sandstones and shales. These strata have been considered river and lake deposits, though there is no proof that at least some were not marine.

Subsurface data from gravity and scattered deep

drilling indicates that a belt of Keweenawan mafic, high-density rocks extends southwest from Lake Superior to northeast Kansas, probably the world's largest such mass (Fig. 10.22). Magmas were erupted along this trend oblique to all older structures. The Keweenawan rocks are only mildly deformed, unlike those of the older mobile belts; therefore, they represent a hitherto unprecedented condition in the central North American crust. This was no mobile belt such as we have been considering. Moreover, basalts are rock types more characteristic of oceanic areas, so are somewhat anomalous within cratons. Profound rifting or fissuring of the continental crust occurred, allowing mafic magmas to well up to the surface from deep within or below the crust (Fig. 10.23). The lavas spread out on the surface of what was otherwise a stable, passive craton. Better-known late Cenozoic rifting and volcanism of similar scale are known in East Africa and in the Pacific Northwest (see Fig. 17.10). The lavas in all three cases rose through fissures and flooded immense areas, burying everything in sight and gradually levelling the landscape; hence they are called flood basalts. Collectively they represent a very important second-order class of tectonic features of the earth's crust. Flood basalts differ from volcanic rocks typical of mobile belts in that: (1) they are almost exclusively basaltic, (2) with few exceptions they occur in cratonic areas, and (3) they are only mildly deformed. Clearly they require a structural explanation different from the mobile belts, namely profound extensional cracking or rifting of the

FIGURE 10.23 Extensional rifting and outpouring of lavas along a broad upwarp in cratonic crust as envisioned for the Keweenawan of north-central United States and the East African rifts. (*After H. Cloos, 1939,* Geologische Rundschau, *v. 30, p. 401; by permission of* Geologische Rundschau.)

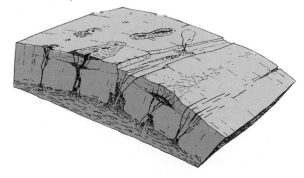

crust; perhaps they are sites of rupturing apart of huge crustal plates as mentioned in Chap. 9. The East African rift valleys (especially the Red Sea; Fig. 9.7) are sites of incipient continental drifting, but in North America, splitting of the continent did not continue for some reason.

SEDIMENTARY EVIDENCE OF THE NATURE OF THE PREPALEOZOIC ATMOSPHERE AND OCEAN

Many Prepaleozoic shales and graywacke sandstones are very dark due to presence of abundant unoxidized carbon and sulfide materials. Also, pebbles of pyrite (FeS_2) and uranium minerals that would be unstable in the presence of free oxygen are known from rocks older than 2 billion years. Iron carbonate ($FeCO_3$—siderite), very rare in younger sediments, is more common in Prepaleozoic ones. If free oxygen had been abundant, elements like iron, manganese, uranium, sulfur, copper, zinc, and vanadium all would tend to be present in their most highly oxidized forms, which is not the case. The facts are in accord with an original oxygen-poor or anaerobic atmosphere as outlined in Chap. 7.

Beginning in Late Prepaleozoic time, iron and silica tended to be chemically separated; the great affinity of iron for oxygen appears to have been largely responsible. Only a small amount of oxidized iron produces red color in sandstone, shale, and soil, and so the appearance of many red sediments should provide an indication of the onset of an oxidizing (aerobic) atmosphere. Red shales are known locally in rocks as old as 2 billion years, but extensive red strata, or "red beds," first appeared in Late Prepaleozoic time. Besides the Upper Keweenawan "red beds" already discussed (Figs. 10.5 and 10.9), prominent examples occur in Glacier National Park, Montana, and in Scandinavia. Red sediments continued to form in all younger rock systems. There can be no doubt that free atmospheric oxygen was abundant by about 1.5 billion years ago, and probably was becoming significant nearly a billion years earlier (see Fig. 7.10).

The banded iron-chert deposits (Fig. 10.6) are known chiefly in Middle and Early Prepaleozoic rocks, which points to some unique chemical (or biochemical) condition. They have long been a great puzzle not only because of their confinement to Precambrian time but also because iron and silica show very different chemical behaviors today. Iron tends to be transported in acid and precipitated in alkaline conditions, whereas silica does just the opposite. Because the peculiar association of iron with silica largely predates widespread "red beds," it would appear that it was related chiefly to oxygen-poor conditions, although there may have been some peculiar biochemical influence, too. With CO_2 relatively much more important, sea water would have been slightly less alkaline than now and could have carried more calcium in solution. Silica apparently reached saturation in much of the Prepaleozoic sea. Because the entire early ocean and atmosphere were anaerobic (oxygen-free), iron could have been abundant in solution throughout the seas as ferrous ion (Fe^{2+}). Depending upon local chemical environments, it combined with sulfur (to produce FeS_2), carbonate (to produce $FeCO_3$), or silica (to form various iron silicate minerals). Locally, however, in shallow agitated environments there was sufficient free oxygen to form iron oxides, even though the atmosphere was not yet oxygen-rich. Once free oxygen became relatively abundant, practically all iron was deposited in oxide form. Iron was fickle, for it ceased its love affair with silica when the attractive newcomer, free oxygen, appeared. It is biologically surprising that free oxygen accumulated so slowly once photosynthesis began. Probably the reason is because there was so much iron present that it could "soak up" all available oxygen for perhaps a billion years.

EVIDENCE OF PREPALEOZOIC GLACIATION

Evidence about Prepaleozoic climates is scant. For the bulk of the record, we have discovered evidence to neither confirm nor deny an assumption of overall climatic conditions similar to later geologic time. Intuitively one suspects that many changes probably occurred. But, though we have little knowledge of the typical Prepaleozoic climate, we do have some evidence of refrigerated climatic extremes.

At the base of the upper Huronian sediments in Ontario, a peculiar, widespread assemblage of rocks called the Gowganda Formation occurs. The most distinctive rock type is a massive, almost completely unstratified and unsorted jumble of large boulders, pebbles, sand, and fine clay surrounded by a finer matrix of dark material (Fig. 10.19, left). Another striking type of deposit in the formation is delicately laminated mudstone that resembles Pleistocene glacial lake clays containing varve laminae interpreted as seasonal layers; some of these contain scattered pebbles (Fig. 10.24). A widespread unconformity marks the base of the Gowganda Formation. At three localities where the sub-Gowganda surface itself is exposed, fine parallel scratches strongly resembling striations made by glaciers are visible.

The peculiarities of the Gowganda Formation argue for glacial tills, river outwash deposits, and muds; some of the deposits probably accumulated in the sea. Today many glaciers extend into the sea along the southern Alaska, Greenland, and Antarctic coasts. Drifting icebergs carry all manner of frozen-in rock debris offshore to be deposited helter-skelter over large areas of the sea floor as they melt (see Fig. 18.4). The pebbles dropped into laminated Gowganda mudstones probably had a similar origin. A major episode of continental glaciation apparently occurred in Canada about 2.2 billion years ago. A large ice sheet spread out from the north into the edge of the sea much as Pleistocene continental ice sheets spread across the North Atlantic continental shelf from New England and eastern Canada. Recognition of ancient glaciation is significant in showing that the well-known Pleistocene refrigeration was not unique in history. Other probable Prepaleozoic glacial deposits exist in the world, but they were not all synchronous.

THE LIFE RECORD BEFORE CAMBRIAN TIME

THE CAMBRIAN CONTRAST

One curious fact has emerged from the exploration for fossils, that of the profound difference between the richly fossiliferous Cambrian rocks and that of older units. While many Prepaleozoic rocks are either igneous or intensely metamorphosed, there are many sedimentary rocks that could retain evidence of life. Yet, almost none bear such evidence. Many of the barren sedimentary rocks of Late Prepaleozoic age are lithologically identical with Cambrian strata re-

FIGURE 10.24 Delicately laminated (varve-like) mudstone with scattered pebbles and sand grains dropped from above (Gowganda Formation, near Blind River, Ontario). Widespread nature and association of this type of deposit with tills suggests dropping of stones from drifting icebergs (see Fig. 18.4).

plete with fossils. Even a relatively rich locality of Eocambrian age in Australia (Table 10.2) would be described as poorly fossiliferous by Cambrian and younger standards. Unfortunately, what is known about life in the Prepaleozoic is not enough to build a satisfactory picture of paleoecology, paleozoogeography, or of the all-important origin of the animal phyla. However, there has been an intensive effort to find fossil and chemical evidence of life in the Prepaleozoic, and some patterns are beginning to emerge.

THE EARLY PREPALEOZOIC RECORD

The earliest known sedimentary rocks are the Onverwacht Group, the Fig Tree Group of South Africa, and the Bulawayo Group of Rhodesia. The rock types plus fossils indicate that the primitive atmosphere (whatever it was) had changed into one containing a dominance of CO_2. This is based, in part, on the presence of limestones by 3 billion years ago. These

TABLE 10.2   Evidences of organisms or related events contained in Prepaleozoic rocks
Partial list of strata containing Prepaleozoic fossils or other evidence of life compiled to show the best-substantiated records of organisms. The stratigraphic arrangement of the formations and systems is only approximate. (Not to scale)

Million years ago	Eras		Systems, formations (or rock units), and locations	Organism, other evidence of life, or events
570	Paleozoic	Cambrian	Lower Cambrian (worldwide)	Coelenterates, protozoans, poriferans, molluscans, worms, echinoderms, trilobites, trilobitomorphs, archaeocyathids, brachiopods. Stromatolites less common
		Eocambrian	Reed Dolomite (California)	*Wyattia* (Mollusca)
			Ediacaran strata (S. Australia)	25 species including medusae (coelenterates); *Rangea, Arborea,* and *Charnia* (frond-type octocorals); *Dickinsonia, Spriggina* (annelid worm); *Praecambridium* (a probable molluscan); and *Tribrachidium* (an echinoderm of the edrioasteroid type)
700			Various localities (worldwide)	Evidence of glaciation dated between about 800–650 m.y. ago
800	Prepaleozoic	Late	Bitter Springs Formation (N. Australia)	Eucaryote cells, evidence that some reproduced sexually; fungi, filamentous blue-green algae, green algae, bacteria
900			Belt Supergroup (Montana)	Stromatolites, eucaryote cells, protozoa
1,100			Beck Springs Dolomite (California)	Stromatolites, possible eucaryote cells (oldest reported)
1,300			Belcher Group (Hudson Bay)	Colonial bacteria, algal tubules, double-walled spheroids
1,700		Middle	Gunflint Formation (Ontario)	12 species of procaryote blue-green algae, some filamentous stromatolitic bioherms, chemical synthesizing bacteria
1,900			(Worldwide)	Stromatolites increasingly common, atmosphere with oxygen component
2,000			Witwatersrand (S. Africa)	Stromatolites, blue-green algae
2,200			Band Iron Formation (Australia)	Microspheres, bacteria
2,400		Early	Soudan Formation (Minnesota)	Blue-green algae, bacteria
2,700			Bulawayo Formation (Rhodesia)	Massive biohermal stromatolites formed by filamentous photosynthetic blue-green algae found in limestones; CO_2 probably a large component of atmosphere
3,100			Fig Tree Formation (S. Africa)	Algal-like spheroids, bacteria
3,300			Kola Peninsula (Finland) and other areas of the world	Oldest known rocks
3,800				

[handwritten annotations:] (photosynthesis)   reef formers

methanogenes. more primitive than prokaryotes

* see M.N.H. magazine Feb. 78 - Gould

African deposits contain bacteria and photosynthesizing blue-green algae (see Table 10.2). Chemical evidence supports the fact that photosynthesis was taking place, because a number of complex organic compounds associated with the process have been found with these occurrences. In fact, most sediments in the Prepaleozoic have yielded chemical "fossils" associated with plant metabolism.

The Bulawayo Limestone is particularly significant, since it contains the earliest known reef-like stromatolites (finely laminated, usually hemispherical algal masses; see Fig. 10.25) which are quite large. Stromatolites gradually became more common in the Prepaleozoic and were havens or communities containing other plants, and probably in Late Prepaleozoic time, animals as well. Certain bacteria also construct stromatolites. If the earliest examples were bacterial, then their presence would not provide

evidence of early oxygen generation as is generally supposed.

THE MIDDLE PREPALEOZOIC RECORD

Widely separated rocks dated between 2.5 and 1.7 billion years ago have yielded a modestly rich and surprisingly diverse marine flora, indicating increasing evolutionary activity. The discovery in 1954 by S. A. Tyler (described by Barghoorn and Schopf) of the first-known diverse fossil occurrence in the Prepaleozoic (Gunflint Formation, dated at 1.8 to 1.6 billion years ago) was largely responsible for a renewed interest in Prepaleozoic fossil studies.

The Gunflint fossils were found in black cherts in a banded iron formation on the north shore of Lake Superior. The microflora occurs between stromatolite

FIGURE 10.25 A microscopic photograph of accretionary layering of stromatolites, from the Gunflint chert of Lake Superior, Canada. (*Photograph by S. A. Tyler.*)

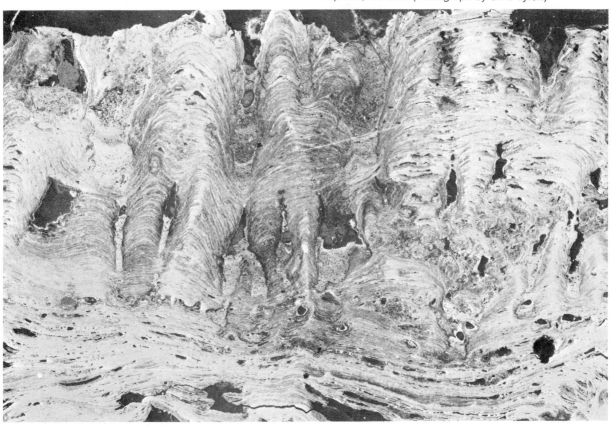

layers (see Fig. 10.25) and consists of over 12 species of mostly blue-green algae, but representing newly evolved groups (see Fig. 10.26). Various bacteria and organisms not yet assigned to any floral family are also present. Similar assemblages of about this age have been discovered in South Africa, Greenland, Hudson Bay, and Australia, and they strongly suggest that by 2 billion years ago there was free oxygen in the atmosphere. This is strengthened by the fact that red beds begin to appear at 1.5 billion years ago.

It is important to note that all known plants discussed so far were procaryotic cells (although a few specimens from the Gunflint might be eucaryotic). Procaryotic cells today include those of very primitive organisms such as bacteria and the blue-green algae and are characterized by the lack of an organized nucleus and certain organelles (tiny units within cells that digest food and have other functions to keep the cell alive and capable of reproducing). Reproduction is accomplished exclusively by a simple cell division. All other cells, from single-celled plants to cells in humans are eucaryotic, with an organized nucleus and organelles. The most important advance in the development of the eucaryotic cell is that the nucleus contains paired strands of genetic material (containing DNA) which make sexual reproduction possible. Sexual reproduction is one of two ways (the other being

mutation) to generate new variation, and it greatly speeded up the evolutionary process leading to the origin of animals. Before sexual reproduction developed, mutations leading to better adaptation or new species were slow to appear because of the very low rate of production of variants in asexual reproduction (simple cell division). Just to emphasize the importance of sex: in an asexually reproducing population, if 10 mutations occur, there will be 11 possible variants. In a sexually reproducing population, 10 mutations could combine to produce 59,049 variants because new genetic information can be passed on so much more rapidly. Sex is not only desirable but necessary!

THE LATE PREPALEOZOIC RECORD

The fossil record between 1.3 billion years ago and the Paleozoic boundary of 700 million years ago is somewhat better documented than the earlier record, with over 16 localities scattered over the world. It reveals very important evidence. In the Beck Springs Dolomite (1.3 billion years old) of eastern California, there has been found the earliest occurrence of green algae which are eucaryotic cells. A very rich microflora was found in the Bitter Springs Formation in the North Territory of Australia (about 900 million years old) and indicates that much evolution had transpired since Gunflint time. There are 56 species of blue-green, green, and red algae, fungi and other plants, and spores, which may indicate that some plants were reproducing sexually.

Probably the most unfortunate gap of the entire fossil record is the 200-million-year span between

FIGURE 10.26 Greatly enlarged microscopic photographs of primitive algae associated with stromatolites in the Gunflint cherts of Canada. (*Photographs by S. A. Tyler.*)

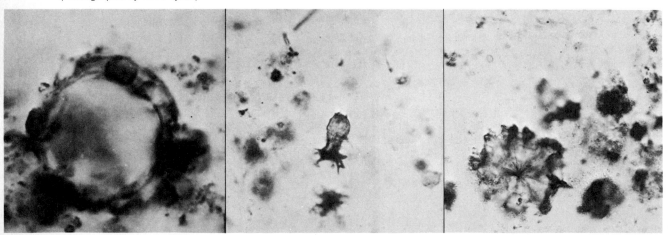

FIGURE 10.27 A few of the complex Eocambrian metazoan fossils from the Ediacaran Hills, South Australia. *Left: Dickinsonia*, a possible worm. *Center: Spriggina*, also thought to be a worm. *Right: Parvancorina*, an animal of unknown affinity (see also Fig. 11.8). (*Photographs by G. R. Adlington.*)

Bitter Springs time and the beginning of the Paleozoic Era. For in that interval, the first multicellular plants and animals had evolved. The multicellular green algae preceded animals and are found in Norway and Australia in rocks dated at about 700 million years. Some of the algae were evidently planktonic (floating), while others were attached to shallow-marine bottoms.

THE EOCAMBRIAN RECORD

Plant communities that must have existed prior to the beginning of the Paleozoic probably provided havens for the earliest animals, which were single-celled planktonic forms. However, our first glimpse of the animal record shows that much evolution had occurred, because quite complex and diverse animals with differentiated tissues existed over 600 million years ago. The first important Eocambrian fauna was found in the Pound Quartzite in South Australia. Since that discovery, the same fauna has been found in scattered localities such as Newfoundland and England. Most of the fauna is composed of jellyfish (coelenterate medusae) of several types, which were probably floaters. The most interesting forms are attached corals (octocorals), several types of worms, possible molluscs, and echinoderms (see Fig. 10.27). This Eocambrian fauna provides us with the first evidence of an animal community from shallow-marine waters. One most important observation is that none of these forms possessed a mineralized skeleton.

Within a startlingly short time interval between the deposition of the Pound Quartzites and the first Cambrian deposits containing fossils (about 570 million years old), two important milestones were reached. A number of animals developed mineralized skeletons, and animal fossils became very common all over the world.

The development of mineralized skeletons was not confined to a single group, but cut across phylum lines, being found in seven phyla at about the same time in the Lower Cambrian. Not all phyla that were to develop skeletons evolved them at the same time; for example, the vertebrate skeleton is first found in the Early Ordovician of Russia (near Leningrad). There have been many theories to explain this "sudden" appearance in the Lower Cambrian. One supposes, for example, that $CaCO_3$ (the dominant mineral in most skeletons) could be more easily precipitated by organisms when a critical point had been reached in the sea water when the necessary elements were sufficiently concentrated. No satisfactory theory, to our minds, has been developed yet.

At least three factors may have been important in the formation of a skeleton, whether it is an outer (exo-) or inner (endo-) skeleton: (1) It would afford protection from ultraviolet rays (in the case of an exoskeleton), allowing animals to move into shallower habitats and also to be less vulnerable to predators (if any were around at that time). (2) Forms attached to the bottom could increase in size if they had a supporting skeleton. (3) Mobile bottom dwellers would have a base for the development of stronger muscles, hence increasing locomotor efficiency (both exo- and endoskeletons).

SUMMARY

In spite of severe handicaps, a workable Prepaleozoic standard chronology was developed in the Great Lakes region. Isotopic dating allowed refinement of the chronology and provides the only valid basis for correlation beyond that region. Continent-wide mapping of isotopic dates shows broad provinces with statistically clustered dates reflecting major periods of metamorphism and granite formation associated with mountain building. The Prepaleozoic basement of the entire continent represents a mosaic of ancient mobile belts delineated both by structural patterns and isotopic dates. The isotopic age provinces show a crude concentric pattern of younger-outward belts, which has been taken as evidence that new continental crust has been added in mobile belts at the margins of a growing craton.

As noted in Chap. 7, timing of continental accumulation is of major interest. The Prepaleozoic record suggests that volcanism was universal prior to about 3.5 billion years ago; only after that time was significant granitic material differentiated.

The oldest North American sediments are both texturally and compositionally immature and reflect

great volcanic activity. But existence of large areas of granitic crust already formed 2.4 billion years ago is indicated by large volumes of Middle Prepaleozoic quartz sands. Features formed by currents provide a basis for reconstructing regional paleocurrent patterns, which help to delineate a rather large, thoroughly weathered cratonic source for the quartz. The Middle Prepaleozoic mobile belts, to a great extent, reworked old crust rather than always producing new crustal accretions, which is consistent with plate-tectonics theory. Indeed, the Prepaleozoic record shows clearly that shield areas are laced with old mobile belts and thus apparently are mosaics of old, obscure lithosphere plate fragments. Exactly when the plate-tectonic mechanism first began to operate is much debated, but it appears that, by no later than 2 billion years ago, clear cratons and mobile belts existed. Prior to that time, perhaps the crust was too thin and too unstable, heat flow perhaps was too great, for permanent rigid plates to persist.

Prepaleozoic rocks are unique in several respects, though in many other ways they are identical with younger ones. Fossils are far less abundant (and of different types) than in younger strata. Early Prepaleozoic strata contain minerals that are very unstable in an aerobic environment, suggesting an oxygen-free atmosphere until about 1.5 to 2.0 billion years ago. Widespread Late Prepaleozoic sediments colored bright red by oxides of iron attest to an oxygen-rich atmosphere by at least 1.5 billion years ago. Banded chert and iron deposits are among the unique and most important trademarks of Early and Middle Prepaleozoic strata. Their depositional history, together with that of carbonate rocks, was the result of special chemical conditions of the sea and atmosphere; oxygen content probably was the chief factor, though biochemical agents also may have been responsible.

Earliest indications of life in some of the oldest known rocks include structures formed by marine algae and bacteria and traces of organic compounds known to be formed only by life processes. Through most of the Prepaleozoic succession, the preserved record of life shows no dramatic change in kind, but it shows an increase in abundance. Unfortunately, no animal fossils are known in Late Prepaleozoic strata; first unquestioned animal fossils appear in Eocambrian rocks. When animal fossils did appear, they were of complex and diverse forms that reflect a great deal of evolution. It is still not clear why animals acquired the habit of secreting skeletons so suddenly in Cambrian time.

After photosynthesis had proceeded for about 2 billion years, free oxygen began to accumulate in sea water and the atmosphere. Only then could efficient animal respiration become possible. Both the sedimentary and fossil records suggest that this did not occur before Late Prepaleozoic time.

Readings

Barghoorn, E., and Tyler, S. A., 1965, Microorganisms from the Gunflint chert: Science, v. 147, pp. 563–577.

Cloud, P. E., 1972, A working model of the primitive earth: American Journal of Science, v. 272, pp. 537–548.

Engel, A. E. J., 1963, Geologic evolution of North America: Science, v. 140, pp. 143–152.

Gastil, G., 1960, The distribution of mineral dates in time and space: American Journal of Science, v. 258, pp. 1–35.

Goldich, S. S., Nier, A. O., Baadsgaard, H., Hoffman, J. H., and Krueger, H. W., 1961, The Precambrian geology and geochronology of Minnesota: Minnesota Geological Survey Bulletin 41.

Patterson, C., 1964, Characteristics of lead isotope evolution on a continental scale in the earth, in Craig, H., et al., Isotopic and cosmic chemistry: Amsterdam, North Holland Press.

Price, R. A., and Doeglas, R. J. W., eds., 1972, Variations in tectonic styles in Canada: Geological Association of Canada Special Paper No. 11.

Rankama, K., 1963–1968, The Precambrian (4 vols.): New York, Interscience.

Schopf, J. W., 1970, Precambrian micro-organisms and evolutionary events prior to the origin of vascular plants: Biological Reviews, v. 45, pp. 319–352.

Society of Economic Geologists, 1973, Precambrian iron-formations of the world (a symposium): Economic Geology, v. 68, pp. 913–1179.

Symposium on geochronology of Precambrian stratified rocks: Canadian Journal of Earth Sciences, v. 5, no. 3, June 1968, pp. 555–772.

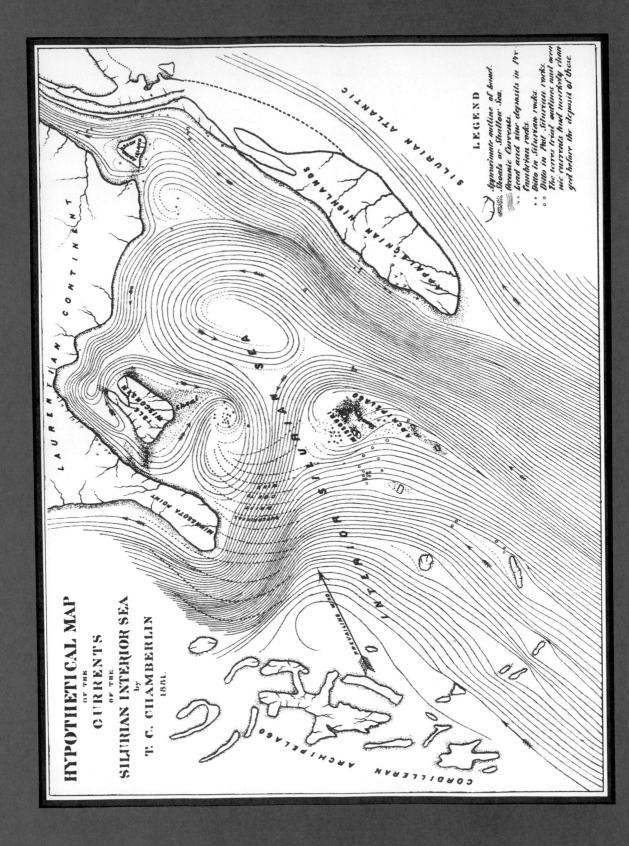

HYPOTHETICAL MAP
OF THE
CURRENTS
OF THE
SILURIAN INTERIOR SEA
by
T. C. CHAMBERLIN
1881.

LEGEND

Approximate outline of Great
Shoals or Shallow Sea.
Oceanic Currents.
Lead and zine deposits in Pre-
Cambrian rocks.
Ditto in Silurian rocks.
Ditto in Pot. Silurian rocks.
The terms trial outlines and oce-
nic currents head westerly than
get before the deposit of these.

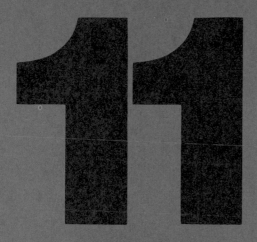

11

EARLIEST PALEOZOIC HISTORY

A STUDY OF THE INTERPRETATION OF EPEIRIC SEAS

In the mud of the Cambrian main,
Did our earliest ancestors dive;
From a shapeless, albuminous grain
We mortals are being derived.

Grant Allen,
The Ballade of Evolution

FIGURE 11.1 Hypothetical paleogeographic map of the United States for early Paleozoic time, one of the first paleogeographic maps ever published. In 1881 the name Silurian often included all of what now comprises Cambrian, Ordovician, and Silurian. This map really refers to the Late Cambrian. *(After T. C. Chamberlin, Geology of Wisconsin, v. IV, p. 530.)*

The beginning of the Paleozoic Era produced a great punctuation mark in the record of earth history because of the sudden appearance in the stratigraphic record of the remains of animals. The Eocambrian fauna described in Chap. 10 was already moderately diverse and included some rather complex forms. Even more puzzling, however, is the sudden appearance of hard, preservable skeletons at the beginning of Cambrian time. It is the appearance of fossil animals, then, that defines the beginning of the Paleozoic Era.

In North America, as in other continents, many areas contain essentially unfossiliferous strata that underlie and are closely related to Cambrian ones. A few animal fossils have been found in strata below typical Lower Cambrian fossils (see Table 10.2), but because such strata generally lack fossils, their true geologic age is somewhat uncertain. In only a few cases have isotopic dates been available to resolve the dilemma. What dates exist suggest that these poorly fossiliferous strata spanned the interval from about 700 to 570 million years ago. In many mobile belts, such strata are found conformable below others with Lower Cambrian index fossils, and a major angular unconformity is found beneath the essentially barren rocks; in eastern North America, the "basement" below these is the Grenville-aged metamorphic complex (see Fig. 10.11). The barren strata clearly are more closely related to Cambrian sediments above than to deformed older Precambrian rocks below. Therefore, it has become customary to refer to such unfossiliferous strata as Eocambrian, meaning the "dawn of the Cambrian" (also called Infracambrian and Riphean). At present, the term Eocambrian is an informal one, but we feel that the advent of the Paleozoic Era should be redefined formally to include the Eocambrian, especially now that fossils are turning up within it.

In this chapter we shall review the rich Cambrian fauna and examine the early Paleozoic history of North America from 700 to 450 million years ago, with emphasis upon the structure of the craton and conditions in shallow epeiric seas that covered it during Late Cambrian time. In order to interpret the lower Paleozoic rocks, it is necessary to introduce several important principles and types of diagrams and maps which, though shown here, apply equally to all periods; *these same principles will be used in subsequent chapters, too.* Comparisons with foreign regions will be noted as well, in keeping with our goal to test continually the working hypotheses of crustal evolution outlined in Chaps. 8 and 9.

CAMBRIAN LIFE

THE ACQUISITION OF SKELETONS

In the last chapter, we observed that by Eocambrian time six phyla of animals had appeared, and we noted that the most significant feature of these was a lack of shells or internal skeletons. We further noted that the sudden appearance of shelled genera is the most singular feature of the earliest Cambrian faunas. Two interesting points can be made in viewing the Cambrian fauna. First, the earliest faunas are not very diverse in terms of kinds of organisms, but, as new faunas appeared, diversity increased until by Ordovician time there was great faunal variety. The second point is that mineralized skeletons gradually appeared in the phyla during the course of the first two Paleozoic periods. The oldest cephalopod skeletons are almost 60 million years younger than the snails and brachiopods. Bryozoans and corals also took that long to develop skeletons after the calcareous (but unrelated) archaeocyathids appeared. Yet, the earliest Cambrian record still is rich in terms of numbers of organisms.

THE TRILOBITES

The fossil record of the Cambrian represents shallow-marine environments with few exceptions. Of this record, the greatest majority of organisms both in kinds and numbers were the trilobites, some 600

FIGURE 11.2 Cambrian faunal provinces on both sides of the North Atlantic Ocean, 2,000 kilometers apart. The "European" fauna characterizes graptolite-bearing shales, while the "American" occurs chiefly in shelly carbonate rocks. Long ago it was believed that land barriers separated and isolated the two faunas, but they are found mixed in the same strata in western Newfoundland and elsewhere. Close similarities of faunas and sedimentary facies on both sides of the North Atlantic evidence a former closer link between the opposing continents, which is discussed in Chap. 12.

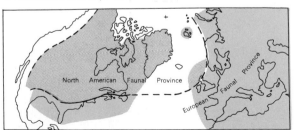

genera of which are known from the Cambrian. These strange animals became highly diversified and widespread during the course of the period and their many evolutionary divergences have permitted us to use the trilobites as time indicators. The early trilobites are thought to have been scavengers or mud-eaters living on the bottom, and most of them probably crawled rather than swam. As a result, they were more sensitive to the type of bottom sediments and tend to be facies fossils. This is reflected in the very interesting distribution of trilobites in the Lower and Middle Cambrian rocks of most of North America and northern Scotland in contrast to those in the maritime provinces of Canada and western Europe (Figs. 11.2 and 11.3), and their close link with sedimentary facies.

There are two evolutionary trends that developed in many groups of trilobites and probably represent grades of evolution; that is, a similar evolutionary level reached by several independently evolving lineages. One was the gradual fusion of posterior segments to form a rigid paddle-like structure called the pygidium (see Fig. 11.4). This usually was accompanied by the reduction in spines along the margins. No one can account for the progressive trend to develop a posterior structure as large and as well formed as the head section or cephalon. Almost certainly the cephalon went through a similar period of fusion during the Late Prepaleozoic when trilobites became more active and mobile. It is a general rule among all animal groups that, with locomotion, the body becomes elongated and sense organs become highly developed and concentrated at the anterior end, while waste is discharged from the posterior.

Another very important trend was for the eyes to migrate away from the central bulb of the head (cephalon). In Early and Middle Cambrian time, they were connected by an eye ridge, but later became separated. In some groups the eyes gradually became smaller and disappeared altogether. Well-preserved specimens show clearly that trilobites had highly developed compound eyes similar to those of the crustaceans and other arthropods.

OTHER ORGANISMS OF THE CAMBRIAN

The trilobites make up more than half of the Cambrian fauna and show by far the most progression in terms of evolution. The rest of the fauna shows a curious mixture of evolutionary "experiments" (organisms that

FIGURE 11.3 Lower Cambrian trilobites. *Left*: *Paedeumias* is found in North America except for the eastern Appalachian belt (i.e., "American" province of Fig. 11.2). *(Photograph courtesy U. S. National Museum.)* *Right*: *Holmia* is a common genus of the "European" faunal province. *(Illustration from* Treatise on invertebrate paleontology; *courtesy Geological Society of America and University of Kansas Press.)*

have a unique appearance, did not give rise to any other group, and were short-lived), along with quite conservative groups that did give rise to the vast faunas to come in post-Cambrian times.

The earliest strata of the Cambrian contain archaeocyathids, a sedentary bottom-dwelling group of solitary, vase-shaped forms with a double skeletal wall (see Fig. 11.5). In fact, they predate the trilobites. Their worldwide distribution when plotted on a map of Pangea suggests that they were tropical, equatorial, much as the coral reefs of today (see Fig. 11.12).

FIGURE 11.4 The morphology of trilobites. *(Courtesy the University of Kansas and Geological Society of America, from the* Treatise on invertebrate paleontology, *part O, 1959.)*

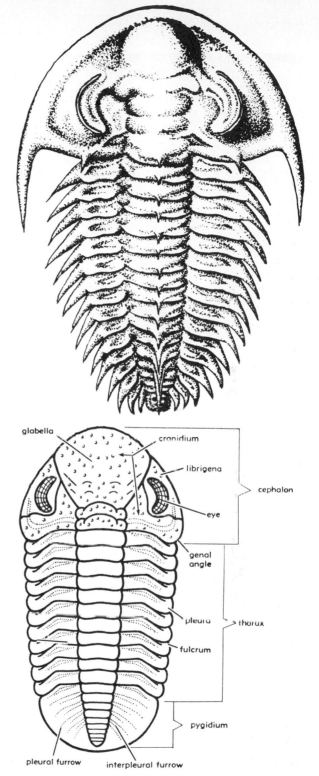

FIGURE 11.5   An archaeocyathid from the Lower
Cambrian of Australia. *(Photograph by G. R. Adlington.)*

They were quite widespread, forming undersea gardens that usually are devoid of other fossils. The genera are short-ranged, thus are used for correlation purposes; they became extinct at the end of the Middle Cambrian.

The brachiopods, which are the next most common group, are dominated by chitinophosphatic inarticulate forms (see Fig. 11.6). The shells are made up of a complex calcium phosphate molecule with much chitin (a plastic organic material) incorporated within the shell layers. These are the earliest brachiopods and were very conservative, showing negligible evolution after the Cambrian. They are considered, therefore, to have been unspecialized and primitive. For example, the modern *Lingula* is scarcely distinguishable from its Paleozoic ancestor. Their longevity may be due to their life in stable ecological niches of the intertidal zone or on the hard substrate of rocks or other shells. A few genera of calcareous inarticulates and articulates made up the rest of the brachiopods. While constituting only 30 percent of the Cambrian fauna, they were destined to dominate the faunas of the post-Cambrian.

Along with the archaeocyathids are some of the most fascinating of the Cambrian animals, the primitive molluscs and mollusc-like forms, which constitute no more than 5 percent of all known groups. In the Early Cambrian, some cap-shaped shells appeared that are so similar to modern day limpets in shape that they almost certainly lived in a similar rocky shore environment. These monoplacophorans (Fig. 11.7) are partly segmented molluscs and have the same body orientation as the chitons, Polyplacophora, which are a conservative group found living today relatively unchanged since the Paleozoic. The monoplacophorans are important because they gave rise to at least two groups of snails that dominated the later Paleozoic, which in turn gave rise to all later snails. By Middle Cambrian time, the clams made their appearance, but they were to remain a rare element until late in Ordovician time. Other molluscs were cone-shaped or limpet-like but had different internal structures; none of these survived the Cambrian.

Although the echinoderms are a very minor part of the entire Cambrian faunal realm, they show many interesting characters indicating experimentation on a level similar to the molluscs. The first echinoderm known is *Tribrachidium* from the Eocambrian fauna (Fig. 11.8, left). It appears to be related to the edrioasteroids of the later periods (see Fig. 11.8); yet nothing like it is known in the Cambrian. In the earliest Cambrian rocks, *Helicoplacus* (Fig. 11.9) has been found, which has spiral rows of polygonal plates and apparently was attached to the substrate. Stalked echinoderms appeared in the Middle Cambrian and are noteworthy in that they had but a few large plates. The echinoderms probably gave rise to the chordates no later than Ordovician time. This conclusion is based primarily on the fact that the larval stages of both are quite similar, but this hypothesis is only one of several concerning the origin of the chordates.

Finally, we must mention the Middle Cambrian Burgess Shale fauna from Field, British Columbia. There in 1910, C. D. Walcott made one of the most exciting fossil finds of the century (Fig. 11.10). The fauna consists of soft-bodied animals mostly belonging to various worms and arthropods. Almost all of these are known only from this one locality, but, due to the unusual preservation, they give rare insight to the soft anatomy of extinct groups. It enables us to have confidence in the correct classification of trilobites and other arthropods because arthropods are primarily classified on the basis of appendages, which are

FIGURE 11.7   Primitive mollusc of the Cambrian Period. *Helcionella* is a simple, cap-shaped monoplacophoran. *(Photograph by G. R. Adlington.)*

FIGURE 11.6   *Lingula,* an inarticulate brachiopod, from the Devonian of Eastern New York. *(Photograph by G. R. Adlington.)*

rarely preserved. Plants and other types of animals present include sponges and jellyfish, all showing beautifully detailed anatomy. Such finds as this give us a rare glimpse of the past, and suggest that perhaps the usual skeletal fossils represent a small percentage of the total fauna. Abundance of animal burrow structures in Cambrian strata over much of the continent attests to a rich soft-bodied element in the fauna everywhere (Fig. 11.11).

RELATIONS OF FOSSILS TO EARLY PALEOZOIC FACIES

Studies of early Paleozoic fossils and their relations to containing sediments have led to many important discoveries about continental histories. In Wales and western England, where such studies began, it was

FIGURE 11.8   *Left*: *Tribrachidium*, earliest known Eocambrian echinoderm. *(Photograph by G. R. Adlington.)* *Right*: Edrioasteroids from the Upper Ordovician, Cincinnati, Ohio, for comparison. *(Courtesy American Museum of Natural History.)*

FIGURE 11.9 *Left*: *Helicoplacus* from the Lower Cambrian. *Right*: *Camptostroma* from the Middle Cambrian also are echinoderms. *(Photographs by J. Wyatt Durham.)*

observed long ago that certain types of fossils tended to be restricted to particular sedimentary facies (Fig. 11.2). A shelly limestone facies is characterized by brachiopod and other shells, whereas a graptolitic shale facies is characterized by the fossils of presumed floating organisms known as graptolites (see Fig. 12.8). Certain peculiar trilobites also occur in the latter facies (Fig. 11.3). Association of particular organisms with certain sediments reflects environmental or ecologic conditions. The shelly fossils commonly show breakage, abrasion, and some sorting as to size of shells, and so must have accumulated in relatively agitated water. But the delicate graptolites apparently were preserved best when deposited in relatively still water. Moreover, the dark color of the typical graptolitic rocks suggests that the environment was poorly oxygenated so that disseminated organic matter was not thoroughly oxidized or decayed. Scavengers probably were excluded by the paucity of oxygen, also favoring the preservation of the delicate graptolites.

Wales lies within the ancient Caledonian mobile belt, and most of England is on the edge of a northern European craton. Some of the earliest studies of sediments in mobile belts were conducted in Wales, and it was noted that the graptolitic shale facies tend to be confined to such belts (especially to the volcanic-rich zones therein), whereas the shelly limestone facies is most characteristic of the craton and the adjacent nonvolcanic edge of the mobile belt. In addition, the dark, poorly sorted graywackes showing graded bedding are characteristic of the graptolitic facies. On the other hand, better-sorted sandstones containing prominent cross stratification are rarely associated with graptolites, but instead are found intimately interstratified with shelly limestones. Both of the latter, then, seem to characterize strongly agitated environments.

Fossil-rich carbonate rocks reflect special conditions, namely a paucity of terrigenous (silicate) detritus, which would suffocate many organisms, and favorable conditions of agitation as are found today especially in shallow tropical seas where invertebrate populations flourish. Countless carbonate-secreting organisms existed wherever extensive bodies of carbonate rock were formed, for skeletal material comprises a sizeable proportion of such rocks. Apparently organisms play the dominant role in extracting calcium carbonate from sea water.

Some fundamental generalizations about relations of faunas and sediments are possible, as was verified in eastern North America by the early finding of graptolitic faunas identical with those of western Britain and in dark mudstone and graywacke rocks (Fig. 11.2). The observation that Cambro-Ordovician faunas are similar on both sides of the Atlantic tends to support hypotheses of continental separation outlined in Chap. 8.

FIGURE 11.10 Examples of soft-bodied animals and arthropods from the Middle Cambrian Burgess Shale at Field, British Columbia. Note the exquisite detail of the arthropod appendages. *(Photographs courtesy U.S. National Museum.)*

EOCAMBRIAN DEPOSITS

EOCAMBRIAN GLACIATION

The most remarkable trademark of Eocambrian rocks is the presence of peculiar unsorted boulder-bearing deposits slightly below fossiliferous Cambrian strata. Since they were first described in northern Norway in 1891, they have been found on practically all continents (Fig. 11.12). Peculiar textures, wide distribution, and local scratched surfaces beneath the conglomerates led most geologists to regard them as tills (Fig. 11.13), representing a major episode of continental glaciation about 700 million years ago. Continental ice sheets spreading onto shallow marine shelves deposited till, which was more or less reworked by gravity sliding and currents, while icebergs dispersed and dropped boulders into normal

FIGURE 11.11 *Scolithus*, an animal burrow tube (vertical dark features), in Upper Cambrian cross-stratified glauconitic sandstone in Wisconsin.

FIGURE 11.12 World distribution of known Eocambrian till-like boulder deposits compared with paleomagnetically determined equator. For comparison brown band shows distribution of Lower Cambrian archaeocyathid fossils thought to have been tropical. Known older Prepaleozoic tills also are shown (triangles). *(Modified from Harland and Rudwick,* The great infra-Cambrian ice age*; Copyright © 1964 by Scientific American, Inc. All rights reserved; Bartholomew's Nordic Projection used with permission.)*

marine sediments beyond glacier margins (see Fig. 10.24).

Eocambrian tills occur over an exceptionally broad latitudinal range, either in terms of present latitudes or of different Eocambrian pole-equator positions (Fig. 11.12). The latter postulated shift is suggested by studies of rock magnetism explained in Chap. 8. The broad latitudinal distribution of the deposits has been rationalized by assuming such a complete chilling of climate that glaciers formed at

EOCAMBRIAN TILL-LIKE BOULDER BEDS

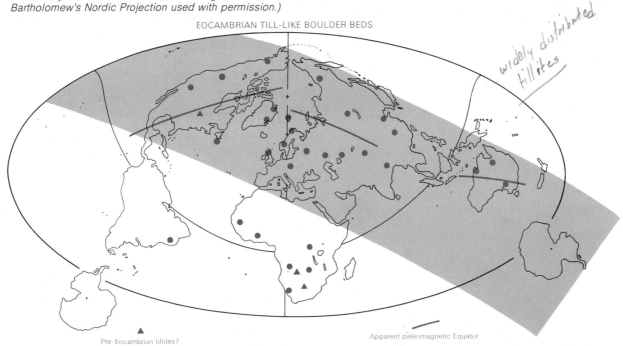

Pre-Eocambrian tillites?

Apparent paleomagnetic Equator

FIGURE 11.13 Eocambrian (Bigganjargga) till and underlying striated rock surface in Varangerfjord, Finmark, northern Norway. Note unsorted texture of the conglomerate and striations parallel to hammers. (Courtesy H. Reading.)

practically all latitudes, or at least that icebergs carried debris into very low latitudes.

EOCAMBRIAN SEDIMENT DISTRIBUTION

Eocambrian strata are well developed in many mobile belts but are nearly absent from most cratons. It is clear that the North American craton was a vast, low land area during latest Prepaleozoic and Eocambrian times. But the margins of the continent received thick Eocambrian sediments with some local volcanic rocks interstratified with them. The sediments represent shallow-marine continental shelf as well as nonmarine deposits.

Figure 11.14 shows an example from the southwestern United States of the relationship of Eocambrian and Cambrian strata to the craton margin. The diagram also illustrates how the geologist constructs cross-sectional diagrams to help interpret thickness, sedimentary facies, and age patterns. In this case we see at once that the thickest strata and most complete age sequence are represented beyond the craton in California and Nevada. Relative thickness of shallow-marine strata provide a direct guide to relative subsidence of the crust. In that region, Eocambrian strata are conformable beneath Cambrian ones, and there, subsidence amounted to thousands of meters. Farther east the Eocambrian and Lower Cambrian are missing entirely, however, and only a few

hundred meters of Middle and Upper Cambrian strata rest unconformably upon the Prepaleozoic basement.

All around the continent we find these same patterns. First, thick Eocambrian and then younger and younger Cambrian and even Early Ordovician marine strata rest upon the Prepaleozoic basement progressively inward toward the center of the craton. Clearly the sea surrounded the continent at first and gradually spread over the craton. This entire transgression spanned about 300 million years—equal to almost the entire remainder of the Paleozoic Era after Cambrian time!

The pattern of facies in Figs. 11.14 and 11.16 show that the clastic sediments were derived from the craton, because the sandy facies always occurred on the inner (craton) side of the shale and carbonate facies. Because the cratonic surface was weathered and eroded for nearly half a billion years before Eocambrian sedimentation began, a huge volume of clastic material must have become available. Undoubtedly it was deposited, eroded, and redeposited countless times during this interval by rivers, wind, and glacial ice. In many cycles of erosion and sedimentation, the compositional maturity of sands and gravels and the separation of clays by winnowing increases. This was the case, because pure, well-sorted and rounded quartz sand is prominent in both Eocambrian and Cambrian strata.

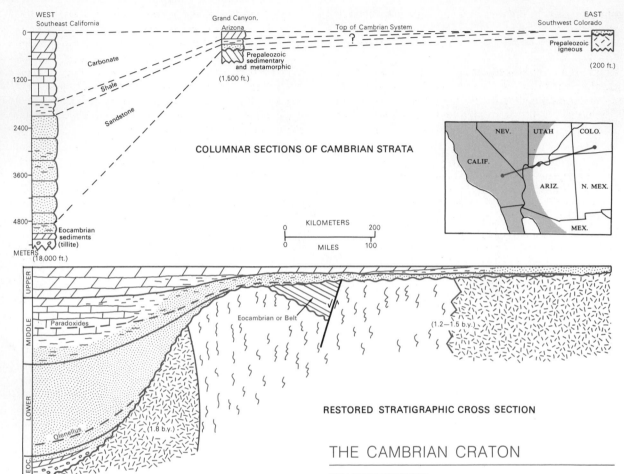

FIGURE 11.14 Cross sections showing Eocambrian and Cambrian strata in the southwestern United States. *Columnar sections* (above) portray factually the character of preserved rocks exposed at several localities. The *restored cross section* (below) was constructed from the columnar sections and ignores subsequent deformation and erosion. It is an interpretive graph in which the vertical axis is stratigraphic thickness, and the horizontal axis is distance between data points. The geologist infers thickness and lithologic facies variations between separate columns of the upper diagram. Such cross sections are helpful in portraying regional relationships among strata in a vertical plane and are invaluable companions to facies maps, which show relationships in the horizontal plane. The vertical scale is exaggerated to show detail. Note the transgressive facies, and how time lines overlap progressively eastward onto the basal unconformity. (See Fig. 11.15 for symbols.)

THE CAMBRIAN CRATON

TRANSGRESSION

Flooding of the craton continued throughout Cambrian time. By the end of the Period, probably two-thirds of the continent was doused beneath a shallow epeiric sea (meaning "a sea over a continent"). The average apparent rate of transgression was about 18 kilometers (10 miles) per million years (or 0.05 feet per year), hardly a tidal wave. Actually the transgression was spasmodic, being interrupted by several partial regressive oscillations of the shoreline. Figure 11.17 is a paleogeographic restoration of the continent's appearance in mid-Late Cambrian time. It was constructed from the evidence contained in the facies map (Fig. 11.16). Considerable interpretation is necessary in preparing any paleogeographic map, especially where the strata have been entirely eroded from a large region in later time (as in northeastern Canada).

STRUCTURAL FRAMEWORK
Around the periphery of the continent, volcanic rocks (chiefly basaltic) interstratified with Eocambrian sediments suggest that deep fissures penetrated the crust there to allow molten mantle material to rise. It has been suggested that such fissures developed when

FIGURE 11.15 Explanation of symbols used on facies maps, paleogeographic and paleogeologic maps, and most cross sections in remainder of book *(Principal sources of data: Alberta Society of Petroleum Geologists, 1964, Atlas of facies maps of western Canada; Clark and Stearn, 1960, Geological evolution of North America; Douglas et al., 1963, Petroleum possibilities of the Canadian Arctic; Eardley, 1962, Structural geology of North America; Kay, 1951, North American geosynclines; Levorsen, 1960, Paleogeologic maps; Martin, 1959, American Association of Petroleum Geologists Bulletin; Donald E. Owen, unpublished maps; Raasch et al., 1960, Geology of the Arctic; Sloss et al., 1960, Lithofacies maps; Rocky Mountain Association of Geologists, 1972, Geologic atlas of the Rocky Mountain region.)*

large, Late Prepaleozoic lithosphere plates were rifted and dismembered—presumably by sea-floor spreading—to form embryonic North America and other continents. As such separation (divergence) of plates proceeded, thick continental shelf sediments accumulated along the new, structurally passive trailing margins of the continents (Fig. 11.18). Apparently the Eocambrian was a time of profound worldwide reorganization of lithosphere plate patterns.

On the southern margin of the craton (chiefly in southern Oklahoma), volcanism and granite intrusion occurred within a narrow zone extending nearly 500 kilometers northwestward into the otherwise stable craton. Very thick early Paleozoic sediments accumulated there within a trough that has been interpreted as an incipient rift structure like the modern East African rift valleys (see Fig. 10.23). Rifts such as this, named aulacogens by the Russians, apparently extend into cratons nearly perpendicular to a new divergent plate boundary. If correct, this feature also reflects plate reorganization.

STRUCTURAL MODIFICATIONS OF CRATONS
As we pointed out in Chap. 8, cratons are only relatively more stable than other regions. Moreover, the cratonic surface in Cambrian time was not perfectly flat topographically. Cratons have undergone structural changes, largely of a mild (epeirogenic)

On the following pages:

FIGURE 11.16 Upper Cambrian sedimentary facies (note symbols in Fig. 11.15). Nowhere does the zero thickness line (heavy brown) represent a Cambrian shoreline; erosional nature of this boundary is proven, especially at the southeastern side of the transcontinental arch and in west-central Canada, where the zero line intersects facies boundaries nearly at right angles. Clearly the facies once extended beyond the zero line. Note southerly orientation of paleocurrent arrows.

FIGURE 11.17 Our interpretation of Late Cambrian paleogeography. Facies maps provide factual evidence for constructing paleogeographic maps, which inevitably are very interpretive. The latter may show land, sea, equator, poles, currents, winds, climatic zones, etc. Wherever we must restore geography beyond the zero limit of facies maps, there is much chance of error. For example, the single large low area shown in northeastern Canada may have had many small islands. (Compare Fig. 11.1.)

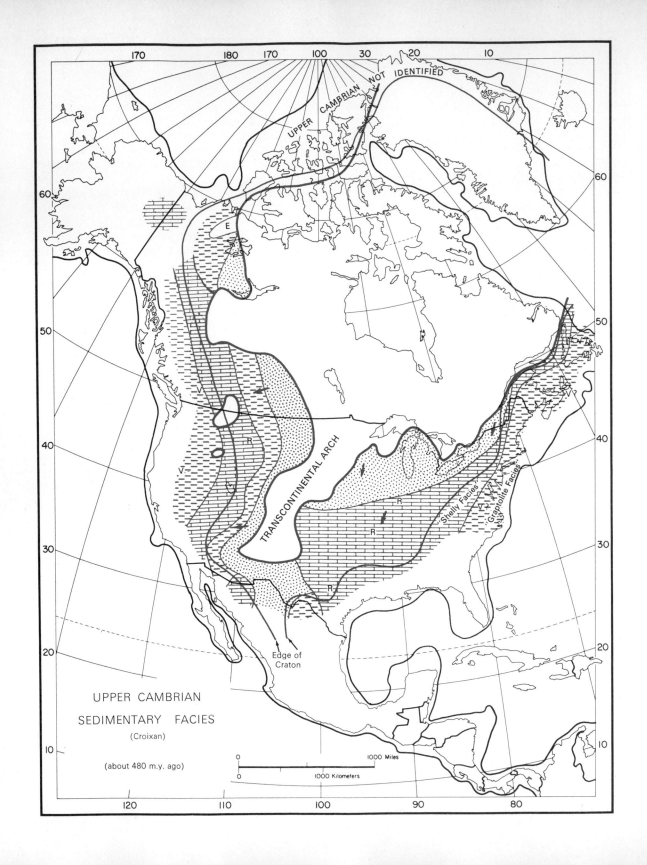

UPPER CAMBRIAN NOT IDENTIFIED

TRANSCONTINENTAL ARCH

Shelly Facies

Graptolite Facies

Edge of
Craton

UPPER CAMBRIAN

SEDIMENTARY FACIES

(Croixan)

(about 480 m.y. ago)

0   1000 Miles

0   1000 Kilometers

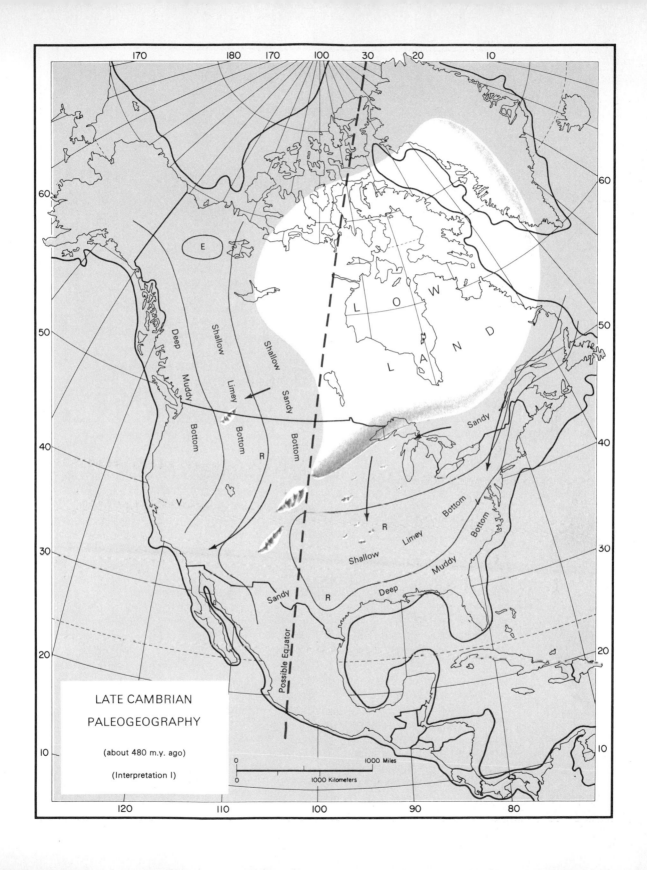

LATE CAMBRIAN

PALEOGEOGRAPHY

(about 480 m.y. ago)

(Interpretation I)

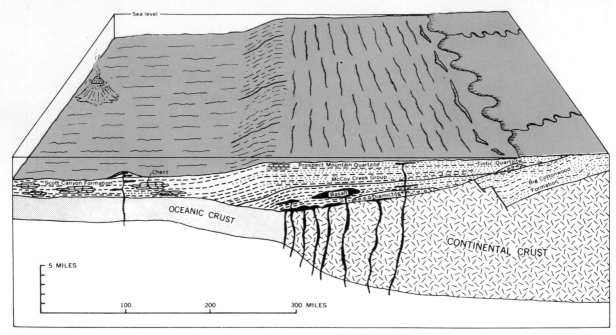

FIGURE 11.18   Diagrammatic restoration of Eocambrian and Lower Cambrian strata, suggesting that western North America was a passive continental trailing edge. Basalts suggest possible rifting of North America away from some other continent by spreading. Typical of trailing edges is the accumulation of thick prisms of continental shelf sediments with mature quartz sandstones and carbonate rocks. *(After J. B. Stewart, 1972, Bulletin of the Geological Society of America, v. 83, pp. 1345–1360; by permission of the Society.)*

sort, and these changes are reflected in the stratigraphic record.

The largest single feature of the Cambrian craton is the so-called <u>transcontinental arch</u> extending from Lake Superior southwest toward Arizona (Fig. 11.16). Along the sides of this trend, Cambrian sandstones lap unconformably against the Prepaleozoic basement, but along its axis, especially in the subsurface beneath the Plains states, Cambrian strata are entirely missing and younger rocks rest upon the Prepaleozoic there. Figure 11.19 shows the present distribution and thickness variations of preserved Cambrian sediments. Are these patterns entirely original ones, or do they reflect post-Cambrian erosion? Finally, what is the age and history of this great arch? Figure 11.16 reveals some clues not apparent from thickness maps alone. We see that dominantly quartz sandy facies border much of the arch, suggesting that it may have been a high feature supplying some of

that sand during Cambrian time; perhaps it was an old Prepaleozoic drainage divide present as a range of low hills converted to islands when the Late Cambrian epeiric sea flooded the craton. But in some areas, boundaries between dominantly sand and the more shaly and carbonate-rich facies are perpendicular to the trend of the arch, suggesting that it did not influence Cambrian sedimentation much there, but was raised and eroded later. Presence of several major unconformities over the transcontinental arch indicates a long history of periodic flooding alternating with erosion. Although it influenced sedimentation somewhat in the Cambrian, much warping of the feature was post-Cambrian.

Broad, gentle warpings of the cratonic crust, apparently in response to small changes of isostatic equilibrium due to density changes in the underlying lithosphere not only have raised areas called arches (domes, if circular), but have depressed areas called

FIGURE 11.19   Cambrian thickness map. Contours connect points of equal thickness of total Cambrian strata. Marked contrasts of thickness are revealed (especially at the margins of the craton), but their causes must be interpreted carefully. For example, subsequent erosion has removed Cambrian strata from large areas. (See Fig. 11.15 for sources.)

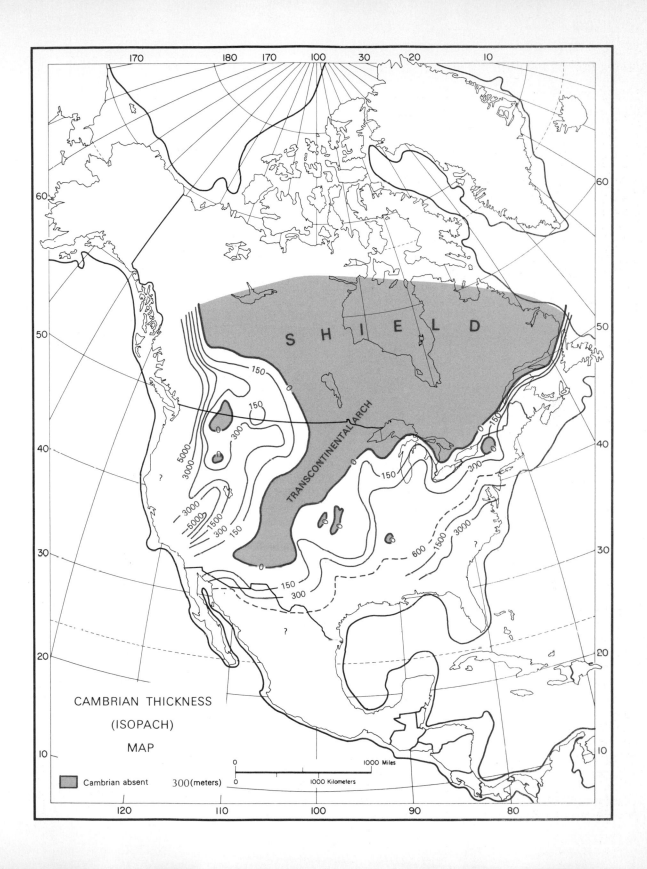

CAMBRIAN THICKNESS

(ISOPACH)

MAP

Cambrian absent   300(meters)

SHIELD

TRANSCONTINENTAL ARCH

150
0
150
300
5000
3000
3000
5000
1500
300
150
0
150
300
0
150
0
600
1500
3000
150
300
0
300

1000 Miles

1000 Kilometers

basins (Fig. 11.20). Faults also are present in cratons, and a few show great displacement. Sedimentation on cratons has been profoundly influenced by all such structural modifications. In addition, basins and arches have greatly influenced the accumulation of oil and natural gas, salt and gypsum, chemically pure limestone, and other economically important resources associated with sedimentary rocks. Therefore, the structure and history of cratons are of more than passing interest to humanity.

Compared with basins, arches typically contain thinner stratigraphic sequences that are interrupted by many unconformities. This implies that basins subsided relatively more rapidly and could therefore receive greater and more continuous sedimentary accumulations than arches. The arches intermittently were shoal areas experiencing little or no sedimentation or were islands undergoing active erosion.

INTERPRETATION OF THE HISTORIES OF ARCHES AND BASINS

In Fig. 11.20 note that thinning of strata over the arch is due both to originally thinner accumulations there and to erosion at several unconformity surfaces. Interpretation of cratonic history rests heavily upon knowledge of the ages of warpings of arches and basins; therefore, it is crucial to make correct interpretations of the causes and timing of the thick and thin accumulations. In Fig. 11.21 we show an identical situation of thinner and thicker strata, but note that a different historical explanation is required in this case, for these structural features were formed entirely *after* deposition of the rocks. In the first case, thinning over the arch was caused by lesser subsidence and smaller accumulation rates as well as by periodic erosion. But Fig. 11.21 shows thinning caused solely by subsequent erosion or secondary truncation (unconformity). The "basin" in this case is likewise purely secondary and is the result solely of downfold-

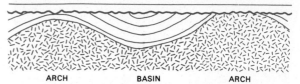

FIGURE 11.21 Differential warping *after* sedimentation to produce arches and basins. Note similarity of gross thickness patterns to those of Fig. 11.20; yet here the contrast resulted solely from erosional truncation *after* warping occurred. Histories of the two cases were very different, though results are superficially similar.

ing and truncation after deposition. If one is exploring for petroleum entrapments, for example, this sort of analysis is of utmost importance because potential structural traps must have formed before petroleum began to migrate through permeable strata if any was to accumulate.

How do we acquire and synthesize the information for studying broad and subtle cratonic structures? Much of the evidence for arches lies in surface outcrops, but only the edges of basins are exposed. Therefore, subsurface information is required to adequately locate basins and parts of some arches. Deep drilling for water and petroleum during the twentieth century has literally revolutionized the study of regional stratigraphy by providing the third dimension to our observations. Where deep drill-hole data are lacking, some insight into subsurface relationships can be gained from various geophysical devices. Magnetic and gravity surveys reveal information about special types and thicknesses of buried rocks. But seismology provides the greatest insight into thickness, structure, and, in some cases, even gross lithology of buried strata.

CAMBRIAN SEDIMENTS

PURE QUARTZ SANDSTONES
To see a world in a grain of sand
And a heaven in a wildflower.   (William Blake)

Let us next interpret the Cambrian sediments deposited in the epeiric sea. That the Upper Cambrian sediments of Fig. 11.16 are marine deposits is proved by presence of fossil groups such as brachiopods

FIGURE 11.20 Contrasts of thickness and unconformities between a cratonic arch and basin being differentially warped sufficiently to influence sedimentation *throughout* their histories.

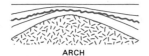

ARCH

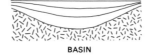

BASIN

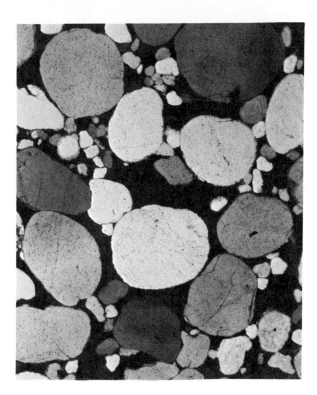

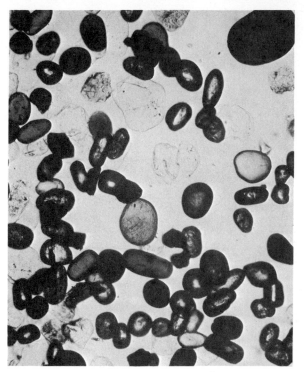

FIGURE 11.22 Microscopic photographs of typical Upper Cambrian sandstones, southern Wisconsin. *Left*: Quartz grains averaging 0.25 millimeter diameter. Rounding is so exceptional for larger grains as to suggest an abrasive history equivalent to rolling by water around the earth 30 or 40 times. *Right*: Concentrated fraction of well-rounded mineral grains of relatively high specific gravity (2.8 to 3.5) from a Cambrian sandstone like that at left. Minerals present average 0.15 millimeter and consist of very durable zircon, tourmaline, and garnet. Such minerals constitute less than 1 percent by weight of typical Cambrian sandstones. *(Right view courtesy J. A. Andrew.)*

and cephalopods known only to have lived in sea water. The green mineral glauconite (a potassium- and iron-bearing mica-like silicate), which is known to form only in sea water, also is widespread in Cambrian strata.

Upper Cambrian sandstones, the dominant cratonic sediment, rank among the most mature in the world. They are unrivaled for perfection of rounding and sorting of grains, and may contain 99 percent quartz with traces of other very stable minerals (e.g., garnet, zircon, and tourmaline) (Fig. 11.22; also see Figs. 10.14 and 10.16). All the clastic minerals point to ultimate metamorphic and granitic sources, but the grains had a long and complex sedimentary history prior to final deposition in the Late Cambrian epeiric sea.

The variation of sand-grain sizes is useful for interpreting their origin. By assuming that observable, present processes are keys to ancient ones, grain size characteristics of modern sands formed in known environments are useful for comparison with the textural properties of ancient examples formed by unknown processes (see Fig. 10.15). Sorting of the

Cambrian sands suggests wind, surf, or vigorous marine currents as the most probable agents. Structures such as ripple marks and cross stratification (Fig. 11.23) prove that they were transported by vigorous currents. Judging from relative average thickness of Cambrian strata in the craton versus the mobile belts (a ratio of 1 to 10), the rate of sedimentation in the craton must have been slow. Very slow addition of new sediments to a strongly agitated environment would be ideal for producing the observed texture and stratification of these rocks.

Rounding reflects the physical durability of

grains and the intensity and duration of abrasion. Experimental studies with transport of sand in laboratory troughs by water and in wind tunnels suggest that wind is more than 100 times as effective a rounding agent as running water. This is because impact between grains is cushioned slightly due to the greater viscosity of water. Cambrian strata are very uniform and widespread, and the presence of scattered marine fossils indicates that most were last deposited in the sea. But the unusually high rounding values suggest a long history of wind transport. How can these facts be reconciled? It is significant that the early Paleozoic pure quartz sands show higher overall grain rounding than do modern sands that have been analyzed. The present does *not* provide a complete key to the past, for we cannot find good examples today of all things found in the ancient record. No definite fossil record of land plants exists in rocks older than Devonian. Without a significant plant cover to hold soil in place, wind would have played a much greater role in erosion and transport of sediment than today. Therefore, the sand grains must have been acted upon by wind for long periods on the Late Prepaleozoic continental surface during the nearly half a billion years before their redeposition in the Late Cambrian sea. Confirmation is provided by the presence locally at the basal Cambrian unconformity of pebbles with peculiar flat sides identical with faces cut on pebbles today through blasting by windborne sand.

Some of the rounding and sorting of the Cambrian sand was probably also accomplished by strong tidal currents, which daily move grains back and forth across sand bars. In only 1 year's time, a grain might travel the equivalent of 2,000 kilometers within a relatively local area. In 1,000 years our little grain would have rolled and bounced 2 million kilometers, equivalent to 50 trips around the earth! Even though it was being moved by water rather than wind, the constancy of strong agitation surely would take its toll by pulverizing less durable grains and rounding durable quartz at a geologically reasonable rate.

Much sand produced by weathering through geologic time has not left the continental plateaus, but has been deposited and re-eroded almost endlessly to become more and more purified. For many years, geologists thought that the most profound purification and concentration of the quartz sand occurred during Early Cambrian time, but as was shown in Chap. 10, much initial concentration actually occurred long before the Paleozoic Era began, as testified by large volumes of Middle and Late Prepaleozoic quartzites (see Fig. 10.19).

Where is all the clay that must have formed by decay of the immense volumes of igneous and metamorphic rocks indicated by the pure quartz sand concentrate? At least half of the original source rock should have weathered to clay; yet shale is rare in Cambrian rocks of the craton. Apparently it was swept from the craton to find its way into deeper, less agitated zones of the sea beyond, for much shale does occur in Eocambrian and Cambrian strata on the outer edges of the craton beyond the Cambrian continental shelf edges. (see Figs. 11.16 and 11.18).

RIPPLE MARKS AND CROSS STRATIFICATION

Cross stratification is one of the most prominent features of the Upper Cambrian sandstones of the craton from Ausable Chasm in northeastern New York to the Canadian Rockies (Fig. 11.23). Ripple marks also are prevalent. Variations in form between wave- and current-formed ripple marks are well known, and the dimensional relations of current ripples differ slightly for the wind and water media due to viscosity differences (Fig. 11.24). It is said that certain cross-stratification types characterize wind dunes, others typify ordinary aqueous currents, while still others form only on beaches. Yet the form of cross strata actually varies greatly within any one modern environment. Sediment volume and coarseness, constancy of currents, and patterns of turbulence control the form of cross strata more than whether the transporting medium is air or water.

Ripples and dunes simply represent different scales of roughening, be they wind- or water-formed. Such phenomena have been studied experimentally for many years, beginning with water tank experiments in Germany in 1825. In those early experiments, brandy and mercury as well as water were used as fluids. Since then hundreds of artificial stream channels called flumes have been constructed to study erosion and transport of sediments by currents and waves (see Fig. 10.18). Similar studies have been made with air transport in wind tunnels. Cross laminae represent the buried lee faces of ripples or dunes (see Fig. 10.17). If the ripple crests are straight, then the cross laminae are simple and

FIGURE 11.23 Upper Cambrian (Galesville) sandstone along the Dells of the Wisconsin River. Lowest exposed stratum displays the simplest geometric type of cross stratification in which inclined laminae are almost perfectly planar, and surfaces truncating their tops are horizontal (see Fig. 10.17).

planar; but experiments show that, due to turbulence, ripple or dune crests more commonly are irregular (Fig. 11.25). As individual dunes migrate, intervening depressions become filled with sand in laminae that dip toward the depression axes. These interdune depression fillings are what we see preserved in most cross-stratified sandstones. The axes of such depressions provide the most accurate index of current direction (Fig. 11.25).

It was long thought that large bodies of cross-stratified sandstone must represent wind-dune sands simply because such dunes were the only readily accessible features that geologists could compare with the rocks. Herein lies a fallacy, however, for as exploration of the sea floor progressed, large *underwater* dunes as much as 65 feet high have been discovered on shallow, agitated sand bank areas. Examples are well known in the English Channel, on the Georges Bank and Nantucket Shoal off New England, and on the Grand Bahama Banks (Fig. 11.26). In these areas, very strong, turbulent currents develop due both to wind blowing over the shallow waters and to tidal fluctuations. The subaqueous dunes rival wind-formed ones in size, are similar in form, and they have essentially identical internal cross stratification. Given the apparently much larger and continuous area of very shallow sea during the Late Cambrian in North America, it seems inescapable that similar subaqueous dunes characterized large areas of that sandy sea floor.

CHANGE TO CARBONATE SEDIMENTS AND THEIR ENVIRONMENT OF DEPOSITION

As the burial of local islands and general submergence of most of the craton progressed, the supply of weathered quartz sand had all been redeposited by the end of the Cambrian Period, leaving little land exposed that could supply further terrigenous clastic material. It is as though a great sand faucet had been turned off after dripping steadily for 100 million years. Consequently, deposition changed to dominantly carbonate sedimentation over most of the craton. Carbonates had been forming already for millions of years on the cratonic margin (Fig. 11.16), and now, with final depletion of the quartz supply, similar

FIGURE 11.24 Profiles of ripple marks of different origins. Ideally, lee-face angle of inclination, wave length (*L*), and amplitude (*A*) vary for wind- versus water-formed current ripples. [*After E. H. Kindle, 1932, in* Treatise on sedimentation *(2d ed.), edited by W. H. Twenhofel; by permission of Williams & Wilkins Co.*]

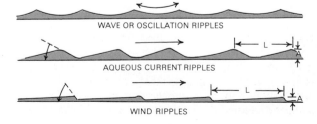

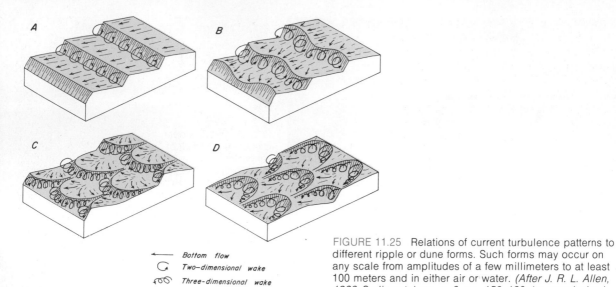

FIGURE 11.25   Relations of current turbulence patterns to different ripple or dune forms. Such forms may occur on any scale from amplitudes of a few millimeters to at least 100 meters and in either air or water. *(After J. R. L. Allen, 1966, Sedimentology, v. 6, pp. 153–190; by permission.)*

FIGURE 11.26   Air view of large submarine sand dunes (or sand waves) on the Bahama Banks, formed of oölite and other carbonate fragments by strong tidal currents at the margin of the banks in water only 6 to 7 meters deep. *(Air photo courtesy R. H. Ginsburg.)*

deposition could spread over most of the shallow epeiric sea floor, which was otherwise favorable for it.

Much limestone proves to consist of macerated shell debris and may even display cross stratification, ripple marks, and other features typical of sandstones formed of terrigenous silicate detritus. Such rocks are called clastic limestones. They make up over half of all limestones and typify the "shelly facies" named long ago in Britain. Several conditions can be inferred from their characteristics. By analogy with living counterparts, most marine invertebrate fossil communities during life required agitated, well-oxygenated water for best growth, and the textures of the "shelly facies" confirm that the material accumulated in sea water with current activity sufficient to roll, abrade, and sort shells. Early Ordovician carbonate rocks imply a minimal influence of mud and sand from eroding lands and also a uniform, relatively shallow marine environment where calcareous-secreting algae and invertebrate animals could thrive. Today such creatures are most abundant and most diversified in the clear and well-lighted shallow, agitated, warm seas of the tropics and subtropics, but important carbonate sediments also form at higher latitudes

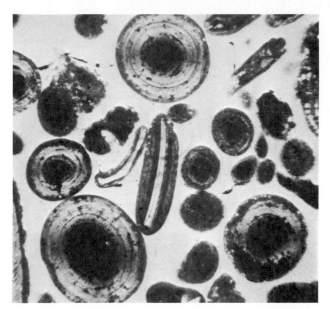

FIGURE 11.27 Microscopic photograph of modern oölites from the Bahama Banks. Note the central grain-nucleus of each sphere and surrounding concentric layers (spheres average about 2 millimeters in diameter). *(Courtesy R. H. Ginsburg.)*

IMPORTANCE OF OÖLITE

Fragmented fossils, scattered quartz grains, sporadic carbonate pebbles, and ripple marks in Late Cambrian–Early Ordovician carbonates point to considerable agitation during their deposition. Also common in these rocks are layers of crowded spherical grains called oölites, consisting of fine, concentric laminae surrounding some nucleus such as a quartz or shell fragment (Fig. 11.27). It is clear that they are accretionary in nature, and where they have been observed forming today (as in Great Salt Lake and on the Grand Bahama Banks), precipitation of carbonate occurs as they are rolled vigorously. Agitation of the water causes loss of carbon dioxide gas to the atmosphere, producing a saturated condition with respect to calcium ions. Calcium carbonate is therefore precipitated on all convenient surfaces such as sand grains. By inductive reasoning from the present to the past, we infer that ancient oölites also attest to strong agitation in shallow, calcium-rich waters; thus they provide an important paleoenvironmental indicator. The occurrence of oölites in thin zones in Late Cambrian and Early Ordovician strata suggests local episodic

changes either of degree of agitation or of chemistry in the epeiric sea.

DEPTH OF THE EPEIRIC SEA

The abundant evidence of agitation strongly suggests that the epeiric sea was shallow. Wavy, laminated, hemispherical stromatolite masses in the early Paleozoic carbonate rocks are thought to have been formed by primitive filamentous marine algae that lived on the sea floor (Fig. 11.28). Some of these stromatolites make up small reef mounds that are more abundant than were their counterparts in Prepaleozoic sediments. As was discussed in the last chapter, algae cannot receive adequate sunlight for photosynthesis in water deeper than 100 to 150 meters because of abrupt absorption of most wave lengths of light by the water. Therefore they provide the best maximum-depth indicator for the Cambrian and Ordovician epeiric sea. Apparently it was not more than 150 meters deep over most of its extent, and probably was much less in many places. Today simple algae that form laminated deposits like those of 500 million

FIGURE 11.28 Lower Ordovician algal stromatolite showing characteristic wavy lamination (Shakopee Formation, Troy, Minnesota) (compare Fig. 10.20). Today marine filamentous ("blue-green") algae form identical laminated carbonate deposits by entrapment of calcium carbonate particles by a mucous coating on the algae. *(Courtesy R. A. Davis.)*

years ago commonly inhabit the intertidal zone near the edge of subtropical seas, so it is inferred that many Cambrian and Ordovician ones occupied a similar environment.

Many of the fine dolomitic sediments associated with algal structures show polygonal networks formed by shrinkage cracking due to drying of fine muds as happens today on exposed tide flats and in drying mud puddles (Fig. 11.29). Polygonal cracks so produced gave rise, through subsequent current-stirring, to flat-pebble conglomerates common in early Paleozoic carbonate rocks. All stages from mere cracking to transport and rounding of the chips exist (Fig. 11.30).

FIGURE 11.29 Desiccation shrinkage cracks in modern lime-mud flats of Florida Bay formed by drying of the flats during low tide. *(Courtesy R. N. Ginsburg; with permission of Society of Economic Paleontologists and Mineralogists.)*

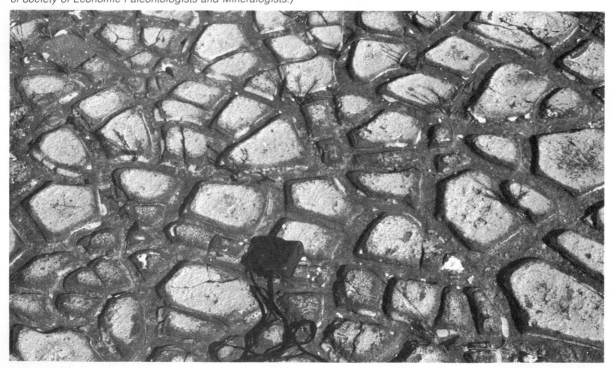

FIGURE 11.30 Flat-pebble conglomerate of fine dolomite chips formed by shrinkage cracking, then torn up by tidal currents or storm waves. Such conglomerates are especially common in Eocambrian Cambrian and Lower Ordovician strata (Upper Cambrian, southeastern Missouri).

NORTHERN GULF OF MEXICO—MODEL FOR EARLY PALEOZOIC EPEIRIC SEAS

The continental shelf of the northern Gulf of Mexico today provides some useful models of environmental conditions in the early Paleozoic epeiric seas. The shelf area is less than 200 meters deep, has moderately strong currents over it (Fig. 11.31), and has varied sediment sources. The Mississippi River supplies the bulk of detritus. On the outer shelf, much clay occurs, but deposition there since the last rise of sea level is nearly negligible. In Florida to the southeast, by contrast, the land is entirely Cenozoic limestone and the climate is humid subtropical. Therefore, essentially no terrigenous silicate clastic debris is supplied from it to the adjacent shelf. Environmental conditions off southern Florida favor carbonate sedimentation, and carbonate sands that contain considerable oölite are prominent on the shelf. At the south end of the peninsula, where nutrient rich waters rise against the shelf edge, ideal growth conditions for organic reefs exist. Behind the Florida Keys reef zone, in the protected, less-than-3-meters-deep waters of Florida Bay to the north, very fine limy muds are forming.

While areas of respective facies differ, Gulf of Mexico sediments present good analogues for major early Paleozoic facies discussed for the craton. Quartz sands along the Alabama-Georgia coasts are texturally similar to those of the Upper Cambrian of the central craton, while carbonate sands and lime muds of the western Florida shelf are roughly similar to the Early Ordovician cratonic carbonate rocks. But, though the Ordovician rocks possess small algal reefs, ancient large animal reefs like those of the present Keys area did not form on the craton until

Silurian time. Likewise, nothing comparable to the Mississippi Delta sediments is known in Cambrian rocks.

On the outer north shelf, clay-rich and slightly glauconitic sediments suggest analogies with finer silty and shaly glauconitic facies of Upper Cambrian strata. West of the Mississippi delta, there are large sandy areas of "nondeposition" where new sediment is not being added and the sandy bottom is only

FIGURE 11.31 Recent sediments around southeastern United States. Much of the sediment, particularly on the outer shelf, is relict from the last fall of sea level. On the average, sand on the outer shelf is moved by storm waves only about once every 5 to 10 years. Note the dominant surface-current movement parallel to the coasts. *(Adapted from F. P. Shepard et al., 1960,* Recent sediments of the northwest Gulf of Mexico; *with permission of American Association of Petroleum Geologists.)*

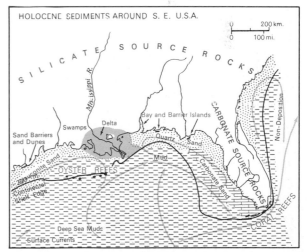

influenced by major storms. Violent hurricanes profoundly influence sedimentation on the Gulf shelf and along the shoreline. Geologically speaking, hurricanes are common, and shorelines that "normally" are very stable are greatly modified within a brief span of time when a violent storm strikes (Fig. 11.32).

Do ancient sediments record largely average conditions or more violent, extreme conditions? As we have seen, vigorous current and wave agitation characterized much of the epeiric sea floor. Doubtless constant currents were present, but some sporadic and more violent processes such as storm waves also were important. For example, layers of conglomerate interstratified with normal quartz sandstone occur near former islands of Prepaleozoic rocks, and flat-pebble conglomerates occur in carbonate rocks (Fig. 11.30). Each coarse layer probably records a single violent storm. It is likely that much, if not most, of the preserved epeiric sea record in such sediments was formed not so much by ordinary or average conditions, but by less-frequent, short-lived, violent events. Cambrian quartz sands may have been only slightly affected by normal currents, but

suffered large-scale transport during brief, violent storms.

PALEOCLIMATOLOGY

HEAT BUDGET OF THE EARTH

Continuing our analysis of paleogeography, we next examine some basic principles for interpreting paleoclimate. The principles elucidated here also apply to post-Cambrian time (see also Chap. 18). The diagnosis of past climates is one of the most complex facets of earth study. Still, certain broad generalizations are warranted, and they serve at least as constraints upon possible interpretations until such time as better guidelines appear. Climate represents a statistical average effect of the many meteorologic changes experienced by any given area, and those changes are related ultimately to atmospheric temperature and motion. Atmospheric motion is induced by differential heating and cooling and by the rotation of the earth. Landmasses, particularly if mountainous, obstruct and complicate movements.

Average temperature of the earth depends upon many factors, including the distance of the sun and tilt of the rotational axis. Also important are the transparency of the atmosphere to both incoming and outgoing radiation as well as the area and height of landmasses. Under any configuration of the earth that we can conceive, there always would have been a

FIGURE 11.32 *Left*: Matagorda barrier island, central Texas coast, immediately after Hurricane Carla, 1961. *Right*: The same coast a few weeks later. (Gulf of Mexico to left, Matagorda Bay to right in both views.) Hurricane-driven waves breached the island, spreading great volumes of sand both seaward and landward, but the breaches closed quickly, restoring equilibrium. *(Courtesy Shell Development Co.)*

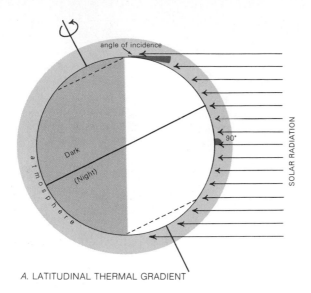

A. LATITUDINAL THERMAL GRADIENT

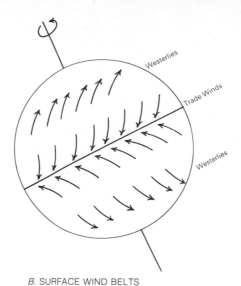

B. SURFACE WIND BELTS

FIGURE 11.33   Chief factors affecting atmospheric circulation. *A*: Solar radiation and the latitudinal thermal gradient of the earth's atmosphere due to varying angle of incidence of sun's rays. *B*: Climatologists believe that the dominant zonal wind belts probably were present throughout geologic time (though variable in breadth and intensity). Spin of the earth causes deflection of winds from north-south paths; this deflection is known as the Coriolis effect.

significant temperature difference between the poles and equator. This results from the geometry of the curved surface of the earth and its atmosphere with respect to solar radiation (Fig. 11.33A). The pole-equator temperature gradient might have been accentuated or ameliorated by variable tilt of the axis, but it could not have been eliminated.

Sea water absorbs much solar radiation, becomes heated, and oceanic circulation distributes the heat over the earth. Land surfaces, however, retain negligible heat; most of the solar energy reaching them is radiated back to the sky (Fig. 11.34). Therefore, the larger the total ocean surface, the warmer is the overall earth temperature, and vice versa, which already was appreciated by Charles Lyell in 1830. Moreover, if the poles lay in open oceanic areas, then circulation of heated sea water from lower latitudes would tend to warm the polar areas. But, if poles lay within or near large landmasses, they would tend to be colder and the pole-equator temperature gradient would be accentuated.

CLIMATIC ZONES

The Coriolis effect resulting from the earth's spin causes air moving north or south to be deflected east or west by rotation; therefore its actual path describes an arc relative to the earth's surface. Near the equator the steady trade winds blow toward the equator

because of the heating and rising of air there, but they are deflected westward (Fig. 11.33B). We assume a trade winds belt on both sides of the equator for all times. Midlatitude poleward-blowing westerly wind belts probably also existed throughout history, but they may have been subject to greater variability. In short, then, regardless of many unknown ancient climatic factors, almost certainly there has been some kind of latitudinal zonation of climate, but more distinct at some times than others.

With small, low continents, there would be little longitudinal (east-west) differentiation of climate, for atmospheric circulation cells could drift easily as the earth spun beneath the atmosphere. But when large and mountainous continents existed, then the atmospheric circulation cells would become more fixed in place by these obstructions (Fig. 11.35). Year after year, seasonal high- and low-pressure cells would be markedly differentiated longitudinally as well as latitudinally as today. Under conditions of fairly large

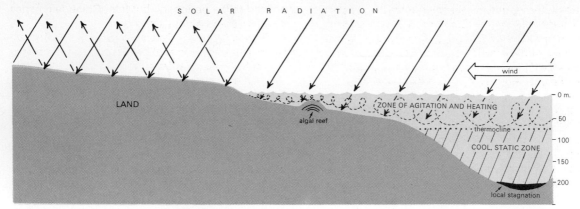

FIGURE 11.34   Contrasting effects of solar radiation on land versus water, and the relative mixing of water layers. The shallow-water irradiated and agitated zone is the habitat of algae and microscopic plankton, which form the base of the aquatic food chain.

continents, we could predict relatively drier areas on the west coasts of lands in subtropical latitudes; east coasts in such latitudes would tend to be wetter. It would be dangerous to extend our "predictions" any further without a great deal of supporting evidence.

GEOLOGIC INDICATORS OF CLIMATE

Certain peculiar sediments and fossil organisms suggest some further constraints upon paleoclimatic interpretations. Fossil plants provide the best paleoclimatic criteria for ancient land areas. For example,

FIGURE 11.35   Effect of a large continent on atmospheric circulation. Latitudinal zonal wind patterns are modified into high- and low-pressure cells locked into position by the land mass.

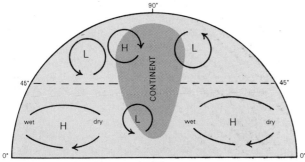

coal forms primarily in humid swampy areas that generally lie near sea level, though peat also forms in cold tundra regions. The nature of the plants present and the more restricted areal extent of peat allow distinction from widespread, lower-latitude coals. Large organic reefs form today only in the shallow, agitated warm seas found between approximately 30°N and 30°S latitudes. Assuming that reef-building organisms had similar ecologic requirements in the past, we can tentatively infer ancient temperature and other conditions roughly analogous to those of present reef areas. Coupled with paleomagnetic evidence for past latitude, reefs can be of great paleogeographic value. Further discussion of this topic will be presented in Chap. 13. Evaporite deposits have climatic significance in that they indicate high evaporation potential, i.e., relative aridity. But this need not indicate excessively hot temperatures, only an excess of evaporation over precipitation. Nonetheless, the setting of most ancient evaporites is consistent with warm temperatures. Proven ancient glacial tills and associated deposits provide evidence of unusual refrigeration of climate. In succeeding chapters, numerous examples of climatic interpretations will be cited and explored further as evidence arises, and we shall try to use these interpretations to help infer ancient paleogeography.

SHALLOW-SEA CURRENTS

Having established the presence of a shallow epeiric sea characterized by persistent currents and periodically stirred by violent storms, we can now speculate about the relation of the craton to overall continental geography. By inductive reasoning from the present seas, we can say that circulation in a broad epeiric sea only a few hundred feet deep would be controlled primarily by winds blowing across the water surface, and secondarily by tidal currents. Islands would, of course, deflect and modify circulation. The orientation of the bottom slope would have little influence in such a sea. Most currents in shallow seas run parallel to the bottom contour (Fig. 11.31), and Cambrian epeiric sea currents apparently roughly paralleled the transcontinental arch.

The Coriolis effect influences ocean water circulation as well as that of the atmosphere. It is known theoretically that water set in motion by wind does not move in exactly the same direction as the air. Due to the Coriolis effect, it is deflected, and, at successively greater depths, is deflected more and more. Obviously the relation may be very complex between surface currents and flow along the bottom, which affects sediments.

POSSIBLE CAMBRIAN CLIMATIC ZONES

In shallow seas one might expect very erratic directions of currents to be indicated in the sediments, and indeed this tends to be so. Nonetheless, a south- to southwest-flowing average bottom-current direction is suggested throughout the central craton by cross-stratification orientations and the concentration of Cambrian shales at the southwest and west edges of the craton (Fig. 11.16). What sort of winds might produce this overall persistence of direction? In spite of difficulties in interpreting bottom-paleocurrent data, trade winds, which today control the tropical surface circulation of the seas, could have produced the observed paleocurrent pattern had the craton then been within the low latitudes. Just such a position is, in fact, suggested independently by paleomagnetic studies (Fig. 11.17)! All evidence points to North America having been a tropical continent in earliest Paleozoic time.

SUMMARY

Eocambrian time is designated as beginning about 700 million years ago after the Grenville orogeny. The periphery of the young continent was a broad continental shelf with thick marine sediment accumulations. Eocambrian and Early Cambrian strata were confined to the periphery, but later Cambrian marine deposits spread across the craton.

Fossils are rare in Eocambrian rocks, but as more are being found, it is realistic to redefine the beginning of the Paleozoic Era to include the Eocambrian. It was during Eocambrian time that a few primitive invertebrate skeletons appeared, but Cambrian strata contain the first truly abundant and complex skeletons. Trilobites dominated the Cambrian fauna, but they show a grouping influenced by bottom environments. The same faunal groups and facies occur in northwestern Europe, suggesting some past closer connection between the two continents. Cambrian trilobites show several prominent structural evolutionary trends, notably reduction of the number of skeletal segments through time. Brachiopods, archaeocyathids, molluscs, and echinoderms make up most of the remainder of the preserved fauna, but there also is evidence that many soft-bodied, rarely preserved creatures existed; in life these may well have exceeded in number the better-preserved animal skeletons.

The continent was unusually stable. The craton, which had been thoroughly weathered and eroded for at least 200 million years, was gradually inundated by Cambrian transgression, forming a huge epeiric sea. Very pure quartz sand was redeposited widely on the sea floor as the shoreline shifted. Much of the sand had been moved ceaselessly over the landscape by wind before being reworked by the Late Cambrian inundation. Final deposition to form cross-stratified, pure quartz sandstones was accomplished by surf and shallow-marine currents, which apparently produced large areas of submarine dunes like those of the modern Georges Bank.

Just before the advent of Ordovician time, carbonate sedimentation superseded that of quartz sand. Vast dolomite deposits characterized this accumulation in seas that could not have been more than 100 to 200 meters deep. Great tidal flats apparently existed at the margin of the sea. Current and storm-wave circulation dispersed sediments in the epeiric sea. In the central craton, prevailing transport by bottom currents was generally southerly and southwesterly, more or less parallel to the transcontinental arch.

Circulation in very shallow seas is almost entirely wind-controlled. Paleomagnetic evidence indicates that the Eocambrian-Cambrian equator lay across central North America in a nearly north-south direction. This in turn suggests that the epeiric sea currents may have been driven by steady trade winds blowing in subequatorial zones that included most of the craton. Occasionally, however, storm winds affected sedimentation profoundly.

It has long been assumed that preserved sedimentary rocks record primarily normal or average conditions for past epochs. But in a region of very slow sedimentation, such as the Cambrian craton and outer continental shelf regions today, normal, day-to-day processes affect sediments very little. Only in the surf zone and on very shallow banks is sediment moved appreciably, except during rare, extremely powerful events such as hurricanes. Recognition of the importance of the so-called rare event requires an important shift of one's time perspective. To paraphrase G. G. Simpson (1952), based upon ordinary human experience, what we may regard by probability as impossible becomes barely possible given millions of years of geologic time; what was regarded as possible becomes probable; and the probable, virtually certain. It follows that catastrophic storms judged rare by human standards must be regarded as common on the geologic time scale, and therefore may have been the principal agents to leave an imprint upon Cambrian sediments rather than more prosaic day-to-day processes.

Readings

Dott, R. H., Jr., 1974, Cambrian tropical storm waves in Wisconsin: Geology, v. 2, pp. 243–246.

Gretener, P. E., 1967, Significance of the rare event in geology: Bulletin of the American Association of Petroleum Geologists, v. 51, pp. 2197–2206.

Harland, W. B., and Rudwick, M. J. S., 1964, The great infra-Cambrian ice age: Scientific American, August, pp. 28–36.

Hayes, M. O., 1967, Hurricanes as geologic agents: Bureau of Economic Geology, The University of Texas, Report of Inv. No. 61.

Holland, C. H. (ed.), 1971, Cambrian of the New World: London, Wiley-Interscience.

Irving, E., 1964, Paleomagnetism and its application to geological and geophysical problems: New York, Wiley.

Kay, M., and Colbert, E. H., 1965, Stratigraphy and life history: New York, Wiley.

Klein, G. deVries, 1970, Depositional and dispersal dynamics of intertidal sand bars: Journal of Sedimentary Petrology, v. 40, pp. 1095–1127.

Kuenen, Ph. H., 1960, Sand: Scientific American, April.

Laporte, L., 1968, Ancient environments: Englewood Cliffs, Prentice-Hall. (Paperback)

McKee, E. D., 1945, Cambrian history of the Grand Canyon region: Washington, Carnegie Institution Publication 563.

Middlemiss, F. A., Rawson, P. F., and Newall, G., eds., 1971, Faunal provinces in space and time: Liverpool, Geological Journal Special Issue No. 4.

Raasch, G. O., ed., 1961, Geology of the Arctic: Toronto, Univ. of Toronto Press.

Rocky Mountain Association of Geologists, 1972, Geologic atlas of the Rocky Mountain region: Denver, Hirschfeld Press.

Schwarzbach, M., 1963, Climates of the past: London, Van Nostrand.

Shepard, F. P., Phleger, F. B., and Van Andel, T. H., 1960, Recent sediments of the northwest Gulf of Mexico: Tulsa, American Association of Petroleum Geologists.

Simpson, G. G., 1952, Probability of dispersal in geologic time: American Museum of Natural History Bulletin, v. 99, pp. 163–176.

Sloss, L. L., Dapples, E. C., and Krumbein, W. C., 1960, Lithofacies maps: an atlas of the United States and southern Canada: New York, Wiley.

Woodford, A. O., 1965, Historical geology: San Francisco, Freeman.

12

THE LATER ORDOVICIAN

FURTHER STUDIES OF TECTONICS AND THE PALEOGEOGRAPHY OF MOBILE BELTS

We crack the rocks
and make them ring,
And many a heavy pack we sling;
We run our lines and tie them in,
We measure strata thick and thin,
And Sunday work is never sin,
By thought and dint of hammering.

Andrew C. Lawson,
formerly of the Geological Survey of Canada and the
University of California

FIGURE 12.1 Anticline in black Ordovician slates (Martinsburg Formation), exposed in an eastern Pennsylvania slate quarry. Slaty cleavage is inclined steeply downward to the right, parallel to axial plane of the fold but discordant with stratification. Such structures typify mobile belts. (Courtesy I. W. D. Dalziel.)

In Chap. 11 we discussed briefly Early Ordovician cratonic sedimentation. Recall that diminution of the quartz-sand supply near the end of the Cambrian Period allowed carbonate deposition, which continued into Ordovician time and resulted in one of the greatest accumulations of dolomite known. In this chapter we consider the history from 450 to about 425 million years ago.

The Ordovician record on the continental margins is clearer than that of the Cambrian. Volcanism was very important both in the Cordilleran and Appalachian regions. In central Nevada, vast volumes of ellipsoidal lavas (Fig. 12.2) associated with stratified chert deposits occur within thick sequences of black mudstone (now slate) that contain many graptolites. Coarse clastic detritus is relatively rare, again suggesting accumulation in still, probably fairly deep water far from land, as on a continental slope or deep ocean floor beyond the shelf edge.

Ordovician graptolitic slates and volcanic rocks also occur in the Appalachian belt, but are accompanied there by coarse clastic terrigenous material. Included are dark graywackes, local conglomerates, and red sandstones and shales, all reflecting elevation of a large land (or islands) in the mobile belt. Faunas in the Appalachian region continued to resemble strongly those of northwestern Europe and eastern Greenland. In Late Cambrian and Ordovician time, mountain-building disturbances and volcanism also affected the Caledonian mobile belt in Britain, underscoring its tectonic similarity with the restless Appalachian belt. The subduction of a now-vanished proto-Atlantic lithosphere plate beneath mobile belts on both sides of the Ordovician ocean basin is thought to have caused the orogenies and accompanying volcanic activity. In this chapter we examine especially the effects of orogenies and the evolution of geosynclines.

ORDOVICIAN LIFE

CONTRASTS WITH THE CAMBRIAN FAUNA

Just as the Cambrian had its unique fauna, which is easily recognized by the predominance of trilobites, so, too, Ordovician organisms are distinct. When one first encounters Ordovician fossils, one is struck by the abrupt change in faunal composition as compared to latest Cambrian faunas (Fig. 12.3). A part of this difference may be related to changes of environment reflected by greater abundance of limestone, dolomite, and shale in the Ordovician sequence.

FIGURE 12.2 Lower Paleozoic ellipsoidal lavas from Toquima Range, central Nevada (Willow Canyon Formation). These lavas and associated chert formed on the sea floor in what probably was deep water beyond the continental shelf margin. *(Courtesy M. Kay.)*

FIGURE 12.3 Ordovician diorama based on an Upper Ordovician fossil site near Cincinnati, Ohio. Note large squid-like nautiloid cephalopod in the foreground. *(Photograph courtesy American Museum of Natural History.)*

The Ordovician was marked by a rather long period of continental submergence climaxed by a Late Ordovician inundation that was the most widespread ever recorded. This great period of submergence was accompanied by mild and uniform climate and a very extensive distribution of bottom-dwelling species. There were many new shallow-marine ecological niches available so it is not surprising to find a much larger number of species than in the Cambrian. By the end of the Ordovician Period, most major groups of animals capable of being preserved as fossils had appeared.

In examining collections from the two early systems, we are struck by several important differences. The trilobites are not as numerous in Ordovician strata as brachiopods, bryozoans, and other groups. The evolution of the trilobites continued as vigorously as before, but along more restricted lines (there are about 1,000 species distributed through only 120 genera). The chief trends of evolution stressed "experimentation" of the eye and development of the central bulb and other gross features of the head section, which, to some researchers, reflects a shift to more active habits (Fig. 12.4).

Probably the most striking change in faunal composition involved the brachiopods. The phosphatic-shelled inarticulate groups (Fig. 11.6) of the Cambrian became diminished in numbers, probably due to ecologic replacement by the calcareous, articulate forms of the Ordovician. Throughout the Ordovician, there was an extensive period of evolutionary expansion involving ovoid forms belonging to several major groups (Fig. 12.5).

The molluscs, represented primarily by the snails, underwent considerable evolution with the innovation of important new groups. But like the trilobites, they were masked by other, more common forms and did not become a substantial part of the fauna again until the Mesozoic Era. However, one very important new group, the nautiloid type of cephalopod, appeared and went through a rapid period of evolutionary expansion during the early and middle parts of Ordovician time when some ten orders arose (they actually are first known as rare forms in the Upper Cambrian). These Ordovician fossils are very common, so apparently the creatures had little competition from other organisms. This is suggested in part by the huge size some of them attained (exceeding 6 meters long; Fig. 12.3); cephalopods were the first large animals to inhabit the seas.

DEBUTS

We must now look at some of the new phyla that appeared during the period. From the Early Ordovician on, the calcareous, colonial bryozoans (see Fig. Al.4) were to become constant companions of the brachiopods. Together they dominated most subsequent Paleozoic faunas. Not only do they occur together, but they seem to have been distantly related because they share several anatomical characters.

FIGURE 12.4 Ordovician trilobites. *Left: Isotelus. Middle: Flexicalymene. Right: Cryptolithus. (Courtesy U.S. National Museum.)*

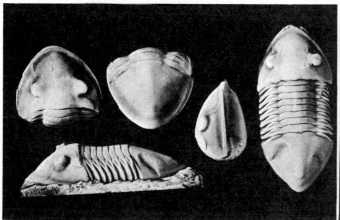

FIGURE 12.5   Strophomenid and ovoid orthid brachiopods from the Trenton Limestone of New York. *(Courtesy U.S. National Museum.)*

We have seen that the floating jellyfish type of coelenterate is found in rocks as old as the Middle Prepaleozoic. Its relative, the rugose coral, first appeared in Ordovician strata as a solitary form (Fig. 12.6). Colonial rugose corals, which also appeared during Ordovician time, developed by gradual elongation and narrowing of the individuals, presumably in response to crowding. When individuals touched, they became compressed and then formed prismatic cross sections, which allowed the tightest possible packing (Fig. 12.7). Colonial dwelling has the advantage of mutual protection and benefit.

THE GRAPTOLITES

Perhaps the most important of the new groups were the graptolites. These, as the Bryozoa, were microscopic individuals grouped together in colonies. They differed in possessing a chitinous rather than calcareous skeleton. A single group appeared in the Late Cambrian, but they radiated rapidly in earlier Ordovician time, spreading all over the world so that many species are shared by widely separated regions. Their very wide distribution is believed to indicate that they were planktonic or floating colonies (Fig. 12.8). Although they are known from many facies, the vast majority of graptolites are found in black shales, particularly in geosynclinal areas, for example on the eastern and western margins of North America, as was discussed at the beginning of Chap. 11. Because they evolved rapidly and became distributed instantly (geologically speaking) all over the world, and then became extinct equally abruptly, they are classic examples of ideal index fossils for stratigraphic zonation (Fig. 12.9).

EARLIEST VERTEBRATE FOSSILS

The earliest known vertebrates are represented by dermal plates from a Lower Ordovician greensand in

FIGURE 12.6 *Streptelasma*, a primitive, solitary rugose coral from the Upper Ordovician of Richmond, Indiana. *(Courtesy American Museum of Natural History.)*

FIGURE 12.7 *Hexagonaria*, a colonial rugose coral, from the Devonian of Michigan. *(Courtesy U.S. National Museum.)*

northern Russia. These plates, which formed an outer armor, are also found in the Harding Sandstone of Colorado. Unfortunately, no complete skeleton has been found thus far, but the plates are thought to be from jawless fishes (Agnatha). Nothing is known of the anatomy of these fishes, but the plates are like those of well-preserved middle Paleozoic jawless fishes. The environment in which the Ordovician creatures lived is not clear. A few marine invertebrates in associated strata suggest that the fish also were marine forms.

LATER ORDOVICIAN OF THE CRATON

MEDIAL ORDOVICIAN REGRESSION
After extensive dolomite deposition for several million years, a widespread disconformity was produced over virtually the entire craton. The sea retreated at least to the marginal mobile belt areas and the interior

was subjected to extensive erosion. Above the disconformity lies a very widespread pure quartz sandstone called the St. Peter. Though the sea had retreated from the craton, the topography was very subdued everywhere. Maximum relief observed in erosional channels is about 100 meters. Remember that only a small relative vertical fall of sea level (or rise of the continental platform) was needed to cause exposure of a tremendous area of the former shallow sea floor.

FIGURE 12.8 *Tetragraptus*, a typical planktonic graptolite colony from the Levis shale of Quebec. *(Photograph courtesy U.S. National Museum.)*

Zones	Great Britain	Eastern North America	Australia	China	
Monograptus nilssoni					Silurian
Monograptus testis					
Cyrtograptus linnarssoni					
Monograptus riccartonensis					
Cyrtograptus murchisoni					
Monograptus crenulatus					
Monograptus greistoniensis					
Monograptus crispus					
Monograptus turriculatus					
Monograptus sedgwicki					
Monograptus gregarius					
Orthograpius vesiculosus					
Akidograptus acuminatus					
Dicellograptus anceps					Ordovician
Dicellograptus complanatus					
Pleurograptus linearis					
Dicranograptus clingani					
Nemograptus gracilis					
Glyptograptus teretiusculus					
Didymograptus murchisoni					
Didymograptus bifidus					
Didymograptus hirundo					
Didymograptus extensus					
Dictyonema "flabelliforme"					

FIGURE 12.9 A diagram of Ordovician and Silurian graptolite zones showing the worldwide distribution of some species. This reflects a planktonic habitat. *(Redrawn from Newell, 1967,* Revolutions in the history of life, Geological Society of America Special Paper 89, *reproduced by permission.)*

As soon as exposure occurred, weathering, solution, and erosion began to remove earlier Ordovician dolomites. In the center of the craton, all dolomite was removed to re-expose Upper Cambrian quartz sandstones (Fig. 12.10). These were reworked and the sand again was spread out (for perhaps the hundredth time since the Middle Prepaleozoic) from the cratonic core to form a veneer over much of the craton. This great sheet of sand totals more than 20,000 cubic kilometers!

Stream and wind dispersal distributed the sand, but it was finally redeposited by the sea as transgression was renewed during the middle part of the Ordovician Period. Much of the preserved sandstone apparently represents shore and nearshore deposits formed at the advancing edge of the sea. Paleocurrent patterns were more complex than in the Late Cambrian sea, but geographic variations of grain size, and lateral facies relationships with other sediments of similar age, all point to derivation of the sand

FIGURE 12.10 Paleogeologic map of the medial Ordovician regional unconformity; transgressive sandstones (St. Peter) were deposited upon this now-buried geology. Such a map helps reveal the history of large features such as the transcontinental arch (see Fig. 11.16) and provides clues to paleogeography. Note that Ordovician strata overlapped beyond Cambrian ones to rest widely upon Prepaleozoic basement in central Canada and on the transcontinental arch (proving the arch had formed *at least* this early). The map is what one would see if he could lift off all younger strata overlying the unconformity. Actual evidence for preparing the map comes only from areas within the zero-thickness boundary of overlying strata. Subjective extrapolations must be made beyond that boundary where such strata are no longer present. (*After Levorsen, 1960, Paleogeologic Maps.*)

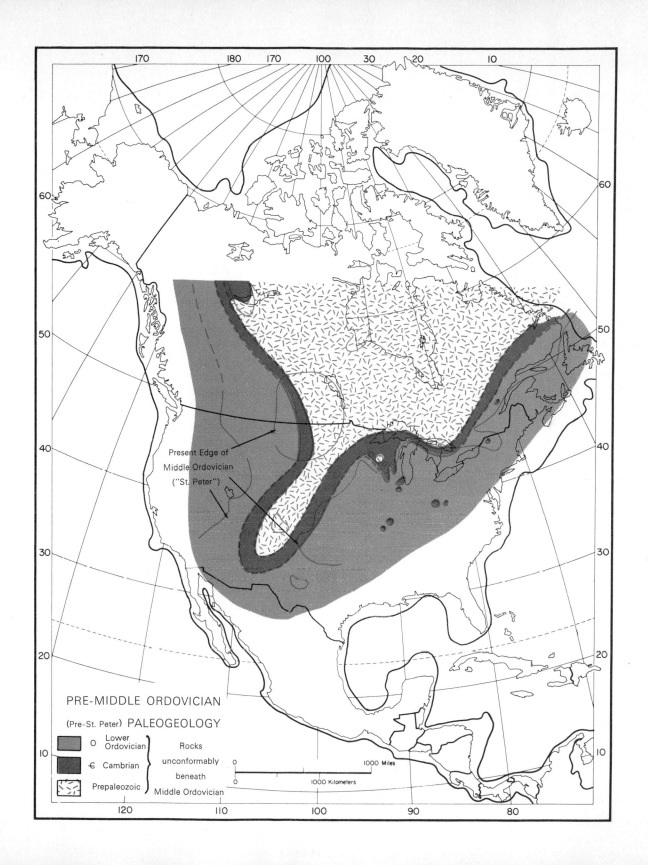

PRE-MIDDLE ORDOVICIAN

(Pre-St. Peter) PALEOGEOLOGY

Present Edge of
Middle Ordovician
("St. Peter")

		Rocks
O	Lower Ordovician	unconformably
∈	Cambrian	beneath
	Prepaleozoic	Middle Ordovician

0
0

1000 Miles

1000 Kilometers

largely from the center of the craton with transport outward. Pure quartz sand drifted southwest into Oklahoma, and even as far west as central Idaho and Nevada. In the latter region, it is interstratified in an unusual association with black, graptolitic mudstones and lavas, the typical early Paleozoic rocks of the western portion of the continent.

It can be shown that the sandstone is older near its outer limits than in the center of the craton (Fig. 12.11). It is a classic transgressive deposit whose deposition, as an educated guess, spanned about 5 million years. As noted in Chap. 11, studies of modern continental shelf sediments show that surprisingly little active sedimentation is now going on over large portions of shelves except in proximity to shorelines and on very shallow bank areas. Most of the silicate sediment on outer shelf surfaces today is relict sand and mud originally deposited during Pleistocene lowerings of sea level and but slightly modified by currents and storm waves since (see Fig. 11.31). On carbonate-producing shelf areas, as off southern Florida, however, new carbonate material is being added more or less constantly as many generations of organisms contribute their skeletons, and as oölites are formed.

How can we reconcile the very limited active clastic sedimentation on shelves with the presence of many ancient, very widespread quartz sandstones that appear to have formed in similar environments? Several disconformities present in the Upper Cambrian and Lower Ordovician successions of the craton reflect minor transgressions and regressions. Assum-

ing that most of the ancient sand was deposited in proximity to the shore zone, then great lateral shifts of the shoreline must have integrated gradually to produce widespread transgressive and regressive sand deposits. Any sandstone formation so deposited must vary in age across the craton as shown in Fig. 12.11. On the shallowest bank areas, which probably were very large in the past, shifting of sand relict from earlier stages of transgression probably also occurred as on sand banks today.

STRATIGRAPHIC SEQUENCE DIVISIONS

In Chap. 4, we referred to regional rock units called *sequences*, which are assemblages of many formations bounded by exceptionally widespread unconformities. The boundaries vary in age from place to place so that the sequences are not universal time divisions with strict age significance. Strata adjoining the bounding unconformities may vary in age several or even tens of millions of years. The unconformity beneath Cambrian rocks across North America and that beneath the St. Peter Sandstone together define the oldest sequence recognized, which has been named Sauk. The St. Peter Sandstone is the basal, transgressive deposit of the second (Tippecanoe) sequence. A total of six unconformity-bounded sequences are widely accepted for North America (Fig. 12.12). The sequence concept applies chiefly to cratonic strata, but increasingly we see evidence that the changes reflected by the great unconformities (and accompanying transgressions and regressions) are probably related to lithosphere plate motions and so also to tectonic events in mobile belts at the plate margins.

LATER ORDOVICIAN EPEIRIC SEA

During the regression of the sea discussed above, mutations and natural selection combined to produce marked changes in the Ordovician shallow-marine

FIGURE 12.11 *Left:* Diagram of medial Ordovician (Chazyan) strata in the southern craton showing northward decrease in age of transgressive quartz sandstone facies (St. Peter). (Vertical scale is exaggerated.) *Right:* Integration of countless successive shoreline deposits formed during transgression to produce a widespread, tabular mass of sandstone. (Vertical scale is exaggerated.)

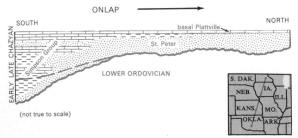

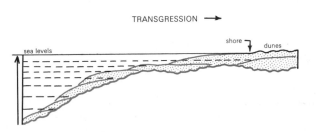

fauna, which was temporarily restricted to a much-reduced shallow-marine area on the borders of the continent. Surely this shrinkage of environment produced unusual selective pressures on the population, tending to accelerate evolutionary diversification. Certainly, the rich later Ordovician fauna, which invaded the craton as the epeiric sea readvanced, was markedly different from the early Ordovician one. It contained the first vertebrate animals as well as the more diversified corals, bryozoa, stromatoporoids, cephalopods, and new groups of brachiopods discussed above.

Upper Ordovician limestones were somewhat less extensively altered to dolomite than were the lower ones, and so their faunas are better preserved; many are 80 or 90 percent shell material. The Upper Ordovician strata of the Ohio-Indiana region are among the most fossiliferous rocks in the world, and they were the inspiration and training ground for a "who's who" of outstanding American geologists. Upper Ordovician fossil-rich carbonate strata also are the most extensive marine deposits on the entire craton; a cratonic basin near Hudson Bay contains a large volume of them and smaller patches occur elsewhere on the shield (see Fig. 12.15). The Late Ordovician epeiric sea represented one of the most complete floods experienced by any continent. A tremendous, uniform shallow sea resulted, the like of which is nonexistent today.

SHALY DEPOSITS OF LATER ORDOVICIAN TIME

In the Appalachian region, much of the later Ordovician sequence is characterized by dark mudstones with sporadic graywacke sandstones and chert layers. The top of the Ordovician sequence farther west also contains thin but persistent dark gray shale (Fig. 12.13). This shale is something of an oddity for the craton and must record some important change of terrigenous source areas. To determine the source, we must examine facies maps and cross sections (Figs. 12.14 and 12.15). The shale becomes thicker toward the east where practically the entire Ordovician sequence in the northern Appalachian region is composed of dark shales. Still farther east, important volcanic rocks occur in the Ordovician succession. Both lavas and pyroclastic rocks are found, which suggest that important volcanic islands were formed here. In summary, the facies patterns point to the

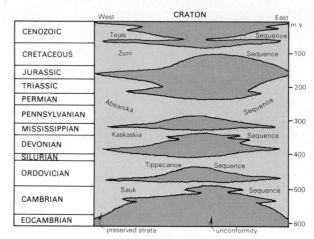

FIGURE 12.12   The six major unconformity-bounded sequences designated by L. L. Sloss for the North American craton. Dark areas represent the major unconformities, that is, space-time without any rock record today. These major unconformities serve to subdivide the cratonic stratigraphic record into widespread "packages" of strata. Note that the more complete depositional records are at the craton margins. The St. Peter Sandstone (SP) lies at the base of the Tippecanoe Sequence; the Oriskany (O) at the base of the Kaskaskia. (After Sloss, 1963, Bulletin of the Geological Society of America, v. 74, pp. 93–114.)

elevation of islands above sea level in the Appalachian belt, which shed sand and much fine mud that was carried westward. Most of this sediment was deposited within the mobile belt itself, but as the volume increased, mud literally spilled out of the mobile belt onto the edge of the craton in New York. Near the end of the period, clay and volcanic ash were carried as far west as Iowa (Fig. 12.15). The distribution of the shale facies suggests southwesterly flowing currents; ripple marks and other features confirm this. By latest Ordovician time, the continent appeared as shown in Fig. 12.16.

VOLCANIC ASH—A CORRELATION TOOL

In the Appalachian Mountain region there is a major lateral facies gradation from dominantly graptolitic shales southwestward to a shelly limestone facies more characteristic of the craton (Fig. 12.15). Because of the differences of environment reflected by these facies, there is also a difference of fauna in

FIGURE 12.13 Ordovician black shales and thin white limestones (Utica Formation) along the New York Thruway, central New York. In this area, mud derived from the east spread beyond the Appalachian belt onto the craton. Differential flow under the influence of gravity on water-laden mud caused contortions.

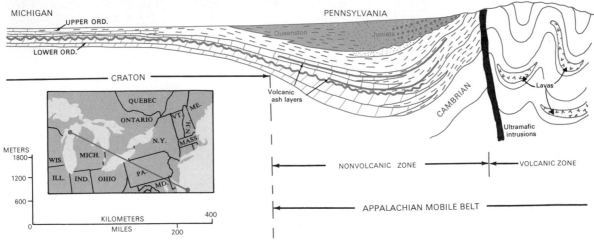

Section without vertical exaggeration

FIGURE 12.14 Restored cross section showing relations of Ordovician facies to the Taconian land. Note the volcanic ash layers that transcend facies boundaries to provide useful time datums for correlation purposes. Note pre-St. Peter unconformity. *(Adapted from M. Kay, 1951, Geological Society of America Memoir 48.)*

FIGURE 12.15 Upper Ordovician sediment patterns for North America, Widely scattered patches of sediments on the Canadian Shield prove the great extent of the Late Ordovician sea. Absence of Ordovician strata on several arches proves subsequent further warping and erosion of them. Note spread of red beds and marine shales westward from the Appalachian region (see Fig. 11.15 for symbols and sources).

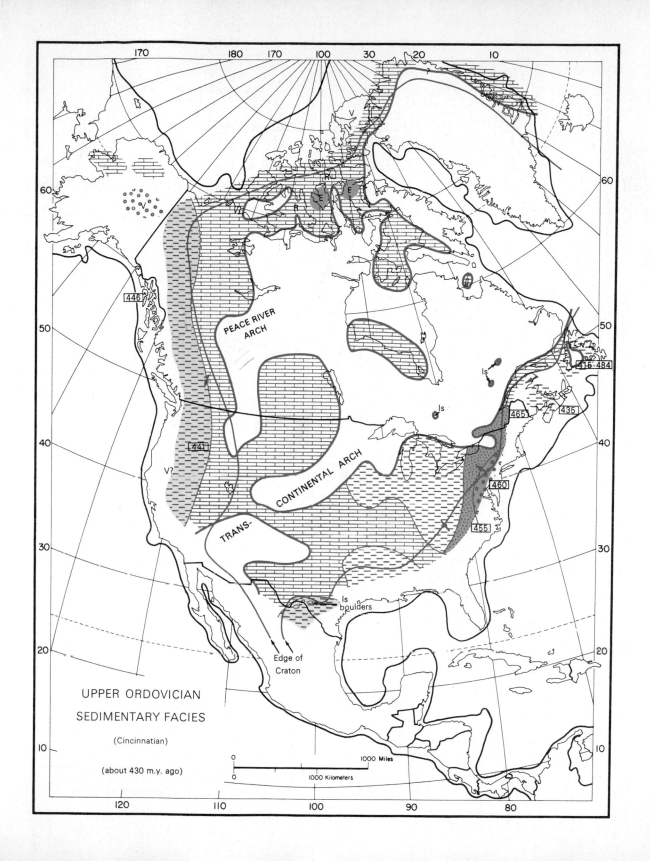

UPPER ORDOVICIAN

SEDIMENTARY FACIES

(Cincinnatian)

(about 430 m.y. ago)

PEACE RIVER
ARCH

CONTINENTAL ARCH

TRANS-

Edge of
Craton

Is
boulders

1000 Miles

1000 Kilometers

446

441

V?

465

460

455

435

415-484

V?

Is

Is

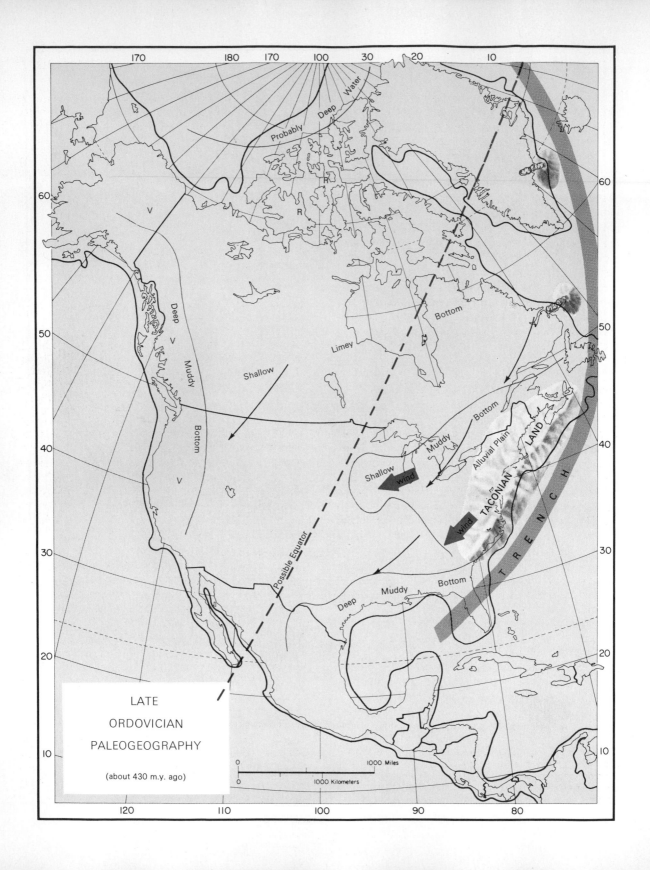

LATE

ORDOVICIAN

PALEOGEOGRAPHY

(about 430 m.y. ago)

each. In turn, the differences between their fossil assemblages make it difficult to correlate between the facies on the basis of index fossils; this illustrates the limitation of "facies fossils" discussed in Chap. 4.

Volcanic ash layers are found in the Ordovician strata of the Appalachian belt. They are represented today by thin, clay-rich layers formed by the alteration of volcanic dust particles. Because each layer was erupted instantaneously in a geologic sense, individual ash strata provide time datum markers. Certain of the ash layers extend into both shaly and carbonate facies, and so they constitute unique datums for correlation between the two facies (Fig. 12.14). Also, those which have not been entirely altered can be used for isotopic dating to provide a numerical date for a sequence of strata much as isotopic dating of glauconite provides for Upper Cambrian strata of the craton.

PALEOWIND PATTERNS

Perhaps of even greater interest is the clue that volcanic ash distribution provides about the possible Ordovician wind pattern in eastern North America. This distribution suggests that the ash was derived from known volcanic centers of the northeastern Appalachian mobile belt (Fig. 12.15) Apparently winds blew from northeast to southwest (in terms of present geographic coordinates; Fig. 12.16), which is the opposite of present winds. Marine currents would also influence ash distribution after it settled into the sea.

PALEOCLIMATIC SPECULATIONS

In spite of uncertainties, it is interesting to note the relation of the ash distribution to the possible position of the Ordovician equator with respect to North America as indicated by paleomagnetic data. In Fig. 12.16 you will note that the apparent equator extended from northeast to southwest—nearly bisecting the continent. This would place the Appalachian mobile belt within the trade wind zone and the ash distribution would be entirely consistent with such an orientation! The vast, richly fossiliferous Ordovician carbonate

FIGURE 12.16 Late Ordovician paleogeography interpreted from Fig. 12.15. Note especially the Taconian land, widespread marine inundation, and position of the equator according to paleomagnetism. Possible wind arrows are interpreted from volcanic ash distribution. (Compare Fig. 12.31.)

deposits of the craton would fall within 40 degrees latitude of the equator. But if the equator then occupied its present position with respect to the continent, those carbonate rocks would span 70 degrees of latitude. As will be shown more fully in the next chapter, such sediments today form chiefly, though by no means exclusively, in low, warm latitudes, and so it would be less surprising to find them at 40°N than at 70°N.

ORDOVICIAN OROGENY IN THE APPALACHIAN MOBILE BELT

EVIDENCES OF INCREASING STRUCTURAL MOBILITY

Volcanic rocks of the mobile belts prove increasing structural mobility, but abundant dark muds deposited widely in the Appalachian belt and far out onto the craton also reflect unrest and uplift in the east. In the southern Appalachians, local red shales appeared in medial Ordovician time, reflecting the elevation of land. In the northern Appalachian region, some spectacular, ill-sorted conglomerates with fragments from mixed Cambrian and Early Ordovician formations occur helter-skelter as local deposits within Ordovician black shale sequences; some of the blocks are many meters long (Fig. 12.17). These unusual deposits occur at several scattered localities. They represent submarine avalanches of debris derived from shallower water environments. The deposits became unstable and slid and rolled into the deeper, adjacent muddy environment. Such bouldery mudstones reflect disturbances of the sea floor heralding much more profound unrest destined to produce major mountain-building episodes later in the period.

In the northeastern United States and southeastern Canada, there is diverse evidence pointing to Ordovician mountain building. In central and western New York and Pennsylvania, Ordovician black shale is succeeded eastward by red shale. The "red bed" facies coarsens eastward in Pennsylvania to sandstone and some conglomerate (Figs. 12.14 and 12.15). The red deposits reflect a major change in depositional conditions. Because they lack fossils and show thorough oxidation of iron, it was suggested long ago that they were nonmarine deposits that bordered a large land somewhere to the east.

FIGURE 12.17 Coarse conglomerate and breccia of Cambrian fragments within Ordovician shales, Trois Pistoles, St. Lawrence River, Quebec. These breccias apparently originated by submarine sliding. *(Courtesy Jean Lajoie.)*

DIRECT EVIDENCE OF MOUNTAIN BUILDING

At a number of localities in the Appalachian region, important unconformities are visible within the Ordovician sequence and between it and Silurian or Devonian strata (Fig. 12.18). Extending from the St. Lawrence River valley south into eastern New York is a zone of major overthrust faulting that originated about the end of the Ordovician, or about the same time that the unconformities were developing. Farther east in New England and maritime Canada, some isolated small granite masses are overlain by Silurian strata and were intruded into older, probable Ordovician ones. Isotopic dates confirm that some granites were formed there during Late Ordovician and Early Silurian orogenesis. Long, narrow ultramafic igneous rock masses were emplaced along the mobile belt (Fig. 12.18). You will recall (Chap. 7) that these unusual rocks represent mantle material faulted into the crust during severe deformation. Finally, Ordovician sandstones and conglomerates are heterogeneous in composition and show derivation from erosion of limestones, sandstones, shales, and rare granitic rocks.

Evidence cited above points to a major interval of mountainous uplift, erosion, and intrusion of some igneous plutons during the Ordovician Period. This event represents the first great Paleozoic mountain building in the Appalachian belt, and is called the Taconian Orogeny for the Taconic Mountains of southeastern New York, where a major unconformity is exposed and where thrust faulting occurred during this upheaval.

PALEOGEOGRAPHY OF MOBILE BELTS AND THE ORIGIN OF GEOSYNCLINAL SEDIMENTS

EARLY IDEAS

Reconstructions of ancient geography of regions from their stratigraphic records is one of the highest goals of the earth historian. We have already demonstrated

FIGURE 12.18 Summary of evidences of the Taconian orogeny.

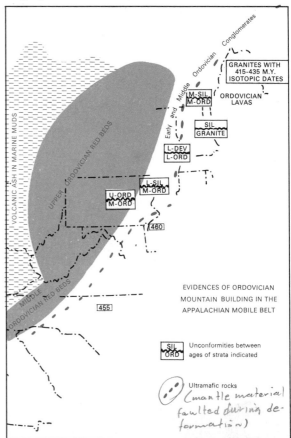

EVIDENCES OF ORDOVICIAN MOUNTAIN BUILDING IN THE APPALACHIAN MOBILE BELT

the reconstruction of cratonic geography for the Cambrian and Ordovician Periods. Now that we have examined some evidence from the Ordovician of the Appalachian region, discussion of paleogeography is appropriate for the mobile belts. The concepts developed in this section will apply as well to other cases.

The American J. D. Dana, one of the first to speculate about paleogeography of ancient mobile belts, believed that more or less permanent ridges of old Prepaleozoic igneous and metamorphic rocks lay at the borders of North America *throughout most of geologic time* (Fig. 12.19). This conclusion was based upon two principal observations: first, that many clastic sediments in the Appalachian geosyncline (such as the Upper Ordovician "red beds") coarsen eastward; and second, to the east of those strata today lie predominantly granitic, gneissic, and schistose rocks considered a century ago to be entirely of Prepaleozoic age. The "Prepaleozoic-looking" rocks, which are particularly prominent in New England where Dana lived, naturally were assumed to be the source of the clastic sediments that had been derived from that direction.

Another American (C. D. Walcott) amplified Dana's concept in 1891 by recognizing the importance of *two* distinct sources of clastic geosynclinal material—the craton as well as a supposed Prepaleozoic borderland beyond the geosyncline on the other side (Fig. 12.19). He suggested that the present Atlantic coastal plain once had been the site of a borderland from which much of the Paleozoic sediments of the adjacent, subsiding Appalachian geosyncline were derived. This concept of persistent borderlands of Prepaleozoic crystalline rocks, which is seen in the old paleogeographic map of Fig. 11.1, was extended to all other North American geosynclines early in our century. Borderlands were assumed to have existed through millions of years along all margins of the continent, but when adjacent geosynclinal tracts ceased to subside, the borderlands disappeared beneath the marginal seas like mythical Atlantis. While important clastic sediments in mobile belts were derived from the craton, the derivation of clastic material from outside the craton (i.e., from the opposite direction) ultimately was of greater volumetric importance.

Inference that the marginal or extracratonic lands were comprised of Prepaleozoic complexes was a natural consequence of the long-held fallacy of age

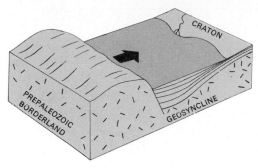

FIGURE 12.19 The hypothesis of Prepaleozoic borderlands as the sources of most geosynclinal sediments. Since 1940, this model, which calls for large highlands *outside* geosynclines, has been replaced by a model of lands raised *within* geosynclinal belts (see Fig. 12.21).

correlation of rocks solely on the basis of metamorphism and deformation (see Chap. 10). In 1930, Harvard University geologists discovered in New Hampshire Silurian and Early Devonian brachiopods in high-grade mica schists long assumed to be Prepaleozoic! The originally designated borderland of "Prepaleozoic" rocks actually was in large part composed of much younger ones. But not only were the supposed borderlands composed mostly of Paleozoic and younger rocks, they could not have been permanent lands during Paleozoic and Mesozoic time, for the fossils found included forms known to have inhabited only marine environments. Intermittently, seas must have occupied almost the entire "borderland" areas as well as the adjacent "geosyncline." Clearly, "borderlands" were neither permanent nor composed wholly of Prepaleozoic rocks. What, then, *was* the nature of the extracratonic lands?

MODERN CONCEPTS OF BORDERLANDS
At the same time that fossils were being discovered in the so-called borderland areas, a German geologist, Hans Stille, and Americans Marshall Kay and A. J. Eardley were drawing the distinction between magmatic or igneous-bearing (eugeosynclinal) and nonmagmatic or non-igneous-bearing (miogeosynclinal) subdivisions of geosynclines. The important observation that ancient mobile belts contain a distinct zone of volcanic and intrusive igneous rocks of diverse ages became the key to a reinterpretation of borderlands.

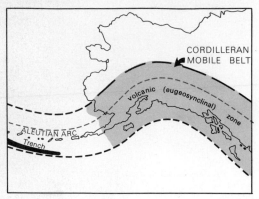

FIGURE 12.20 Continuity of ancient mobile belt with a modern volcanic island arc-submarine trench system, which, together with stratigraphic evidence, has led to the paleogeographic comparison of many ancient belts with modern arcs.

It is important to realize that the original conception of a geosyncline by Hall, Dana, and Schuchert included only the nonmagmatic (miogeosynclinal) portion of the expanded geosynclinal conception of Stille and Kay. *The original Appalachian example included only half of the mobile belt, and the less disturbed half at that!* Synthesis and comparison of the characteristics of the entire Appalachian belt with other examples had led Kay and others to reject the old borderland concept in favor of a series of temporary islands periodically raised *within the "magmatic" portion of the belts themselves*. Details of stratigraphy in the latter zone show, besides abundant volcanic outpourings, many angular unconformities, coarse conglomerates, varying ages of igneous plutons, and metamorphism, as we have seen in our discussion of the Taconian Orogeny (Fig. 12.18). In short, there is ample evidence of the persistence of great structural mobility throughout much of their histories. The rocks, moreover, attest to formation of both volcanic and nonvolcanic islands intermittently within the mobile belt itself. Some old mobile belts even extend seaward into still-active volcanic arc-trench systems (Fig. 12.20). Clastic detritus deposited in the so-called geosynclines has been derived from such islands, not from borderlands wholly composed of Prepaleozoic rocks and lying outside the belt. The rocks exposed to erosion in the islands were, for the most part, but slightly older geosynclinal deposits augmented periodically by additions of igneous extrusions and intrusions.

SOURCE MODELS FOR THE SEDIMENTS IN MOBILE BELTS

Having examined some of the stratigraphic record for Prepaleozoic and early Paleozoic mobile belts, we can now formulate generalizations about the sources of sediments found therein. Mobile belt deposits are so enormously variable, both in types and relative abundances, that simplifying models are helpful.

Prior to onset of structural mobility in mid-Ordovician time, the Appalachian region was a broad, tranquil continental shelf receiving clastic sediments only from the cratonic source (Fig. 12.21). Mature quartz sand continued to enter from the craton when the region became a mobile belt in later Ordovician time. But it soon became subordinated to less mature sands from other sources (Fig. 12.22).

The second, or volcanic-source model (Fig. 12.21) is exemplified by some of the Ordovician of the eastern Appalachian belt. As indicated earlier, during medial Ordovician time extensive volcanism occurred there, and volcanic islands were formed as evidenced by volcanic-rich clastic sediments associated with lavas and by the fine volcanic ash blown far to the south and west. Apparently structural mobility had accelerated sufficiently so that volcanic accumulation exceeded subsidence and built islands similar to those of modern volcanic island arcs.

A third, or tectonic-land model (Fig. 12.21), which was emphasized particularly by Marshall Kay, is exemplified by Upper Ordovician and Lower Silurian strata of the Appalachian region. As shown above in Figs. 12.14 and 12.15, these sediments coarsen toward the east, and the uppermost Ordovician red beds apparently represent deposition on the western margin of a large land. Microscopic study of Ordovician sandstones shows that they belong to feldspathic and lithic graywacke clans, for they contain quartz, feldspar, and fragments of older sedimentary and minor metamorphic rocks (Fig. 12.22); volcanic detritus is subordinate. Their composition and great volume indicate that a large land comprised chiefly of older, partly metamorphosed sedimentary and some igneous rocks was the principal source of these sediments. Such a land was raised structurally during the Taconian orogeny, and that tectonic land was the dominant source of Ordovician terrigenous clastic sediments in eastern North America. How large might it have been? Considering that Upper Ordovician clastic sediments of the western Appalachian belt comprise approximately 100,000 cubic kilometers

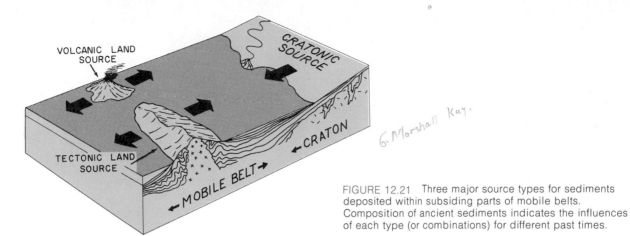

G. Marshall Kay.

FIGURE 12.21 Three major source types for sediments deposited within subsiding parts of mobile belts. Composition of ancient sediments indicates the influences of each type (or combinations) for different past times.

and assuming symmetrical deposition of sediment on both sides, erosion of a land approximately 1,000 kilometers long by 50 kilometers wide, and an average of 2 kilometers high, would be adequate. Of course the land was elevated as it was eroded, thus it need not have been as high at any one time as these dimensions suggest. But comparison with some of the modern island arc-trench systems shows that such dimensions are by no means excessive. Apparently we can account for the great volume of Ordovician geosynclinal sediments by envisioning a dynamic mobile belt that was elevated and eroded in one part, while the resulting sediment was deposited in another, subsiding part. In this manner, many mobile belts have been filled to form geosynclines.

SEA LEVEL, SEDIMENTATION, AND SUBSIDENCE

In Chap. 8 we noted that American geologists assumed from the outset that all geosynclinal sediments were deposited either in shallow-marine water or on land just above sea level. This view was based upon the strata of the western nonvolcanic portion of the Appalachian Mountain belt, where many rocks contain algal stromatolites and other marine fossils, whereas others are clearly river-delta deposits. Meanwhile, European geologists were arguing that typical geosynclinal sediments were deep-marine in origin—perhaps even trench deposits. It is now clear that many different environments have existed in mobile belts. As shown in Fig. 12.23, the relation of sedimentation to sea level is the most critical factor.

Very thick sediments, whether beneath continen-

FIGURE 12.22 Microscopic photograph of a typical Ordovician (Normanskill) graywacke from eastern New York. Grains include quartz (white), feldspar, and varied sedimentary rock fragments (dark) derived from erosion of earlier Paleozoic strata in the ancient Taconian tectonic land to the east. Note the very poor sorting and rounding (grains average about 2 millimeters in diameter). Overall sedimentary immaturity indicates that the material suffered rapid erosion, transport, and dumping.

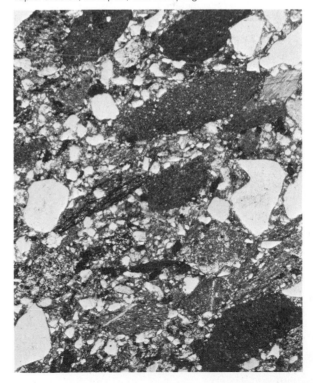

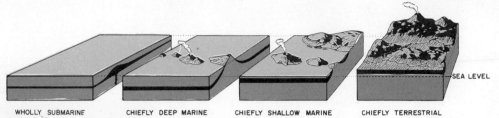

WHOLLY SUBMARINE   CHIEFLY DEEP MARINE   CHIEFLY SHALLOW MARINE   CHIEFLY TERRESTRIAL

FIGURE 12.23 Models of different possible sedimentary environments within mobile belts. The difference is relative position of sedimentary and volcanic accumulations (black) with respect to sea level, which determines degree of agitation by waves and currents as reflected in sediment textures. The relative size and height of land and climate also influence the types of sediments, especially determining whether carbonate rocks will form. Note in the example to the right that very thick, wholly nonmarine sequences also may accumulate within mobile belts.

FIGURE 12.24 Regional crustal subsidence produced by lateral as well as vertical accumulation of shallow marine and nonmarine alluvial plain sediments. Through sedimentation over a large area, the crust is warped down a limited amount; the locus of active accumulation constantly shifts, spreading the load over an increasing region and producing further subsidence. This appears to be what happened along the eastern margin of the craton during Late Ordovician time and on the Gulf of Mexico shelf more recently.

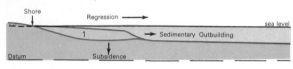

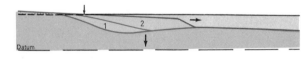

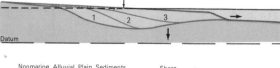

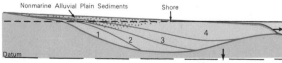

ACCUMULATION OF THICK SHALLOW MARINE
AND ALLUVIAL PLAIN SEDIMENTS

tal shelves, or in geosynclines, or in cratonic basins, imply subsidence of the crust during deposition in order to provide room for them to accumulate vertically. Only if deposition began in deep water, as the Europeans believed, could considerable accumulation occur without any subsidence. If the sediments are all shallow-water types, then subsidence must just balance accumulation to maintain a constant shallow environment. This is what many early Americans believed happened in geosynclines. But what causes the subsidence?

James Hall of New York argued over a century ago that a large mass of sediments must depress the crust in accord with Archimedes' principle (see Chap. 8). We know that ice caps 3,000 meters thick have depressed the crust 1,000 meters. But Hall neglected the fact that, because sediments like ice are less dense than the solid earth beneath, they have to pile up on the crust to a great thickness to cause only modest crustal subsidence. Unlike ice, however, sediments cannot accumulate much above sea level without being spread laterally by wind, rivers, or waves. Apparently shallow-marine sediments cause subsidence only when large volumes are spread over broad areas, for example on the continental shelf around the Mississippi delta. Because sea level inhibits vertical accumulation, deposition must spread laterally, and over time it surely does depress the crust isostatically (Fig. 12.24). The early Paleozoic continental shelf of eastern North America apparently behaved in this fashion.

Where deposition begins in deep water, a great thickness can accumulate, and it will depress the crust significantly. But once sea level is reached, currents and surf will spread the sediments laterally. If the rate of sedimentation continues to exceed subsidence, then deltas will build outward, literally crowding the sea away to produce regression. Onshore, rivers will build a broad alluvial plain slightly above sea level, which will migrate seaward as the deltas advance. Thus we might expect to find vertical transitions from deep to shallow-marine and then to nonmarine sediments as "holes" were filled. In fact, exactly such sequences are common in old mobile belts; for example, the Ordovician of the Appalachian belt was capped by the red beds deposited at the height of the Taconian Orogeny.

While sediment loading isostatically must cause some subsidence, it is not the primary factor in the profound subsidence evidenced by very thick geosynclinal sediments deposited in active mobile belts. Other possible causes include phase changes that thin the crust or lithosphere (see Chap. 8) and forcible downbuckling. A century ago, Dana, you may recall, invoked shrinkage of a cooling earth to cause downbuckling beneath geosynclines, but there is no evidence that the earth is contracting. Later, thermal convection currents in the mantle were thought to buckle down the overlying crust. Still more recently, subduction of one lithosphere plate beneath another (probably also driven by convection) has been postulated to explain most of the subsidence evidenced in mobile belts.

ENVIRONMENTS OF THE APPALACHIAN REGION

EARLY PALEOZOIC SHELF

Let us now apply the preceding discussions to the Appalachian region. From Eocambrian to mid-Ordovician time, there was no mobile belt there. Instead, as we have noted, a broad, structurally passive continental shelf faced a proto-Atlantic Ocean basin, which was completely destroyed before the present Atlantic formed (Fig. 12.25A). Sedimentation closely balanced subsidence so that thick, shallow-marine Eocambrian and Lower Cambrian sands followed by later Cambrian and Ordovician carbonates accumulated on the shelf, while fine muds and occasional limestone slide blocks accumulated on the continental slope beyond. Because no high lands existed nearby and the entire shelf was located within the subtropics, carbonate deposition was favored. Organisms flourished to produce a shelly carbonate facies. With the beginning of subduction of the proto-Atlantic lithosphere beneath North America, the Appalachian mobile belt was born (Fig. 12.25B, C). The shelf deposits eventually became deformed,

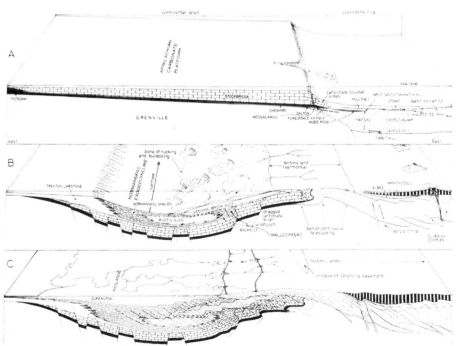

FIGURE 12.25 Restorations of eastern North America showing A. Cambrian-Early Ordovician passive carbonate shelf; B. Mid-Ordovician (early Taconian) disturbances as subduction began to the east (note beginning of graptolitic-graywacke facies deposition); C. End of Ordovician upheaval of Taconian land and the deposition of red clastic wedge (molasse) to the west. (From J. F. Dewey and J. M. Bird, 1970, Bulletin of the Geological Society of America, v. 81, pp. 1031–1060.)

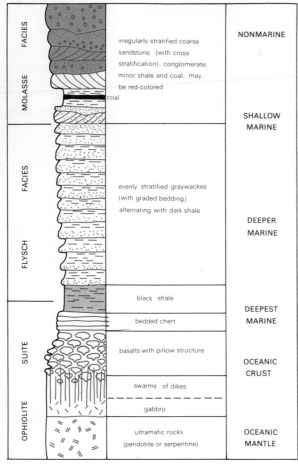

MOLASSE FACIES	irregularly stratified coarse sandstone (with cross stratification), conglomerate, minor shale and coal; may be red-colored	NONMARINE
	coal	SHALLOW MARINE
FLYSCH FACIES	evenly stratified graywackes (with graded bedding) alternating with dark shale	DEEPER MARINE
OPHIOLITE SUITE	black shale	DEEPEST MARINE
	bedded chert	
	basalts with pillow structure	OCEANIC CRUST
	swarms of dikes	
	gabbro	
	ultramafic rocks (peridotite or serpentine)	OCEANIC MANTLE

FIGURE 12.26 Idealized vertical sequence from oceanic igneous rocks upward through graptolitic-graywacke (flysch) strata to red clastic (molasse) deposits. Such successions are commonly seen in orogenic belts and represent a progression from deep marine to nonmarine conditions, as long ago argued by Europeans.

together with the younger, truly geosynclinal strata as the mobile belt developed.

THE GRAPTOLITIC FACIES

With the onset of structural mobility, the region became engulfed in black muds and dark sands (graywackes). Because of the large organic content of the muds and few obvious current features in the sands, it was assumed long ago that the region had subsided in mid-Ordovician time to produce quiet, deep-water environments lacking oxygen and current

agitation. Floating graptolite colonies drifted over the area, and their delicate remains sank to the bottom and were preserved in the muds.

Understanding of the origin of the important graptolitic shale facies was slow in coming. In the Alps, early Mesozoic dark shales and cherts that lie upon serpentine and basalt with pillow structure all together form what the Europeans have long called an ophiolite suite (Fig. 12.26). The igneous members of the suite were interpreted as oceanic crust, and the overlying fine sediments as deep-marine deposits. The Alpine sequence became more sandy in Cretaceous and Eocene times, and the alternating, very evenly stratified sandstones and shales (called *flysch* in Switzerland) resemble the graptolitic facies of the early Paleozoic. But even if the ophiolites are deep marine in origin, how could the overlying sands be so? A deep-water origin was proposed for the similar early Paleozoic graptolitic shale facies of Wales in 1930. The immature dark graywackes, which contain graded bedding so characteristic of that facies (Fig. 12.27), were interpreted as rapidly "poured into" a deep, tranquil zone of the sea floor. Since then, graded bedding has been widely recognized in similar sequences all over the world, including the Alps. Presence of ripple marks led many to believe that these sequences were shallow deposits, because ripples were thought to form only there, the deep seas being assumed static. With the recognition in 1950 of the importance of episodic turbidity currents (see Fig. 10.7) in transporting relatively coarse sediments into otherwise quiet environments and the discovery of ripples on the deep-sea floor (Fig. 12.28), a reinterpretation of many sandy geosynclinal strata as relatively deep-water deposits became possible. The turbidity current revolution finally settled the controversy between the American and European views of geosynclinal sedimentation.

Returning to the Appalachian belt, we find Ordovician ophiolite sequences (most clearly in Newfoundland) overlain by graptolite or flysch facies (Fig. 12.26). Basalts extruded onto the deep-ocean floor were buried first by cherts and black muds containing manganese concentrations. By mid-Ordovician time, however, immature sands showing graded bedding were introduced rapidly by turbidity currents and were apparently deposited on deep-sea fans (Fig. 12.29). Volcanic rocks continued to be erupted intermittently. Apparently volcanic and other islands were being raised nearby as a result of subduction of the

FIGURE 12.27   Graded bedding in Ordovician graywacke (Martinsburg Formation) near Middletown, New York; hammer handle at right provides a scale comparison. These sandstones are interpreted as turbidity current deposits. *(Courtesy Earle F. McBride.)*

the turn of the century, an American paleontologist noted that many graptolites in eastern New York had been oriented by bottom currents. More recently, current-sculptured sole marks found on the bottoms (soles) of sandstone strata where interstratified with shale (Fig. 12.30) have provided much additional paleocurrent data to help us interpret the paleogeography. As shown in Fig. 12.25, much of the sediment derived from rising lands was deposited in a small ocean basin between the continent and the island arc to the east. This basin was probably similar to the modern Sea of Japan between the Japanese arc and mainland Asia, which has an oceanic crust. During the Taconian Orogeny, the Ordovician small ocean basin was closed as the rising island arc and the basin's sediments were crushed against the rigid continent.

The evenly stratified dark Alpine (*flysch*) strata pass upward into nonmarine light-colored irregularly stratified coarse conglomerate, cross-stratified sandstone, minor shale, and coal. These deposits (called *molasse* in Switzerland) reflect the final uplift of the

proto-Atlantic lithosphere (Fig. 12.25). Although relationships are not as clear farther south, the presence of identical rock types there suggests similar conditions. Throughout the mobile belt, facies patterns and paleocurrent criteria indicate that sands of the graptolitic facies were derived from the east, where the Taconian volcanic and tectonic lands were rising. At

FIGURE 12.28   Deep-marine ripple marks on a fine sand bottom at a depth of 3,500 meters in the South Pacific Ocean near the Antarctic Circle. Prior to the development of submarine photography around 1950, it was assumed that the deep seas were static and currentless; therefore ripple marks could form only in shallow water. *(Official NSF photo, USNS Eltanin, Cruise 15; courtesy Smithsonian Oceanographic Sorting Center.)*

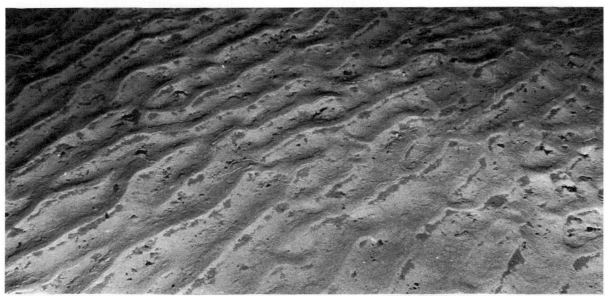

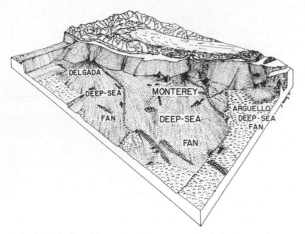

FIGURE 12.29 Diagrammatic portrayal of deep-sea fans west of California (view looking east toward continent; vertical scale exaggerated). Such fans are fed largely by turbidity currents, with submarine canyons that cut across continental shelves acting as conduits for movement of sediment to the fans. Most sediment transported by turbidity currents is deposited on such fans. Note the gradual burial of topographic features. *(From H. W. Menard, 1960,* Bulletin of the Geological Society of America, *v. 71, pp. 1271–1278; used by permission.)*

Alps above sea level. Some of the molasse is red-colored, and is in every way analogous to Late Ordovician–Early Silurian red beds of the Appalachian belt (Fig. 12.14).

RELATIONS BETWEEN EUROPE AND NORTH AMERICA

In Chap. 11 we noted a long-recognized similarity of early Paleozoic faunas between eastern North America and northwestern Europe (see Fig. 11.2), which has suggested some kind of connection. The tectonic history of the two sides of the North Atlantic also suggests a former physical relationship, for sequences of rock types and major orogenies are almost identical. The presence of an early Paleozoic mobile

FIGURE 12.30 Flute structures on the bottom (sole) of an Ordovician graywacke (Martinsburg Formation), near Staunton, Virginia; the bulbous ends (right) point up-current. Turbulent bottom currents scoured elongate flutings in cohesive muds and deposited sand within them. Long after lithification, erosion exposed the structures on the now more-resistant sandstone. Such sole structures appear only at sandstone-shale interfaces because wet clay particles bond together immediately upon deposition to produce sufficient cohesion so that the mud surface can be fluted. Sand surfaces do not have this property; they are essentially cohesionless. *(Courtesy Earle F. McBride.)*

belt in Britain and Norway that has an almost perfect mirror image in east Greenland and Newfoundland is especially compelling. A few ardent advocates invoked continental drift many years ago to explain these similarities, but their restorations had Europe and North America joined throughout the Paleozoic Era (see Fig. 8.1). A plate-tectonic interpretation, however, postulates that they were separated until Devonian time. During the Ordovician and Silurian Periods, these two continents and Africa were moving toward each other as the intervening proto-Atlantic plate was being consumed by subduction on both sides (Fig. 12.31). Such subduction is postulated in order to explain volcanic and orogenic activity on both continents. Such activity began earlier in the northern region between Greenland and Scandinavia (Early Ordovician) than in the Appalachian region, where it began in mid-Ordovician time and culminated with the Taconian Orogeny around the end of the period.

Based upon the above concept, the map of Fig. 12.31 has restored Prepaleozoic and lower Paleozoic strata of northwestern Scotland, western Norway, and Svalbard to their apparent original positions as part of the ancient continental shelf along eastern Greenland and western Newfoundland. The lower Paleozoic rock sequences and paleocurrent directions (as well as the faunas already noted; Fig. 11.2) match perfectly. Normal cratonic quartz sandstones of northwestern-most Scotland with obvious derivation from the north-west—now a deep ocean—were especially hard to explain without drift. Opponents had to invoke a sunken Atlantis continent for a source, but there can be little doubt that cratonic North America was that source. Eastern Newfoundland and a portion of the mainland farther south apparently were then either part Africa or Europe (Fig. 12.31). All these tectonic elements were crushed together by continental collision in Devonian time, as detailed in the next chapter. When the present Atlantic basin began forming in Mesozoic time, irregular separation of the two continents left these fragments of one attached to the other.

SUMMARY

A major regression of the sea from the entire craton occurred in early medial Ordovician time, due either to a fall of sea level or a small rise of the entire continent. This change may have resulted from the beginning of subduction of the proto-Atlantic lithosphere plate. Strata just deposited in the previous 30 million years or so were laid bare to erosion. Thin Lower Ordovician dolomites were quickly stripped off the central craton to expose Upper Cambrian quartz sandstones. Sand was eroded and redeposited during retransgression of the sea in medial Ordovician time to form a very widespread sheet-like unit demonstrably older at the margins of the craton (where transgression commenced) than near the center. The quartz sand supply dwindled again as the sea advanced to flood the entire craton. In later Ordovician time, epeiric-sea fossiliferous carbonate deposits were laid down everywhere until the end of the period when dark muds began to encroach from the east.

Ordovician faunas differ considerably from Cambrian ones. The medial Ordovician transgression brought an especially distinctive fauna, for the new epeiric sea teemed with brachiopods and bryozoans. Cephalopods, gastropods, diverse echinoderms, and corals also were prominent. Vertebrate animals, represented by jawless, armored fish, appeared for the first time. Of greatest importance for dating and correlating Ordovician strata, however, were the rapidly evolving graptolites, whose tiny colonial fragments were widely dispersed over the ocean floor.

In the mobile belts, very thick Eocambrian, Cambrian, and Ordovician marine sedimentary and volcanic rocks accumulated. Most of the lavas were submarine, except in the Appalachian belt, where volcanic islands appeared. Near the end of Ordovician time, a large, mountainous tectonic land was raised there during the Taconian Orogeny. Whereas terrigenous clastic sediments deposited in the mobile belts during earliest Paleozoic time had been derived largely from the craton, during Ordovician time in the Appalachian belt, more and more clastic material was derived from volcanic and tectonic lands raised *within the belt itself*. As presently interpreted, those lands were composed of only slightly older geosynclinal rocks and igneous masses rather than ancient Prepaleozoic granitic and metamorphic rocks lying outside the belt to the east. Mobile belts have been structurally disturbed more or less continuously throughout their histories. Uplift of one portion by mountain building produced land sources for sediments that were deposited in another, subsiding portion of the same belt. The weight of the thick

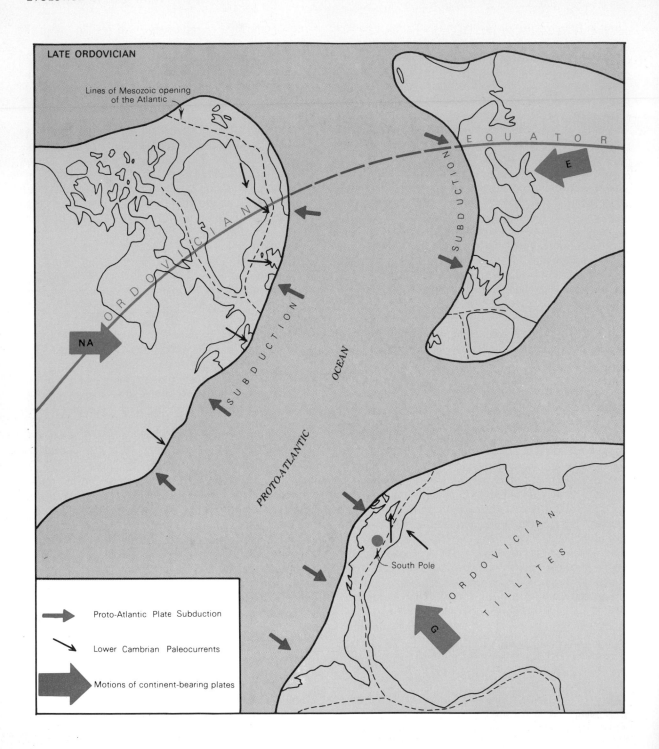

LATE ORDOVICIAN

Lines of Mesozoic opening
of the Atlantic

EQUATOR

ORDOVICIAN

SUBDUCTION

NA

OCEAN

SUBDUCTION

PROTO-ATLANTIC

E

G

South Pole

ORDOVICIAN TILLITES

→ Proto-Atlantic Plate Subduction

→ Lower Cambrian Paleocurrents

→ Motions of continent-bearing plates

sediments themselves depressed the crust somewhat, but more profound structural disturbances—chiefly vise-like compression between two lithosphere plates—must have been the primary cause of subsidence that made possible the accumulation of thousands of meters of sedimentary and volcanic rocks.

During times of relative structural serenity, carbonate rocks were deposited in shallow parts of many mobile belts, but at other times, subsidence produced deeper water environments in which muds were deposited. Where sedimentation began on oceanic crust, sequences can be seen with ultramafic igneous rocks (presumably from the upper mantle) overlain by thick pillow basalt (presumably oceanic crust), bedded chert, and black shale. When sand was introduced intermittently by turbidity currents into relatively deep water, the very evenly stratified graptolitic (or flysch) facies resulted. During the most active time of uplift of new mountains, so much sediment was produced by erosion that a great terrigenous clastic wedge or prism of red strata (the molasse facies) accumulated so that the sea was literally crowded away and the shoreline migrated cratonward.

The patterns summarized here were forerunners of things to come. We shall see that the early Paleozoic tectonic and sedimentary framework, and the concepts developed in this and the preceding chapter for interpreting them, will be applicable to much of the later record as well. In subsequent chapters, therefore, we shall not dwell at length upon similar patterns, but rather shall emphasize new ones as they appear. Keep in mind that details of the stratigraphic record are not ends in themselves, but rather provide vehicles for illustrating *how geologic history can be*

interpreted and how restorations are made. Our objective is not to bury the reader in a mass of murky detail, as it were, but rather to exhume from the strata some principles of historical reasoning and broad insights into the evolution of the earth.

Readings

Beuf, S., et al., 1971, Les gres du paleozoique inferieur au Sahara: Paris, L'Institut Francais du Petrole, Special Publication No. 18.

Bird, J. M., and Dewey, J. F., 1970, Lithosphere plate-continental margin tectonics and the evolution of the Appalachian orogen: Bulletin of the Geological Society of America, v. 81, pp. 1031–1060.

Dapples, E. C., 1955, General lithofacies relationship of St. Peter Sandstone and Simpson Group: American Association of Petroleum Geologists, v. 39, pp. 444–467.

Dott, R. H., Jr., and Shaver, R. H., eds, 1974, Modern and ancient geosynclinal sedimentation: Tulsa, Society of Economic Paleontologists and Mineralogists Special Publication No. 19.

Eardley, A. J., 1962, Structural geology of North America: New York, Harper & Row.

Kay, M., 1951, North American geosynclines: Geological Society of America Memoir 48.

Kay, M., and Colbert, E. C., 1964, Stratigraphy and life history: New York, Wiley.

Ketner, K. B., 1966, Comparison of Ordovician eugeosynclinal and miogeosynclinal quartzites of the Cordilleran geosyncline, in U. S. Geological Survey Professional Paper 550-C, pp. C54–C60.

King, P. B., 1959, The evolution of North America: Princeton, Princeton.

Levorsen, A. I., 1960, Paleogeologic maps: San Francisco, Freeman.

McBride, E., 1962, Flysch and associated beds of the Martinsburg Formation (Ordovician), central Appalachians: Journal of Sedimentary Petrology, v. 32, pp. 39–91.

Schenk, R. E., 1971, Southeastern Canada, northwestern Africa, and continental drift: Canadian Journal of Earth Sciences, v. 8, pp. 1218–1251.

Sloss, L. L., Dapples, E. C., and Krumbein, W. C., 1960, Lithofacies maps: an atlas of the United States and southern Canada: New York, Wiley.

Swett, K., and Smit, D., 1972, Paleogeography and depositional environments of the Cambro-Ordovician shallow-marine facies of the North Atlantic: Bulletin of the Geological Society of America, v. 83, pp. 3223–3248.

FIGURE 12.31 Hypothetical Ordovician paleogeographic map showing the subduction of a proto-Atlantic Ocean plate as North America (NA), Europe (E), and the southern supercontinent Gondwanaland (G) began their approach. Relative latitudes are revealed by paleomagnetic data, but exact relative positions cannot otherwise be determined (Gondwanaland undoubtedly is shown too close to the others). All that remains of the proto-Atlantic today are mafic oceanic igneous rocks and deep-sea sediments found within the old mobile belts on either side of the present Atlantic. *(Data from J. F. Dewey and J. M. Bird, 1970, Bulletin of the Geological Society of America, v. 81, pp. 1031–1060; P. E. Schenk, 1971, Canadian Journal of Earth Sciences, v. 8, pp. 1218–1251; and K. Swett and E. Smit, 1973, Bulletin of the Geological Society of America, v. 83, pp. 3223–3248.)*

13

THE MIDDLE PALEOZOIC

TIME OF REEFS, FORESTS, AND SALT DEPOSITS

He who with pocket-hammer
smites the edge
Of luckless rock
or prominent stone, disguised
In weather-stains
or crusted o'er by Nature.
The substance classes by
some barbarous name,
And thinks himself enriched,
Wealthier, and doubtless
wiser than before.

William Wordsworth,
The Excursion (1814)

FIGURE 13.1 The Gilboa forest of the Devonian in the Catskill Mountains, eastern New York. *(Courtesy American Museum of Natural History.)*

The tectonic configuration established in Ordovician time also characterized the middle part of the Paleozoic Era, that is, from 425 to 355 million years ago. Continued subduction of the proto-Atlantic lithosphere plate resulted in volcanic activity and deformation, on both sides of the old ocean, which shrank and finally was destroyed by the collision of northwestern Europe and northeastern North America in Devonian time. As the deep-ocean barrier disappeared, the faunas and floras of both continents became more and more alike. The edges of each colliding plate were intensely deformed and uplifted to form a single great mountain chain between two cratons. Granite batholiths formed within this great intercratonic mobile belt, and huge prisms of red clastic sediments were shed from the mountains onto both cratons. The cratons themselves also were affected by the continental collision, as indicated by the most extensive warping ever of arches and basins.

Apparently organisms first invaded the land in middle Paleozoic time, indicating that an atmospheric ozone shield from deadly ultraviolet radiation existed by now (if not earlier). The oldest proven land plants are latest Silurian, and coastal forests were commonplace by Late Devonian time. The earliest definite air-breathing animals known were Early Devonian spider-like creatures, whereas the oldest land vertebrates were Late Devonian amphibians. Epeiric seas were widespread over all mid-Paleozoic cratons. Unusually rapid evolution produced many new faces among both marine invertebrates and vertebrates. Especially important was the first development of large animal reefs, whose locations were largely determined by warping of cratonic basins. Major evaporite deposits also occurred within many of the basins rimmed by reefs.

Microorganisms produce the raw materials of petroleum after burial in sediments. Their organic compounds undergo chemical changes that lead to petroleum. If favorable permeable rocks overlain by impermeable ones are present at the right time and place, oil and gas can be trapped in significant quantities. Buried middle Paleozoic organic reefs are among the most prolific of such traps, but sandstones are also important. The famous Drake well drilled near an oil seep in 1859 at Titusville, Pennsylvania, tapped a petroleum-saturated Devonian sandstone. Exploration for elusive hidden traps requires sophisticated geologic and geophysical tools. As a byproduct of practical oil-seeking, science has been rewarded with a wealth of otherwise inaccessible subsurface geologic data, which adds greatly to the accuracy of our historical analysis, especially for the deeply subsided basins. And such greater accuracy is itself essential if we are to find new petroleum reserves for this energy-hungry world of ours. Even with the best possible skill and luck, however, it is a losing battle, for petroleum—like all mineral resources—is a nonrenewable resource in terms of the human time scale. Our best estimates indicate that in less than 100 years from now the world's petroleum reserves will have been exhausted. Long before that, however (certainly by the year 2000), we must have other major energy sources to supplement petroleum. Coal and nuclear fuels with all their environmental hazards are most likely to be the first supplements, but they are nonrenewable, too, and will probably be exhausted by the year 3000.

MIDDLE PALEOZOIC LIFE

INVERTEBRATE ANIMALS

With the spread of the great Upper Ordovician sea, many new creatures appeared, and favorable conditions permitted vast populations to develop. We have a rich heritage of these in exceptional fossil accumulations around Cincinnati, Ohio, just as rich, middle

FIGURE 13.2 *Mucrospirifer,* a common Devonian spiriferid brachiopod, from Arkona, Ontario. *(Courtesy U.S. National Museum.)*

Paleozoic faunas occur farther west at the Falls of the Ohio. At first glance, the Silurian fauna has much of the aspect of later Ordovician life in the predominance of brachiopods and bryozoans, but corals and echinoderms increased in importance. The brachiopods of the middle Paleozoic changed, with elongated and deep-bodied forms replacing many of the dominant thin, ovoid Ordovician types (Fig. 13.2).

By Silurian time, two groups that had undergone such rapid evolution before—the nautiloid cephalopods and graptolites—had virtually disappeared. A few important graptolite genera are useful in correlating Silurian strata. Other more primitive and long-ranged graptolites persisted until Early Mississippian time, but are very rare. Thus far we have seen that, since the beginning of the Cambrian Period, there was a rapid development and decline to near extinction of three groups, the trilobites, nautiloids, and graptolites, which were important elements in the early Paleozoic faunas and are so useful as index fossils. We do not know the cause of this pattern, but the most common hypothesis is that some predator group evolved and either ravaged the species themselves or ate their principal food source, thus disrupting the food chain. It is possible, for example, that trilobites fell prey to large, predatory nautiloids. On the other hand, before the height of nautiloid diversity, the trilobites were well on their way to extinction, and thus may not have been affected by the nautiloids.

By Devonian time (Fig. 13.3), coiled ammonoids had evolved from a nautiloid stock. They were to become the dominant invertebrate group of the Mesozoic Era. Even when they first appeared, they were distinctive enough to be useful for intercontinental correlation, particularly of Upper Devonian strata. It is felt by most authorities that they were dominantly swimmers and floaters, which made them ideal index fossils as were the graptolites before. First, they could be geographically dispersed instantaneously (by geologic standards), and, secondly, they could cross sedimentary facies boundaries. They might flourish in well-circulated waters above a foul, stagnant mud bottom on which no benthonic organism could survive, and after death sink to the bottom, where chances of preservation would be increased in an oxygen-poor environment.

The most obvious change in the Silurian marine fauna was an increase in the number of colonial rugose and tabulate corals. Some of their colonies

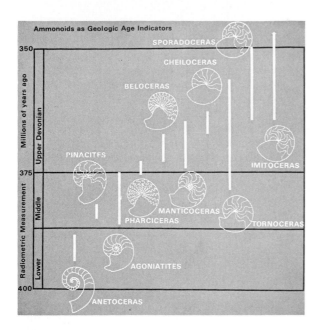

FIGURE 13.3   Evolution of Devonian ammonoid cephalopods, showing changes in the sutures (which separated the internal chambers) with time. In Paleozoic ammonoids, suture patterns are called *goniatite sutures*. *(Redrawn from a figure in* Bursts of evolution *by M. A. House in* The Advancement of Science, *1963, p. 501; by permission of British Association for the Advancement of Science.)*

became so large as to form the first coral reefs in several parts of the world; Devonian reefs were even more widespread (Fig. 13.4). A companion to these coral-reef formers was the calcareous sponge Stromotoporoidea group, which originated in the Ordovician. It, too, formed Silurian and Devonian reefs and "gardens," and so was a common rock-former.

The trilobites still were present, and in some

FIGURE 13.4   A diorama based on a fossil site in the Middle Devonian of central New York; note the rugose coral colony in the left foreground. *(Photograph courtesy American Museum of Natural History.)*

facies they are fairly abundant, but are represented by only a few groups. A distant arthropod cousin appeared in fair numbers during the Silurian—the sea scorpions (eurypterids; Fig. 13.5). These strange forms superficially resemble the air-breathing scorpions, but probably were not related. They were monsters of the sea during the Silurian, and some attained lengths of 3 meters before becoming extinct in late Paleozoic time.

THE FISHES

We turn next to a group that adds immeasurably to our knowledge and is important historically, though not abundant, namely the fish (Fig. 13.6). In the Silurian, and more particularly the Devonian, many complete skeletons of the jawless and armored fish have been preserved. These primitive fish, the Agnatha, were heavily armored and had flattened anterior shields. They undoubtedly were poor swimmers and in all probability were filter feeders living on the bottom. The internal skeleton was never preserved, and so we suspect that they had a cartilaginous skeleton like primitive fish of today.

FIGURE 13.5 A sea scorpion (*Eurypterus*) from the Silurian Bertie Water Lime at Litchfield, New York. (*Courtesy U.S. National Museum.*)

Sometime during the Silurian Period, though perhaps earlier, the first vertebrate jaw developed. It was modified from several gill arches, which are anterior equivalents of ribs with the specialized function of supporting part of the respiratory system. Probably at first they were imbedded in muscle, but early in the evolution, the upper jaw became fused to the head shield, thus forming a rigid and far more efficient eating machine. Primitive jawed fish are called placoderms (Fig. 13.6), and, while some of them retained the ancestral flattened body, most became streamlined and armor was reduced. These changes presumably were in response to selective pressure for better swimming. Some of the primitive jawed fishes, such as *Dunkleosteus* (Fig. 13.7), became large and voracious predators. One other carnivorous group worthy of mention appeared. The sharks, which are a primitive group that lost the heavy armor (though vestiges occur as dermal plates), developed highly efficient muscles for swimming. The first of the sharks probably were marine forms, but several important groups very early invaded the fresh waters.

The development of a jaw in the fishes is yet another example of how a major new character enables a group to undergo adaptive radiation into an unoccupied adaptive zone. The fishes rapidly expanded after the Devonian and are some of the most common pelagic (swimming) organisms of today.

There has been a great deal of controversy regarding the original environment of the vertebrates. Professor Romer of Harvard University felt that the evidence favors a fresh-water habitat because marine invertebrates are not found associated with the primitive fish fossils. But Dr. Denison of the Field Museum in Chicago, after reviewing newer evidence, has presented a strong case for the marine origin of fishes.

The Osteichthyes are the most successful of all fish because they developed bony skeletons and reduced the rigid armor to thin scales, thus greatly increasing their agility and speed. The development of paired fins, gill slits protected by bony opercula, and bony rays to strengthen the fins, all served that end. Of the two recognizable subgroups of the bony fishes, the ray-finned fish (Actinopterygii) has greatly exceeded all other kinds in numbers and diversity today. The other ray-finned group (Choanichthyes), which includes the lobe-finned fish (see Figs. 13.6 and 13.8) and the lungfish, developed internal nos-

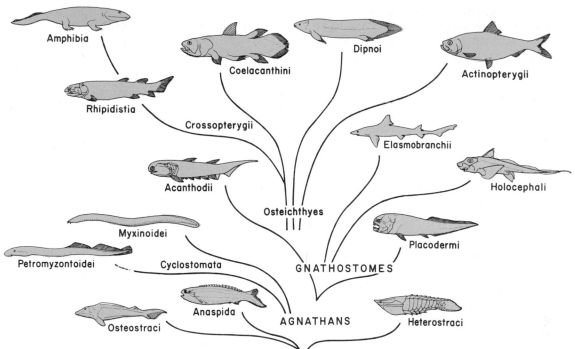

FIGURE 13.6 Evolution of the fishes. *(Drawing courtesy American Museum of Natural History.)*

trils that enabled them to respire with their mouths closed. It is because of the internal nostrils that some of the forms gradually began to breathe air, probably out of necessity when trapped in drying tidal or fresh-water ponds. The Choanichthyes are represent-

ed today by the lungfish and the "living fossil," the coelacanth, considered extinct since Mesozoic time until 1938, when the first living specimen was caught off East Africa. Other important characteristics of this group are that the fins are lobed and muscular and have articulated rays that allowed some of them to "walk" on the bottom. The amphibians are assumed to have derived from this group. Current thinking among specialists is that the Rhipidistia of the Upper Devonian are likely the ancestral group (see Fig. 13.6).

FIGURE 13.7 Restored head shield of *Dunkleosteus,* a large placoderm, from the Devonian of northern Ohio. *(Courtesy American Museum of Natural History.)*

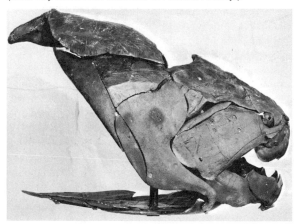

FIGURE 13.8 A lobe-finned coelacanth fish (Choanichthyes) from the Triassic of New Jersey. *(Courtesy American Museum of Natural History.)*

INVASION OF THE LAND

The very first amphibian made its appearance during the Late Devonian and is called a labyrinthodont (because of the labyrinthine infolding of tooth enamel). It was an awkward model, looking like something a committee put together. Its limbs were nothing more than jointed lobed fins; its head and tail were fish-like, too. Nonetheless, it did breathe air. Our record of them comes from some scraps in eastern Canada and several good specimens from northeastern Greenland. They would hardly be classified as common

FIGURE 13.9   Devonian coral showing a wide annual band; the white lines bracket that band; the fine lines are presumed to be daily growth lines. *(Photograph by G. R. Adlington.)*

fossils, yet they are of enormous importance as the evolutionary stem of all air-breathing, vertebrate land animals. Further discussion of early land vertebrate development and the rise of land plants, which appeared at least as early as Silurian time, is deferred to the next chapter.

We shall see that a relatively few characters that were "preadapted" for life on land in both the plants and animals were strongly selected for terrestrial environments. Consequent elaboration of these characters, along with the development of additional functional anatomy, permitted a very fast invasion of the land.

FOSSILS AS CALENDARS

We have seen that isotopic dating gives us the best estimate of "absolute" time in terms of years. But the smaller units of time—the month and day—are so brief geologically that they cannot be resolved by any isotopic method known. As we have seen, fossils were used to establish the "relative" time scale, and now it appears that they also will prove useful for limited "absolute" chronologies. A very ingenious line of investigation initiated by Professor John Wells of Cornell University has helped to bridge the gap between years and the smaller units of time by the use of fossils.

Biologists have observed that modern corals deposit a single, very thin layer of lime once a day. It is possible, with some difficulty, to count these diurnal (day-night) growth lines and to determine how old the coral is in days. More important, seasonal fluctuations will cause the growth lines to change their spacing yearly so that annual increments can also be recognized, much as in growth rings of trees.

Out of curiosity, and because he is a paleontologist, Wells began looking for diurnal lines on fossil corals. He found several Devonian and Pennsylvanian corals that do show both annual and daily growth patterns. But he was astonished to find that the Pennsylvanian forms had an average of 387 daily growth lines per year-cycle, and that the Devonian corals had about 400 growth lines (Fig. 13.9)!

By making counts between annual marks, Professor Wells found an average of 360 growth lines per year on modern corals he examined. He then constructed a graph (Fig. 13.10) based on the latest isotopic dating of the periods back to the beginning of the Cambrian, which suggested a systematic decrease in the number of days per year through geologic time. Several years later, Pannella and his associates (1968), using many more organisms from Recent to Cambrian, have shown that while a systematic decrease is probably valid, their points on a graph formed an S-curve rather than a straight line. This is probably caused by the differing effects of tidal action accompanying the changes of continents, ocean basins, and epeiric seas.

Meanwhile, geophysicists and astronomers estimated that there has been a deceleration (presumably due to tidal friction) of the earth's rotational velocity in recent centuries amounting to 2 seconds per 100,000 years. If one extrapolates this rate (which may or may not be valid), the Cambrian day would have been 21 hours and the year 420 days long. By the same reasoning, the length of a Devonian year should have been 400 days long. Thus, the astronomic extrapolations and Wells's coral data for the Devonian are remarkably similar. More recent studies using both corals and brachiopods have shown that Wells may have undercounted the growth lines. Some estimates place the length of the year in Early Silurian time at 421 days. Much more sampling and control will have to be done before this method can be used with accuracy.

Algal stromatolites also look promising for giving information on lengths of ancient years. It has been found that modern stromatolites in Bermuda waters lay down a daily growth increment, and study of Cambrian algal stromatolites from Washington shows that there are from 400 to 420 second-order increments in a series of growth bands.

Much research is being done to explore this exciting new development in the use of fossils to understand earth history. For example, stromatolites similar to types that today form chiefly in the intertidal zone existed 2.7 billion years ago. They were several centimeters high, suggesting that the moon was present at least by then to influence tides. But it is interesting to note that the height of such algal colonies increased to a maximum of 6 meters in late Prepaleozoic time. If these stromatolites *were* restricted to the intertidal zone, their increase in height would suggest a gradual decrease in the distance of the moon from the earth as the cause of increased tides. Since Prepaleozoic time, however, seemingly the tides have decreased while the length of the day has increased, which changes presumably were due

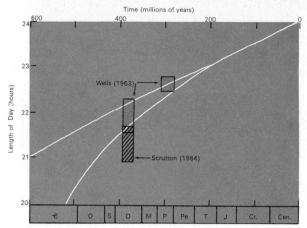

FIGURE 13.10 Changing length of the day during Paleozoic time on the basis of coral growth bands (two curves show limits of uncertainty). If correct, the data suggest the moon was 3 to 3.5 percent nearer the earth and the equilibrium tide was nearly 10 percent greater in early Paleozoic time than now. *(Data from Wells, 1963, Nature; Scrutton, 1964, Palaeontology.)*

to movement of the moon away from the earth (conservation of angular momentum requires that the earth must accelerate if the moon approaches it, and vice versa). Perhaps we might be able to learn more about the history of the earth-moon system from fossils than from actual moon exploration (see the Wade reference, 1969).

THE SILURIAN CONTINENT

AFTERMATH OF THE TACONIAN OROGENY

In the Appalachian mobile belt, erosion of tectonic islands raised during the Late Ordovician Taconian disturbance was already advanced by Early Silurian time. Deposition in the western part of the belt continued more or less uninterrupted from the Ordovician, but an unconformity marks the base of Silurian strata farther east. Gradual eastward encroachment of the sea over the erosional surface is clearly documented. Coarse sandy and gravelly deposits gradually shifted eastward as transgression proceeded.

Early Silurian clastic sediments clearly were derived from the core of the mobile belt east of New

York. This is shown by the coarsening of sandstones and increase of conglomerate eastward (Fig. 13.11) and by orientation of current-formed features. Silurian quartz sandstones are widespread in the Appalachian Mountains, making up roughly 100,000 cubic kilometers (25,000 cubic miles). This staggering volume of sand with scattered quartz pebbles represents weathering, winnowing, and concentration from a tremendous volume of source rocks in the Taconian Mountains.

Unusual Silurian iron-rich sedimentary deposits in the southern Appalachian Mountains provide ore for the important Birmingham steel industry; a similar ore used to be mined in Newfoundland, too. Apparently iron was introduced by rivers in unusual concentrations to a somewhat restricted, marginal-marine environment. The sediments forming there, including marine shells, became replaced and cemented with red hematite.

NEW MOUNTAIN BUILDING IN THE NORTH

In the eastern part of the mobile belt, Silurian volcanism was important in New England and Maritime Canada. Many of the eruptions were submarine, for lavas are found interstratified with richly fossiliferous limestones, sandstones, and shales. Only local, small islands existed.

In Newfoundland, northeastern Greenland, and northern Canada, there were new mountain-building disturbances during the Silurian Period. Possibly structural upheaval began first in the south during the latest Ordovocian (the Taconian Orogeny) and then spread progressively northward during the Silurian. Late Silurian terrigenous clastic sediments occur in the Arctic (Franklin) mobile belt, and especially in northern Greenland. Locally, an angular unconformity separates Ordovician and Silurian strata, and another, even more profound one, marks the boundary between Late Silurian and Early Devonian ones.

GENERAL PALEOGEOGRAPHY OF THE CRATON

Silurian strata are widespread over large parts of the craton, but in the south-central and southwestern portions, they are conspicuously absent (Fig. 13.11). Scattered distribution of Silurian strata led to a long-held view that the Silurian epeiric sea was restricted in area and was studded with many low islands. Such

an interpretation was strongly influenced by the hazardous assumption that *present limits of marine strata closely approximate their original distribution.* As we have seen, one must be careful to evaluate the importance of unconformities in cratonic sequences that may account for a great deal of erosion of formerly more extensive strata. Such a discontinuity within the Devonian System is known to occur over all of the craton!

If we examine the sedimentary facies for the Middle Silurian, we see that most of the preserved rocks are carbonates and some shale that must have formed under a minimum influence of lands. Moreover, the deposits and their fossils suggest a former shallow sea with uniform conditions over an immense region. Therefore, Fig. 13.12 shows our preferred interpretation of Middle Silurian paleogeography in which it is assumed that marine Silurian strata originally covered practically all of the craton.

ORGANIC REEFS

GENERAL CHARACTERISTICS

Although animals inhabit all portions of the seas, the greatest number and diversity of marine animals and plants live in warm, shallow waters within the zone of light penetration and strong agitation. The floating microscopic plankton forms the beginning of the long marine food chain. Lowly microscopic plants begin the chain, followed by microscopic animals, which feed on the plants and each other. Bottom-dwelling invertebrates in turn feed upon all of the plankton and upon each other. Various swimming invertebrates, fishes, and marine mammals in turn are predatory on the other creatures. Because plants provide the ultimate basis of all animal diets, the food chain must begin in shallow, lighted water where photosynthesis is possible. Furthermore, agitation tends to make shallow waters well oxygenated for animal respiration. But nutrients must also be available.

Optimum growth conditions for reef-building organisms exist where the upwelling of deep, fertile

FIGURE 13.11   Middle Silurian lithofacies; note large patches of carbonate rocks over the Canadian Shield region and widespread organic reefs, R. Also note limestones in the outer volcanic zones of the mobile belts (see Fig. 11.15 for symbols and sources).

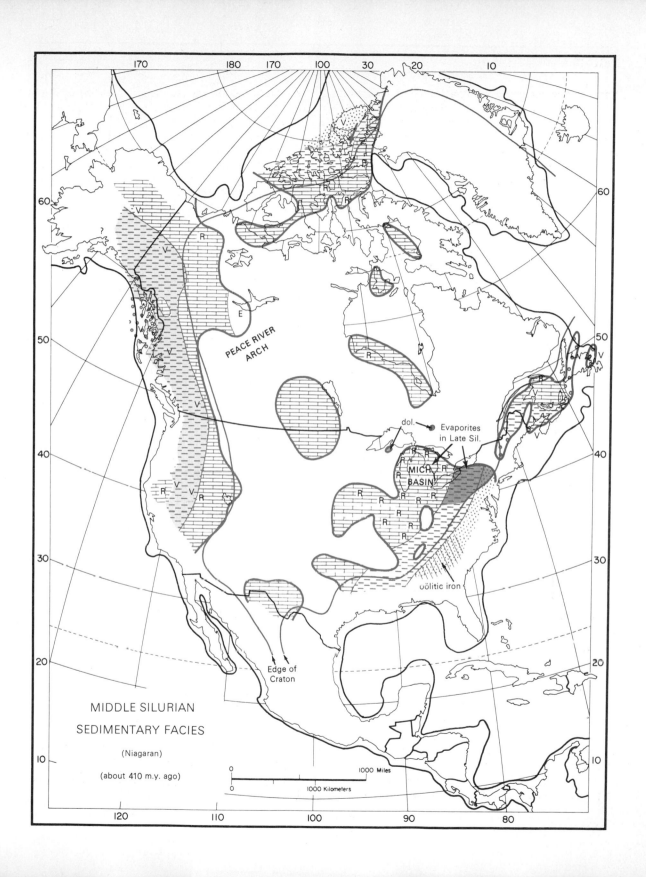

MIDDLE SILURIAN

SEDIMENTARY FACIES

(Niagaran)

(about 410 m.y. ago)

PEACE RIVER ARCH

MICH. BASIN

dol.

Evaporites in Late Sil.

oölitic iron

Edge of Craton

0   1000 Miles

0   1000 Kilometers

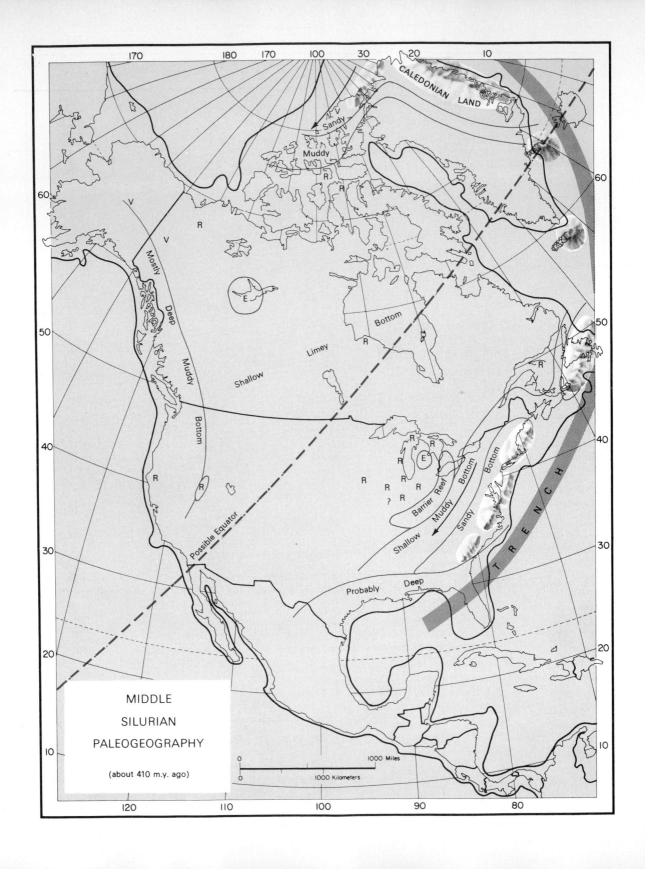

MIDDLE

SILURIAN

PALEOGEOGRAPHY

(about 410 m.y. ago)

waters occurs. Upwelling characterizes many areas of the sea, but is especially important where currents impinge against submerged slopes or escarpments and are deflected upward, as along the northwest side of the Florida Straits and the eastern, windward edge of the Bahama Banks (Fig. 13.13). At such locations in tropical and subtropical latitudes, crowded and diverse marine animal and algal communities develop, and the growth of skeletal organisms is so great that wave-resistant, mound-like masses of calcium carbonate are built up on the seafloor to sea level. These are called organic reefs (Fig. 13.14). Reefs are the most "urbanized" and therefore most ecologically complex areas of the sea floor.

Reefs are geologically as well as biologically important, for large masses of carbonate rocks are built by reef-forming organisms. Many ancient reefs exist and are of special interest because of their significance both environmentally and as petroleum traps. The main reef core rock commonly is characterized by a distinctive fabric of interlocking skeletal material. Typically it forms a massive, lenticular body surrounded by clearly stratified sediments (Fig. 13.15).

Today corals and calcareous algae are the most prominent reef builders, but in the past many other organisms contributed as well. These included algae alone (especially prior to the Silurian Period) and stromatoporoids, sponges, bryozoans, crinoids, brachiopods, and certain molluscs. Many ancient reefs have been more or less converted to dolomite, causing considerable modification of original textures. Organic reefs, both modern and ancient, vary greatly in size and form. Some occur as long, linear barrier reefs at the edges of shelf or bank areas. Examples include the Florida Keys and Bahaman reefs (Fig. 13.13), and the largest of all modern examples, the Great Barrier Reef of Australia (see Fig. 13.34), 1,700 kilometers long. Also important are the more circular fringing reefs developed around islands, and the Pacific atolls built on submerged prominences. Small, isolated reefoid masses called patch reefs or knolls are very common in a variety of settings.

Reefs encompass many subenvironments characterized by differences in the organic community

FIGURE 13.12 Middle Silurian paleogeography. Note that only small peripheral lands are inferred; most of the continent lay beneath an epeiric sea.

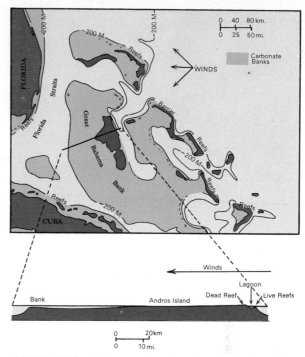

FIGURE 13.13 Modern reefs and carbonate banks of the Bahama Islands–Florida Keys–Cuba region. The Bahama Islands and the Keys are emerged dead Pleistocene reefs; living reefs lie seaward of them. Reefs have grown almost continually in this region since late Mesozoic time. Carbonate sands cover the banks. Contour line indicates 200 meters below sea level. *(After Newell and Rigby, 1957, Society of Economic Paleontologists and Mineralogists Publication 5.)*

and sediments. The windward side is generally rather steep-faced and is constantly battered by waves; it is the zone both of most active growth and of destruction. Fragments of reef rock periodically are torn loose to slide down the reef front into deep water. Thus an apron of coarse, poorly sorted, angular reef debris characterizes the fore reef facies. The debris shows crude stratification inclined as much as 30 degrees away from the reef front. Back reef facies consists chiefly of stratified clastic carbonate sand derived from the reef, though oölite also may be present as well as evaporite layers.

SILURIAN REEFS

One of the most interesting aspects of Silurian time is the great development of large organic reefs involv-

FIGURE 13.14 Underwater photo of a modern reef top near Key Largo, Florida. Living brain coral (center) and staghorn coral (lower) are surrounded by sand produced by wave erosion of reef. The water is only a few meters deep, allowing photography under ordinary sunlight. *(Courtesy George Lynts.)*

ing for the first time important contributions from animal skeletons as well as from algae. Silurian examples extend from Tennessee to the Arctic and from Ohio to Alberta (Fig. 13.11), but the prevalence of reefs is by no means confined to North America. The original Silurian strata studied and named in eastern Wales contain important reef masses along the eastern side of what we call the Caledonian mobile belt. Similar reefs also are well known elsewhere in Europe, but Devonian reefs were the most widespread ever.

Small reefs had been formed by algae since Early Prepaleozoic time. Primitive filamentous algae, today most characteristic of the intertidal zone, have built laminated structures throughout history (see Figs. 10.20 and 11.28), but beginning in early Paleozoic time, more complex types became important reef contributors. Animals began building reefs in Cambrian and Ordovician times, but why there was such a sudden burst of highly complex animal and algal reef building in Silurian time is not clear. Probably it was due to a combination of evolutionary changes and existence of shallow-marine conditions that gave some selective advantage to "urbanized" living.

FIGURE 13.15 Air view of inaccessible reef-like limestone lens in Upper Devonian strata exposed in the Canadian Rocky Mountains near Mount MacKenzie, Alberta. Massive light-colored rock extends as tongues into adjacent, stratified material, which dips away from the lens-shaped mass. Light-colored masses at right edge presumably are slide blocks derived from the reef. *(Courtesy W. S. MacKenzie, Geological Survey of Canada.)*

Figure 13.16 shows a Silurian reef exposed by quarrying operations. Various facies of this complex are revealed clearly both by lithology and fossil communities. Stromatoporoids and corals built the main, wave-resistant core, while crinoids, molluscs, and brachiopods dominated the back reef and flanking deposits. Because shoal reefs grow in the surf zone only a meter or so below sea level, the depth of surrounding water can be estimated accurately in some fossil examples by tracing a reef front stratum down into inter-reef deposits and noting the vertical difference of level. The Thornton reef at Chicago apparently was surrounded by water at least 60 meters deep. Thus we see how significant ancient organic reefs are in providing clues to past environments.

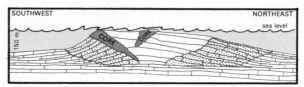

FIGURE 13.16   Middle Silurian reef at Thornton, Illinois, south edge of Chicago. Various facies have been exposed by quarrying, thus allowing detailed study. Continuity of dipping reef flank into flat inter-reef strata allows determination of approximate water depth. *(After Ingels, 1963, Bulletin American Association of Petroleum Geologists, v. 47, pp. 405–440; by permission.)*

ORIGIN OF MARINE EVAPORITE DEPOSITS

THE MICHIGAN BASIN
Another prominent group of Silurian sediments is the important evaporite deposits found particularly in New York, Ohio, Michigan, and in western Canada (Fig. 13.11). These deposits are the basis for major plasterboard and chemical industries. Potassium, important as fertilizer, also may be obtained from evaporites, for example in Saskatchewan. Evaporites characterize chiefly Upper Silurian strata that formed immediately after the maximum reef development shown in Fig. 13.12. In Michigan, especially, subsidence accelerated in a circular area (Fig. 13.17). Subsidence commenced rather suddenly far from any rising arch or mountainous uplift. An increase in density or a decrease in thickness of the crust beneath Michigan must have upset local isostatic equilibrium. Probably there was some indirect relation to structural movements in the Appalachian mobile belt.

In the Michigan Basin, up to 1,500 meters of sediments were deposited, chiefly dolomite rock [$CaMg(CO_3)_2$], but with as much as 750 meters of rock salt ($NaCl$) and anhydrite-bearing ($CaSO_4$) strata. These represent sediments precipitated from sea water under conditions such that evaporation exceeded total replenishment of water by rainfall, river flow, and inflow from the open sea. Evaporite strata required concentration of brines from an immense total volume of sea water. If the Silurian sea were as saline

as that of today, it would have required evaporation of the equivalent of a column of normal sea water nearly 1,000 kilometers deep (about 600 miles) to deposit 750 meters of evaporite strata! Certainly the water was never 1,000 kilometers deep over Michigan; rather it was apparently shallow at all times. Therefore, we must postulate continual replenishment of water as evaporite sediments were precipitated over a subsiding basin floor.

RESTRICTED CIRCULATION
A presumably analogous situation is seen today in central Asia in the Gulf of Kara-Bogaz-Gol, an embayment of the Caspian Sea. The Gulf's waters are replenished more or less continuously across a shallow bar, but because the region is very arid, evaporation causes continuous precipitation of salts from very

FIGURE 13.17   Two thickness maps of Michigan showing that the circular Michigan basin originated in Late Silurian time (contours in meters). The basin suddenly subsided as much then as it had throughout all of earlier Paleozoic time. *(Adapted from Cohee, 1948, U.S. Geological Survey Oil and Gas Chart 33; Alling and Briggs, 1961, Bulletin American Association of Petroleum Geologists, v. 45, pp. 515–547; by permission.)*

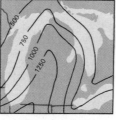 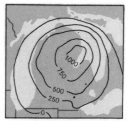

A. CAMBRIAN THROUGH MID-SILURIAN   B. UPPER SILURIAN

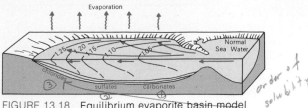

FIGURE 13.18 Equilibrium evaporite basin model illustrating the restricted circulation hypothesis of precipitation. Normal sea water flows continually into the restricted basin at a rate closely balanced by evaporation, which produces dense brine that sinks and precipitates different evaporite minerals according to concentration; numbers indicate water density. *(After Briggs, 1957; by permission of Michigan Academy of Science.)*

concentrated, dense waters at the bottom of the Gulf. These brines cannot escape to the Caspian Sea because of the restricting bar. As a result, a dynamic equilibrium[1] exists between precipitation of evaporite sediments at the bottom and replenishment and evaporation of water at the top. As long as this equilibrium persists, evaporite sedimentation continues. The Gulf of Kara-Bogaz-Gol provides a kind of model to help us understand evaporite deposition in ancient sedimentary basins (Fig. 13.18).

Circulation of sea water within many ancient basins was restricted by a variety of causes so that dense brines sank and precipitation of salts occurred. Though relative aridity of climate is indicated, excessively hot temperatures are not necessarily required, for dry winds could accomplish the evaporation effectively. But if we assume both moderately warm temperatures *and* dry winds, then evaporites can be explained readily wherever oceanic circulation in shallow seas was impaired. In Michigan, the fringing reef complexes that began developing around the basin in Middle Silurian time apparently caused restriction of circulation leading to evaporite precipitation (Fig. 13.19). Also, a slight lowering of sea level

may have occurred near the end of Silurian time, which would have produced more islands and shoals around the basin margin, further restricting circulation. In New York, circulation was restricted not only by shoals on the west and south (Fig. 13.19), but also by land to the east. Red-colored fine clastic sediments are found intimately interstratified with the marine evaporites there.

Ordinary sea water today contains about 3.5 percent dissolved salts, of which sodium chloride (NaCl) is the most familiar and most abundant (Table 13.1). Theoretically, complete evaporation of sea water should produce sequential deposition of a series of evaporite minerals in reverse order of their relative solubilities in water. Laboratory experiments performed as early as 1849 showed the theoretical salt precipitation sequence to be expected. In reality, however, actual precipitation sequences depend upon a variety of possible events that may disturb the evaporation process; it is the rule to find incomplete sequences. For example, cessation of precipitation or solution of earlier salts results from seasonal temperature (and humidity) changes or from destruction of circulation barriers to allow dilution of brines by normal sea water. Furthermore, a lateral sequence of evaporite mineral facies is frequently encountered, as shown in Fig. 13.18, with a continuous flow of water undergoing constant evaporative precipitation of carbonate first, sulfates farther along, and chlorides at the farthest end of the flow.

A fruitful new line of research in evaporite (and dolomite) formation suggests that supratidal deposition also may be of major importance. For example, salt flats several kilometers wide border intertidal lagoons of the Persian Gulf. The flats are but a few centimeters above average high tide, but occasionally are flooded. Salt water fills the pores of sediments beneath the salt flats, but it is abnormally saline due to the high evaporative potential of arid climates. Gypsum and dolomite are precipitated within the salt flat muds (Fig. 13.20). Under conditions of balanced subsidence and deposition, significant thicknesses of evaporite-bearing strata could accumulate, and lateral shifts of shoreline would cause transgressive or regressive migration of the evaporite facies.

POSTDEPOSITIONAL CHANGES
Evaporites undergo many changes after deposition. These involve chiefly reactions with ground water because of the great solubility of the evaporite miner-

[1]Equilibrium commonly is thought of only in static terms, but in nature practically nothing is static. Development of the concept of equilibrium (or a *steady-state* condition) in dynamic systems was a major advance for geology. In an open system, there may be a constant flow of material and energy, as in an evaporite basin, but the system looks much the same from one time to the next. A beach also exemplifies dynamic equilibrium or steady state, for energy is constantly being expended in the system and sand is constantly in motion, yet the beach does not change in form significantly through long spans of time (except for severe storms, which temporarily upset the equilibrium). Most large rivers with steady discharges also illustrate dynamic equilibrium, at least over modest time spans.

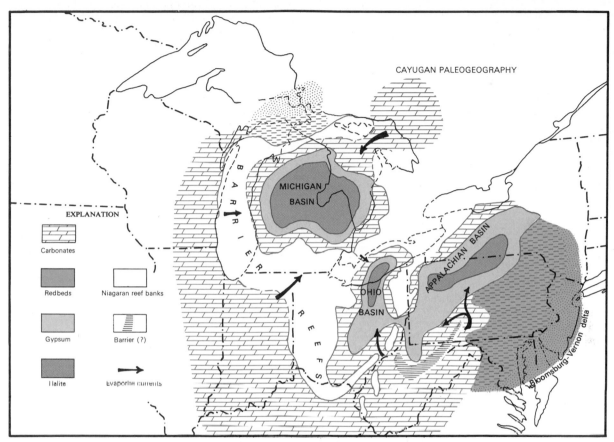

FIGURE 13.19 Late Silurian paleogeography of the Michigan–New York evaporite basins. Barrier reefs restricted marine circulation into the basins; evaporites occur in basin centers in Michigan and Ohio, but not toward the landward margin in New York. *(After Alling and Briggs, 1961, Bulletin American Association of Petroleum Geologists, v. 45, pp. 515–547; by permission.)*

als. The most common change is simple solution of evaporite layers, causing subsidence or collapse of overlying and interstratified insoluble sediments into solution caverns. Collapse causes fragmentation of nonevaporite rocks, producing a jumbled breccia of angular fragments. Another typical change is the

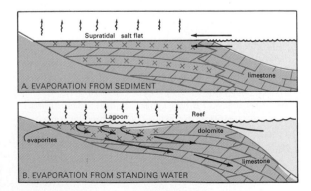

FIGURE 13.20 *A*: Evaporites and dolomite formed due to infrequent washovers onto supratidal salt flats called sabhka and some seepage from normal marine waters at right. *(Adapted from Illing et al., 1965, Society of Economic Paleontologists and Mineralogists Special Publication 13.) B*: Dolomization of limestone ($CaCO_3$) by seepage reflux of very saline brines derived from restricted lagoon (left) where evaporation increases brine density. Dense, magnesium-rich brines are "pumped" through limestone toward right; calcium is carried away and replaced by magnesium to produce dolomite [$CaMg(CO_3)_2$]. *(Adapted from Adams and Rhodes, 1960, Bulletin American Association of Petroleum Geologists, v. 44, pp. 1912–1920; by permission.)*

TABLE 13.1 Proportions of principal salts dissolved in normal sea water

Salts	Amount dissolved, %	Thickness if all were precipitated from 2,000 meters of water
NaCl (salt)	2.72	47.5
MgCl$_2$	0.38	5.8
MgSO$_4$	0.16	3.9
CaSO$_4$ (anhydrite)	0.12	2.3
K$_2$SO$_4$	0.08	
CaCO$_3$ (calcite)	0.01	0.6
MgBr$_2$	0.01	
	3.48%	60.1 meters

conversion of anhydrite to gypsum near the surface by hydration, that is, the combination of two molecules of water with each molecule of calcium sulfate. As a result of such changes, we only rarely see normal, original evaporite sequences in surface outcrops. Undisturbed, natural evaporite deposits are largely confined to the subsurface below the depth of penetration of fresh ground water. Indeed, many evaporite deposits were unknown until deep drilling penetrated hitherto unexplored basins.

ORIGIN OF DOLOMITE ROCKS

CALCIUM AND MAGNESIUM IN THE SEAS

Magnesium is about three times as abundant in sea water as calcium; yet calcium is added six or seven times faster by rivers (Table 13.2). At present rates of input, calcium abundance should be doubled in only 1 million years, and magnesium in 18 million years. Because the composition of sea water seems to have remained fairly constant, at least over later geologic

TABLE 13.2 Relative abundance of four chief bases dissolved in river and sea waters

	River water, %	Sea water, %
Calcium	73	3
Magnesium	11	10
Sodium	9	84
Potassium	7	3
	100%	100%

time, both calcium and magnesium must be constantly removed from sea water so that some sort of balance of their abundances—a dynamic equilibrium—is maintained. The figures also show that calcium is less soluble than magnesium, so is more readily extracted. The carbon dioxide content of sea water largely controls deposition both of calcium and magnesium, but carbon dioxide is in turn sensitive to temperature. With an increase of temperature, carbon dioxide escapes from water to the atmosphere. This causes dissolved calcium to become less soluble, and inorganic precipitation of calcium carbonate may ensue, especially in warm climates. The reaction is as follows, in which loss of CO$_2$ drives it to the right:

$$Ca^{2+} + 2HCO_3^- \;\rightleftharpoons\; CaCO_3 + H_2O + CO_2 \uparrow$$

A decrease of temperature accompanied by an increase of pressure, as in the deep seas, increases both calcium and magnesium solubilities. Therefore, many of the skeletons found more than 5,000 meters below sea level are composed of silica, which is less soluble there than calcium carbonate.

Given an abundant supply of calcium and magnesium from weathering of rocks on land, the quantity of carbonate sediments that can form in the shallow seas is a function of the carbon dioxide content of sea water, which, in turn, is controlled ultimately by the amount of that gas in the atmosphere. Sea water is more nearly saturated with respect to calcium carbonate than more soluble magnesium carbonate, *even though magnesium is more abundant*. It follows that special conditions are required to precipitate magnesium carbonate to form dolomite.

IMPORTANCE OF SALINITY

From the above discussions, it is clear that some carbonate sediments can form by direct, inorganic precipitation from sea water through evaporation. Dolomite [CaMg(CO$_3$)$_2$] is the most common evaporitic carbonate mineral. Experimental as well as observational evidence indicates that dolomite forms in an alkaline medium such as sea water with slight excess salinity and slightly elevated temperatures. Under such conditions, magnesium as well as calcium becomes insoluble, with magnesium substituting for some calcium ions to form either impure calcite or dolomite.

Although a case can be made for direct precipitation of dolomite under evaporative conditions, there are immense volumes of dolomite rock that show no clear evidence of having so formed. Many Paleozoic carbonate rocks initially formed as accumulations of calcareous skeletons; yet no known organism secretes the mineral dolomite. Therefore, much dolomite rock must have been converted after deposition by the following reaction:

$$2CaCO_3 + Mg^{2+} \rightarrow CaMg(CO_3)_2 + Ca^{2+}$$

Recrystallization of limestone to dolomite rock is called dolomitization, but it occurs in a variety of ways and at varying times after initial deposition. Magnesium must have been introduced, and, from previous discussions, percolation of high-salinity brines through limestone seems the most plausible mechanism. In recent years, extensive search of modern sediments (using x-ray identification techniques) has shown that modern dolomite, long thought nonexistent, is common in supratidal muds. Evaporative increase of salinity in pore waters, as in the example of supratidal evaporites cited above (Fig. 13.20), causes magnesium to be precipitated. The brines undergo continuous seepage reflux, the reality of which has been confirmed by experiments as well as field observations of modern sediments. This makes possible dolomitization of large volumes of limestone.

COMPARISONS OF MODERN AND ANCIENT CARBONATE SEDIMENTATION

We should next seek suitable environments in the modern seas that might match those of early and middle Paleozoic carbonate deposition. We should remember, however, that there is little assurance that *all* ancient environments have modern counterparts (and vice versa). But it is reasonable—even essential at first—to assume that the more persistent ancient environments are at least approximated somewhere today.

We look to the modern shallow seas for analogies with ancient epeiric seas for reasons outlined in Chap. 11. Broad, smooth continental shelves, together with the Baltic Sea, Black Sea, and Hudson Bay, represent small contemporary epeiric seas. From paleogeographic evidence, it is clear that total land area today is many times larger than during the first two-thirds of Paleozoic time. Therefore, we can only hope for small modern areas analogous to past

carbonate-forming epeiric seas. Figure 13.21 shows the chief areas where carbonate sediments are forming today. Disregarding the pelagic foraminiferal oozes of the deep-sea floor, other carbonates shown are predominantly shallow, tropical sea products. A minimum of contaminating terrigenous clastic detritus is the most important factor in precipitation either organically or inorganically of significant carbonate strata. Nonetheless, there is no doubt that warm, well-lighted water provides optimum conditions. It taxes the imagination little to translate these conditions to the Paleozoic carbonate-producing epeiric seas. Lands were relatively small and low until late Paleozoic time, and shallow seas were large. Calcareous-secreting organisms thrived, and their skeletons contributed the bulk of carbonate material (Fig. 13.22). The wide distribution of ancient carbonates with abundant and diverse invertebrate faunas long has been taken to indicate wider-than-present Paleozoic tropics.

Note that so far we have assumed relatively constant sea-water chemistry and other ecologic requirements for most marine organisms since Eocambrian time. The assumption of near-constancy of ecologic requirements of respective types of organisms is based upon simple probability. Many fossil communities show the same associations of types of organisms and of sediments as do their modern counterparts. It is improbable that *all* members of very complex communities could have changed requirements to the same degree *and* at the same rate through time. Therefore, the finding of ancient communities with two or three dozen organisms showing mutual relations closely duplicating their modern counterparts argues for similar requirements through time. Sulfur isotope variations in evaporites suggest some changes of sea-water composition, but probably they were small.

PALEOCLIMATE AND PALEOGEOGRAPHY

Abundance of iron oxides in certain Silurian and Devonian sediments indicate a strongly oxidizing atmosphere. Important evaporite sediments prove a relatively high evaporation potential over large parts of North America, and so one is tempted to conclude that the climate was relatively warm. Over Devonian land in eastern North America apparently there was

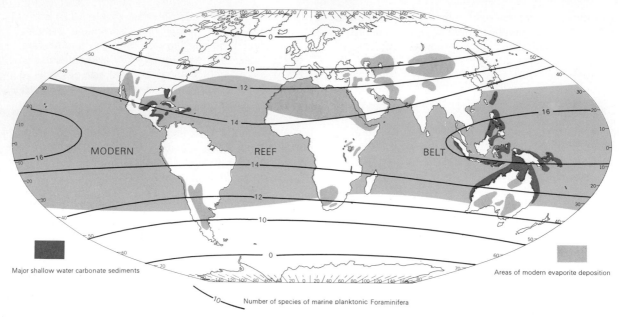

Major shallow water carbonate sediments

Areas of modern evaporite deposition

-70- Number of species of marine planktonic Foraminifera

FIGURE 13.21 Distribution of modern organic reefs, major shallow-marine carbonate deposition, and evaporites (chiefly nonmarine). *(Adapted from Lowman, 1949,* Geological Society of America Memoir 39; *Rodgers, 1957,* Society of Economic Paleontologists and Mineralogists Special Publication 5; *Goode's world atlas, 1964.)* Also shown are contours indicating latitudinal diversity (number of species) of modern marine planktonic Foraminifera as a function of temperature. With few exceptions, both marine and nonmarine organisms show similar patterns of increasing diversity in warm, tropical latitudes. *(After F. G. Stehli,* Science, *v. 142, 22 November 1963, pp. 1057–1059; copyright 1963 by American Association for the Advancement of Science.)*

moderate humidity as suggested by luxuriant forests that cloaked the region (Fig. 13.1). Presence of great organic reef complexes and rich, diverse marine fossils in both Silurian and Devonian marine strata suggest warm, shallow, agitated seas by analogy with the restriction of modern reefs to shallow tropical seas (Fig. 13.21). We also find today that the greatest diversity of marine organisms occurs in the warm subtropics and tropics (Fig. 13.21), where optimum (though not exclusive) ecologic conditions for many organisms exist.

Devonian land plants are similar the world over, suggesting that climate was rather uniform. Wide distribution of richly fossiliferous middle Paleozoic marine carbonate rocks, and especially the great latitudinal spread of fossil reefs (Fig. 13.23) suggest

subtropical conditions for North America, Europe, Siberia, and Australia. It has long been felt that the average climate of the earth through time has been milder and more homogeneous than it is today. The present certainly is not a very good key to the past in terms of climate!

When compared with modern distributions, middle Paleozoic reef and carbonate rock patterns pre-

FIGURE 13.22 Microscopic photograph of a limestone composed of abraded and sorted fossil skeletal debris (grains average about 0.5 millimeter). Such clastic limestones are common in most Paleozoic sequences (lower Pennsylvanian, northern Nevada).

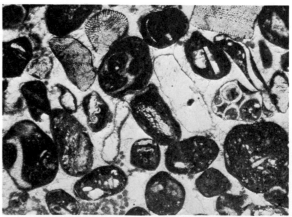

DEVONIAN LATITUDINAL DATA

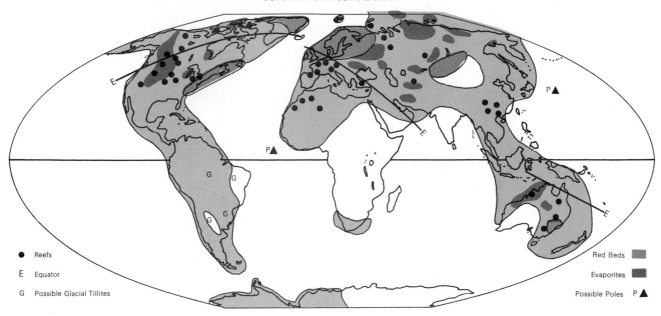

- ● Reefs
- E Equator
- G Possible Glacial Tillites

Red Beds

Evaporites

Possible Poles P ▲

Approximate maximum transgression

sont a serious paleogeographic problem because they are found at 70°N latitude with respect to the present equator, 30° higher than reefs now grow! Three hypotheses can be presented to explain this seemingly anomalous latitudinal spread: (1) The past warm-water (subtropical) oceanic belt was so wide as to extend to about 70 or 80°N, thus bathing the entire North American continent in warm seas (Fig. 13.24A); (2) the subtropical belt, whether wider than today or not, was differently oriented with respect to North America and the present equator; i.e., the continent or

FIGURE 13.23   Presumed paleoclimatic and latitudinal indicators for Devonian time compared with paleomagnetically indicated equator position. Coral reefs, evaporite deposits, and red bed facies all fall near and parallel to the paleomagnetically indicated equator; Devonian tills in South America lie far from the equator near the paleomagnetically indicated pole for South America. *(Adapted from Schwarzbach, 1963; Irving, 1964; McElhinny, 1967, UNESCO Symposium on Continental Drift, Montevideo.)*

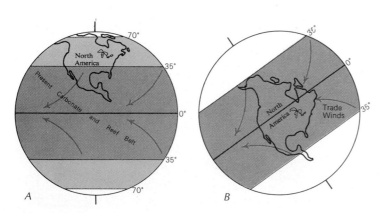

FIGURE 13.24   Two alternate hypotheses to explain distribution of middle Paleozoic reefs and associated richly fossiliferous carbonate strata containing great diversity of species. *A*: Wider tropics (up to 70° latitude) then than now (darkest area) due to overall warmer average climate, which is indicated by much other evidence. *B*: Different relative orientation of the tropics with respect to North America as is suggested by paleomagnetic evidence.

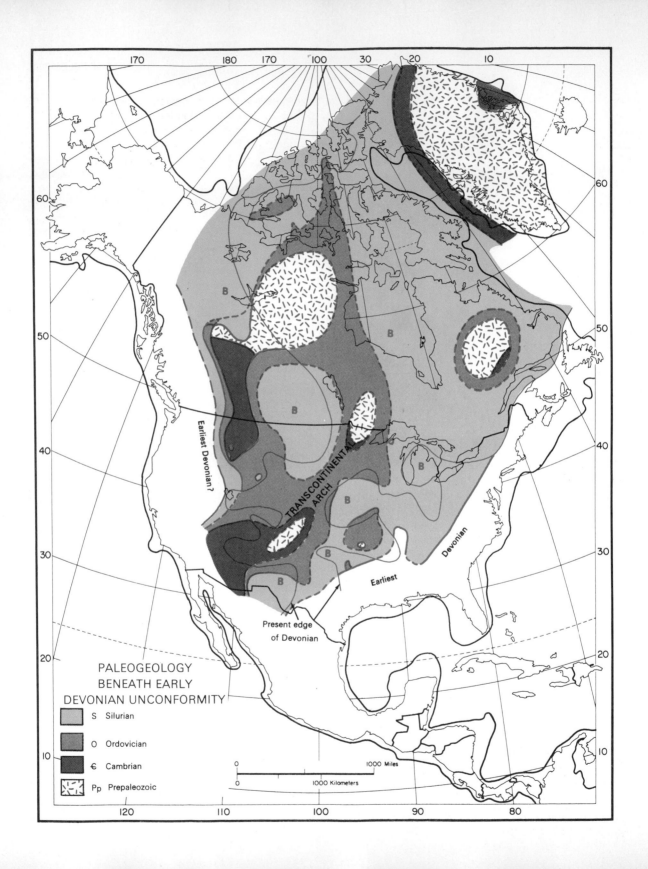

PALEOGEOLOGY
BENEATH EARLY
DEVONIAN UNCONFORMITY

S Silurian

O Ordovician

€ Cambrian

Pp Prepaleozoic

Earliest Devonian?

TRANSCONTINENTAL ARCH

Present edge
of Devonian

Earliest

Devonian

0   1000 Miles
0   1000 Kilometers

the pole-equator system (and climatic zones) have somehow shifted relative to each other (Fig. 13.24B); or (3) reef-forming organisms have changed their ecologic adaptations slowly through time so as to become restricted gradually in latitudinal range.

The third possibility above seems least likely for reasons already discussed, and it is essentially untestable. Choice between the first and second hypotheses is not easy. An argument in favor of the second comes from paleomagnetic studies, which suggest that the middle Paleozoic magnetic field was oriented very differently than now with respect to North America. This relationship would bring the reefs within a subtropical belt parallel to that equatorial position. Comparing Fig. 13.23 we see that worldwide reef distribution and paleomagnetic data concur with other geologic evidence summarized before that points to a very different past arrangement of lithosphere plates.

DEVONIAN STRATA OF THE CRATON

REGRESSION AND TRANSGRESSION

In Late Silurian and Early Devonian time, marine deposition became restricted to a few basins and the marginal mobile belts. In the latter regions, marine Upper Silurian and Lower Devonian strata are perfectly conformable. Over most of the craton, however, regressive facies developed, and finally most of the craton emerged as a low land for a few million years. Younger, transgressive Devonian strata rest unconformably upon a variety of older rocks, including even Prepaleozoic ones on several arches (Fig. 13.25). The Early Devonian unconformity is one of those rare cratonwide breaks like that beneath the Ordovician St. Peter Sandstone. It is the boundary between the second (Tippecanoe) and the third (Kaskaskia) cratonic sequences illustrated in Fig. 12.12. Pure quartz

FIGURE 13.25 Paleogeologic map showing rocks beneath the widespread Early Devonian unconformity. A great deal of warping affected the craton prior to Middle and Late Devonian transgression, but Prepaleozoic basement was exposed widely only in Canada (B represents basins). Note that youngest strata beneath a regional unconformity tend to be less extensive than successively older ones; Lower Devonian strata are least widespread, Upper Silurian slightly more so, Middle Silurian and Ordovician ones, respectively, still more so. (*After Lorersen, 1960,* Paleogeologic maps, *and D. E. Owen, unpublished maps.*)

sandstone south of the eastern Great Lakes (Oriskany of Fig. 13.30) resembles the lower Paleozoic quartz sandstones in every respect, and, like them, it was deposited at the margin of a transgressive sea. It was derived from erosion of those same older sands, reflecting the repetitive nature of early cratonic history. Apparently the Ordovician and Silurian carbonate rocks previously deposited were dissolved by ground water, and any shale (of which there was very little) was weathered, with the resulting clay carried beyond the edges of the craton.

CRATONIC BASIN DEPOSITION

Warping of the North American crust was widespread in Early Devonian time, and practically all of the craton and even portions of the mobile belts were above sea level and being eroded to produce profound changes on the continent. Basins, arches, and

FIGURE 13.26 Effects of Devonian regression and transgression on cratonic arches and basins. Note differential erosion beneath the unconformity and differential deposition above. Subsurface information is mandatory to make such reconstructions in basin areas.

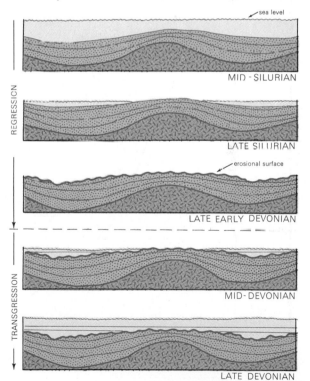

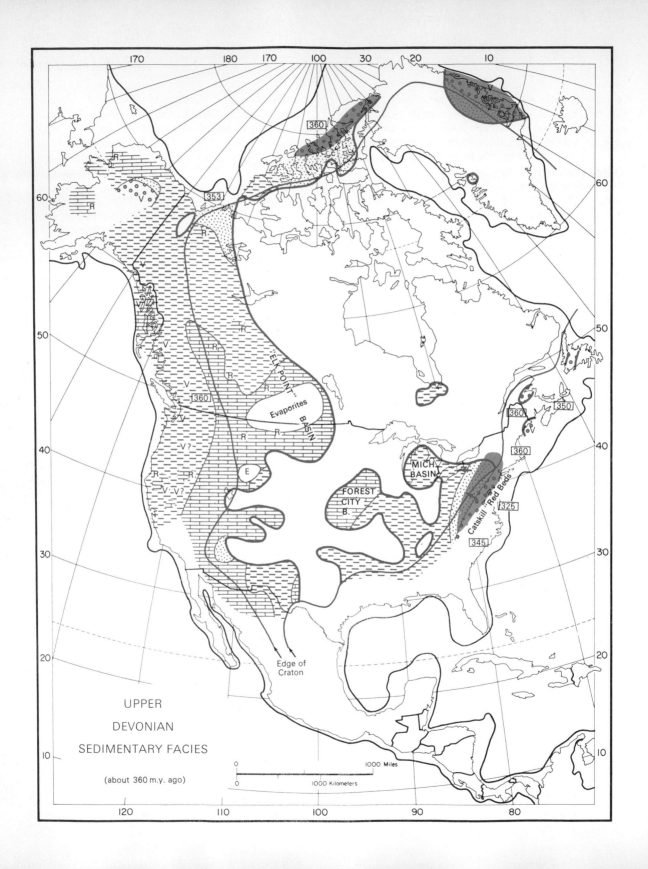

UPPER

DEVONIAN

SEDIMENTARY FACIES

(about 360 m.y. ago)

360

353

360

350

360

360

325

345

"ELK POINT"

Evaporites BASIN

E

MICH BASIN

FOREST CITY B.

Catskill "Red Beds"

Edge of Craton

1000 Miles

1000 Kilometers

domes became more sharply delineated than ever before. The net result of warping and regression was much differential erosion (Fig. 13.25). As transgression occurred near the middle of the Devonian Period, marine deposition resumed first in basins and gradually encroached upon arches and domes (Fig. 13.26). As a result, Upper Devonian marine strata are most widespread (Fig. 13.27).

The Michigan Basin continued to subside and more evaporites were deposited. The Williston or Elk Point Basin farther west also received important evaporites. An immense barrier-reef complex developed around its margin and extended far northwest in Canada (Fig. 13.28). Apparently it formed along a zone of deeper water upwelling against a shallow carbonate shelf. Devonian reefs long had been known in the Canadian Rocky Mountains (Fig. 13.15), and petroleum had been produced from some since 1920, but reefs beneath the plains were unknown until about 1947, when drilling encountered phenomenal petroleum reserves trapped therein. The discovery triggered one of the continent's greatest oil booms and provided a wealth of information about buried rocks previously unknown. Indeed, existence of the basin itself was hardly appreciated before that time.

DOMES AND ARCHES
Because the stratigraphic record on domes and arches is less complete than in basins, and because some of the unconformities there are very subtle (Fig. 13.29), it is more difficult to date their upwarpings. At least some Cambrian and Ordovician strata are present on most, but Silurian and Early Devonian rocks are absent over all the arches and domes (Fig. 13.25). This situation naturally tempts us to guess that warping and erosion began in Silurian time, but there are other clues for us to consider, too. If shoals or islands were present in Silurian time, sandy or muddy facies would be expected around them; yet only Silurian carbonate rocks are found (Fig. 13.11). Moreover, the paleogeologic map of the mid-Devonian unconformity reveals that present Silurian rock distribution on the craton is due largely to Early Devonian erosion. Therefore we conclude that many of the domes were not upwarped until Devonian time. Because of the

number of such features, we also conclude that this was a time of unusually severe cratonic deformation.

THE ACADIAN OROGENY IN THE APPALACHIAN BELT

EVIDENCE OF DEVONIAN OROGENY
James Hall's original observations leading ultimately to the concept of a geosyncline were based largely upon comparisons of Devonian strata of the craton with those of the northern Appalachian Mountains. Whereas the entire Devonian sequence averages less than 300 meters thick in the craton, it is nearly 1,800 meters in the Appalachian region of eastern Pennsylvania and New York (Fig. 13.30). In New England and southeastern Canada, Devonian rocks include considerable lava and volcanic ash; and throughout most of the Appalachian belt, later Devonian strata include red sandstone, conglomerate, and shale (Figs. 13.27 and 13.31), a molasse facies named for the Catskill Mountains of southeastern New York.

Like the older red clastic succession of Late Ordovician age in the same region, Catskill sediments also coarsen toward the east, and so must reflect elevation and erosion of another prominent land there. We know that this land was composed largely of only slightly older fossiliferous Paleozoic rocks. The Devonian tectonic land encompassed most of the area of the older eroded Taconian land (Fig. 13.28). Thus the major Devonian orogeny was superimposed upon the older one. In the maritime region of southeastern Canada and in northern New England, granites of Devonian age and unconformities related to this orogeny are well displayed. Therefore, this mountain-building episode has been named the Acadian Orogeny for the old French colonial name for that region.

More granite and regional metamorphism (Fig. 13.32) developed during the Acadian than the Taconian event, and it is probable that some thrust faults along the present St. Lawrence valley were formed at this time. The Acadian Orogeny, therefore, was a more severe disturbance of the earth's crust and represents the greatest orogeny for the Appalachian mobile belt.

DATING THE OROGENY
At several scattered localities in the Acadian region, angular unconformities with Mississippian strata rest-

FIGURE 13.27 Upper Devonian sedimentary facies. Note importance of reefs and evaporites in western Canada, and isotopic dates for widespread granitic rocks (see Fig. 11.15 for symbols and sources).

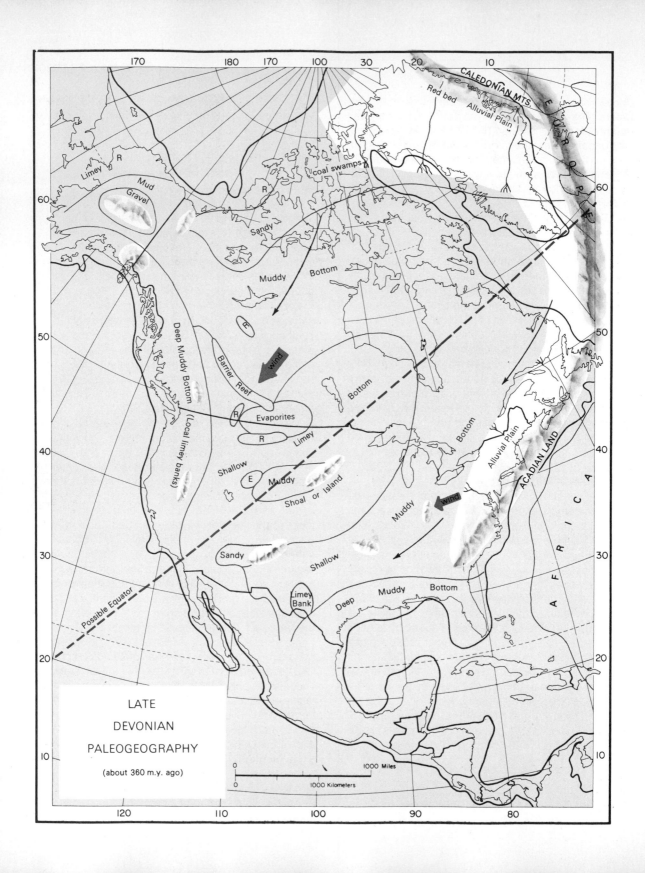

LATE

DEVONIAN

PALEOGEOGRAPHY

(about 360 m.y. ago)

ing variously upon deformed Lower Devonian or older rocks intruded by granitic plutons serve to date the orogeny rather closely. Also, Acadian granites have been extensively dated by isotopic methods (Fig. 13.27). In Nova Scotia and New England, many yield dates of from 330 to 360 million years ago. In Maryland a large mass known as the Baltimore Gneiss has been dated by the K-Ar method, using biotite mica, as from 300 to 350 million years (Early Mississippian). But zircon from the same rock yielded U-Pb dates of from 700 to 1,100 million years (Late Prepaleozoic)! These were among the first discordant isotopic dates discovered for a single rock (See Chap. 6). In recent years such discordances have proven to be more common than was supposed; indeed, they are now more or less expected in complex mobile belts that suffered multiple, superimposed orogenies.

Discordant dates reflect two major events in the history of the Baltimore Gneiss. This rock apparently was formed first as a granite during the widespread Grenville Orogeny (700 to 1,000 m.y.), which affected the entire southeastern margin of present North America and produced the complex metamorphic and igneous rock basement of the Paleozoic Appalachian mobile belt (see Fig. 10.11). The Maryland granite was again heated and deformed during the Acadian Orogeny, at which time biotite was recrystallized so that it shows an isotopic age of 300 to 350 million years. Zircon is very resistant to temperature changes; therefore it has retained isotopic ratios of uranium and lead that reflect the date of original crystallization of the granite. Biotite is very sensitive to heating in excess of 200°C during metamorphism, and argon 40, being a gas, tends to leak from the mica crystal lattice. The result is resetting of the mica's isotopic clock; all the ^{40}Ar now present in the biotite of the Baltimore Gneiss has accumulated from decay of ^{40}K since Acadian metamorphism.

Isotopic evidence, unconformities, and the thick Middle and Late Devonian Catskill clastic rocks all together indicate that Acadian mountain building occurred during the latter part of the Devonian and Early Mississippian Periods through an interval of perhaps 25 to 30 million years (360 to 330 m.y. ago).

FIGURE 13.28 Late Devonian paleogeography. Note the importance of marginal tectonic lands in the east and north. Wind direction in the east is inferred from volcanic ash distribution; in the west, from reef patterns.

FIGURE13.29 Outcrop on the northeast side of the Ozark Dome near Louisiana, Missouri, showing Ordovician, Silurian, and Mississippian strata. C is the Late Devonian–Mississippian Chattanooga Shale, which overlies a subtle but widespread regional unconformity; presence of the unconformity is apparent here only from the absence of Devonian fossils.

THE CATSKILL CLASTIC WEDGE
Conditions under which the Late Devonian clastic sediments accumulated in the eastern United States were almost identical to those under which the Late Ordovician sediments of the same region had formed. But the total volume and coarseness of the Catskill rocks exceeds those of the Ordovician, and the pebbles and sand grains of the Catskill are chiefly composed of metamorphic and granitic rock fragments, feldspar, mica, and quartz (Fig. 13.31). The red color is due to the presence of a small percentage of iron oxide between the grains.

EVOLUTION OF THE EARTH

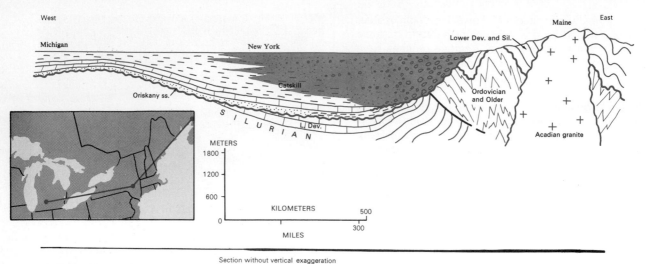

FIGURE 13.30 Restored cross section of Devonian rocks in eastern United States showing effects of the Acadian orogeny and the upper Lower Devonian (pre-Oriskany) unconformity. Note superimposing of Acadian on older Taconian folding.

Most of the Catskill sediments were deposited on a vast alluvial coastal plain that sloped gently westward from the eroding Acadian mountains to the epeiric seashore. This plain was on the order of 300 to 500 kilometers wide at its maximum extent. At its eastern side, it was built of gravels and sands debouched by rivers from the foot of the mountains. Many channel deposits are evident in the eastern Catskill facies, and cross stratification is the rule. Along the western shoreline, however, finer sediments were deposited on river deltas and beyond the deltas to the west; black marine muds and some thin sands spread onto the craton (Fig. 13.27).

In southeastern New York, the red sediments contain remains of a spectacular buried fossil forest (Fig. 13.1). Tree stumps about 30 centimeters in diameter are found buried in their rooted positions, but some trees were rafted down rivers to the epeiric sea, there to be buried in black shales. Identical land plants occur in Devonian red beds in eastern Greenland, and coal is present in Arctic Canada. The oldest known vertebrate land animals also occur in the Greenland strata.

A MODERN ANALOGUE

The volume of preserved Catskill sediments is on the order of 288,000 cubic kilometers, as contrasted with about 105,000 cubic kilometers of Ordovician ones. The Devonian clastic wedge must represent the erosion of at least half of a mountainous Acadian landmass with dimensions shown in Fig. 13.33. But assuming that uplift was more or less continuous over 20 to 30 million years, the land need not ever have been more than about 2 kilometers high.

For comparison it is instructive to note that a modern, analogous tectonic land exists in New Guin-

FIGURE 13.31 Microscopic photograph in polarized light of an Upper Devonian (Catskill) sandstone from southeastern New York. Immature sand composed largely of metamorphic rock fragments (e.g., quartzite and schist) was derived from erosion of the Acadian tectonic land in present New England. (Larger grains are about 2 millimeters in diameter.)

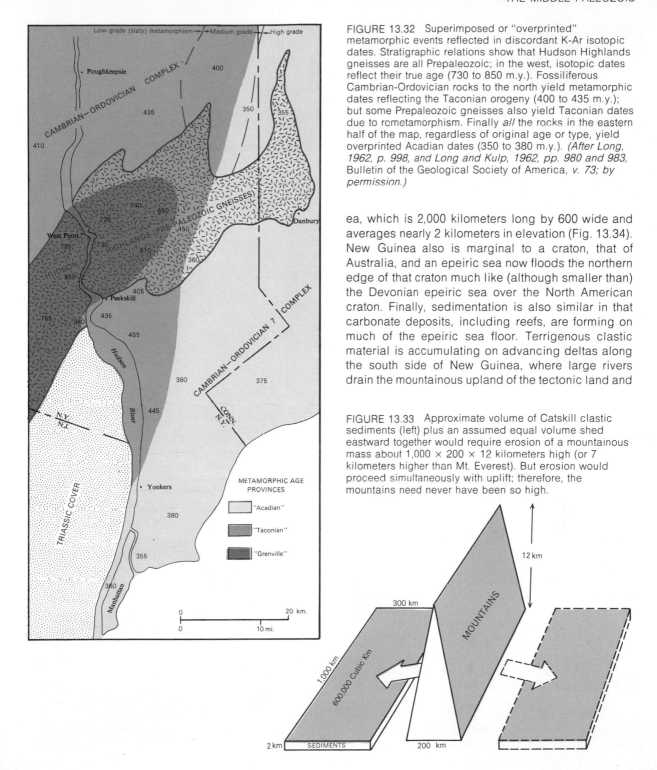

FIGURE 13.32 Superimposed or "overprinted" metamorphic events reflected in discordant K-Ar isotopic dates. Stratigraphic relations show that Hudson Highlands gneisses are all Prepaleozoic; in the west, isotopic dates reflect their true age (730 to 850 m.y.). Fossiliferous Cambrian-Ordovician rocks to the north yield metamorphic dates reflecting the Taconian orogeny (400 to 435 m.y.); but some Prepaleozoic gneisses also yield Taconian dates due to rcmetamorphism. Finally *all* the rocks in the eastern half of the map, regardless of original age or type, yield overprinted Acadian dates (350 to 380 m.y.). *(After Long, 1962, p. 998, and Long and Kulp, 1962, pp. 980 and 983, Bulletin of the Geological Society of America, v. 73; by permission.)*

ea, which is 2,000 kilometers long by 600 wide and averages nearly 2 kilometers in elevation (Fig. 13.34). New Guinea also is marginal to a craton, that of Australia, and an epeiric sea now floods the northern edge of that craton much like (although smaller than) the Devonian epeiric sea over the North American craton. Finally, sedimentation is also similar in that carbonate deposits, including reefs, are forming on much of the epeiric sea floor. Terrigenous clastic material is accumulating on advancing deltas along the south side of New Guinea, where large rivers drain the mountainous upland of the tectonic land and

FIGURE 13.33 Approximate volume of Catskill clastic sediments (left) plus an assumed equal volume shed eastward together would require erosion of a mountainous mass about 1,000 × 200 × 12 kilometers high (or 7 kilometers higher than Mt. Everest). But erosion would proceed simultaneously with uplift; therefore, the mountains need never have been so high.

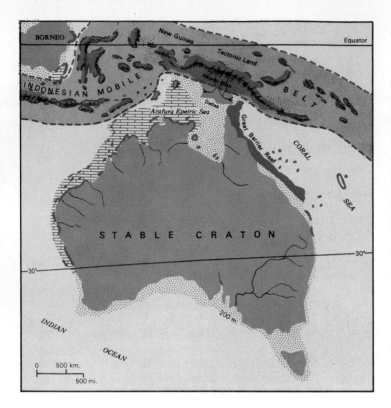

FIGURE 13.34 Relations of modern tectonic lands of New Guinea and Indonesia to the Australian craton, a close analogue to Late Devonian conditions in North America. New Guinea is a complex tectonic and volcanic borderland with dimensions comparable to the Acadian land. Rivers carry clastic sediments southward from the mountains toward the craton where they are deposited in deltas at the margin of an epeiric sea, which, like its larger Devonian counterpart, has carbonate sedimentation and active reefs (dark brown areas) over much of its area (rotate page 90° clockwise for a closer comparison with Fig. 13.28). *(Analogy originally suggested to writers by T. S. Laudon.)*

flow south across a broad, heavily forested alluvial plain. Southward (cratonward) regression of the sea appears to be happening today in southern New Guinea due to sedimentation just as it did in North America 350 million years ago.

COLLISION OF EUROPE AND AFRICA WITH NORTH AMERICA

THE CALEDONIAN MOBILE BELT

An early Paleozoic mobile belt, called the Caledonian, has long been recognized in Britain and Norway along the northwestern margin of the European craton. It was from studies of this belt that James Hutton formulated his revolutionary eighteenth century views of the earth. The Caledonian belt is a twin of the northern Appalachian–East Greenland system, and it contains similar rocks, which reveal a similar history. Episodes of volcanism and deformation are

recorded in the Caledonian belt at least as early as Cambrian time; more severe ones in the Ordovician correspond to the Taconian, whereas the culminating Caledonian orogeny (although slightly older) corresponds closely to the Devonian-Acadian event of America. Even major cratonic events show some correlation between the two continents. The warping of basins and arches as well as major transgressions and regressions correspond to a remarkable degree (see Fig. 12.12), suggesting that cratonic structure is also controlled by plate movements.

The first geologists to study eastern Greenland were Europeans, which was a happy circumstance, because what they found was so familiar that they might easily have forgotten they had left home. (It was in East Greenland that Alfred Wegener, father of the continental drift theory, lost his life in a blizzard in 1930). We noted in Chaps. 11 and 12 the great similarity of lower Paleozoic strata and fossils there with those of northwestern Scotland (see Fig. 12.31). Devonian and Carboniferous sediments are also virtually identical with those of northwestern Europe (as well as with those of the Appalachian belt), and species of fossil plants and fishes in these rocks are the same on both continents.

STRUCTURAL SYMMETRY

Structurally the two margins of the ocean are mirror images. In East Greenland as in the Appalachian belt, thrust faulting carried thick mobile-belt rocks westward against the craton. In Norway, thrusting was eastward against the European craton. In both cases intense metamorphism and large granitic batholiths developed in what is now the seaward side of those belts. Locally a definite bilateral symmetry still can be seen across the mobile belts today. In northwestern Scotland, Caledonian thrust faulting carried metamorphic rocks northwest over flat, lower Paleozoic carbonates, but eastward thrusting occurred nearby in Norway (Fig. 13.35). In Newfoundland, a central zone of intensely deformed Ordovician and Silurian oceanic rocks lies between two zones of mildly deformed continent-margin rocks. Finally, across the southern Appalachian belt of the United States, there is a central zone of most intensely deformed and metamorphosed Paleozoic and older rocks bounded on both sides by less deformed ones. From the above clues, as well as from paleomagnetic data, we can reconstruct the continents as they must have looked before and after the Late Devonian collision.

CATSKILL–OLD RED SANDSTONE
MOLASSE FACIES

The Devonian red bed deposits of eastern Greenland and northwestern Europe, like those of the eastern United States, reflect the rapid erosion of high mountains (Fig. 13.36). In Fig. 13.37 we see that the Old Red Sandstone of Europe was the mirror image of the American (Catskill) red beds. The sediments were transported by rivers in both directions away from a single great Caledonian-Acadian mountain range. Approximately equal volumes occur on both sides. The identity of land plant and fish fossils contained therein is readily explained as well.

INTERCRATONIC MOBILE BELTS

Now we realize that the intercratonic type of mobile belt noted in Chap. 8 results from continent-continent collision. Moreover, it follows that some apparently marginal mobile belts (e.g., the Appalachian) actual-

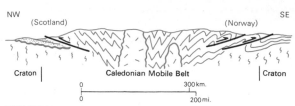

FIGURE 13.35 Simplified composite cross section of the Caledonian mobile belt, northwestern Scotland to southern Norway. Rocks of the mobile belt were thrust toward *both* margins; the belt apparently was symmetrical, being an *intercratonic* one.

ly were intercratonic at some time and have been subsequently dismembered by continental drift. It is only since the development of plate-tectonic theory, however, that such belts could be adequately explained. While the similarities of northwestern Africa with eastern North America were not as striking as those of Europe, the bilateral symmetry of the southern Appalachian belt in light of plate theory suggested collision also with Africa. Comparative studies were initiated, and based upon the similarities found, we infer that much of the southeastern edge of North America was originally part of Africa, as shown in Fig. 12.31. Figure 13.37 summarizes the probable evolution of the Appalachian-Caledonian system in terms of lithosphere plate movements; the scheme shown also applies to other intercratonic belts in a general

FIGURE 13.36 Coarse, angular Devonian ("Old Red") conglomerate, Solund District, western Norway, one of several areas of thick, nonmarine sediments deposited in downfaulted basins following the Caledonian orogeny. Note heterogeneous composition representing varied older metamorphic rocks. *(Courtesy Tor H. Nilsen.)*

EVOLUTION OF THE EARTH

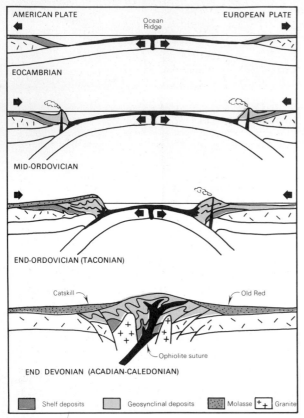

FIGURE 13.37 Hypothetical evolution of northern Appalachian–Caledonian mobile belts during closure of the postulated proto-Atlantic Ocean basin, which culminated in continental collision in Devonian time. Note that both continents had passive trailing edges with carbonate shelf zones facing the proto-Atlantic until mid-Ordovician time when subduction began. The old ocean basin completely disappeared later except for a mafic-rock suture zone (bottom). Note the symmetry of red Catskill and Old Red clastic wedges.

way. Of particular importance in unravelling these complex belts is the identification of suture zones, which are crushed oceanic rocks (typically the mafic ophiolites) that mark the old boundary between converging plates. Such a zone is generally all that remains of the ancient ocean basin that was consumed between two converging continents. Besides the suture, the central zone characteristically also has the deformed volcanic arc rocks, granitic plutons, and shows high-temperature metamorphism. Bound-

ing the central zone on both sides are intensely deformed sedimentary rocks thrust outward toward adjoining cratons (Fig. 13.37). Marginal mobile belts formed between converging oceanic and continental crust do not show such bilateral symmetry.

MOUNTAIN BUILDING IN THE ARCTIC AND CORDILLERA

In the Franklin mobile belt of Arctic Canada, northern Alaska, and (possibly) adjacent Siberia, Silurian sandstone and volcanic rocks (Fig. 13.11) and Devonian conglomerate (Fig. 13.27) reflect mountain building there essentially simultaneous with Caledonian-Acadian orogenesis farther east; the Franklin and Caledonian belts even merge in northeastern Greenland. Some granitic rocks of latest Devonian and earliest Mississippian ages are widely scattered in the Arctic (Fig. 13.27). Local ultramafic and volcanic rocks of early and middle Paleozoic ages present on the northernmost tip of Canada apparently reflect oceanic and island-arc rocks crushed against the northern edge of the continent, causing the upheaval of mountains and thrust faulting of Paleozoic strata southward against the craton during Late Devonian or Mississippian time (the Innuitian Orogeny; Fig. 13.38).

Disturbances also began to affect the Cordilleran region of western Canada and the United States for the first time. Important conglomerates were deposited in eastern and southern Alaska during the Silurian to Pennsylvanian Periods (see Figs. 13.11 and 13.27). In Nevada and Idaho, at least 1,000 meters of Mississippian and Early Pennsylvanian sandstone and chert-pebble conglomerate (Fig. 13.39) lie unconformably upon folded early Paleozoic oceanic sediments and pillow basalts (see Fig. 12.2). Together with the granites yielding Late Devonian ages in western Canada (Fig. 13.27), these rocks indicate a mild episode of mountain building (the Antler Orogeny). Unlike the orogenies discussed before, only low islands were elevated by this event, and no red bed alluvial plains resulted. Here most of the clastic sediments eroded from the tectonic lands were deposited at or below sea level. There is not yet enough evidence for as detailed an interpretation of the middle Paleozoic disturbances in the Franklin and Cordilleran belts as for the Acadian-Caledonian belts.

FIGURE 13.38   Devonian nonmarine strata in eroded anticlines and synclines of Parry Islands fold system of the Franklin mobile belt, Arctic Canada (large plunging anticline at left, syncline in middle, anticline at right). Folding occurred in Mississippian time. [*Courtesy National Air Photo Library (Canada), Surveys and Mapping Branch, Department of Energy, Mines and Resources; Photo No. T419R-132.*]

FIGURE 13.39   Evidence of the Antler orogeny in northern Nevada. Deformation began in Late Devonian and continued through Mississippian time; note westward overlap of conglomerate as lands were eroded (C and R denote index fossil zones). *(After R. H. Dott, Jr., 1964, Kansas Geological Survey Bulletin 169.)*

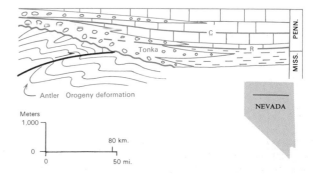

SUMMARY

Middle Paleozoic time was one of continued extensive carbonate sedimentation in shallow epeiric seas on the craton, but with increasingly important influences of domes and arches. There was a great increase both of organic reefs and evaporite deposits. Obstructions to marine circulation, especially through activities of reef-building organisms, repeatedly enhanced evaporative precipitation.

Important oscillations of relative land and sea

levels occurred, producing unconformities and attendant changes of fossil faunas. The most profound unconformity is within the lower part of the Devonian succession, rather than at a system boundary. Extensive warping affected almost the entire continent; probably it was related to the beginning of severe Acadian and Caledonian disturbances in marginal mobile belts. In the northern Appalachian region and eastern Greenland, as well as in northwestern Europe, severe middle Paleozoic mountain building produced large tectonic lands. Erosion of the lands resulted in the deposition of immense volumes of coarse, red clastic sediments adjacent to them.

The middle Paleozoic was a time of unusual evolutionary diversification among practically all organisms—marine and land; animal and plant; invertebrate and vertebrate. Corals and crinoids became extremely important, and ammonoids appeared. Fishes enjoyed their greatest evolution ever, culminating in exploration of land by a lobe-finned hopeful amphibian. Land plants, which appeared at least by the end of Silurian time, developed rapidly so that diverse forests with large trees cloaked the Devonian lowlands. Primitive land plants were dependent upon water to complete their reproduction, so probably could not inhabit drier uplands yet.

Distribution of organic reefs, richly fossiliferous carbonate rocks, evaporites, land plants, and also paleomagnetic evidence suggest that the relative position of the continent with respect to the axis and equator of the earth, and to climatic zones was different from that of today.

Geologic, paleontologic, and paleomagnetic evidence all point to a Devonian collision of northwestern Europe and Africa, with eastern North America as the culmination of consumption of proto-Atlantic lithosphere by subduction beneath those continents. Plate-tectonic theory explains the mountain building in the Appalachian–East Greenland and the Caledonian mobile belts as the result of that subduction and collision. Orogenies affecting the Cordilleran and Arctic regions cannot yet be explained fully in terms of plate theory, however.

Readings

Alberta Society of Petroleum Geologists, 1964, Geological history of western Canada: Calgary. (An atlas of stratigraphic maps)

Alling, H. L., and Briggs, L. I., 1961, Stratigraphy of Upper Silurian Cayugan evaporites: Bulletin of the American Association of Petroleum Geologists, v. 45, pp. 515–547.

American Association of Petroleum Geologists, 1973, Arctic geology: Tulsa, American Association of Petroleum Geologists Memoir 19.

Bird, J. M., and Dewey, J. F., 1970, Lithosphere plate-continental margin tectonics and the evolution of the Appalachian orogen: Bulletin of the Geological Society of America, v. 81, pp. 1031–1060.

Boucot, A. J., 1974, Evolution and extinction rate controls: Developments in palaeontology and stratigraphy, v. 1: Amsterdam, Elsevier.

Kay, M., ed., 1969, North Atlantic geology and continental drift: American Association of Petroleum Geologists Memoir 12.

Ladd, H. S., ed., 1957, Treatise on marine ecology and paleoecology: Geological Society of America Memoir 67, v. 2, Chaps. 10, 11.

LeBlanc, R. J., and Breeding, J. G., 1957, Regional aspects of carbonate deposition: Tulsa, Society of Economic Paleontologists and Mineralogists Special Publication No. 5.

Millot, J., 1955, The coelacanth: Scientific American, December.

Oswald, D. H., ed., 1968, Proceedings of the international symposium on the Devonian System: Calgary, Alberta Society of Petroleum Geologists.

Pannella, G., MacClintock, C., and Thompson, M. N., 1968, Paleontological evidence of variations in length of synodic month since Late Cambrian: Science, v. 162, pp. 792–796.

Pray, L. C., and Murray, R. C., 1965, Dolomitization and limestone diagenesis: Society of Economic Paleontologists and Mineralogists Special Publication 13.

Raasch, G. O., ed., 1961, Geology of the Arctic: Toronto, Univ. of Toronto Press.

Scrutton, C. T., 1964, Periodicity in Devonian coral growth: Palaeontology, v. 7, pp. 552–558.

Sloss, L. L., 1972, Synchrony of Phanerozoic sedimentary-tectonic events of the North American craton and the Russian platform: Montreal, 24th International Geological Congress Proceedings, sec. 6, pp. 24–32.

Wade, N., 1969, Three origins of the moon: Nature, v. 223, pp. 948–950.

Wells, J. W., 1963, Coral growth and geochronometry: Nature, v. 197, pp. 948–950.

Woodford, A. O., 1965, Historical geology: San Francisco, Freeman.

14

LATE PALEOZOIC HISTORY

A TECTONIC CLIMAX AND RETREAT OF THE SEA

Here about the beach I wander,
Nourishing a youth sublime
With the fairy tales of science,
And the long results of time.

Alfred Tennyson

FIGURE 14.1 Contorted fine sediments overlain by unsorted, heterogeneous conglomerate near Squantum Head, Boston, Massachusetts. The conglomerates were originally interpreted as glacial deposits, but more probably represent submarine sliding of gravels during Late Devonian and Mississippian orogenic disturbances.

We have shown that during early and middle Paleozoic time, North America experienced repeated and widespread transgressions and regressions by epeiric seas. Except at times of considerable mountain building, carbonate rocks tended to form widely, even in the mobile belts. But accompanying major orogenic episodes, clastic wedges engulfed mobile belt areas and even influenced cratonic sedimentation.

The North American Paleozoic stratigraphic record had been dominated by marine conditions before, but during Carboniferous and Permian times (355 to 230 m.y. ago), it became increasingly influenced by tectonic disturbances that raised much of the craton above sea level. Early Carboniferous or Mississippian strata represent a transition from middle Paleozoic marine to later nonmarine conditions. By the end of the era, practically the entire continent stood above sea level and was inhabited by cosmopolitan plants and animals. What had been a paradise for denizens of the sea was to become one for luxuriant swamp forests, lurking reptiles, and giant insects. Wholesale extinctions of large groups of shallow marine invertebrates and explosive proliferation of land organisms accompanied the shrinkage of shallow seas.

Naturally, general increase of structural unrest resulted in drastic changes in sedimentation in North America and Europe. Near the end of the Mississippian Period (roughly the Early Carboniferous of Europe),deposition of typical carbonate rocks almost ceased on the craton. Pennsylvanian (or Late Carboniferous) and Permian strata contain great volumes of terrigenous clastic sediments derived increasingly from peripheral mountains, but also from large uplands within the craton itself. Pennsylvanian strata contain our most important coal deposits, which formed under special climatic and topographic conditions favoring luxuriant growth of swamp forests. Identical conditions and floras existed in Europe and North Africa, as well. The coals occur in repetitive strata showing a conspicuous pattern of cyclic alternations of rock types that perhaps is of worldwide significance. The Late Paleozoic, therefore, represents a time of revolutionary change.

Uplift of the Caledonian and Franklin mobile belts ceased in late Paleozoic time. Erosion reduced the mountains there so that marine deposition resumed in Pennsylvanian and Permian times. Meanwhile, farther south, Africa and South America, which were fused together as part of the supercontinent Gondwanaland, completed their collision with North America and Europe. This last compression resulted in extensive thrust faulting and uplift in the southern Appalachian belt, and even more profound orogeny across Europe. Simultaneously, the collision of Siberia with eastern Europe formed the Ural Mountains, while several microcontinents collided and fused with southern Siberia to become China. Thus by the end of the Paleozoic Era, all the continents were more or less interconnected and total land area was very large. This situation greatly influenced the evolution of land life by facilitating dispersal of the organisms and providing a great diversity of ecologic niches.

MISSISSIPPIAN ROCKS OF THE CRATON

LAST WIDESPREAD CARBONATES

The Mississippian Period was characterized by the last widespread carbonate-producing epeiric sea in North America. Carbonate sediments spread over most of the craton and large portions of the mobile belts (Fig. 14.2). Limestones rich in crinoid fragments predominated (Fig. 14.3). They display conspicuous clastic textures. Breakage and sorting of fossil fragments, oölite, cross stratification, ripple marks, and scoured structures are common. These features point to close analogies with the modern Bahama Banks (see Fig. 13.13). Organic reefs occur widely in Mississippian strata from southern United States to northern Alaska, but they are smaller than the great barrier reef complexes of middle Paleozoic time. Evaporite deposits are less important as well.

SEDIMENTARY AND TECTONIC CHANGES

During Late Mississippian time, distinct changes in cratonic sedimentation commenced. There was a change in the composition of terrigenous sediments. Prior to Late Mississippian time, only very pure quartz sandstones and shales composed overwhelmingly of only one clay mineral (illite) were deposited. Beginning in later Mississippian time, heterogeneous sands bearing feldspar and mica, as well as quartz, began to appear in the southeastern craton, and they became more widespread during subsequent periods. Simultaneously, the mineralogy of associated shales became more complex.

Most of the above changes indicate exposure by erosion of new, more heterogenous source rocks. Facies and paleocurrent patterns (Fig. 14.2), together with mineralogy, indicate that the source of much of the clastic material lay in eastern Canada and was composed chiefly of igneous and metamorphic shield rocks long buried beneath early Paleozoic strata. Material also was derived from lands in the Appalachian belt. Erosion during various periods of regression gradually had stripped most of the Paleozoic veneer from eastern Canada, laying bare significantly large areas of old, crystalline basement. Comparison of Mississippian and Pennsylvanian facies with earlier lithofacies and paleogeographic maps indicates that the entire continent gradually was tilted up in the east. As a result, epeiric seas of late Paleozoic time covered less and less of the eastern craton (Fig. 14.4), and erosion cut deeper and deeper. Over the western craton and adjacent eastern Cordilleran mobile belt, pure quartz sands still were deposited intermittently throughout late Paleozoic time (Figs. 14.5 and 14.7). They were derived largely from erosion of earlier Paleozoic sediments on the west-central craton in Canada, where the Prepaleozoic basement was not yet widely exposed.

At the end of the Mississippian Period, a major regression carried the sea entirely out of the craton, and the earliest Pennsylvanian seas are confined only to marginal mobile belts. As a result, a major discontinuity occurs beneath Pennsylvanian strata over the craton (Fig. 14.6). A great deal of warping and faulting occurred before the sea returned to the craton in mid-Pennsylvanian time (Figs. 14.7 and 14.8).

LATE PALEOZOIC REPETITIVE SEDIMENTATION—AN ENIGMA

SEDIMENTARY CYCLES

Beginning in Late Mississippian and continuing through Early Permian times, the strata deposited over the craton and inner parts of the mobile belts displayed a striking repetitive pattern, which is present to varying degrees in late Paleozoic strata on other continents. Upper Mississippian deposits in the southeastern craton show clear repetitions of sandstone-shale-limestone sets repeated several times vertically. Illinois geologists have shown that

FIGURE 14.2 Middle Mississippian paleogeography. Note emerging lands in eastern North America and Antler lands in Cordillera.

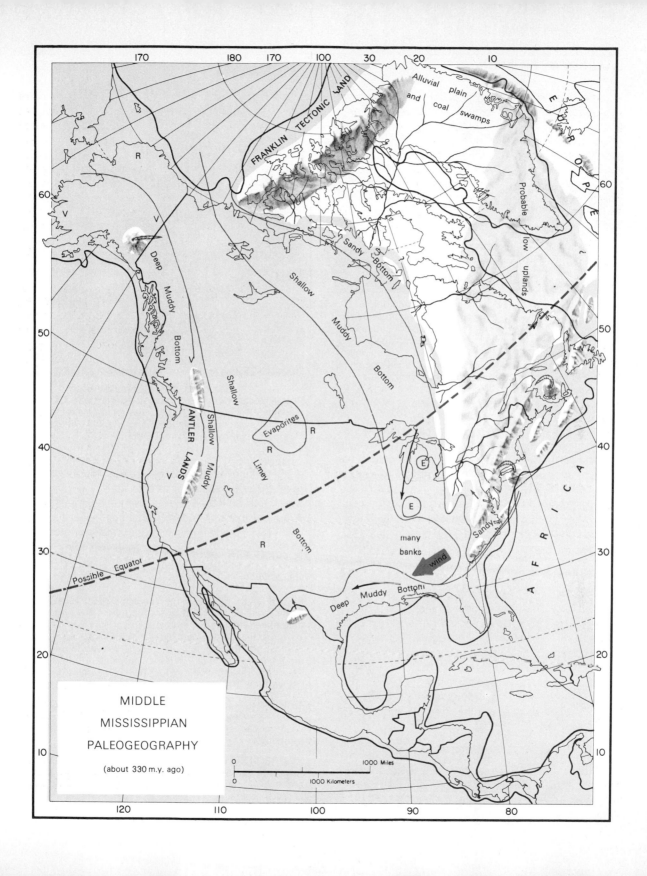

MIDDLE

MISSISSIPPIAN

PALEOGEOGRAPHY

(about 330 m.y. ago)

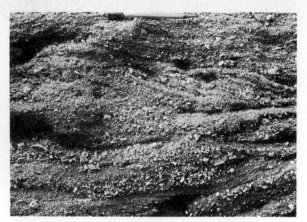

FIGURE 14.3 Cross stratification in pure crinoidal Mississippian Burlington Limestone near Hannibal, Missouri. This formation, with a total volume of about 300×10^{10} cubic meters, contains the skeletal remains of approximately 28×10^{16} individual crinoid animals. After death, the skeletons were disarticulated and the fragments dispersed like sand and gravel as shown here; it is a typical clastic limestone.

FIGURE 14.5 Large-scale cross stratification in Pennsylvanian pure quartz sandstone, Tensleep Canyon, Wyoming. Although commonly interpreted as wind-dune deposits, association with fossiliferous marine limestones suggests a shallow-marine origin for many of these sands.

the sandstones and shales represent in part deltaic deposits formed by river systems flowing from the southeastern Canadian region. In Pennsylvanian time, influxes also came from the Appalachian region (Fig. 14.8).

Practically all Pennsylvanian strata on the continent show some kind of repetitive pattern, but the most striking occurs in coal-bearing sequences. At least 50 late Paleozoic cycles are known, many of which can be traced widely over the southern craton. A typical cycle (Fig. 14.9) commences at the base with cross-stratified sandstone and conglomerate

FIGURE 14.4 Effects of uptilting of the eastern craton beginning in middle Paleozoic time.

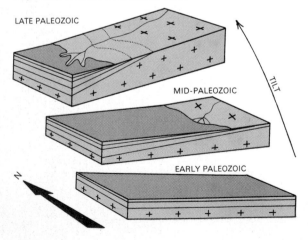

On the following pages:

FIGURE 14.6 Paleogeology beneath widespread Early Pennsylvanian unconformity showing great warping and differential erosion in the craton. Old arches and basins were reactivated and new ones formed. Features in Canada are inferred because they lie beyond the present eroded (zero) edge of Pennsylvanian strata. *(From A. I. Levorsen,* Paleogeologic maps, *W. H. Freeman Co., Copyright © 1960; and unpublished maps by D. E. Owen.)*

FIGURE 14.7 Middle Pennsylvanian sedimentary facies; note eastern coal-bearing regions and western coarse clastic areas reflecting uplifted regions (see Fig. 11.15 for symbols).

FIGURE 14.8 Middle Pennsylvanian paleogeography. Note enlargement of land area in eastern North America and many local lands in the southwest as compared with Mississippian time.

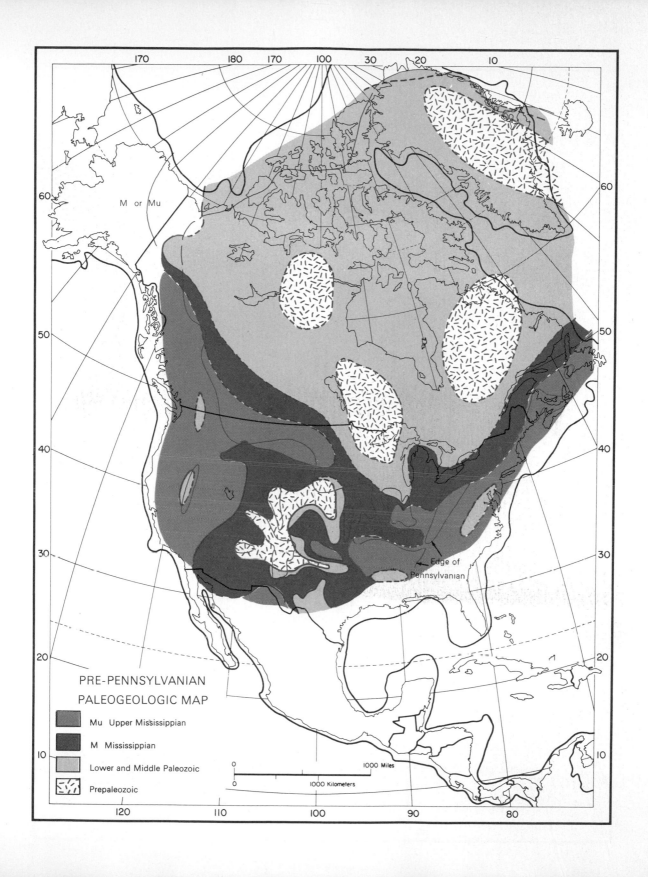

PRE-PENNSYLVANIAN
PALEOGEOLOGIC MAP

Mu Upper Mississippian

M Mississippian

Lower and Middle Paleozoic

Prepaleozoic

M or Mu

Edge of
Pennsylvanian

0   1000 Miles

0   1000 Kilometers

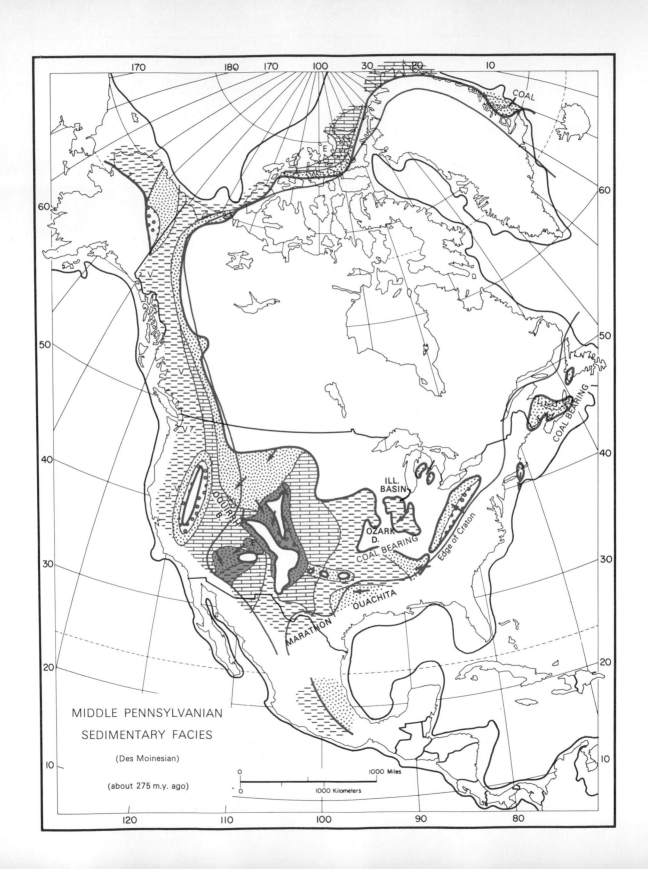

MIDDLE PENNSYLVANIAN
SEDIMENTARY FACIES

(Des Moinesian)

(about 275 m.y. ago)

COAL

COAL BEARING

ILL.
BASIN

OZARK
D.

COAL BEARING

Edge of Craton

OUACHITA

MARATHON

0   1000 Miles

0   1000 Kilometers

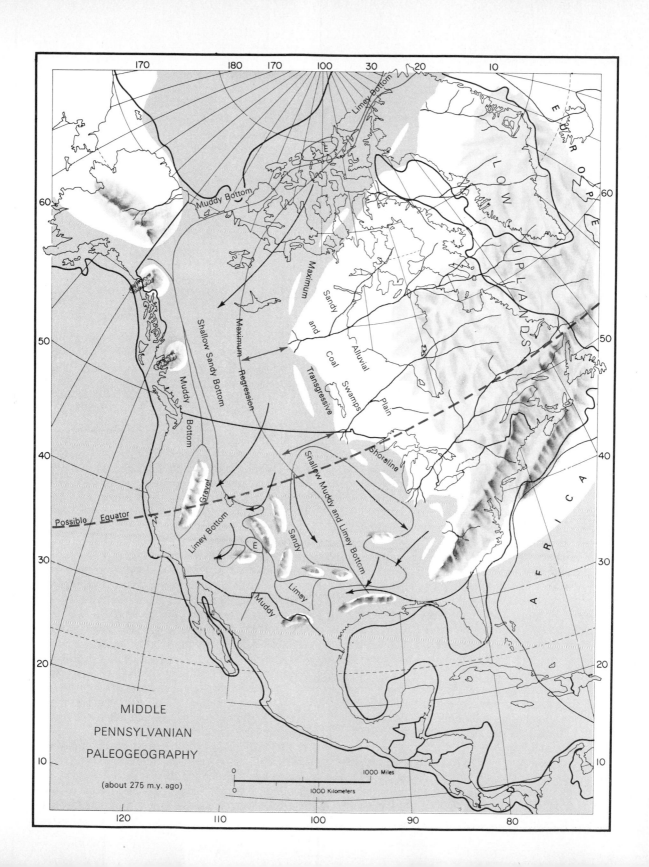

MIDDLE
PENNSYLVANIAN
PALEOGEOGRAPHY

(about 275 m.y. ago)

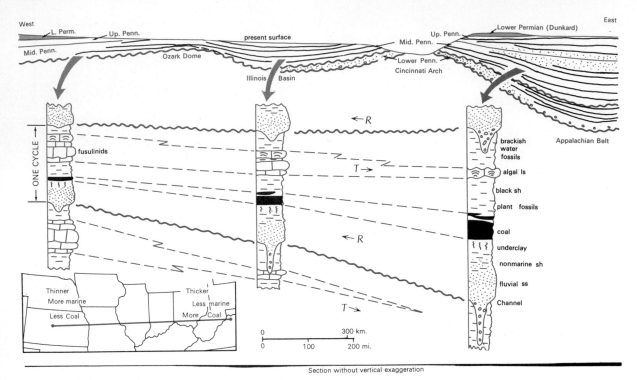

Section without vertical exaggeration

FIGURE 14.9 Idealized cross section showing lateral and vertical relations of one Pennsylvanian depositional cycle (*cyclothem*) reflecting major transgressions *T* and regressions *R* of the sea and shoreline.

↳ *sequence of repetitive strata*

resting unconformably upon older strata; channel structures, fossil logs, and relatively poor sorting of grains indicate these were formed by river processes. The middle of the cycle contains coal and plant-bearing shales, while the upper part generally contains marine or brackish-water fossiliferous shales and limestones. Many such cyclic sets of strata occur vertically stacked upon one another. Lateral variations within the sets are just as important as vertical alternations, for both record drastic environmental shifts through time.

In the southwestern craton, the upper marine rocks of the cycles are best developed, whereas the coal-bearing, nonmarine rock types predominate in the east. Alternation of marine and nonmarine deposits points to many transgressions and regressions of the sea over wide areas. With more than 20 distinct cycles representing about 60 million years, however, it is apparent that geologically these were rapid oscillations, occurring on a wholly different time scale than earlier Paleozoic transgressions and regressions. Apparently they require a different explanation. Possible causes of these oscillations include: (1) worldwide rise and fall of sea level due either to advance and retreat of continental glaciers or to large-scale warping of the deep-sea floor; (2) spasmodic tectonic up-and-down motions of the entire continent; or (3) fluctuations of clastic sediment supplied to deltas, perhaps in response to cyclic climatic changes affecting erosion in the uplands; as more sediment was supplied, deltas would advance, but transgression would occur when the supply decreased. The last cause, while an important process, is too localized to explain Pennsylvanian oscillations. The second (a kind of Saint Vitus's dance hypothesis) overtaxes credulity because so many short-term oscillations were involved. We prefer the first mechanism, with changes in known late Paleozoic glaciers on Gondwanaland as the ultimate cause.

Pennsylvanian strata provide perhaps the clearest examples of complex interactions of land-sea relationships. A long-term rise of all eastern North America is evident (Fig. 14.4), but more localized differential warping of the craton also occurred on an intermediate time scale and accentuated many arches and basins. Finally, superimposed on both of those effects were the short-term transgressions and regressions presumably caused by sea-level oscillations. The unique rate of each process was superimposed upon that of the others to produce a very complex stratigraphic record.

COAL SWAMPS

Basal sandstones and shales of a Pennsylvanian cyclic set are interpreted as river and delta deposits, while the coals formed in vast coastal swamps containing jungle-like vegetation. The Dismal Swamp (2,500 square kilometers), Dutch lowlands (15,000 square kilometers; Fig. 14.10), Florida Everglades (25,000 square kilometers), and comparable-sized south Lousiana swamps all provide useful modern comparisons.

The ancient swamps were dominated by large, scaly-barked trees (lycopsids) that had evolved first in Late Devonian time. Fossil seeds show adaptations for dispersal by floating on water. Other plants also were present, and amphibians, primitive reptiles, air-breathing molluscs, and insects inhabited the soggy forests as well. Land floras and faunas of Europe and North Africa were identical. Presence of cold-blooded animals and the nature of the vegetation over the three continents indicate that the climate must have been very humid and probably warm. The Carboniferous trees lack growth rings, which together with the presence of large, thin-walled cells in the tree trunks suggest a lack of distinct annual seasonal changes either of temperature or humidity, and that growth was rapid. This flora is considered to have been a tropical one.

As trees and shrubs died and fell to the swamp floor, much of their debris was protected from decay by rapid submergence and burial, thus excluding oxygen and preventing attack by all but anaerobic bacteria. With time, it was compacted by the weight of subsequent sediments to form peat. The typical ratio of thicknesses of uncompacted peat and coal is about 10 to 1. To form a sizeable commercial coal seam (Fig. 14.11) therefore required a fantastic quantity of vegetation (especially when we know that much of it must have decayed completely). Through gradual escape of more volatile hydrocarbon compounds from plant tissues, peat changed to coal, which is more compact and contains a higher percentage of carbon; increased burial and compaction over time

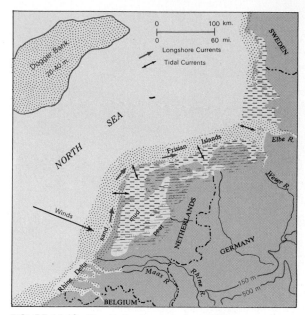

FIGURE 14.10 The Netherlands before diking began—model for ancient, low, swampy coasts. Note peculiar, nonprotruding delta of the Rhine River, which carries negligible sand; vigorous tidal and longshore currents due to winds sweep sand from the shallow (epeiric) North Sea floor northeastward to form long sand spits and barrier islands.

produced better grades of coal. Principal areas of commercially important Pennsylvanian coal are the Illinois basin, the central Appalachian Mountains, and the Maritime Provinces of Canada (Fig. 14.7).

PALEOGEOGRAPHIC RECONSTRUCTION

Ultimately the sea flooded the low, coastal swamps, and the marine upper phase of each cyclic set was formed during such transgressions, thus burying the peat. As with sediments previously discussed, the understanding of Pennsylvanian strata has been facilitated greatly by studies of modern sediments. Much of the craton must have been exceedingly flat and very near sea level so that the coal swamps then, like the southern Florida Everglades and Dutch lowlands now, were practically at sea level when the peat originally accumulated. Therefore, only a small rise of sea level or sinking of land could cause very widespread inundation of former swamps. Conversely, a very small fall of sea level would cause an equally widespread regression, enlargement of land area,

FIGURE 14.11
Pennsylvanian coal above
light-colored sandstones
deposited in a channel that
crossed an immense coastal
swamp. Note lenticular cross
section of its main channel
deposit (bottom); differential
compaction of coal and
shales accentuated the lens
form. (Cagles Mill Dam, near
Terre Haute, Indiana.)

and a great expansion again of rivers, deltas, and swamp forests. These are exactly the conditions recorded in the cyclic strata, namely many geologically rapid vertical fluctuations of relative sea and land levels, causing widespread oscillatory transgressions and regressions over nearly one-fourth of

FIGURE 14.12   The Mississippi delta—model for certain Pennsylvanian deposits. *Left*: Modern protruding or "birdfoot" delta (A.D. 1500 to present) and swampy deltaic plain integrated from three older deltas, now partially submerged. Sand from delta feeds Chenier beach ridges and barrier islands farther west (see Fig. 11.31). *Right*: Submerging effect of a 5-meter rise of sea level for comparison with the transgressive phase of Pennsylvanian cycles (no part of map area is more than 10 meters above sea level today). *(Adapted from R. Leblanc and H. Bernard, 1954, in* The Quaternary of the United States; *and H. N. Fisk and J. McFarlan, 1955,* Journal of Sedimentary Petrology.*)*

the craton. Between low mountainous uplands along the Appalachian mobile belt and a persistent epeiric sea over the western craton, the shoreline oscillated across a region 800 to 1,000 kilometers wide (Fig. 14.8). Overall sedimentary facies, paleocurrent patterns, and mineral composition of sandstones indicate that large rivers flowed west from the Appalachian region and southwest from eastern Canada to debouch into the epeiric sea along its eastern margin.

Much of the Pennsylvanian sandstone and shale in the central United States represents environments associated with deltas (Fig. 14.12). As the shoreline oscillated, so did the deltas (Fig. 14.13). River-deposited channel sandstones pass laterally into black shales and coals formed in swamps between channels. Channel sands grade westward into silty

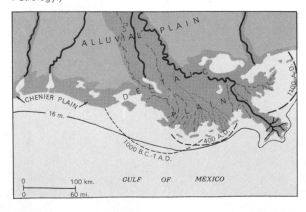

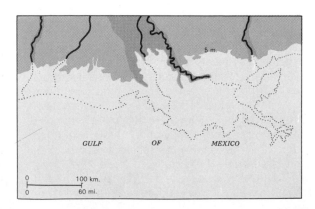

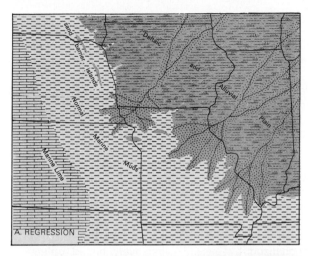

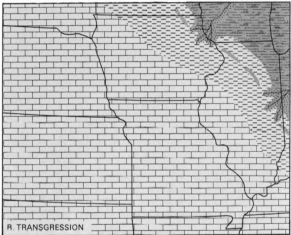

FIGURE 14.13 Regressive and transgressive fluctuations of environments for a typical Pennsylvanian sedimentary cycle in the central United States (compare Fig. 14.12). *(Suggested by data from H. R. Wanless et al., 1963, Bulletin of the Geological Society of America.)*

shales containing marine and brackish-water fossils. Frequent pulsations of deposition occasionally produced sedimentary features of great variety (Figs. 14.14 and 14.15).

TERMINAL PALEOZOIC EMERGENCE OF THE CONTINENT

PERMO-TRIASSIC GEOGRAPHY

The tendency of the eastern side of North America to tilt upward continued from late Paleozoic into Meso-

FIGURE 14.14 Contorted Pennsylvanian sandstones west of Tulsa, Oklahoma. Water-saturated fine sands were deposited on a gently sloping delta fringe and were so unstable that they slipped and became contorted. Such deformation is common in rapidly deposited sand-shale sequences.

zoic time. Recall that changing facies and paleogeographic patterns indicated that normal marine deposits became progressively more restricted, and westward encroachment of terrigenous detritus increased manyfold (Fig. 14.16). These tendencies

FIGURE 14.15 Sole marks preserved on the bottoms of sandstones deposited upon shale, here displayed in the walls of Gilcrease Art Museum, Tulsa, Oklahoma. Muds develop sufficient cohesion soon after deposition that subsequent currents can scour or score the surface and deposit sand in the depressions. These formed in Pennsylvanian coastal and deltaic deposits, but such marks form in a wide variety of other environments (e.g., Fig. 12.30).

culminated in Permo-Triassic time, and so there was more land area, of greater average elevation, than ever before during the Paleozoic Era (Fig. 14.17).

By Early Permian time, a sea comparable in size to the Black Sea occupied the south-central craton, but final stands of the Permian epeiric sea were in present southwestern Texas and the northern Rocky Mountains. In the former area, a great Late Permian organic reef complex flourished around the edge of a basin that was more than 300 meters deep (Fig. 14.18), but as the sea retreated farther, the reefs died. Thick evaporite and red bed deposits rapidly filled the basin and buried them. The reef complex had swarmed with an amazing variety of life. Brachiopods, including bizarre forms adapted to coral-like growth, were prominent, together with sponges, bryozoans, and advanced forms of algae; corals were less important than in middle Paleozoic reefs.

At the west edge of the craton, phosphate and chert formed with black shales, possibly as a result of upwelling of nutrient-rich cold waters at the edge of the craton. Meanwhile, evaporites and red beds accumulated widely in western United States and evaporites and carbonate rocks in northern Canada where the mid-Paleozoic mountains had stood (Fig. 14.16). Undoubtedly evaporites and red beds also were deposited across central Canada, but erosion has removed them. This vast expanse of Permian evaporites records the slow drying up of epeiric seas as the land area of the earth enlarged to nearly an all-time maximum.

THE RED COLOR PROBLEM

Origin of the red color of many sedimentary rocks has been a subject of controversy for generations. Red beds are the trademark of Permian and Triassic strata on five continents; these two systems account for nearly half all red strata. The vivid color adds immeasurably to the beauty of many landscapes, especially throughout western United States. Such deposits long have been considered to bear witness to unique paleoclimatic conditions, but exactly what those conditions might be is still a heated controversy known widely as the red bed problem.

Abundance of Permian evaporite deposits indicates high evaporation rates over epeiric seas and adjacent lowlands. Wind-dune sands around the Grand Canyon have been cited as evidence of a desert, but dunes alone are no climatic proof, for coastal dunes are common today in humid as well as arid regions. Fossil Permian plants in Arizona, however, do suggest some aridity because they have small, thick, hair-covered leaves and spines typical of modern dry-climate floras. But farther east, sporadic Permian and Triassic coals as well as aquatic plants, fresh-water molluscs, fish, amphibians, and some aquatic reptile fossils suggest moderate humidity. Late Triassic floras suggest gradations from mild-temperate conditions in East Greenland through wet tropics in the Appalachian region, tropical savannas (alternately dry and wet) in southwestern United States, and wet tropics in southern Mexico. From paleontological evidence, it would appear that red-colored sediments developed in *both* arid and humid areas, but some of the seeming paradox could be rationalized by assuming relatively humid upland areas blanketed with abundant vegetation and more arid lowlands. Upland vegetation first became possible with the appearance of coniferous trees in late Paleozoic time, owing to a capability for complete reproduction without abundant water. But trees grow even in true deserts along large, permanent rivers, and animal carcasses and logs float long distances from uplands down rivers through arid regions and even out to sea.

Red color indicates, first and foremost, thorough oxidation of iron in sediments. Only a trifling 1 to 6 percent of iron is required because it is such a potent coloring agent and is so unstable in the presence of free oxygen. Sole agreement is that the chemical environment favoring red color must be strongly oxidizing and slightly alkaline. Though most appear to have been nonmarine, some red beds interstratified with marine limestones probably were deposited in the sea.

Recent studies suggest that oxidation of iron minerals to produce red oxides disseminated among sediment grains occurs largely after deposition. Progressive breakdown of iron in sand grains has been observed in late Cenozoic sediments that are now turning red. In finer, clay-rich sediments, it appears that brown, hydrated iron minerals "age" to red hema-

FIGURE 14.16 Middle Permian lithofacies and isotopic dates of granitic plutons. Note the importance of evaporites and red beds in the western craton and phosphate and widespread volcanic rocks with associated limestones in the Cordilleran belt. (See Fig. 11.15 for symbols and sources.)

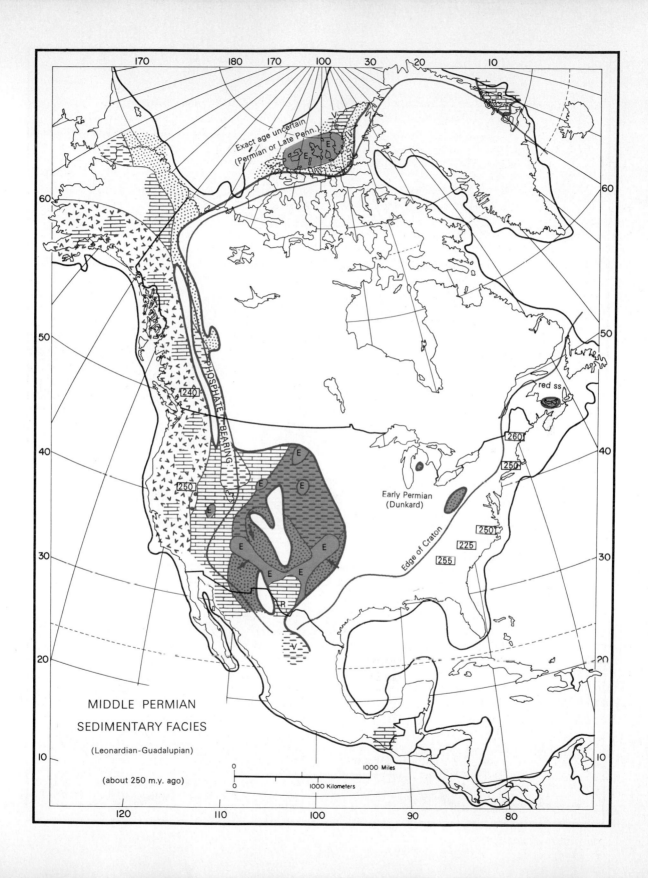

MIDDLE PERMIAN

SEDIMENTARY FACIES

(Leonardian-Guadalupian)

(about 250 m.y. ago)

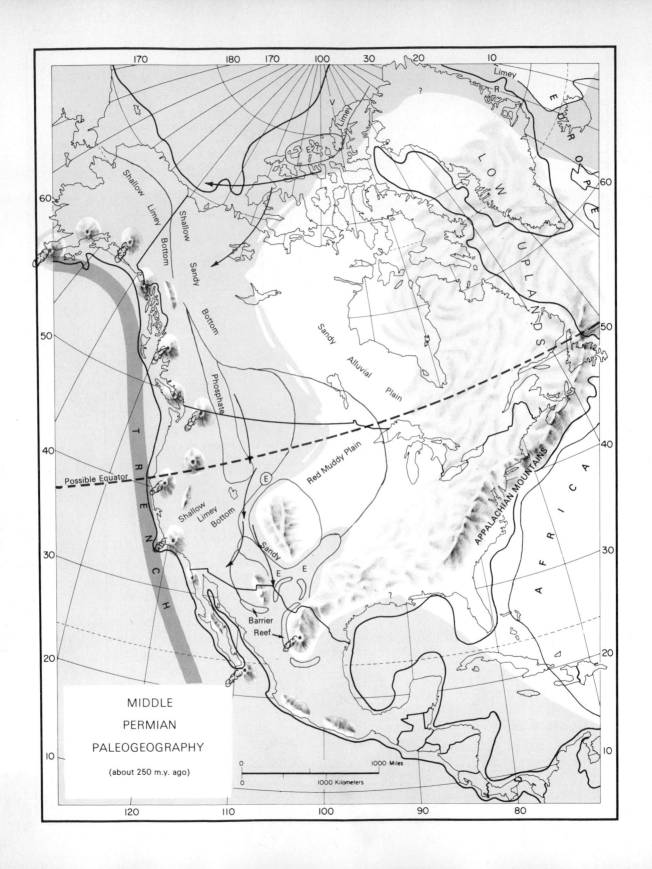

MIDDLE

PERMIAN

PALEOGEOGRAPHY

(about 250 m.y. ago)

tite by dehydration after burial. The bulk of the evidence points to development of red color through intense oxidation of iron in sediments under relatively warm conditions.

Once established in sedimentary rocks, indestructibility of red color is legendary, as evidenced by dirt roads and river waters in regions of red beds, or in the dust that reddened skies during the "dust bowl" days of the 1930s. Few distant recipients appreciated that their skies and eyes were being reddened by 200-million-year-old dust blown hundreds of miles. Red color even survives cooking, as pioneer vertebrate paleontologist E. C. Case (University of Michigan) found on collecting trips into Oklahoma and Texas around 1900. Occasionally Case mixed biscuits for his field parties, but they were so brightly colored by the local water that only the chef was enthusiastic about his ferruginous creations.

PERMO-TRIASSIC PALEOCLIMATE

The position of the equator indicated by paleomagnetic evidence for late Paleozoic time is noteworthy in relation to distribution of various sediments, fossil types, and paleocurrents. Most significant is that coal-bearing deposits would lie at or within 20 or 25 latitudinal degrees of the apparent paleoequator (Figs. 8.11 and 14.8), thus in tropical to subtropical zones, where mild, humid, nonseasonal climatic conditions would be expected. Land apparently was low enough and small enough during Mississippian and Pennsylvanian time that rainfall was uniform. During Permian and Triassic time, position of the possible paleoequator seemingly was not greatly different, but both area and elevation of land had increased. This would cause greater differentiation of climate (Fig. 14.19); therefore, it is plausible to postulate relatively humid Permo-Triassic uplands, but drier lowlands. Much of North America would fall in the late Paleozoic tropics and thus in the Trade Winds belt. The western side of the continent would be leeward of the Appalachian Mountains, which could account for aridity there.

FIGURE 14.17   Middle Permian paleogeography. Note the almost complete exclusion of the sea from the craton and eastern North America and the volcanic island system in the Cordillera.

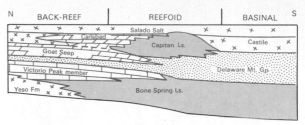

FIGURE 14.18   Complex Permian facies around the Capitan reef, southwestern Texas. *(After P. B. King, 1948, U.S. Geological Survey Professional Paper 215.)* "Those of us who grew up with West Texas Permian geology . . . learned facies the hard way . . . as we looked into the interior of the range, we saw all of our fine units [formations] dissolve before our eyes, merging into a monotonous sequence of dolomite." *(P.B. King, 1949.)*

The presence of land areas in Europe and Africa east of the old Appalachian Mountains would affect climate significantly. The subtropical high-pressure cell over Africa would be weaker and drier than over open ocean. Moisture, however, was available by evaporation from Permian and Triassic epeiric seas over central Europe and northern Africa. Such moisture would have been carried westward to the windward side of the Appalachians, where humid-forest plants have been found.

TECTONICS

CRATONIC DISTURBANCES

As noted above, the craton suffered more severe deformation in Pennsylvanian time than ever before. Many basins and arches were warped and eroded again, but the most intense disturbances occurred in Oklahoma and the nearby Rocky Mountain region. Erosion of mountains more than 1,000 meters high in both areas produced thick, coarse, red gravels, sandstone, and shales, which grade laterally into normal marine strata. As shown in Fig. 14.20, the Colorado Mountains probably were large fault blocks. Whatever their ultimate cause, which is not known, it is most unusual for such high mountains to be raised within cratons.

FINAL COMPRESSION OF THE APPALACHIAN BELT

Until late Paleozoic time, the southern margin of the continent was a stable shelf and continental slope.

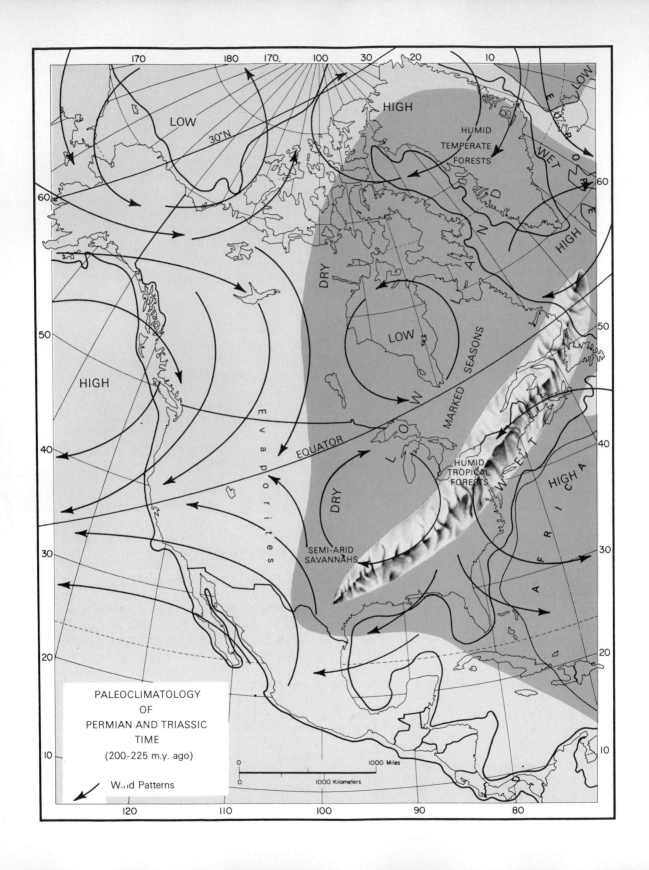

PALEOCLIMATOLOGY
OF
PERMIAN AND TRIASSIC
TIME
(200-225 m.y. ago)

Wind Patterns

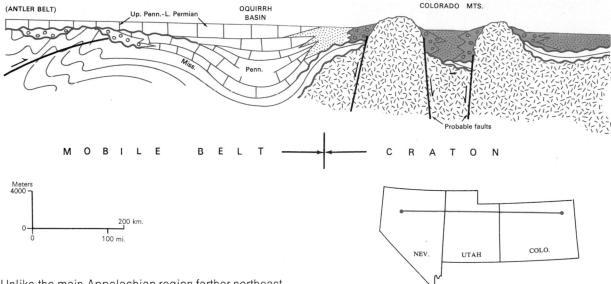

FIGURE 14.20 Colorado Mountains elevated in Pennsylvanian time and the coarse, red facies deposited adjacent to them. Lower Paleozoic strata were so thin that erosion quickly exposed the Prepaleozoic basement. Solution karst topography developed on Paleozoic limestones adjacent to main uplifts. Farther west, the Antler orogenic belt became buried, but the Oquirrh basin subsided abruptly and received 7,000 meters of Pennsylvanian strata, underscoring the new instability of the crust.

Unlike the main Appalachian region farther northeast, tectonic disturbances did not affect the Marathon-Ouachita region (Fig. 14.7) until the Mississippian Period. Only thin, fine, deep-water shale and chert had accumulated there earlier. But now sand derived from the eroding mountains farther east was introduced by turbidity currents, and volcanic ash entered from the south (Figs. 14.21 and 14.22). A volcanic island arc had formed, implying that subduction had begun. During the Pennsylvanian Period, the trough between the arc and the shelf to the north filled with clastic sediments, and by Permian time, the entire Marathon-Ouachita sector was uplifted and deformed (Fig. 14.21).

Farther northeast in the main Appalachian region, folding and thrust faulting of Pennsylvanian and older strata toward the craton also occurred (Fig. 14.23). While results of late Paleozoic deformation are the most obvious, they were not as profound as the effects of the earlier orogenies. Evidence for dating the final compression is as follows: (1) in southwestern Texas and south-central Oklahoma, folded Penn-

sylvanian strata are overlain unconformably by Early Permian ones; (2) in the central Appalachian Mountains, Pennsylvanian rocks were folded and faulted, together with older ones; (3) from Nova Scotia to Florida, nearly flat-lying Upper Triassic strata rest with unconformity upon folded Paleozoic rocks (Figs. 14.24 and 14.25); and (4) isotopic dating of scattered granitic bodies formed during this final upheaval indicates an average age between 200 and 250 million years. It follows that the final deformation of the mobile belt occurred between Late Pennsylvanian and Late Triassic times, the exact timing varying somewhat with locality. Figure 14.26 summarizes the Paleozoic history of the Appalachian belt.

Along the axis of the mobile belt, Paleozoic and Prepaleozoic rocks were most intensely deformed and metamorphosed by the various orogenies. Carbonate rocks became marble, and other strata became schists or quartzites. In the western and eastern

FIGURE 14.19 Hypothetical paleoclimatic map for late Paleozoic and early Mesozoic time. Accepting the paleomagnetically indicated equator position, we see that most of North America, Europe, and North Africa would have been in the Trade Winds belt, whereas today they lie chiefly in the Westerly Wind belt (see Fig. 11.33B). The restoration explains fossil and sediment evidence very well. The Tethys seaway (see Fig. 15.16) apparently supplied moisture to eastern North America.

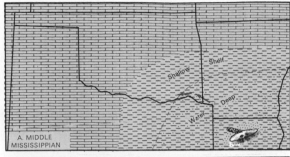

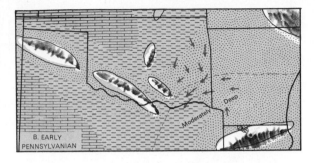

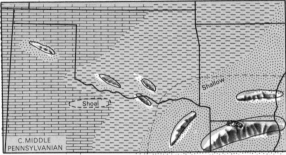

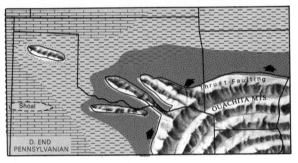

FIGURE 14.21 Changing late Paleozoic geography of the Oklahoma region. The Ouachita region was a deep-water area starved of sediments until Late Mississippian time; turbidity currents then introduced large volumes of sand, which gradually filled the deep troughs. By Late Pennsylvanian time, uplift produced lands in the region from which much red clastic sediment was shed, partially burying some uplifts. *(From published and unpublished work by L. M. Cline and associates.)*

FIGURE 14.22 Steeply tilted Pennsylvanian sandstone and shale, Marathon Mountains, southwestern Texas. Identical sequences of rhythmically alternating sandstone and shale in Europe (named *flysch*) are established as deep-water turbidity current deposits; these and like ones in the Ouachita region are similarly interpreted. *(Courtesy L. M. Cline.)*

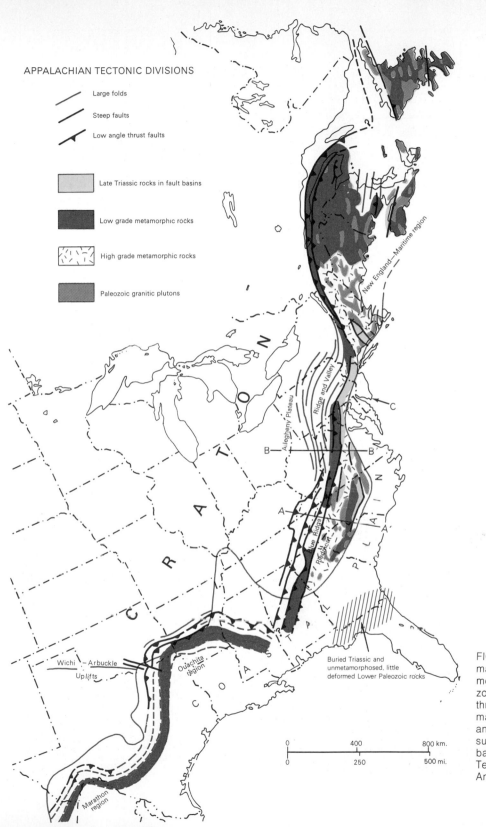

APPALACHIAN TECTONIC DIVISIONS

Large folds

Steep faults

Low angle thrust faults

Late Triassic rocks in fault basins

Low grade metamorphic rocks

High grade metamorphic rocks

Paleozoic granitic plutons

New England—Maritime region

Allegheny Plateau

Ridge and Valley

C R A T O N

Blue Ridge region

Piedmont

A P P A L A C H I A N   C O A S T A L   P L A I N

Buried Triassic and unmetamorphosed, little deformed Lower Paleozoic rocks

Wichi – Arbuckle Uplifts

Ouachita region

Marathon region

0		400		800 km.
0	250		500 mi.	

FIGURE 14.23 Tectonic map of the Appalachian mobile-belt system. Note zonation across the belt from thrust faulting at the cratonic margin to a metamorphic and granitic pluton belt with superimposed Triassic fault basins. *(Adapted from Tectonic map of North America.)*

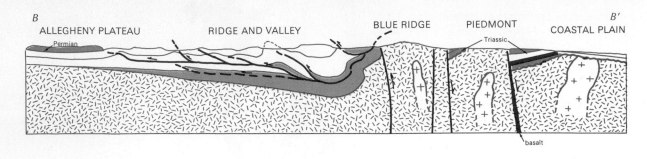

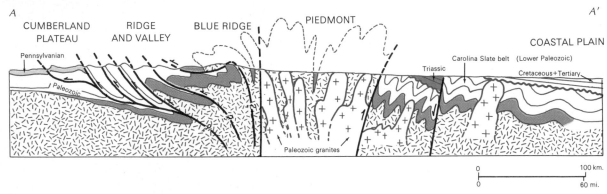

FIGURE 14.24 Two cross sections across the
Appalachian belt (see Fig. 14.23 for locations). Note the
apparent bilateral structural symmetry of the belt in A-A'
and current interpretation of the western thrust faults as
flattening downward so that they involve only the
superficial sedimentary rocks. Note also Triassic
block-faulted basins in both. *(Bottom adapted from P. B.
King, 1950, Bulletin American Association of Petroleum
Geologists. Top adapted from V. E. Gwinn, 1964, Bulletin
of the Geological Society of America.)*

Appalachians (see Fig. 14.24, A-A'), the older shales
were converted to slate and some coals to anthracite
(hard coal). Ultramafic (serpentine) rocks and old
basalts in the central zone represent a suture of old
oceanic material crushed between colliding continents (e.g., Fig. 13.37).

Recall that the rocks of the eastern, high-grade

FIGURE 14.25 Angular
unconformity between
intensely folded
Mississippian strata (Horton
Group) below and Upper
Triassic red bed deposits
(Wolfville Formation) at Rainy
Cove, Nova Scotia. This
unconformity reflects the late
Paleozoic to Middle Triassic
Appalachian orogeny.
(Courtesy G. de Vries Klein.)

metamorphic belt of New England and the Atlantic coast states long were considered to be entirely Prepaleozoic because they "looked old." Later they were shown to be largely Paleozoic, and so are simply more intensely disturbed equivalents of the strata of the Ridge-and-Valley region to the west. As is typical of the deep root zones of mobile belts, it is difficult to differentiate the deformation and metamorphism produced by each successive orogeny because the effects have all been superimposed. Granitic rocks are also hard to differentiate; some are Prepaleozoic, but some also formed during each episode of Paleozoic orogeny. Isotopic dating shows that the largest share of the granite—as well as metamorphism—was of Acadian age, however.

As noted before, the Appalachian belt is bilaterally symmetrical in the southeastern United States, as it is in Newfoundland. Deformation and metamorphism decrease eastward toward the Coastal Plain just as they do northwestward toward the craton (Fig. 14.24). This situation is like that of the Caledonian belt as discussed in the last chapter, and is taken to be additional evidence that this also was an intercratonic belt squeezed between colliding North America and Gondwanaland.

SIGNIFICANCE OF THRUST FAULTING

The Appalachian region was among the first mobile belts to be studied in detail, and the fold structures long ago were taken to indicate a shortening of the circumference of the earth (see Fig. 8.6). In Chap. 8 we noted that this idea was consistent with the long-held assumption of a hot origin for the earth followed by cooling and shrinkage, and early laboratory experiments using vise boxes to squeeze layers of sand and clay or wax produced amazingly faithful replicas of the natural folds (see Fig. 8.8). The recognition of thrust faults lent much weight to crustal shortening across mobile belts, because lateral displacements on some thrusts could be shown to amount to several kilometers. At the end of the last century, however, some European geologists suggested that thrust faulting and much of the associated folding might involve only the weak, superficial sediments and not the more rigid basement. That is, the sediments might wrinkle and slide over the basement like a rug pushed across a floor (Fig. 14.27). One of the main reasons for such a suggestion was the apparent weakness of the rocks involved in the thrust

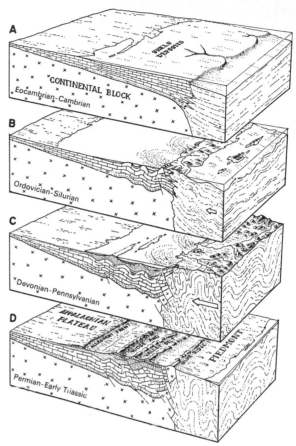

FIGURE 14.26 Diagrammatic summary of Appalachian history showing the gradual development of tectonic borderlands *within* the mobile belt through repeated orogenies and final culminating structural upheaval of the belt by apparent cratonward compression. *(Modified from Dietz and Holden, 1966, Journal of Geology, v. 74, p. 581; by permission of University of Chicago Press.)*

sheets. How could they be forcibly pushed so far? It was suggested instead that all the crumpling was simply caused by large-scale sliding of weak strata off an isostatically upwarped basement ridge merely due to gravity (Fig. 14.27). If this were the usual case in nature, as some argued, then folding and thrust faulting might not really indicate significant real shortening of the crust at all.

Figure 14.28 highlights the critical issue. It shows two contrasting interpretations of the structure of the Jura Mountains on the French-Swiss border. The upper,

Subsidence and Sedimentation

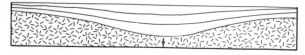

Uplift and Sliding

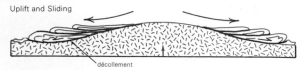

décollement

FIGURE 14.27 Downwarping and geosynclinal sedimentation followed by isostatic upwarping causing folding and thrusting by *gravity tectonics* (i.e., sliding laterally of the superficial sedimentary sequences). According to this hypothesis of mobile-belt structures, the basement rocks were not involved in deformation and no lateral shortening of earth circumference occurs (compare Figs. 14.24 and 14.28).

or *décollement* hypothesis, requires no significant shortening of the crust at all, while the lower one does because basement blocks are also involved in thrust faulting. Comparison of the cross sections of the Appalachian belt (Fig. 14.24), which are based in part

FIGURE 14.28 Two contrasting interpretations of the deep structure of the Jura Mountains on the French-Swiss border based considerably upon tunnel exposures. *Upper*: Involves superficial slippage of strata over the basement along a flat shear surface or *décollement. (From Buxtorf, 1908,* Nat. Carte Geol. Suisse.) *Lower*: Involves faulting of the basement as well as superficial strata along steeper faults. *(From Aubert, 1949; by permission of* Geologische Rundschau.) Conflicts between interpretations illustrated in this classic region have great bearing on whether or not the crust has been compressed and shortened in mobile belts.

upon drill-hole data, suggests that most of the thrust faults there actually flatten downward and do not involve the basement rocks very much. While most of the more obvious thrusts apparently are superficial, there still is compelling evidence in parts of this and other mobile belts of large slabs of both granitic and mafic oceanic basement rocks having been shoved up over younger sediments for considerable distances; that is, more like the lower section of Fig. 14.28. These cases provide inescapable evidence that indeed there has been net shortening across mobile belts. It is clear, however, that the superficial weak sediments become more intricately contorted and maybe shoved farther along thrust faults than the stronger basement rocks.

LATE PALEOZOIC MOUNTAIN BUILDING IN EURASIA

The end of the Paleozoic Era was a time of unusually widespread mountain building, the result largely of several continental collisions. Besides North America, central Europe was the site of great mountain building in the Hercynian mobile belt (Figs. 14.29 and 14.30). It appears that the Hercynian Orogeny was caused by the collision of northern Africa with Europe (Fig. 14.29). Similar-aged deformation around the southern perimeter of the Siberian craton may have resulted from the collisions of several small Chinese cratons (Fig. 14.30). Upheaval of the Ural belt at this same time is thought to have resulted from collision between lithosphere plates carrying Siberia and Europe. Late Paleozoic mountain building affected all other continents, too, but discussion of that is deferred to the next chapter, on Gondwanaland.

Both the Hercynian and Ural belts experienced

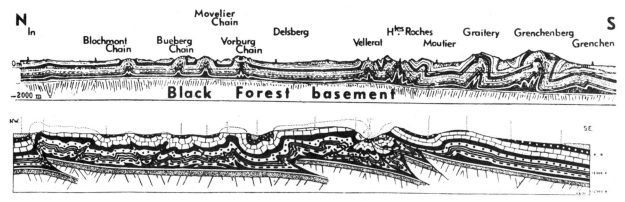

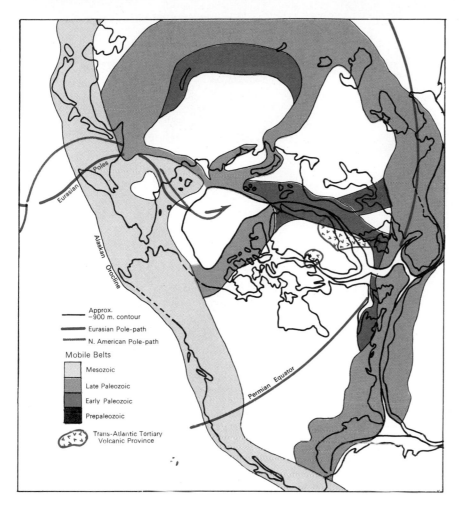

FIGURE 14.29 Northern continent reconstruction assuming continental drift and based upon reconciliation of geologic, paleontologic, and paleomagnetic evidence. Eastward rotation of North America re-fits disjunctive mobile belts and shorelines, closes most of the Arctic Ocean. It also places eastern North American fossil assemblages closer to nearly identical European cousins, yet keeps equally similar western North American and Asiatic ones in proximity. Note rotation also brings together Eurasian and North American apparent relative paleomagnetic polar-wandering paths. *(Adapted from many sources, especially Carey, 1959; Irving, 1964.)*

earlier deformation, especially in Devonian time, following which Europe was widely covered by Early Carboniferous (Mississippian) carbonate rocks. But this was a lull before an orogenic storm, for the major Hercynian mountain building characterized by widespread metamorphism and granitic plutonism began in Late Carboniferous (Pennsylvanian) time. In northwest Africa as in Europe, erosion of large tectonic lands produced large quantities of sand and mud, which were spread out into the surrounding epeiric sea to form huge deltas and coal swamps just like those of eastern North America at the same time. The resulting sediments show characteristics of repetitive or cyclic sedimentation, too. Most of Europe's mineral wealth of coal and ores was formed at this time. As in

North America, the last epeiric sea deposition occurred in restricted basins of later Permian time. Evaporites and nonmarine red beds became more and more widespread, and by the beginning of the Triassic Period, all of Europe was land. Late Paleozoic plants, fresh-water fish, amphibians, and reptiles, as well as the sediments, were almost identical with their counterparts in North America and northern Africa.

Siberia is much like North America in possessing a central craton completely encircled by mobile belts (Fig. 14.30). Marine Eocambrian strata are well represented in marginal mobile belts, and most of the craton was inundated by Cambrian and later epeiric seas. Great volcanism characterized the early Paleo-

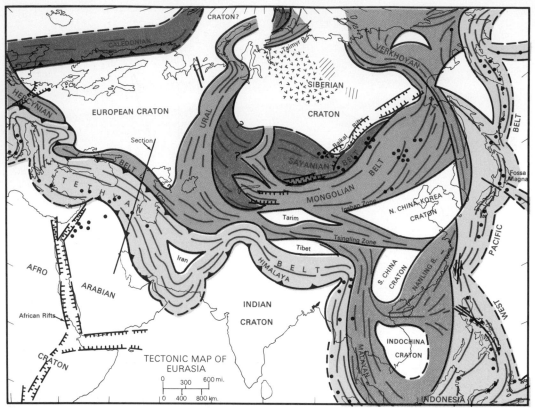

FIGURE 14.30 Tectonic map of Asia and adjacent regions, showing Paleozoic and younger mobile belts, including western Pacific oceanic island arcs, rifts, and very recent volcanoes. (See Fig. 17.29 for key to symbols.) *(After* Tectonic map of Eurasia, *U.S.S.R. Academy of Sciences, 1:500,000, 1966; Beloussov, 1962; Huang, 1960; Hseih, 1962.)*

zoic mobile belt on the south, and Ordovician, Silurian, and Devonian systems all include important red beds there. By Carboniferous time, coal-bearing deposits became widespread both in mobile belts and upon the craton itself. Although it is widely believed that the Ural belt resulted from Late Paleozoic collision of Siberia with Europe, it is at present difficult to interpret the relations of the other late Paleozoic mobile belts of Asia with plate motions. Relations of late Paleozoic deformations to older, Caledonian-aged ones of the southern Asiatic belts and to several zones of intense shearing tens of kilometers wide and extending eastward across China are impossible to interpret satisfactorily without much new data.

PERMO-TRIASSIC FAULTING

Almost as characteristic of Permo-Triassic time as red beds was vertical faulting accompanied by basaltic eruptions on all the continents discussed so far. In northern Siberia during Late Permian and Early Triassic time, an extensive basalt lava plateau was formed (Fig. 14.30). Because basalt is not generally associated with cratons, its eruption here indicates unusual rupturing of the thick continental crust to allow magma to rise from the lower crust or mantle. Explosive diamond-bearing igneous pipes (slender, vertical bodies) may have formed at this time, too. Much Triassic faulting occurred in the Ural belt. Permian basaltic and other magmas were erupted from fissures at Oslo, Norway, within the stable European craton. Triassic or Jurassic basalts were erupted from similar fissures in northwestern Africa.

In the Maritime region of eastern Canada, vertical faults created local basins within the Appalachian mobile belt during the late Paleozoic (Fig. 14.31),

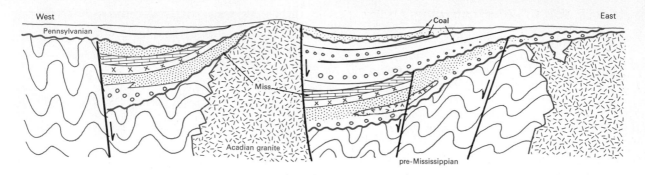

West   East

Pennsylvanian

Coal

Miss.

Acadian granite

pre-Mississippian

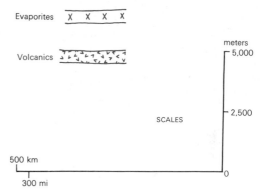

Evaporites   X X X X

Volcanics

meters
5,000

SCALES

2,500

500 km

300 mi   0

FIGURE 14.31   Block-fault uplifts in the Maritime Provinces, southeastern Canada. Faulting was most active in Mississippian time and waned in the Pennsylvanian as basins filled with nonmarine coal-bearing sediments. *(Adapted from M. Kay, 1951, Geological Society of America Memoir 48.)*

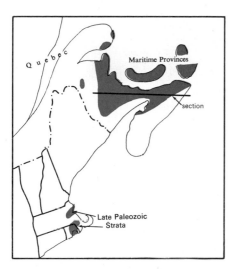

Quebec

Maritime Provinces

section

Late Paleozoic
Strata

apparently reflecting the beginning of a new tectonic behavior that was to become widespread in Triassic time. Maritime Mississippian rocks include coarse conglomerate (Fig. 14.32), red beds, evaporites, limestone, and basalt. By Pennsylvanian time, coal was deposited in the same basins. From Nova Scotia south to Florida, very thick Upper Triassic nonmarine red beds and interstratified basaltic flow rocks and sills also occur in downfaulted basins (Fig. 14.23). Both the faulting and the basaltic igneous activity here point to a new, extensional behavior after 400 to 500 million years of compressional tectonics. It is generally felt that all of this basaltic eruption and faulting around the end of the Paleozoic Era on at least four different continents is the first hint of the beginning of breakup of the continents and the opening of the present Atlantic basin.

FIGURE 14.32   Coarse Mississippian conglomerate and red sandstones associated with basalt and with marine carbonate and evaporite deposits in northern Nova Scotia.

LATE PALEOZOIC ANIMAL LIFE

GENERAL PATTERNS

In many respects, life late in the Paleozoic Era was quite different from that of the earlier Paleozoic. At first, some groups did not change very much, but quickly the entire aspect of the fauna became distinctive. New environments, such as vast coal swamps, provided a uniqueness not seen before nor fully duplicated after.

Slow, westward withdrawal of the epeiric sea was accompanied by spread of nonmarine sediments, which provide a more complete record of nonmarine environments than at any previous time. Thus we can gain much information on land organisms, for example, the excellent record of Pennsylvanian plants associated with coal deposits.

INVERTEBRATE ANIMALS
Arthropods

After the Devonian Period, the trilobites became very rare, and few species survived until ultimate extinction in Late Permian time. The ostracods (see Fig. 14.33), while fairly abundant from Ordovician time onward, reached a degree of diversity permitting them to be used for stratigraphic zonation. The Pennsylvanian has been called, somewhat whimsically, the age of cockroachs by allusion to occurrences of insects in lake and lagoonal deposits of coal-bearing regions. The record is sufficiently well known to gain a picture of the first animals to invade the air. Although insects first appeared in the Silurian, nothing

FIGURE 14.34 *Knightites*, a bellerophontid gastropod from the Middle Pennsylvanian near Gunsight, Texas. This is a bilaterally symmetrical, coiled form; the group is the most primitive of the snails. *(Courtesy U.S. National Museum.)*

FIGURE 14.33 An ostracod (*Trachyleberis*), a calcareous bivalved crustacean from the Pliocene of Italy. *(Photograph courtesy P. A. Sandberg.)*

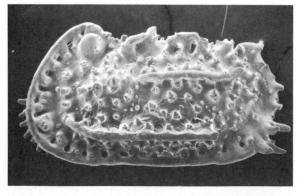

is known about them until the above occurrence. They must have undergone rapid evolution, since such diverse forms as dragonflies with wing spans of 30 inches and Texas-sized cockroaches 4 inches long are known by Pennsylvanian time.

Mollusca

The gastropods and pelecypods again are relatively minor parts of the fauna, although some primitive groups are abundant locally (Fig. 14.34). Fresh-water clams, for example, are used in Europe for correlating nonmarine deposits. Most of these molluscs became extinct by the end of Permian time, but a few survivors gave rise to advanced Mesozoic molluscan faunas.

We have seen that during Late Devonian time the ammonoids became widespread, and evolutionary expansion occurred rapidly in many groups. This trend continued and the ammonoids are of great service for upper Paleozoic correlation (Fig. 14.35).

Brachiopoda

The whole aspect of upper Paleozoic brachiopod assemblages is different from that found in Devonian faunas; however, differences are primarily in propor-

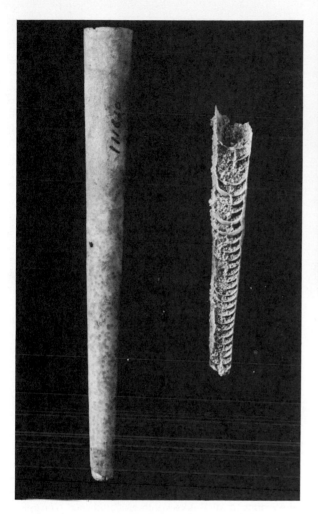

FIGURE 14.35 *Left*: A straight nautiloid cephalopod from the Permian of West Texas. *Right*: A goniatite cephalopod (*Imitoceras*) from the Mississippian of Indiana. *(Courtesy U.S. National Museum.)*

Mesozoic time. Thenceforth, brachiopods were a very minor part of the fauna.

Bryozoa

Up to Carboniferous time, the dominant group of Bryozoa was the massive or branching colonial trepostomes. During late Paleozoic time, the fenestellid Bryozoa overshadowed other groups. Fenestellids, which began in Ordovician time and gradually became important, tended to have delicate, lacy colonies. One bizarre genus *(Archimedes),* an index fossil of the Mississippian, had a corkscrew-shaped colony, which was simply a twisted cone (Fig. 14.36).

Echinodermata

We have seen that in the Silurian and Devonian periods organic reefs became increasingly important and widespread. In Mississippian time, this trend continued in many regions, but the composition of reef communities was much different from previous coral-hydrozoan-bryozoan-brachiopod reef assemblages. Commonly, the younger reefs and "gardens" possessed a crown of crinoids (Fig. 14.37), for massive reefoid limestones in many cases are composed

tions. Spiriferid brachiopods are common (some being as large as 12 inches), but they were exceeded in numbers by productid brachiopods, which are rarer in older rocks. The productids (see Fig. Al.6) invaded a host of environments not available to bivalved forms before because of remarkable adaptability provided by spines on their shells. Spines permitted them to either fasten to a hard surface or to serve as pilings in soft mud. They also served as a food-straining sieve and for protection. The productids continued to dominate until the end of the Permian, when they became extinct. Only four brachiopod groups (rhynchonellids, terebratulids, spiriferids, and ultraconservative inarticulates) survived into

FIGURE 14.36 *Archimedes*, a unique, screw-shaped colony of fenestellid Bryozoa; a commonly encountered Mississippian fossil. *(Courtesy American Museum of Natural History.)*

almost entirely of crinoid stems. Reef paradises resulted in evolutionary expansion of crinoids (Fig. 14.38), and during the Mississippian they reached their climax, constituting the largest portion of the fauna, if not in numbers of species certainly in bulk (Fig. 14.3). Although crinoids are living today, they were never again as diverse as during Mississippian time.

Corals

Some accumulations of colonial corals in Carboniferous rocks qualify as reefs, but in general, conditions did not favor coral-reef development. Colonial coral genera as well as solitary corals are fairly abundant. In Europe the corals, together with goniatites, are used to subdivide stratigraphic units, but no dramatic evolutionary developments are discernible.

Protozoa

The protozoans, though the simplest organisms in terms of organic structures, have undergone an amazingly complex and varied evolution, suggesting that all are not the primitive *Amoeba*-like forms that we have come to think of from elementary biology classes. Protozoan fossils are first encountered in Lower Cambrian rocks (possibly also in the Prepaleozoic), but are rare below the Upper Devonian. It was not until Mississippian time that we see the first active diversification, when Foraminifera became common. The Foraminifera possess calcareous skeletons and delicate, thread-like pseudopods used for locomotion and food gathering.

During Pennsylvanian and Permian times, fusulinid Foraminifera (Fig. 14.39) underwent important evolutionary expansion and proliferated so that by Permian time many genera had evolved. The evolutionary pattern is unusually well documented (see Fig. 5.6). Long-ranged genera tend to be conservative, while short-ranged forms are likely to be highly specialized and have had lower survival potential. Many of the short-ranged genera are extremely useful as correlative tools because they are known throughout the world and are nearly everywhere in the same stratigraphic order. Permian fusulinids are found in two distinct faunal realms. One is best developed in central Asia and Japan but also is recognized along the coast from Alaska south to California. The other realm includes all of North America east of the latter zone. Differences between the two are striking, particularly when comparing faunas as close together as California and Nevada. All fusulinids vanished by the end of the period.

EVOLUTION OF EARLY LAND LIFE

INVASION OF THE NEW HABITAT

Because the remains of countless plants gave rise to the great Pennsylvanian coals just discussed, it is appropriate to consider next the rise of land plants. Aquatic (marine?) plants are dominant elements in

FIGURE 14.37 Mississippian diorama. Most stalked forms are crinoids; note the screw-shaped colonies of *Archimedes*, a bryozoan, in left foreground. *(Courtesy American Museum of Natural History.)*

FIGURE 14.38 Camerate crinoid *Dichocrinus* from Lower Mississippian of eastern Iowa. Crinoids began during the Ordovician with two or three circles of skeletal plates on the head (below the arms). As evolution proceeded, more plates were added and there were many changes and elaborations involving the head and arms. *(Photograph courtesy U.S. National Museum.)*

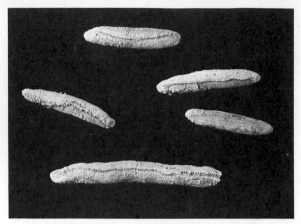

FIGURE 14.39   Fusulinids (*Parafusulina*) from the Permian of West Texas. *(Courtesy U.S. National Museum.)*

the Prepaleozoic record as compared to animals. The blue-green calcareous algae formed small reefs in the Prepaleozoic, and examples are also abundant throughout younger rocks. As far as we know, land surfaces were barren of any plant life until Late Silurian time, though primitive ground-hugging forms likely appeared earlier as noted in Chap. 10. The first known appearance of plants on land predates that of land vertebrate animals by about 40 million years.

Both plants and animals faced formidable problems in adapting to the land habitat. Mechanisms had to be developed: (1) to prevent desiccation, (2) to provide a strong supporting structure, and (3) to permit reproduction and facilitate dispersion of species in the new environment. Many obstacles were solved by plants through development of a vascular system, which assured all cells from top to bottom of the plant receiving moisture and nourishment; photosynthesis also could work more efficiently. As a part of the total system, more elaborate roots increased absorptive surface areas for acquisition of nutrients from soil and provided stability for large trees. The key to the system was the origin of long tracheid cells, which allow liquid to rise by capillary action. In addition, these cells provide woody supportive tissue to strengthen the main stem. Both the vascular system and the development of durable spores occurred in some water plants, thus they were preadapted for life on land. Spores are the simplest device for reproduction and dispersal because they are moved readily by air currents. Desiccation was alleviated by develop-

ment of protective plant epidermal cells and analogous protective cells around seeds. Triradiate spores can be produced by algae and they are known in the Cambrian. The tetradiate spores produced by vascular plants have been reported from Early Silurian rocks of New York, but the first definite vascular plant remains were found in the Late Silurian of Wales. It is thought that the vascular plants developed in water.

PALEOZOIC LAND PLANTS

By Early Devonian time, the first diversification of the primitive psilophytes occurred (Fig. 14.40), and very rapid evolution ensued so that, as mentioned in Chap. 13, by Late Devonian time lowland forests were in existence. The origin of the early psilophytes is unknown, but they probably were leafless vascular plants, which likely evolved from a filamentous green-algal stock.

The famous Gilboa forest in the Upper Devonian red beds of the Catskill Mountains of New York (see Fig. 13.1) contains five groups of plants derived from the Psilophytales, which were eclipsed at the end of the Devonian by hardier descendants. Trees in this forest are estimated to have been over 10 meters high and inhabited a predominantly low-lying, river-margin environment. These new plant groups spread rapidly over a large portion of the Northern Hemisphere so that by Carboniferous time they were the chief contributors to swamp vegetation and peat. The reservoir of energy stored in the vast, buried coal swamps of Europe, Asia, and eastern North America was destined to power the industrial revolution. Paleozoic plants have contributed both directly and indirectly to the housing, production, and general well-being of Western civilization for over 200 years. They still constitute one of the greatest reservoirs of stored energy.

The true ferns (Filicineae) first appeared in Late Devonian time, and they were extremely abundant through the Carboniferous. Some were at least 70 feet in height, but, despite their prominence, they probably were not principal contributors to the accumulation of peat, until the Late Pennsylvanian. For many years they were confused with the similar seed ferns (pteridosperms), which reproduced, as the name implies, by seeds rather than spores. The seed ferns were a dominant group in the Gilboa forest, some species reaching 30 to 40 feet in height. The seed ferns are the possible stock from which more advanced cycads and flowering plants (angiosperms) arose during the Mesozoic Era.

A visit to any coal strip-mine in the eastern part of the United States likely will yield, within a few minutes, examples of the lycopsids and sphenopsids, the most important coal-forming groups of the Carboniferous. The lycopsids were relatively diverse in the late Paleozoic. They are represented today by the temperate-forest ground pine, a very small, bushy plant, pale by comparison with its towering ancestors, which grew over 100 feet high (Fig. 14.41). The lycopsid trees were strange-looking because they were without branches except at the very top of the trunk. The leaves were long and spike-like, similar to the individual palm leaf of today. As the trees grew, leaves were replaced from the top, and as old leaves dropped off, scars were left in rows or spirals. Reproduction was by spores contained in cones. The other common plant group found in association with coal is the Sphenopsida. They are characterized by being jointed, and by having prominent horizontal grooves (Fig. 14.41). In life, a circle of small branches protruded from each joint, and these bore a circlet of leaves and occasional cones. They undoubtedly formed thickets not unlike the biblical Nile River bulrushes (papyrus), and were probably ensconced in a similar habitat *sans* Moses. They averaged about 15 to 20 feet tall. Like the lycopsids, they were never very diversified, and are represented today by the curious little *Equisetum* (or horsetail), a sort of "living fossil" found along swamp banks and railroad embankments.

NORTHERN VERSUS SOUTHERN FLORAS

The fossil record of Carboniferous plants is the earliest to be complete enough to recognize three floras in different climatic belts. The tropical to subtropical flora was dominated by the lycopsids until Late Pennsylvanian time when the lycopsids were overshadowed by the ferns. This flora (Eurameric) is found in North America, Europe, North Africa, and China. A less widely distributed warm temperate flora is found in eastern Siberia, consisting of seed ferns, some lycopsids, and primitive conifers with seasonal growth rings. By Permian time these northern floras had changed to conifer-dominant types, with the Chinese and Eurameric separated into distinct units. The Permian record is not as good as that of the

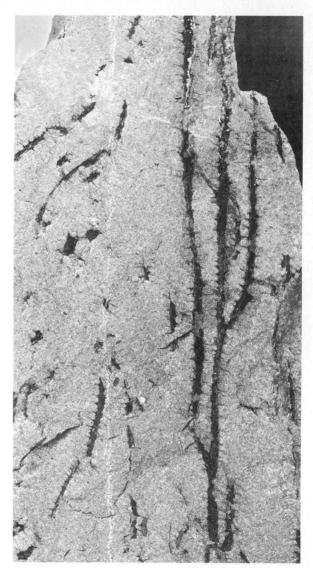

FIGURE 14.40 An example of a Devonian primitive vascular plant (*Psilophyton*) from the Lower Devonian of Gaspé, Canada. *(Photograph courtesy F. M. Hueber and U.S. National Museum.)*

Carboniferous because the low-lying swampy environments favorable for plant preservation were replaced by terrain of higher elevation and drier climates. Primitive conifers were better adapted than other plants for drier conditions.

The southern Gondwana or *Glossopteris* flora, named after the dominant genus (a large seed fern;

FIGURE 14.41 Diorama of a Pennsylvanian coal forest; large trees with scar patterns are lycopsids, single tree (*Calamites*) in the right foreground with the horizontal grooves is a jointed sphenopsid. *(Courtesy American Museum of Natural History.)*

FIGURE 14.42 Typical fossil leaf of the seed fern *Glossopteris*, principal genus and namesake of the southern late Paleozoic flora. Note tongue-like shape, central rib, and fine venation. Lower Karroo strata, Waterburg Coal Field, Transvaal, South Africa. *(Courtesy E. P. Plumstead.)*

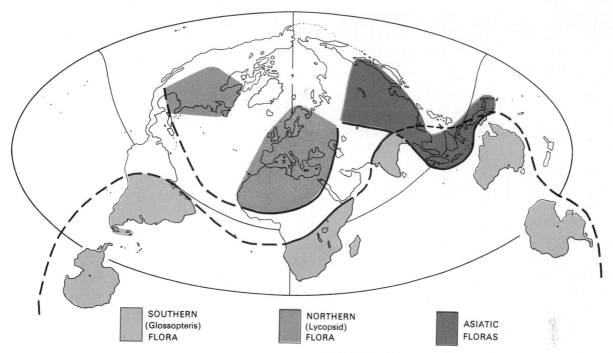

SOUTHERN
(Glossopteris)
FLORA

NORTHERN
(Lycopsid)
FLORA

ASIATIC
FLORAS

FIGURE 14.43   Distribution of late Paleozoic land floras of the world. Note the separation of lycopsid and *Glossopteris* floras. Some *Glossopteris* elements did mix with the intermediate Asiatic flora during Carboniferous time in New Guinea and east Africa, in Triassic time in southern China, and possibly in Jurassic in southern Mexico. Climatic zones presumably formed the barriers between the floras. *(Adapted from Just et al., 1952, National Research Council Report on Paleobotany N22; and Gothan and Weyland, 1954, Lehrbuck der Palaobötanik.)*

Fig. 14.42), is relatively undiversified, but is widespread (Fig. 14.43). It is associated with glacial deposits and contains tree trunks with prominent seasonal growth rings (see Fig. 14.44). Therefore, it is assumed that, unlike its northern contemporary, the southern plant lived under cooler, temperate climatic conditions. It, too, produced immense coal deposits. The *Glossopteris* flora is critical because of its bearing on continental drift and the evidence it provides of worldwide climatic zonation. We shall examine it further in Chap. 15.

VERTEBRATE ANIMALS

Amphibians

Water, besides being the elixir of life, has certain properties making it possible for organisms living in it to have an easier way. For example, because of its density and buoyancy, it helps support an animal's weight. This is critical to our understanding of the skeletal support essential for a relatively large animal to live in air—a factor all-important to the primitive amphibian. Many evolutionary modifications were involved in developing strong skeletons. In the amphibians, "experimental redesign" of the vertebral column provided both protection of the spinal cord and support for leg and back muscles. In several ways, the vertebral elements were interlocked to allow flexibility of the column for movement, but at the same time to achieve greater support than in fishes (Fig. 14.45).

The earliest amphibians (Late Devonian) were fish-like in appearance, but very quickly in some forms the body and head became flattened and limbs shortened. Eyes shifted to the top of the head, suggesting that the creatures may have spent much time in very shallow water.

Development of adaptive skeletal mechanisms

FIGURE 14.44   Seasonal growth rings in petrified *Dadaxylan* wood (Permian or Triassic), Senekal, Orange Free State, South Africa. Such rings may reflect hot–cold, wet–dry, or dark–light seasons. (Diameter of log is about 40 centimeters.)

for land dwelling solved only part of the problem presented by the new habitat. The amphibian had inherited a fish-type egg, one which encased the embryo in a membrane that allowed oxygen to enter and wastes to pass out into surrounding water. This type of egg is simple, but effective in a water medium, but amphibians had to remain near the water or at

least return to it to lay eggs. The problem was solved by reptiles. The reptilian egg has an outer, leathery case or shell that allows oxygen to enter, and various membranes within the egg prevent liquids from escaping through drying in air. In addition, such eggs contain food for the embryo (the yolk) and a storage area for wastes. It is a completely self-contained unit.

If we are to discriminate between a fossil amphibian and a reptile, we need more information, for eggs are very rarely preserved. The skulls have many diagnostic features; most important is the presence in

FIGURE 14.45   A portion of a primitive amphibian (*Eryops*) vertebral column (model) showing the interlocking nature of the vertebrae. *(Courtesy American Museum of Natural History.)*

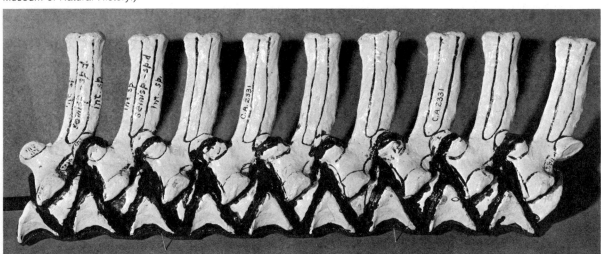

FIGURE 14.46 *Seymouria*, a lower Permian amphibian that developed many reptilian characteristics and has been used as an example of transitional "experimentation." *(Courtesy American Museum of Natural History.)*

amphibia and fish of grooves called lateral lines that mark the presence of a sensory device in aquatic forms. The reptiles lack this system, seemingly because of selection against this (for them) useless device. It goes without saying that the oldest reptile fossils, which are encountered in Lower Pennsylvanian strata at Joggins, Nova Scotia, are so close to their amphibian forebears that, lacking skulls, we would be hard pressed to be sure of their identity (Fig. 14.46).

Reptiles

The first reptiles (cotylosaurs) left a poor record, but there is no doubt that they experienced some Pennsylvanian diversification. As the seas regressed during Permian time, large, low lands appeared, and in Oklahoma, Texas, and New Mexico we find a rich and varied reptilian fauna first studied in detail by E. C. Case. Rivers meandered across broad flood plains dotted with lakes and swamps. A diverse fauna of bizarre reptiles together with amphibians and forms transitional between them (Fig. 14.46) inhabited these environments.

The cotylosaurs gave rise to all other reptiles, two groups of which were of particular note during the Permian Period. One group, the pelycosaurs (Fig. 14.47) was quite diversified, including several genera which had long, elaborately branched dorsal spines. Undoubtedly these spines had skin stretched between them. The most plausible function of this contraption was regulation of body-heat budget. Some students believe that such weird adaptive structures mean that the climate was very hot. In any event, these pelycosaurs were so specialized for a narrow environmental niche that, when conditions changed at the close of Permian time, they became extinct. The pelycosaurs possessed teeth differentiated into incisors, canines, and molar teeth. This contrasts with other reptiles, which had teeth all of the same or very similar shape, varying only in size. Most reptiles do not masticate food, but swallow it whole. No molars or other specialized teeth are needed.

The other group, which is in the same subclass (Synapsida) as the pelycosaurs, is the mammal like reptiles (Therapsida). These, together with the pelycosaurs, came to dominate Permian lands, for by this time there was a sharp decline in amphibians. The mammal-like reptiles underwent very rapid expansion and there is much variation in morphology. For example, in contrast to the waddle of other reptiles and amphibians of late Paleozoic time, they must have enjoyed a faster, smoother gait because a small pelvis served as an abutment for legs that had rotated to be more parallel to the body. This characteristic and the nature of their teeth suggest that some of these forms were effective predators. The therapsids had a secondary palate, a mammalian feature; however, they had small braincases and other clearly reptilian characters. Mammal-like reptiles survived the crisis of terminal Paleozoic extinctions, and some time during Triassic or Jurassic time, gave rise to true mammals. Having spawned them, the therapsids—like the homing salmon—expired (See Fig. 15.17).

Not much can be said regarding the zoogeographic distribution of land vertebrates during the Late Permian and Triassic. Both the therapsids and

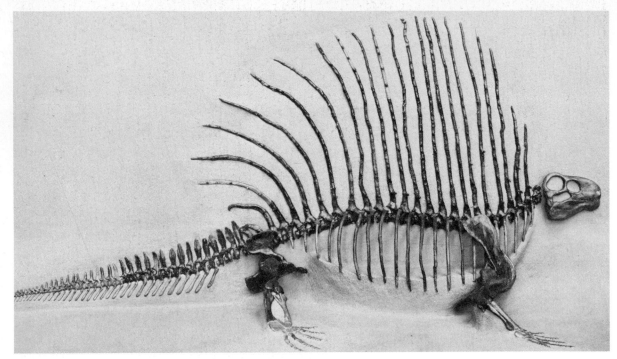

FIGURE 14.47 *Edaphosaurus*, a pelycosaur from the Permian of Texas. *(Courtesy Field Museum of Natural History.)*

non-mammal-like reptiles are found in widely separated localities. In addition, the therapsids are dominant in the Upper Permian (170 genera), in which the other reptiles are represented by 15 genera. The therapsids were physiologically adapted to warm humid conditions, which apparently were common in the Late Permian, whereas the other reptiles apparently were more suited to higher temperatures with seasonal droughts. The latter conditions are thought by some to have been more widespread by Late Triassic time when the other reptiles came into dominance.

SUMMARY

The close of the Paleozoic Era originally was designated on the basis of a major discontinuity in development of marine life. On the inorganic side, we find that increasing tectonic activity had elevated much

new land. Land life evolved and expanded rapidly as new terrigenous habitats multiplied, not only in North America but worldwide. Conversely, marine life suffered great extinctions as shallow epeiric sea environments shrank almost to nothing.

Remarkable correspondence of explosive evolution of Devonian land plants and land animals with the marked increase of land area by mountain building hardly can be sheer coincidence, and provides an outstanding example of mutual relations between the physical and organic evolution of the earth. Epeiric seas reached their maximum sizes in later Ordovician, Silurian, and Devonian times, but beginning in the Devonian in eastern North America and Europe, mountain building provided diverse new habitats for invasion by neophyte land organisms, which were quick to respond to fill the ecological vacuums. But why did they not invade the Late Ordovician Taconian land, which apparently was similar to the Devonian ones? It has been speculated that for a considerable time after the atmosphere became oxygen-rich, there was lethal ozone at ground level. Much oxygen presumably had to accumulate before the ozone layer rose to a higher level safe for advanced forms of life. Possibly this did not occur until Devonian time. Clearly the development of life was closely tied to physical history at this time.

During late Paleozoic time, tectonic unrest increased in tempo all over the continent, including even parts of the formerly tranquil craton. Accompanying the tectonic revolution were marked changes of sedimentation. First, there was the introduction of much more heterogeneous clastic material, reflecting erosion of earlier Paleozoic strata to expose Prepaleozoic basement in Canada. Second, nonmarine and deltaic sediments became more widespread as the continent was tilted up and the epeiric sea was crowded farther westward, causing regression and spread of clastic sediments on a much greater scale than during previous orogenies. As the continent was tilted, black and gray-colored nonmarine and brackish-water Pennsylvanian sediments accumulated in oxygen-poor swampy environments near sea level in repetitive (cyclic) patterns, which evidence many oscillations of the shoreline amounting to several hundred kilometers. During Permian and Triassic time, however, as land became still more elevated, red sediments accumulated widely in oxygen-rich environments.

The greatest activity was in the Appalachian and Hercynian mobile belt, where complex folding, thrust faulting, and local granite formation occurred as a result of collision of northwestern Africa and southern Europe with southeastern North America. Simultaneously, other lithosphere plate collisions were occurring, for late Paleozoic mountain building affected nearly every continent. Siberia collided with Europe, southeast Asia with Siberia, and some adjustments occurred also in South America, Africa, Australia, and Antarctica. By Permian time, the continents were grouped together in one huge mass called Pangaea. Land area was extremely large at this time so that land life had little difficulty dispersing over the entire supercontinent.

The possible position of the equator as suggested by paleomagnetic evidence during late Paleozoic time is noteworthy in relation to distribution of various sediments, fossil types, and paleocurrents. Most significant is that coal-bearing deposits would lie at or within 20 or 25 latitudinal degrees of the apparent paleoequator, thus in tropical to subtropical zones, where mild, humid, nonseasonal climatic conditions would be expected. Land apparently was low enough

and small enough during Mississippian and Pennsylvanian time that rainfall was uniform. During Permian and Triassic time, position of the possible paleoequator seemingly was not greatly different, but both area and elevation of land had increased. This would cause greater differentiation of climate; therefore, it is plausible to postulate relatively humid Permo-Triassic uplands, but drier lowlands. Much of North America lay in the late Paleozoic Trade Winds belt. The western side of the continent would be leeward of the Appalachian Mountains, which might account for aridity there.

Early Mesozoic faulting and basaltic eruptions in eastern North America, northwestern Africa, and even in Norway and in far-off Siberia apparently attest to the initial breakup of Pangaea, which was soon to lead to the birth of the present Atlantic basin. It is significant that after Triassic time, similarities among North America, Africa, and Europe are much less obvious than they were before.

Readings

Alberta Society of Petroleum Geologists, 1964, Geological history of western Canada: Calgary. (An atlas of stratigraphic maps)

Andrews, H. N., Jr., 1961, Studies in paleobotany: New York, Wiley.

Beloussov, V. V., 1962, Basic problems in geotectonics: New York, McGraw-Hill. (English translation)

Douglas, R. J. W., et al., 1963, Geology and petroleum possibilities of northern Canada, in Proceedings Sixth World Petroleum Congress (sec. 1): Frankfurt, pp. 519–571.

Hatcher, R. D., Jr., 1972, Developmental model for the southern Appalachians: Geological Society of America Bulletin, v. 83, pp. 2735–2760.

Hseih, C. Y., 1962, On the geotectonic framework of China: Scientia Sinica, v. XI, pp. 1131–1140.

Huang, T. K., 1960, The main characteristics of the structure of China: preliminary conclusions: Scientia Sinica, v. IX, pp. 492–544.

Kay, M., ed., 1969, North Atlantic—geology and continental drift (Memoir 12): Tulsa, American Association of Petroleum Geologists.

Klein, G. de Vries, ed., 1968, Late Paleozoic and Mesozoic continental sedimentation, northeastern North America, Boulder: Geological Society of America Special Paper 106.

Ladd, H. S., ed., 1957, Treatise on marine ecology and paleoecology: Geological Society of America Memoir 67, v. 2, Paleoecology.

McKee, E. D., Oriel, S. S., et al., 1967, Paleotectonic investigations of the Permian system: U.S. Geological Survey Professional Paper 515.

Nalivkin, D. V., 1960, The geology of the U.S.S.R.: London, Pergamon.

Newell, N. D., et al., 1953, The Permian reef complex of the Guadalupe Mountains region, Texas and New Mexico: San Francisco, Freeman.

Raasch, G. O., ed., 1961, Geology of the Arctic: Toronto, Univ. of Toronto Press.

Roberts, R. J., et al., 1958, Paleozoic rocks of north-central Nevada: Bulletin of the American Association of Petroleum Geologists, v. 42, pp. 2813–2857.

Robinson, P., 1971, A problem of faunal replacement on Permo-Triassic continents: Palaeontology, v. 14, pp. 131–153.

Seyfert, C. K., and Sirkin, L. A., 1973, Earth history and plate tectonics: an introduction to earth history: New York, Harper & Row.

Spencer, E. W., 1962, Basic concepts of historical geology: New York, T. Y. Crowell.

Stokes, W. L., 1973, Essentials of earth history (3d ed.): Englewood Cliffs, Prentice-Hall.

Trewartha, G. T., 1954, An introduction to climate (3d ed.): New York, McGraw-Hill.

Wanless, H. R., et al., 1963, Mapping sedimentary environments of Pennsylvanian cycles: Bulletin of the Geological Society of America, v. 74, pp. 437–486.

15

GONDWANALAND

India

And so these men of Indostan
Disputed loud and long,
Each in his own opinion
Exceeding stiff and strong,
Though each was partly in the right
And all were in the wrong!

The Blind Men and the Elephant (Ancient Hindu fable, John Godfrey Saxe)

FIGURE 15.1   Striated glaciated surface or pavement beneath Carboniferous Dwyka Tillite, Nooitgedacht, near Kimberley, South Africa; surface was cut on Prepaleozoic metamorphosed volcanic rocks. *(Courtesy J. C. Crowell, 1967.)*

In Chap. 8 we presented a brief introduction to the concept of large-scale continental displacements, and we noted some of the evidence from the five continents that today lie south of North America and Eurasia. Now we are ready to examine that evidence more fully and to trace the Paleozoic and early Mesozoic history (700 to 180 m.y. ago) of the southern supercontinent Gondwanaland as we have already done for North America and Eurasia.

The five southern or Gondwana continents have several peculiarities as compared to the northern or Laurasian ones, as the latter are sometimes called. Much of the formers' perimeters are represented not by marginal mobile belts active sometime since Cambrian time as in the north, but rather by truncated cratonic edges (Fig. 15.2). Most of the coastlines look like places where the continents were somehow broken off—pieces of a jigsaw puzzle that once fit together to form a single gigantic craton. The perfect match of Africa and South America was, of course, a major clue that inspired the first continental reconstruction proposed in 1858 by the European, Antonio Snider. For this and other equally compelling reasons, the southern continents have long provided a stronger case for continental displacements than have the northern ones. As with the blind men trying to describe the elephant, the way continents developed depended greatly upon where one was standing.

As we shall see, largely nonmarine rocks of late Paleozoic and early Mesozoic ages are strikingly similar on all five southern continents. They also contain nearly identical fossils, which implies a close relation among the five until the late Mesozoic when widespread flooding of the sea occurred and physical histories as well as organisms began to diverge. Important differences between late Paleozoic floras of Laurasia and Gondwanaland suggest a barrier of some kind between, but it must have been a climatic rather than a physical one because nonmarine fossil reptiles show very strong similarities on *all* continents (Fig. 8.11). Finally, the distribution of late Paleozoic glacial deposits also suggested a polar grouping of southern continents (Fig. 15.1). The evidence points to physical connections among all the present continents in Permo-Triassic time when total land area of the earth was about the same as today—a rare event in earth history with tremendous implications for both climate and the evolution of land life. Paleomagnetic evidence and isotopic dating accumulated since 1955 added great strength to the old continental-drift idea, and the still newer knowledge from the deep seas has clinched major continental displacements through time. Both rock paleomagnetism from conti-

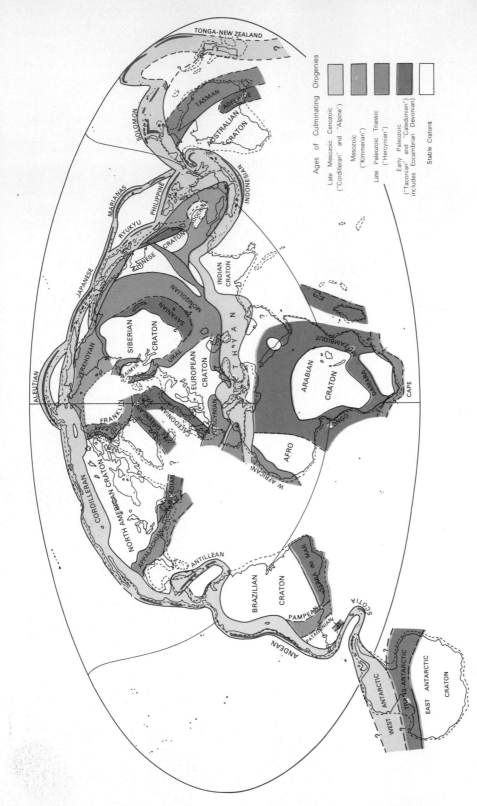

FIGURE 15.2 World mobile belts of the past 700 million years (Eocambrian and younger); note that early Paleozoic belts also appear on Fig. 10.13. Note disjunctive belts (those cut off at continental margins) and their apparent matching counterparts on opposite continents in several cases. Note also modern island arcs and oceanic deep trenches. (Bartholomew's Nordic Projection; used with permission.)

Ages of Culminating Orogenies

Late Mesozoic Cenozoic ("Cordilleran" and "Alpine")

Mesozoic ("Kimmerian")

Late Paleozoic Triassic ("Hercynian")

Early Paleozoic ("Taconian" and "Caledonian" includes Eocambrian Devonian)

Stable Cratons

nents and sea-floor magnetic anomalies have provided independent time scales for studying those displacements.

Ambiguities and incompleteness of the geologic and paleontologic data for years assured a lively and colorful debate between contending opinions on drift. Advocates sometimes have been drugged by the most flimsy evidence, such as the seasonal flight path of the Arctic tern as it zigzags from side to side of the Atlantic Ocean in its migration between the northern and southern polar regions. It has been alleged that this weird habit reflects the indecision of a poor confused species whose ancestors began migrating before the continents separated and not knowing what else to do, continued to try to keep both shores in sight over the ages as the continents continually sailed apart. Perhaps, after all, the most compelling biologic evidence for drift is the distribution noted in Chapter 8 of the modern species of the lowly earthworm, a creature totally incapable of transoceanic travel.

are many criss-crossings of old mobile belts as well as many overprintings of successive metamorphisms. Some of the oldest rocks known (3.0 to 3.5 b.y.) occur in southern Africa, and Russian geologists report an isotopic date approaching 4.0 billion years from the Antarctic craton. Old dates also occur in central and western Africa (2.7 to 2.9 b.y.), southwestern Australia (2.7 b.y.), southwestern India (2.5 b.y.), and northeastern Brazil (2.1 b.y.) (see Fig. 10.13). Banded iron formations formed widely (Fig. 15.3), the world's largest gold deposits formed in Precambrian conglomerate in South Africa, and important copper deposits accumulated in Eocambrian strata in Africa (Fig. 15.4).

At present it is impossible to interpret the relationships among these continents for Prepaleozoic time; apparently they were more or less separate units until the early Paleozoic. The first event that can be detected clearly on most of them was the Eocambrian glaciation (see Fig. 11.12). The glacial evidence, however, tells us little about the relative positions of the continents.

GONDWANA HISTORY

PREPALEOZOIC

As in the north, the patterns of Prepaleozoic rocks in the southern five continents are very complex. There

FIGURE 15.3 Flat-lying Prepaleozoic banded iron deposits and shales (Brockman Iron Formation) in strata about 2.1 billion years old; Colonial Asbestos Mine, Wittenoon Gorge, Hammersley Ranges, Northwest Australia. *(Courtesy D. E. Ayres.)*

FIGURE 15.4 Tightly folded strata of Eocambrian age in Katanga arc of the Damara mobile belt, northern Zambia, south-central Africa (see Fig. 15.2 for location of early Paleozoic or "Pan-African" Damara belt). Photograph is of open-pit Chambishi copper mine of Roan Selection Trust Ltd.

PALEOZOIC

Regardless of continental positions, between 800 and 700 million years ago, important new tectonic patterns developed on all the southern continents. A series of mobile belts began that were to persist through early Paleozoic time in much of Africa and in southern Australia (Figs. 15.2 and 15.5). Presumably these belts were formed by the convergence of lithosphere plates, which culminated in the collision of a mosaic of continental cratons to form the Gondwana supercontinent by Ordovician or Silurian time (Fig. 15.6). Apparently a sliver of present eastern North America was then a part of northwestern Africa and South America, and apparently that margin of Gondwanaland was experiencing subduction of the old proto-Atlantic plate (see Fig. 12.31). In late Paleozoic time, recall, Gondwanaland collided with North America and Europe, resulting in the final Appalachian and the Hercynian Orogenies (Fig. 14.29). Meanwhile, a microcontinent (small, isolated sliver of continental crust) made up of present southern South America, part of West Antarctica, and probably also New Zealand apparently continued to converge upon southern Gondwanaland where the Pampean, Cape,

FIGURE 15.5 Basal unconformity at road level between Ordovician-Silurian Table Mountain Sandstone of the Cape mobile-belt sequence and underlying Cambrian Cape Granite (520 m.y.), 5 kilometers south of Cape Town, South Africa. The granite formed during widespread early Paleozoic "Pan-African" orogenesis. Paleozoic Table Mountain Sandstone has thick, cross-stratified quartz sand derived from the southern African craton and dumped into the subsiding Cape belt.

and Tasman mobile belts remained active throughout the Paleozoic (Fig. 15.2). The South Africa drift advocate, Alexander Du Toit (mentioned in Chap. 8), used these zones of late Paleozoic orogeny in perfecting his now-famous reconstruction of Gondwanaland (Fig. 15.6). His Samfrau geosyncline was finally crumpled by collision of the microcontinent with Gondwanaland proper in Permo-Triassic time (Fig. 15.7).

Early Paleozoic epeiric seas covered most of the northern Afro-Arabian, northern Indian, western South American, and much of the Australian and Antarctic cratons. Devonian seas were the most widespread (see Fig. 13.23), and their faunas in the south were distinct from those of the north. During Late Ordovician time, a large ice sheet covered much of North Africa, which was then located at the south pole according to paleomagnetic data (Fig. 12.31). Silurian and Devonian till-like deposits also occur in central South America (see Fig. 13.23). By late Paleozoic time, much of the Gondwana supercontinent had become land, and collisions with the North American–Eurasian supercontinent (commonly called Laur-

asia) produced land connections among all the present continents. In Devonian time, Gondwanaland shifted northward suddenly about 60 degrees of latitude so that the pole lay in central Africa (Fig. 15.8). Then it shifted about 40 degrees laterally to bring Australia near the pole by Permian time; in the Triassic Period, the supercontinent moved northward away from the pole.

Because much land area lay at high latitudes in late Paleozoic time, large ice sheets formed on up-

FIGURE 15.6 Du Toit's reconstruction of the *Gondwanaland supercontinent* during late Paleozoic time based upon congruence of shorelines, matching of structural features in ancient basement rocks (lines), and late Paleozoic mobile belts—here combined in what Du Toit called *Samfrau geosyncline* (from the combination of letters from the names of three continents). Apparently a microcontinent compressed the Samfrau belt against main Gondwanaland in Permo-Triassic time. After reconstructing, Du Toit plotted distributions of Gondwana rocks and especially glacial phenomena; these seemed far more intelligible on such a reconstruction. Many of the structural trends recently have been defined better by isotopic dating (compare Figs. 10.13 and 15.9). *(Adapted from Du Toit, 1937.)*

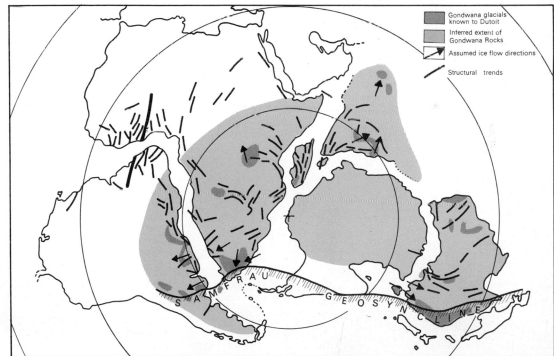

Gondwana glacials known to Dutoit

Inferred extent of Gondwana Rocks

Assumed ice flow directions

Structural trends

EVOLUTION OF THE EARTH

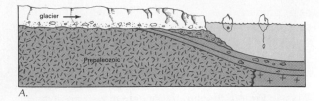

A.

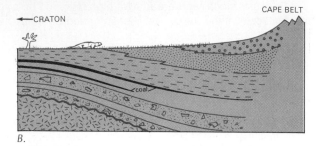

B.

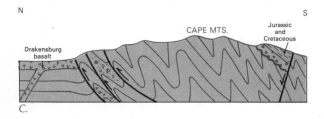

C.

FIGURE 15.7 Diagrammatic portrayal of development of the Cape mobile belt. Prior to *A*, the Cape Granite formed in Cambrian time (lower right; see also Fig. 15.5). Then Table Mountain Sandstone and other middle Paleozoic marine strata accumulated in the mobile belt; the craton was low land. *A*: During late Paleozoic time, continental glaciers spread south to the belt where thick till (Dwyka) was deposited. *B*: After glaciation, widespread Mesozoic ("Gondwana") nonmarine strata and basalts accumulated. *C*: In Triassic and Jurassic time, the belt was upheaved with cratonward thrusting.

land areas. Late Paleozoic glacial deposits are known on all five of the southern continents. Together with the similarity of present continental margins, the distribution of these deposits provided one of the criteria used for early reconstructions of Gondwanaland (Fig. 8.11). Their presence on either side of the present equator and the absence of similar-aged tillites at high present latitudes in North America and Europe was baffling, to say the least, in terms of present geography. Moreover, in several places the ice seemed to have moved onto the continents from present deep-ocean basins (e.g., southeastern Aus-

FIGURE 15.8 Paleomagnetic reconstruction of southern continents. *A*: Present positions of South America and Africa and the relative positions of the south pole as determined paleomagnetically from rocks on each continent. *B*: Note close match of restored zig-zags in respective apparent Paleozoic "polar wandering" paths for Australia and South America with Africa's. It is more convenient to draw *relative* pole shifts, but actually the continents moved. *C*: Latitudes indicated for Australia and India do not agree exactly with geologic data (compare Fig. 15.16). *(After McElhinny, 1967, UNESCO Symposium on Continental Drift, Montevideo.)*

A. PRESENT

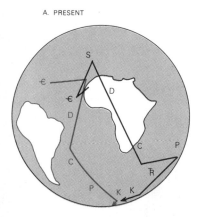

B. PALEOZOIC FIT

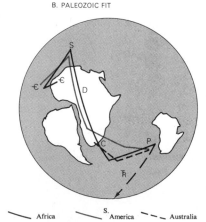

C. MESOZOIC LATITUDES

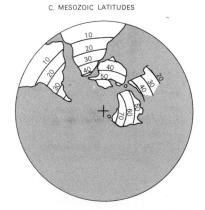

——— Africa ——— S. America – – – Australia

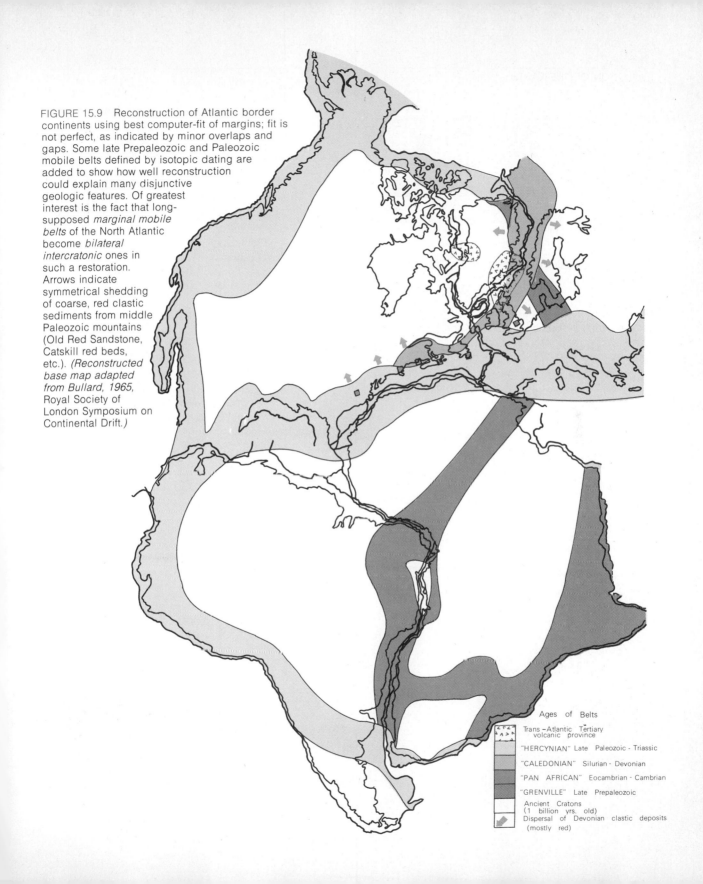

FIGURE 15.9 Reconstruction of Atlantic border continents using best computer-fit of margins; fit is not perfect, as indicated by minor overlaps and gaps. Some late Prepaleozoic and Paleozoic mobile belts defined by isotopic dating are added to show how well reconstruction could explain many disjunctive geologic features. Of greatest interest is the fact that long-supposed *marginal mobile belts* of the North Atlantic become *bilateral intercratonic* ones in such a restoration. Arrows indicate symmetrical shedding of coarse, red clastic sediments from middle Paleozoic mountains (Old Red Sandstone, Catskill red beds, etc.). *(Reconstructed base map adapted from Bullard, 1965,* Royal Society of London Symposium on Continental Drift.*)*

Ages of Belts

Trans–Atlantic Tertiary volcanic province

"HERCYNIAN" Late Paleozoic - Triassic

"CALEDONIAN" Silurian - Devonian

"PAN AFRICAN" Eocambrian - Cambrian

"GRENVILLE" Late Prepaleozoic

Ancient Cratons (1 billion yrs. old)

Dispersal of Devonian clastic deposits (mostly red)

FIGURE 15.10 Diagrammatic cross section showing how largely nonmarine Gondwana rocks were dated in India by relating them to a few interstratified marine tongues of Himalayan (Tethyan) facies containing fossils referable to European standard geologic time scale shown at right (UC—Upper Carboniferous).

tralia and eastern South America; Fig. 15.6). The Dwyka Tillite of South Africa contains boulders with Lower Cambrian archaeocyathid fossils (see Fig. 11.5) that are unknown in Africa but which could have come from Antarctica (Fig. 15.6). To these criteria, Du Toit added structural features (Fig. 15.6). Then in the

FIGURE 15.11 Distribution of the largely nonmarine Gondwana rocks on the southern continents and their remarkable similarity on all five continents. (Bartholomew's Nordic Projection used by permission.)

1960s, the new clues for refitting the jigsaw puzzle provided by extensive paleomagnetic studies (Fig. 15.8) and isotopic dating (Fig. 15.9) gave new precision and confidence to the restorations.

THE GONDWANA ROCKS

Late Paleozoic and early Mesozoic strata of the southern cratons occur in a distinctive sequence that is amazingly similar on all five now widely separated continents. These rocks are collectively known as the Gondwana rock succession. The name Gondwana is derived from an ancient tribe in India. When Gondwana rocks were first discovered a century ago, their age was unknown. The principal fossils belonged to a then strange, and as yet undated flora. Because the relative geological time scale was developed in Europe in largely marine strata, the Gondwana succession could be dated only by relating its largely nonmarine fossils to some more familiar, well-dated marine index fossils. This first became possible in northern India where lower Gondwana tillites are interstratified with Upper Carboniferous and Lower Permian fossiliferous marine strata (Fig. 15.10). Elsewhere, upper Gondwana strata are associated with

DISTRIBUTION OF GONDWANA ROCKS

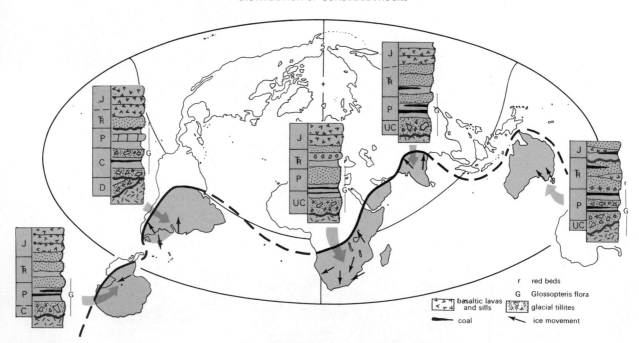

FIGURE 15.12　Lower Gondwana (Dwyka) till, Port St. Johns, South Africa. This, the most famous ancient glacial deposit in the world, formed both on the African craton and in the north edge of the Cape mobile belt. *(Courtesy J. C. Crowell, 1967.)*

fossiliferous marine Jurassic ones, and so the total sedimentary succession spans Carboniferous through Jurassic time. Once Gondwana fossils were dated in terms of the standard marine index fossils of the European standard time scale, they themselves could then be used to correlate among different Gondwana localities soon to be explored on all southern continents (Fig. 15.11).

On all five southern continents the Gondwana succession has prominent glacial tillites in the lower part (Fig. 15.12). These distinctive, unsorted layers overlie scratched surfaces on old rocks abraded by boulders frozen in ice (Fig. 15.1). Many erratic boulders in the tillites also show scratches characteristic of glacial materials (Fig. 15.13). Such deposits in

FIGURE 15.13　Gondwana glacial deposits (Talchirs Tillite), India. Striated cobble of a quartzite whose possible source is only known today 200 kilometers from the resting place of this erratic. *(Courtesy P. K. Ghosh, Geological Survey of India.)*

FIGURE 15.14 Fruits of *Glossopteris* within which seeds matured much as in an apple today. The fruits attached near bases of leaves. Preserved in association with coal seams in South Africa (scale is in centimeters). *(Courtesy E. P. Plumstead; with permission of South African Association for Advancement of Science.)*

India and South Africa were the first pre-Pleistocene tills to be recognized—and only two decades after proposal of the theory of Pleistocene continental glaciation in Europe in 1840. Although some tillites occur within marine sequences, most of them are interstratified with predominantly nonmarine strata containing important coal seams (Fig. 15.11). Associated with the coals are many fossil plants belonging to an assemblage collectively referred to as the *Glossopteris* flora mentioned in Chap. 14, so named for a seed fern which bore elaborate fruits (Fig. 15.14).

The younger Gondwana rocks include chiefly nonmarine sediments, including conspicuous red beds and wind-blown sands. In some areas such as eastern Africa, northern India, and South America, they intertongue with marine deposits. Finally, on all continents except Australia, the Gondwana succession is capped by thick basalt flow-rocks and sills (Fig. 15.11). On the southeastern Australian island of Tasmania, Late Triassic or Jurassic basalt dikes are present, but no extrusive rocks. In South Africa and Antarctica, the Gondwana basalts are of Jurassic age; in South America they are chiefly Early Cretaceous; while in India 3,000 meters of basalt are Late Cretaceous and early Cenozoic in age. All these rocks form immense basalt plateaus deep within cratons where such rocks are not normally encountered. Like Prepaleozoic (p. 177) and Triassic ones in North America and Permian ones in Siberia (p. 300), these basalts are somewhat peculiar, being poor in olivine and sometimes containing minor quartz unlike oceanic basalts. They were fed by countless fissures and now are marked by dikes.

The famous diamond mines of South Africa have been excavated in a very unusual igneous rock called kimberlite (for Kimberley, South Africa), which also occurs in Arkansas, Siberia, Brazil, and Canada. The rock is a magnesium-rich ultramafic species akin to eclogite (see p. 101). It is typified by a brecciated chaotic texture, and contains a great variety of inclusions of rocks through which the pipe-like masses of kimberlite rose as they blasted their way up through the crust. Especially significant are very dense constituents indicative of high pressure, including the diamonds, whose densities are 3.1 to 3.5. These dense materials point to derivation at depths below the crust and rapid transport to the surface, which is borne out by textures suggesting explosive emplacement. In Africa, diamonds occur in Cretaceous sandstones, and some pipes penetrated Late Jurassic rocks; therefore the kimberlites are assumed to have been emplaced in Early Cretaceous time, that is, soon after the formation of the nearby Gondwana basalts. Seemingly they provide further evidence of unusual Mesozoic crustal disturbances.

THE *GLOSSOPTERIS* FLORA

How could a uniform land flora be dispersed across wide oceans? Twenty species of leaves found in Antarctica are common also to India across the equator and far away; yet *Glossopteris* seeds were too large to be windborne. The distribution of the *Glossopteris* flora does not overlap that of the late

FIGURE 15.15 Erect petrified tree stump 32 centimeters in diameter and showing 36 growth rings; found in Gondwana strata on Terrace Ridge, the Ohio Range of Horlick Mountains, Antarctica (a few hundred miles from the South Pole). Together with Mesozoic and Cenozoic plant fossils, they attest to an intermittently milder past climate than now. *(Photo by W. E. Long; furnished by J. M. Schopf.)*

Paleozoic lycopsid flora of North America, Europe, and North Africa (see Fig. 14.43). Because of its intimate association with glacial deposits, the presence of well-developed seasonal growth rings in petrified tree trunks (Figs. 14.44 and 15.15), great abundance of leaves at most fossil localities, the *Glossopteris* assemblage is interpreted to have been a temperate-climate deciduous flora. The lycopsid flora, on the other hand, apparently was a tropical-climate one, which is consistent with paleomagnetic data (Fig. 15.16). The *Glossopteris* flora flourished over wide, swampy low areas on the perimeter of glaciated regions. Interlayering of *Glossopteris*-bearing strata with several tillites indicates considerable climatic fluctuation and implies that the flora was adapted to a wide range of ecologic conditions. The flora persisted into Triassic time long after glaciation had ceased, but by the Jurassic Period, *Glossopteris* had vanished to be replaced by gingkos, cycads, and various new seed ferns, which reflects a Mesozoic

warming of climate as Gondwanaland moved away from the pole and began to break up.

ANIMAL FOSSILS
An unexcelled variety of Permo-Triassic amphibian, reptile (including the best-known mammal-like forms; Fig. 15.17), fish, and invertebrate fossils occur in Gondwana strata. Some fossil leaves even show evidence of having been eaten by insects (Fig. 15.18). A small fresh- or brackish-water reptile called Mesosaurus is unique to South Africa and Brazil and marks the Carboniferous-Permian boundary. Presumably it could not have crossed the wide Atlantic, and

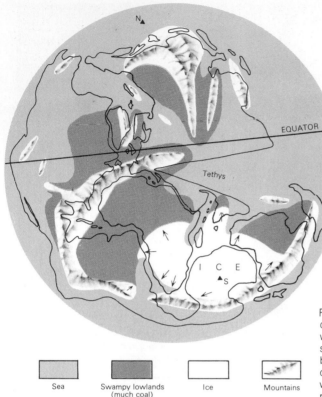

FIGURE 15.16 Late Paleozoic reconstruction of all continents—a compromise of all available data. The fit was recently improved by discovery in deep-sea drilling of submerged continental basement, which fills the hole between South Africa–Antarctica–South America. Ice never covered the entire white area at once. Note epeiric seas, which must have furnished moisture for glaciers. (S—south pole; N—north pole.)

so was taken years ago as additional evidence for existence of the reconstructed Gondwanaland. While many other Gondwana land vertebrates also are common to the southern continents, they have much in common with their *northern* counterparts. Unlike the *Glossopteris* flora, vertebrates clearly could move easily between the northern and southern supercontinents. For example, African Permo-Triassic vertebrates have about as much in common with Eurasian, North American, and Siberian ones as with those of other Gondwana continents. And Jurassic dinosaurs in eastern Africa were identical with those of Wyoming! Clearly land animals managed to disperse north and south more readily than did plants, which perhaps is not surprising in view of the mobility of animals. Their wide distribution also requires that the herbivorous animals were adaptable to a diet of either lycopsid or glossopterid vegetation.

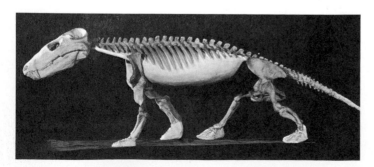

FIGURE 15.17
Cynognathus, a therapsid (mammal-like) reptile from the Karroo series, South Africa. *(Courtesy American Museum of Natural History.)*

FIGURE 15.18 *Glossopteris* leaf apparently chewed by insects; leaves as well as fruits must have provided food for animals. *(Courtesy E. P. Plumstead with permission of South African Association for Advancement of Science.)*

ANTARCTICA—A TRIUMPH OF PREDICTION

Two-thirds of Antarctica (East) is a craton, whereas a youthful mobile belt marks its Pacific (West) margin (Fig. 15.2). An early Paleozoic Trans-Antarctic mobile belt underlies the western edge of the present craton. Since 1909, when Cambrian fossils were discovered at the west edge of the craton, early Paleozoic strata of the Trans-Antarctic belt have been studied intensively. Isotopic dating shows that the belt was deformed near the end of Cambrian time (about 500 million years ago). Granitic batholiths were emplaced and extensive metamorphism occurred as in Australia, Africa, and South America.

Early Paleozoic mountains were leveled by erosion, and then a Devonian epeiric sea flooded the region as in the Brazilian and Australian shields. Devonian marine fossils present have close affinities to those of Australia, South America, and the Cape

belt of South Africa. A very widespread and distinctive sequence of cross-stratified sandstones with important coal seams overlie the Devonian rocks (Figs. 15.11 and 15.19). Important fossil plants were discovered only 100 kilometers from the pole during a 1901–1904 British expedition, and by the ill-fated Scott expedition of 1910–1912. On the return down Beardmore Glacier from the South Pole, Scott's men collected many rock samples, which were found with their frozen bodies at their last camp. Specimens were returned to London and studied eagerly, but the plants were misidentified as lycopsids. Years later, re-examination proved that the fifth southern continent also contained the *Glossopteris* flora! Among the specimens were samples also of basalt described as forming thick, black bands in the sandstones along

FIGURE 15.19 Gondwana strata (Beacon Sandstone) with black basaltic sills (Ferrar Dolerite) exposed along glaciated walls of Finger Mountain near McMurdo Sound at west edge of the East Antarctic craton. *(Courtesy R. F. Black.)*

the walls of Beardmore Valley (Fig. 15.19). Yet another characteristic of the Gondwana rock clan was present.

Naturally, early geologic revelations from the frozen continent were greeted with eager anticipation, especially by Southern Hemisphere geologists and by botanists, who were puzzled by presence of coal and fossil trees at such high, hostile latitudes. What conceivable circumstances could have allowed growth of an ancient land forest within the Antarctic Circle, a region of six months' darkness?

Relatively few new discoveries occurred until after World War II. The International Geophysical Year or I.G.Y. (1957–1958), an imaginative venture in worldwide scientific cooperation, gave impetus to a new wave of scientific exploration on a scale so large that Antarctica appeared in danger of overpopulation. As a result of more extensive explorations, many more Gondwana plant localities from Carboniferous through Triassic ages have been found. Gondwana rocks have recently been found beyond the craton where they were caught in mountain building (Fig. 15.20). Mesozoic basaltic rocks (Ferrar) also have

FIGURE 15.20   Spectacular fold structures in Paleozoic strata, Sentinel Mountains, West Antarctic mobile belt. This area represents westernmost known occurrence of Gondwana rocks; beneath them is a thick marine sequence much as in the South African Cape mobile belt (see Fig. 15.7). *(Courtesy Campbell Craddock.)*

been found over an immense area, although their main concentration is around Beardmore Glacier. Isotopic dating indicates that these igneous rocks are about 160 million years old (Jurassic), or about the same as those on surrounding continents.

Even by 1930, nearly all elements typical of Gondwana sequences elsewhere were known save evidence of glaciation and nonmarine vertebrate fossils. Similarities seemed strong enough that the great South African geologist, A. L. Du Toit, surmised that Antarctica was a full-fledged Gondwana continent, and so it must have been glaciated and probably also had been inhabited by land animals. It was 30 years before his prediction was proven true. During the I.G.Y., New Zealand geologists found a probable till, and in 1960, Ohio State University personnel discovered another example in a complete sequence of Gondwana rock types, including coal and *Glossopteris* remains, only 200 miles from the South Pole. In 1967 an amphibian jaw bone, and in 1969 reptilian bones, both of Triassic age, were found, proving that *cold-blooded land animals as well as forest trees inhabited the continent 210 million years ago!* At least one of the land-dwelling reptiles was identical with Triassic forms long known on other southern continents. The similarity of East Antarctica to the other four Gondwana continents is now firmly established (Fig. 15.11).

CRITICISMS OF DRIFT

Early continental-drift theorizers postulated a single huge polar ice cap, which spread radially outward from the center of Gondwanaland (Fig. 15.6). In Chap. 8 we noted also that the Wegener-Köppen late Paleozoic restoration of the continents placed the north pole in the north Pacific Ocean, which would make it a warmer pole (Fig. 8.11). They drew the equator through the northern (lycopsid) coal belt, which already was considered tropical on paleobotanical grounds (see p. 285). Based upon evaporite deposits and presumed fossil dune sands, midlatitude deserts like today's were postulated for both hemispheres (see Fig. 8.11). Everything seemed to fit, and for drifters the most heartwarming recent development has been the almost perfect agreement of paleomagnetic results for ancient latitude with the old 1920s restorations (compare equators in Figs. 8.11 and 15.16).

The scientist should try to maintain a healthy skepticism, even of his or her own pet hypothesis, and should never stop testing his or her assumptions and prejudices. American geologist A. A. Meyerhoff has pointed out that the total area covered by Gondwana glaciation was three times as large as the North American Pleistocene ice cap. A single landmass so large would have been too dry in its interior for the nourishment of such a huge cap. In Pleistocene time, for example, Siberia was certainly cold enough, but it was too dry to have had a significant ice cap. Meyerhoff questions the entire drift concept for this and other reasons. He believes, instead, that ice caps formed individually upon five always-separate continents that were located where they are today, thus having water surrounding all of them to provide adequate moisture by evaporation to feed the glaciers. In addition, he notes that in southwestern Africa and in Brazil, tillites occupy ancient valleys as much as 1,000 meters deep. This indicates that considerable topographic relief existed and that the glaciers might have formed on high, cool plateaus even if the continents were located at low latitudes just as glaciers are found today on high equatorial mountains in South America, eastern Africa, and New Guinea.

How might one test such criticisms? First, stratigraphic evidence indicates that glaciation of all of Gondwanaland was not simultaneous. It began first in South America in Silurian or Devonian time and continued there into Carboniferous time. It was confined to the Carboniferous in Antarctica and Africa, but spanned part of Carboniferous and part of Permian time in India and Australia (Fig. 15.11). Paleomagnetic evidence confirms also that Gondwanaland was drifting as a unit across the South Pole (Fig. 15.8); thus the centers of glaciation shifted as this occurred. Therefore, we need not postulate a single ice cap much larger than Pleistocene ones. The glacier nourishment problem is further alleviated by the recognition in recent years that epeiric sea embayments existed within Gondwanaland (Fig. 15.16). These areas of water would have been important sources for evaporation of moisture to feed the interior ice cap areas.

Northern Australia provides proof that even the northern fringe of late Paleozoic Gondwanaland really was cold. Permian brachiopod shells from a marine layer just above a tillite have been subjected to a geochemical technique for determining the temperature of the sea water in which the shells were formed.

The analysis involves precise determination of the ratio of two isotopes of oxygen contained in the calcium carbonate of the shells. That ratio is related to the original water temperature, as explained more fully in the next chapter. The results indicate a Permian sea-water temperature of only 7°C, which supports the Gondwanaland restoration. Meanwhile, Permian coral reefs grew in present Indonesia now only 1,200 kilometers to the north; however, we know from paleomagnetic evidence that Australia was about 5,000 kilometers away then.

Another climatic puzzle raised by Meyerhoff is the presence of Permian evaporites near tillites in Bolivia and Brazil. Although these two deposits suggest opposite temperature conditions, there is no reason to suppose that they formed at exactly the same time. Instead they probably reflect marked changes of climate within Permian time, and such changes would neither prove nor disprove continental drift.

The *Glossopteris*-bearing coal deposits in Gondwanaland, like the northern lycopsid-bearing coals, formed largely in broad, coastal swamps as indicated by interstratification with fossiliferous marine and brackish-water strata. Because of their higher latitude, the Gondwana swamps were simply more temperate and more variable in climate than the tropical lycopsid ones. Clearly the glaciers flowed down to sea level and out onto continental shelves at many places, such as the northern Australia locality mentioned above (Fig. 15.16). Presence of apparent warm-temperature indicators such as evaporites need not be alarming, for today tree ferns grow within a mile of glaciers in New Zealand, and ice extends down to areas with apple orchards in Norway. The point is that if glaciers flow from their cool source areas into warm areas more rapidly than melting occurs, they will persist there in spite of warmth.

Some paleontologists have argued that the geographic distributions of certain fossil marine invertebrates and land plants, which have been claimed to be climatic indicators, show a closer parallelism with the present equator than with paleomagnetically indicated ones. Such evidence allegedly discredits both continental drift and polar wandering. Arguments both for and against continental separations based upon either fossil or modern organisms tend to be very ambiguous, however. Apparently the evidence provided by organisms simply does not have the resolving power necessary to provide definite proof or disproof. The single example of penguins living today at the equator around the Galápagos Islands (see Fig. 17.44) simply because they followed the cold Humboldt Current all the way north along the western margin of South America makes one wonder how many of the past biogeographic distribution patterns reflect similar climatic anomalies.

In spite of the fact that the flora and fauna are not very helpful in resolving the history of drifting continents, there are some comparisons that can be made. Certain fresh-water fishes, birds, turtles, frogs, worms, and several insect groups are known only from South America and Africa. It is assumed that they evolved prior to the Cretaceous separation of the two continents and that their similarity therefore supports drift. Large flightless birds such as the ostrich and the flying parrots are known from South America, Africa, Australia, and New Zealand (parrots recently migrated northward to Central America). Their distributions seem also to date from prior to the breakup of Gondwanaland.

EPISODIC PALEOCLIMATIC CHANGES

At least six distinct tillite zones are known in Africa, and more than two dozen have been claimed in Australia. Clearly the glaciers advanced and retreated repeatedly, but it is impossible to tell how many times. Apparently the *Glossopteris* flora was adapted to a variable cool-temperate climatic zone surrounding the ice caps. As the glaciers retreated, the flora expanded poleward, and vice versa. Considering the long time during which large glaciers existed on Gondwanaland (and by analogy with the Pleistocene), we conclude that worldwide sea level must have fluctuated up and down many times as the ice shrank and expanded. All coastlines around the world—even in the tropics—would have experienced transgressions and regressions, and it seems more than coincidental that the most conspicuous repetitive or cyclic sedimentation in the entire geologic record occurs in strata of Carboniferous and Early Permian ages in many parts of the world (see Chap. 14). While late Cenozoic glaciation has lasted about 20 or 30 million years so far, late Paleozoic Gondwana glaciation spanned about 100 million years.

Antarctica shows the most dramatic evidence of climatic changes through time of any place on earth.

Today only lichens, algae, and a few puny grasses (and no truly land animals) grow there, but dense forests with sizable trees (Fig. 15.15) have grown there at several times in the past. Moreover, in Permo-Triassic time, cold-blooded vertebrate animals roamed the landscape only 100 miles from the present South Pole. Clearly the continent had a much milder climate during the past than now, even though glaciers as well as forests also existed there during the late Paleozoic. Major temperature fluctuations characterized late Paleozoic time, as indicated by the presence of cold-blooded animal remains and coal seams interstratified with tillites.

The late Paleozoic polar region suffered much greater oscillations of climate than did the equatorial region, just as in Pleistocene time. Whereas in much of Gondwanaland the tillites alternate with coal and reptile-bearing strata, North America and Europe show no evidence of significant climatic fluctuations—only a gradual long-term increase of aridity through Permian and Triassic times.

THE GREAT GONDWANA SEPARATION

After more than 200 million years of peaceful union, the outbursts of middle Mesozoic basalt forewarned of trouble in Gondwanaland. Before the end of the Jurassic Period, serious separation proceedings were underway, and by the end of the Cretaceous, so many links had been broken that an ultimate complete split became inevitable. Strife and unrest even extended to that northern family, where North America and Eurasia also split up before the end of the Mesozoic Era. Antarctica maintained ties with Australia and South America longer, but even these links had broken by mid-Cenozoic time. Of all the former continental unions begun in Paleozoic time, only that of Europe and Asia has survived to the present. Recently Africa and India have joined with Eurasia, but the fate of this bond is as yet uncertain, and meanwhile there is trouble brewing within Afro-Arabia where the Red Sea rift began forming about 30 million years ago. And so it goes, although drawn mysteriously together many different times, continents seem to have difficulty maintaining stable associations for long periods.

Besides the Mesozoic basalts of eastern North America, northwestern Africa, and parts of interior

Gondwanaland, all of which reflect pulling apart of lithosphere plates along countless fissures to allow mantle material to rise, early-warning signals of impending separation also include the formation of elongate marine embayments along the sites of eventual new continental margins. In Middle Jurassic time, transgression began in eastern Africa and Madagascar, and by Cretaceous time, transgressions were occurring on all sides of Africa, South America, and at least three sides each of India and Australia. Important evaporite deposits discovered by recent exploration for petroleum indicate that most of these embayments did not at first have free access to the open ocean (Fig. 15.21). But the fact that in all cases normal-marine Late Cretaceous or early Cenozoic strata succeeded the evaporites seems to reflect broadening of the basins as the present continents broke apart and new deep-ocean basins such as the Atlantic were formed between them by sea-floor spreading. Block faulting in Late Triassic time in eastern North America, in the Cretaceous in eastern Brazil and in western Africa, and in early Cenozoic time in western India contributed to the early development of these basins. Sea-floor magnetic anomalies (see p. 143), age patterns of deep-sea sediments, and paleomagnetic data from the continents together provide the best clues to the initial opening and subsequent spreading histories of the new ocean basins, and thus of continental separation. Such analysis suggests that North America and Gondwa-

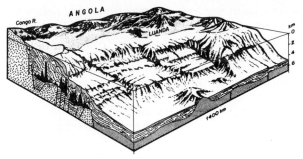

FIGURE 15.21 Diagram of west Africa showing faulted margin resulting from opening of the present South Atlantic basin beginning in Cretaceous time. Evaporites deposited initially in the early Atlantic, when later deeply buried, formed intrusive salt ridges and mounds (domes) shown in black. Similar features are being found beneath the margins of most continents facing the Atlantic and Indian Oceans. *(From Beck, 1972, in* Australian Petroleum Exploration Association Journal, *v. 12, pp. 7–28.)*

naland began separating to form the central Atlantic basin in Late Triassic time. India probably began separating from Africa in Jurassic time, while South America split from Africa in the Cretaceous Period (Fig. 15.22). An Australia-Antarctic unit probably

FIGURE 15.22 Rio de Janeiro harbor and Sugar Loaf Mountain, Brazil. The picturesque rounded stone monoliths that characterize Rio de Janeiro were eroded from early Paleozoic granites very similar to many along the west coast of Africa; all apparently formed during Pan-African orogenesis (e.g., Cape Granite, Fig. 15.5). *(Courtesy Pan American Airways.)*

began separating from its other Gondwana partners in Jurassic or Cretaceous time, but remained together as a kind of siamese-twin continent until early Cenozoic time when Australia split for warmer latitudes. At least a slender connection existed until early or middle Cenozoic time between southernmost South America and West Antarctica. At that time, the formerly continuous Andean orogenic belt was dismembered by large-scale faulting to form the Scotia island arc between the two continents (Fig. 15.2).

SUMMARY

Impressive in the south is evidence of widespread early Paleozoic (Pan-African) mountain building (Fig. 15.22), which is little known in the northern continents. This event produced much land area, followed in middle Paleozoic time by epeiric sea transgressions of Australian, North African, South American, and part of the Antarctic cratons. India, two-thirds of Africa, and probably much of East Antarctica remained land. Plateaus or mountains may have characterized large regions, as indicated by buried ancient, deep valleys. Continental glaciation apparently began earliest in South America (middle Paleozoic), and South Africa (early Carboniferous), and continued latest in Australia (Late Permian) (Fig. 15.11). Glaciation on such a scale probably provided the mechanism for simultaneous cyclic sedimentation in the northern continents as a result of worldwide fluctuations of sea level as ice expanded and contracted. During Triassic and Jurassic times, the southern continents were practically all land. Immense outpourings of basalt accompanied faulting and the formation of diamond-bearing intrusions; in India and South America, eruptions continued into Cretaceous time. Late Jurassic–Cretaceous transgression affected all five continents, but to varying degrees.

The most significant relation among southern continents is the striking similarity of largely nonmarine Paleozoic and Mesozoic rock sequences on all five of the continents now so widely separated by deep seas (Fig. 15.11). Why were there basaltic eruptions and intrusions more or less simultaneously on all continents in cratonic regions where such rocks are considered unusual? How could the *Glossopteris* land flora be dispersed from one continent to another across 5,000 kilometers of salt water? And how could cold-blooded animals and luxuriant forests grow within the Antarctic circle? Even if somehow the climate were warm enough there, could trees adapt their photosynthesis to 6 months of light and 6 months of darkness? And how could continental glaciers develop at low elevations near the present equator (and on both sides of it), especially when simultaneously North America and Europe, now at high latitudes, clearly were enjoying warm, humid conditions and lacked any glaciers? Also, why do several important structural zones seem to vanish at coastlines?

The evidence that prompts the above questions suggests a different past relation among southern continents and between them and the equator and poles, which is supported by paleomagnetic data. It demands large displacements of formerly contiguous land masses—both northern and southern—after early Mesozoic time.

Readings

Adie, R. J., ed., 1972, Symposium on Antarctic geology and solid earth geophysics: Oslo, Universitetforlaget.

Boucot, A. J., Johnson, J. G., and Talent, J. A., 1969, Early Devonian brachiopod zoogeography: Boulder, Geological Society of America Special Paper 119.

Briden, J. C., Smith, A. G., and Sallomy, J. T., 1971, The geomagnetic field in Permo-Triassic time: Geophysical Journal of Royal Astronomical Society, v. 23, pp. 101–117.

Brown, D. A., Campbell, K. S. W., and Crook, K. A. W., 1958, The geological evolution of Australia and New Zealand: London, Pergamon.

Bushnell, V. C., and Craddock, C., 1970, Geologic maps of Antarctica: Antarctic Folio Series, Folio 12, 23 plates; New York, American Geographical Society.

Childs, O. E., and Beebe, B. W., 1963, Backbone of the Americas: Tulsa, American Association of Petroleum Geologists Memoir 2.

Clifford, T. N., 1967, The Damaran episode in the Upper Proterozoic-Lower Paleozoic structural history of southern Africa: New York, Geological Society of America Special Paper 92.

Doumani, G. A., and Long, W. E., 1962, The ancient life of the Antarctic: Scientific American, September, pp. 169–184.

Furon, R., 1963, Geology of Africa: Edinburgh, Oliver & Boyd. (English translation by A. Hallam and L. S. Stevens)

Jenks, W. F., ed., 1956, Handbook of South American

geology: New York, Geological Society of America Memoir 65.

Kummel, B., 1970, History of the earth (2d ed.): San Francisco, Freeman.

McElhinny, M. W., 1973, Palaeomagnetism and plate tectonics: Cambridge, Cambridge.

Meyerhoff, A. A., 1970, Continental drift: implications of paleomagnetic studies, meterology, physical oceanography, and climatology: Journal of Geology, v. 79, pp. 1–51.

Plumstead, E. P., 1962, Fossil floras of Antarctica, Geology 2, in British trans-Antarctic expedition 1955–1958: Science Reports.

Smith, A. G., and Hallam, A., 1970, The fit of the southern continents: Nature, v. 225, pp. 139–144.

United Nations, 1971, Tectonics of Africa: Earth Sciences Series, v. 6, Paris, UNESCO.

van Bemmelen, R. W., 1954, Mountain building: The Hague, Martinus Nijhoff.

Wadia, B. N., 1953, Geology of India (3d ed.): New York, St. Martin's.

16

THE MESOZOIC ERA

AGE OF REPTILES AND CONTINENTAL BREAKUP

Lives of great men all remind us
We can make our lives sublime,
And, departing, leave behind us
Footprints on the sands of time.

Longfellow,
Psalm of Life, (1838)

FIGURE 16.1 Triassic reptile tracks unearthed in 1966 during excavation for a new state building in Hartford, Connecticut. The discovery was so phenomenal that the site for the building was changed and the excavation reserved as a public park (Dinosaur State Park)—a rare yielding by "progress." *(Courtesy John Howard.)*

Marked changes in life occurred after the great extinctions in late Paleozoic time. Between 230 and 65 million years ago, on land, especially, new innovations appeared, including dinosaurs in Late Triassic time, true feathered birds and skinned, flying or gliding reptiles, as well as primitive mammals, in the Jurassic Period. Lycopsid trees declined, while gingkos, cycads, and conifers flourished. Flowering plants appeared in late Mesozoic time, with ancestors of *Magnolia, Eucalyptus, Sassafras,* oak, poplar, fig, and willow prominent in Cretaceous forests. Insects underwent a new burst of evolution at about the same time, suggesting a cause-and-effect relationship with plant evolution. Pollen-transporting insects short-circuited the plant reproductive cycle by their selectivity of flowers. This reduced the gene pool size for specific plants, accelerating evolution and adaptive specialization of both plant and insect. In the seas, too, new faces were evident, and many ecologic niches were vacated by extinction of hordes of Paleozoic invertebrates. Only a few corals had survived, but they diversified and expanded greatly. Molluscs underwent great evolutionary diversification as they adapted to a wide spectrum of habitats. Fish continued to flourish, but were challenged by ammonoids and a host of swimming reptiles that readapted to the aquatic habitat.

In 1822 an English physician and amateur paleontologist, Gideon Mantell, called upon a rural patient in Sussex. His wife discovered some large Jurassic bones in the yard while she was waiting. After considerable study, a skeleton was assembled and named *Iguanodon.* The discovery aroused much attention, but in spite of its fame, the zoological affinity of *Iguanodon* was not recognized quickly. None other than Georges Cuvier, leading comparative anatomist, finally guessed that it was a member of an extinct group of huge reptiles, later named dinosaurs (thunder lizards). Since then, dinosaurs have stirred the human imagination more than any other geologic phenomenon. Indeed, it is probable that their bones, together with those of giant Pleistocene mammoths, inspired early myths about giants. Today their powers of stimulation still are great, as evidenced by browsing in any well-stocked children's library. Finds over the years, such as fossil dinosaur eggs, polished stones thought to represent gizzard gravel, and, most recently, petrified dinosaur stomach contents, all have served to prevent extinction of interest in ancient monsters.

In 1835, only 13 years after the first discovery of the bones of *Iguanodon,* yet still 7 years before the term "dinosaur" was coined, giant footprints were found on Upper Triassic flagstones being laid on the streets of Greenfield, Massachusetts. The tracks immediately drew attention and triggered a rash of speculation and fantasy. Discovery of bones of the barely extinct giant Moa bird in New Zealand lent credence to the most popular belief that the tracks were made by some overnourished Triassic fowl (as yet, no one knew that birds appeared later). Dramatic support for the existence of such creatures was provided at a scientific meeting by famous English theologian-geologist W. E. Buckland, who "exhibited himself as a cock on the edge of a muddy pond, making impressions by lifting one leg after another." So lively was debate over footprints that it provided a great stimulus to American interest in fossils. By the end of the nineteenth century, a dozen localities in the Connecticut Valley had yielded Triassic bones that clearly were reptilian, including some of the earliest dinosaurs. The controversy was settled, with birds the

FIGURE 16.2 Map restoring past paleomagnetic south pole positions for all continents except Asia.
Medium-weight dashed arrows trace pole migrations for Gondwana and Euramerica in early Paleozoic; S-CI records their collision in middle and late Paleozoic time. Heavy arrow traces pole migration for all of Pangaea until breakup in early Mesozoic time (M). Light dashed arrows record separation of the different continents since then. *(After McElhinny, 1973,* Palaeomagnetism and plate tectonics, Cambridge; *used with permission.)*

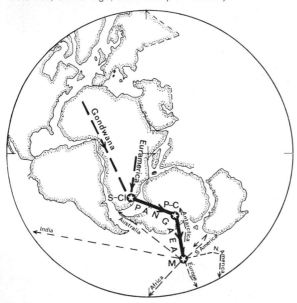

losers, though the oldest true fossil birds (Jurassic) soon were to be discovered in Germany. What was a lively controversy is now so long forgotten that reminiscent vestiges are to be found only in nineteenth century literature, for example the famous passage quoted above.

Collection and study of fossil vertebrate animals have given rise to some of the best romantic lore in the history of geology. The most colorful period was around the turn of the twentieth century when exploration of the American West was revealing untold wealth of virgin fossil localities, many of which yielded hitherto completely unknown ancient monsters. The men who collected them were at least as colorful as their finds. It was a heroic drama on the American geologic stage, and the stars were cast well for their roles.

Two of the most flamboyant participants were E. D. Cope and O. C. Marsh, who scoured arid western hills for reptile and mammal bones. At first, these two pioneers were harrassed only by occasional unfriendly Indians, but later—and far more dangerously—by each other in a jealous feud that eclipsed even the earlier Sedgwick-Murchison altercation in Wales. They falsified collecting locations to keep them secret, and telegraphed descriptions of newly discovered skeletons back east in order to beat the other to publication. Cope even bought a scientific journal in order to publish his articles uninhibited, and to have a ready medium for editorial diatribes against Marsh and the U.S. Geological Survey.

BREAKUP OF EUROPE AND NORTH AMERICA

At the end of the last chapter we reviewed the breakup history for the Gondwanaland supercontinent. Now we must do the same for the northern supercontinent Laurasia. Figure 16.2 shows the positions of the South Pole *relative to the continents* for the past 500 million years based upon paleomagnetic data from all continents except Asia. This map conveniently summarizes paleomagnetic evidence both for the times of supercontinent assembly and for breakup. Formation of the supercontinent Pangaea by the collision of Laurasia (or Euramerica) with Gondwanaland is shown by the convergence of their respective apparent polar wandering paths in mid-Paleozoic time (the S-C1 pole position of Fig. 16.2). The coincidence of pole paths forming the heaviest line reflects the existence of Pangaea moving as a single unit until the early Mesozoic (M pole position), and its path shows the relative movement of the pole across Gondwanaland during late Paleozoic time as discussed in the previous chapter in connection with glaciation. Note that the cessation of glaciation in Late Permian time coincided with the relative movement of the south pole to a warmer oceanic position. The light dashed arrows reflect the breakup and divergent relative motions of the separate continents going their own ways (but beginning at different times) in the Mesozoic.

We already have noted that Late Triassic faulting and basaltic igneous activity in Europe, North America, and northwestern Africa forewarned the opening of the central Atlantic basin. Ocean-floor magnetic anomalies and deep-sea-sediment ages indicate that actual opening of the central Atlantic between North America and Africa began in Early Jurassic time (about 180 m.y. ago). On the other hand, magnetic anomalies indicate that Europe did not separate from Greenland until the beginning of the Cenozoic Era (65 m.y. ago), which is consistent with the presence of widespread basalts of that age in northern Britain (including Ireland's famous Devil's Causeway) and southern Greenland. Greenland, however, began separating from the North American mainland in latest Cretaceous time (about 80 m.y. ago). Small Jurassic-Cretaceous igneous bodies, which include some eroded volcanic necks in New Hampshire, near Montreal, Quebec, in Newfoundland, and a series of now-submerged volcanic seamounts extending east from New England on the Atlantic sea floor, apparently formed along fractures activated during the early opening of the ocean basin. We see, therefore, that the breakup of Laurasia like that of Gondwana had a complex history, which was anything but synchronous.

The relationship between Europe and Africa in early Mesozoic time was very different than today's. Part of Iran may have been attached to Arabia, and Turkey and Greece apparently were part of North Africa. Italy, Sardinia, and Corsica were all attached to southern France and eastern Spain (Fig. 16.3). Northwestern Africa and Spain remained more or less attached until Jurassic time, but an oceanic basin called the Tethys lay between Mesozoic Eurasia and

the remainder of Afro-Arabia and India (Fig. 16.3). The present Mediterranean Sea has long been considered a remnant of the old Tethyan Ocean basin, which shrank as India and Arabia collided with Asia. As we shall see, however, the Mediterranean was formed instead by spreading behind the Italy-Yugoslavia-Greece-Turkey blocks as they were torn from Africa and driven north to collide with Europe. The floors of the Black and Caspian Seas may be the only remnants of the original Tethys that have not been crushed and shredded by continent-continent collisions. Cyprus apparently is a portion of an old Tethyan ocean ridge squeezed and elevated during Cenozoic motions of Africa toward Europe.

The Tethys seaway, extending from southern Europe and North Africa all the way east to present Indonesia, was originally recognized and named because of a remarkable homogeneity of Mesozoic marine faunas. The rich assemblages include a greater diversity of species of corals, sponges, ammonoids, and other molluscs than to the north or south. Some forms, such as stromatoporoids, the exotic rudistid clams (see p. 366), and certain microfossils, were confined to the Tethyan region. All these fossils occur within a vast expanse of thick carbonate rocks, which were deposited in shallow shelf environments on the margins of the supercontinents facing the Tethys. (The prolific oil fields of Arabia produce petroleum from Tethyan limestones of Jurassic age.) The diverse marine fauna, the presence of widely scattered evaporites, and the types of fossil plants that lived on adjacent lands all point to a warm climate; apparently the entire Tethyan realm lay in the subtropics and tropics.

The shelves on both sides of the Tethys ocean basin were originally connected (Triassic-Jurassic) at the west end through Spain and northwestern Africa as well as over the *restored* positions of Yugoslavia, Greece, Italy, and Turkey as shown in Fig. 16.3. At least by Jurassic time, there was also a shallow marine connection westward to the present region of the northern Gulf of Mexico and Central America. These interconnections of shallow seas allowed the wide migration of a distinctive Tethyan fauna. Deep, wide oceans pose almost insurmountable barriers to the dispersal of most shelf faunas; bottom-dwelling molluscs in particular have a larval stage too short to survive long transatlantic voyages. Therefore, it does not seem coincidental that early Mesozoic Tethyan

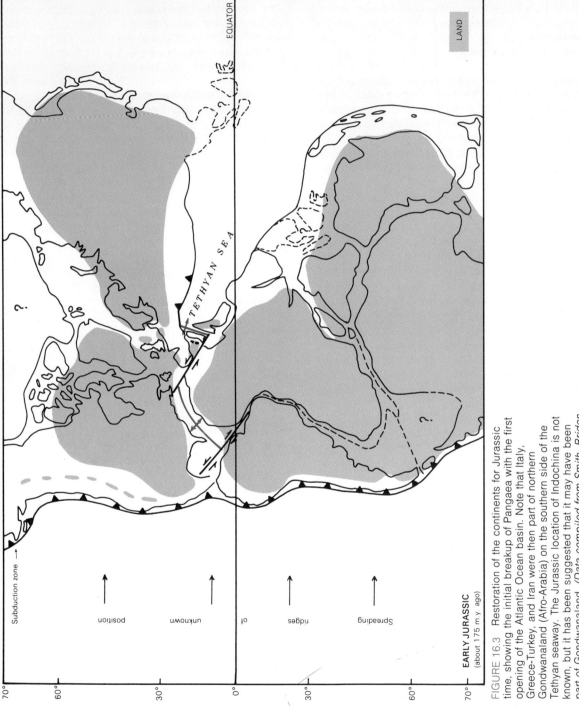

LAND

EQUATOR

TETHYAN SEA

Subduction zone →

Spreading ridges of unknown position

EARLY JURASSIC
(about 175 m.y. ago)

70° 60° 30° 0° 30° 60° 70°

FIGURE 16.3 Restoration of the continents for Jurassic time, showing the initial breakup of Pangaea with the first opening of the Atlantic Ocean basin. Note that Italy, Greece-Turkey, and Iran were then part of northern Gondwanaland (Afro-Arabia) on the southern side of the Tethyan seaway. The Jurassic location of Indochina is not known, but it has been suggested that it may have been part of Gondwanaland. *(Data compiled from Smith, Briden, and Drewry, 1973, Phanerozoic world maps, Palaeontological Association; Smith, 1971, Geological Society of American Bulletin; Dewey et al., 1973, Bulletin of the Geological Society of America; Larson and Chase, 1972, Bulletin of the Geological Society of America.)*

faunas are very homogeneous from southern North America eastward to southern Asia. As Africa and North America separated, however, marine faunas became more distinctive, apparently due to a widening deep Atlantic barrier.

As Laurasia became restive and began to break up, Triassic volcanism not only occurred in eastern North America and northwestern Africa, but also on the northern (European) side of the Tethys. In Early Jurassic time, when the central Atlantic had actually begun opening, the western Tethys also became fragmented. Because Europe remained coupled with northern North America, a major transform-fault system must have developed between Africa and Europe. Apparently it passed through northwestern Africa and southern Spain, perhaps slicing that region into several elongate blocks (Fig. 16.3). A spreading center or centers apparently formed simultaneously in the western Tethys and began to break up the former continental shelf there (Fig. 16.4). First,

Yugoslavia, Greece, and Turkey (and perhaps also Iran) were torn from the shelf to begin their eastward journeys as microcontinents. Soon the Italian block was also torn away and sent eastward. These fragments in their turn were transported northeastward to collide eventually with southern Eurasia (Fig. 16.5). The culmination of these collisions occurred in mid-Cenozoic time with the great Alpine-Himalayan orogenic upheaval resulting from the collisions of Afro-Arabia and India with Eurasia, to be discussed in Chap. 17.

FIGURE 16.4 Diagrammatic evolution of the north African margin of the Tethys Seaway during Jurassic time. Note progressive fragmentation of this edge of Gondwanaland as sea-floor spreading broke up the Tethyan margin. Former shallow-water carbonate rocks became downfaulted, and deep-water Jurassic sediments were deposited in the troughs. *(From Bernoulli and Jenkyns, 1974, in R. H. Dott, Jr., and R. H. Shaver, eds.,* Modern and ancient geosynclinal sedimentation; *used with permission of Society of Economic Paleontologists and Mineralogists.)*

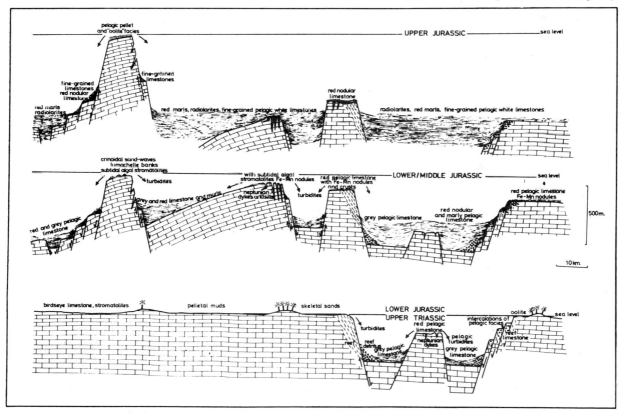

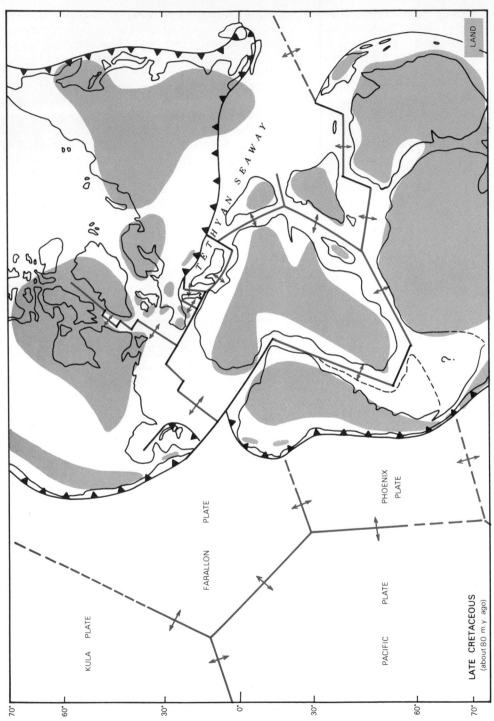

LATE CRETACEOUS
(about 80 m.y ago)

FIGURE 16.5   Restoration of the continents for Late Cretaceous time. Note the further breakup of former supercontinents, including the separation of Greenland from mainland North America and of Italy, Greece-Turkey, and Iran from Gondwanaland. (Data from same sources as Fig. 16.3.)

During late Mesozoic time, several new small-ocean basins formed in the western Tethys by the rifting of the microcontinental blocks from the main continents. Deep-water sediments were deposited in them—the very same deep-water sediments originally regarded a century ago by European geologists as typical geosynclinal strata (see Chap. 8). These sediments were deposited upon ophiolite igneous sequences considered by most geologists now as having formed originally at several spreading sites in the new micro-ocean basins within the Mesozoic Tethys. Some of the small-ocean basins were closed through subduction followed by collision between continental fragments as suddenly as they had started opening—just like so many ice cubes jostling about in a punch bowl.

The Mesozoic history of the Tethys was so complex as to be perhaps forever indecipherable in a detailed way. To some extent, this complexity must be due to the fact that such a large mass of continental crust was fragmented into so many microcontinents by transform faulting and that local sea-floor spreading then shifted those continental fragments to and fro as overall Atlantic spreading directions and rates changed.

GENERAL SETTING OF MESOZOIC NORTH AMERICA

Although Mesozoic life was greatly changed, sedimentation and tectonism in North America were no different from the preceding Permian Period. Marine deposition was continuous in much of the Cordillera, and red sediments continued to accumulate over much of the craton. In the Appalachian belt, as we already have seen, structural disturbances continued through the Triassic Period. In the Cordilleran region, disturbances were intensifying after the relatively mild Antler Orogeny. Extreme volcanism broke out late in the Permian and continued in the Triassic Periods all up and down the western part of the Cordilleran belt from southern Alaska to Central America (Figs. 14.17 and 16.6). Resulting islands represented a clear parallel with modern volcanic island arcs. Immediately following this outburst, mountain building occurred in many parts of the western Cordillera. This event (Cassiar Orogeny) was followed by relative tectonic quiescence until whole-sale upheaval of the entire belt began in late Mesozoic time and continued into the early Cenozoic Era. The culminating event, for simplicity called the Cordilleran Orogeny, terminated subsidence and marine deposition in most of the belt. In Cretaceous and Cenozoic time, patterns of subsidence and sedimentation changed; a long, deeply subsided zone in the present Rocky Mountains developed just east of the former subsiding belt and *within the edge of the craton*! Farther west, complex local basins and mountainous blocks dotted the Cordillera itself; near the Pacific Coast, marine embayments persisted until late Cenozoic time. In the Arctic, marine Cretaceous sedimentation occurred in northern Alaska and over the northern Franklin belt of Canada.

PERMO-TRIASSIC OF THE CORDILLERAN BELT

Late Permian volcanism reflected onset of profound structural disturbances that culminated in the Cassiar Orogeny, named for a mining district in central British Columbia, Canada. From Oregon to Alaska, Lower and Middle Permian rocks were deformed and intruded both by granitic and ultramafic bodies, and are now overlain unconformably by Middle and Upper Triassic strata. Upper Triassic rocks contain thick conglomerates with many pebbles of Permian and older types. Therefore, we conclude that a large, mountainous tectonic land or, more likely, a somewhat discontinuous series of islands occupied the western part of the mobile belt. A gulf lay between it and the craton, which was a broad, low land (Fig. 16.7).

A Permian volcanic belt was recognized in the western Cordillera in the 1940s, and evidence of the Cassiar Orogeny was recognized in the 1950s. In 1970 a plate-tectonic interpretation was proposed in which the Permo-Triassic volcanic belt is seen as a

On the following pages:

FIGURE 16.6 Upper Triassic sedimentary facies. Environments and rock types were extremely variable in the western Cordillera, where up to 7,000 meters of Triassic strata are known. Note that much limestone formed in the volcanic zone of the Cordilleran belt. (See Fig. 11.15 for symbols and sources.)

FIGURE 16.7 Late Triassic paleogeography.

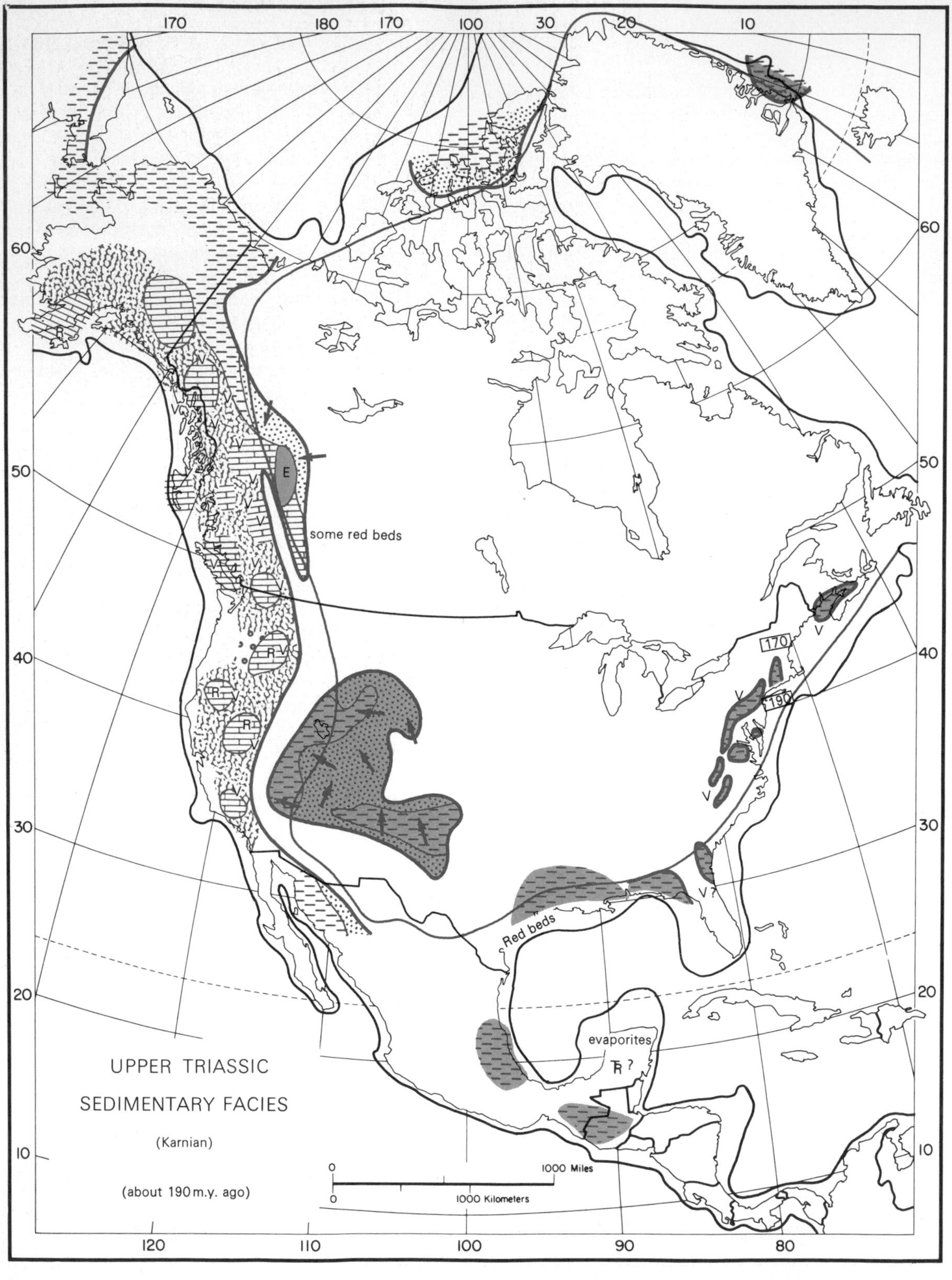
170
180
170
100
30
20
10
60
50
40
30
20
10
some red beds
Red beds
evaporites
Ʀ ?
170
190
0
1000 Miles
0
1000 Kilometers
120
110
100
90
80

UPPER TRIASSIC SEDIMENTARY FACIES

(Karnian)

(about 190 m.y. ago)

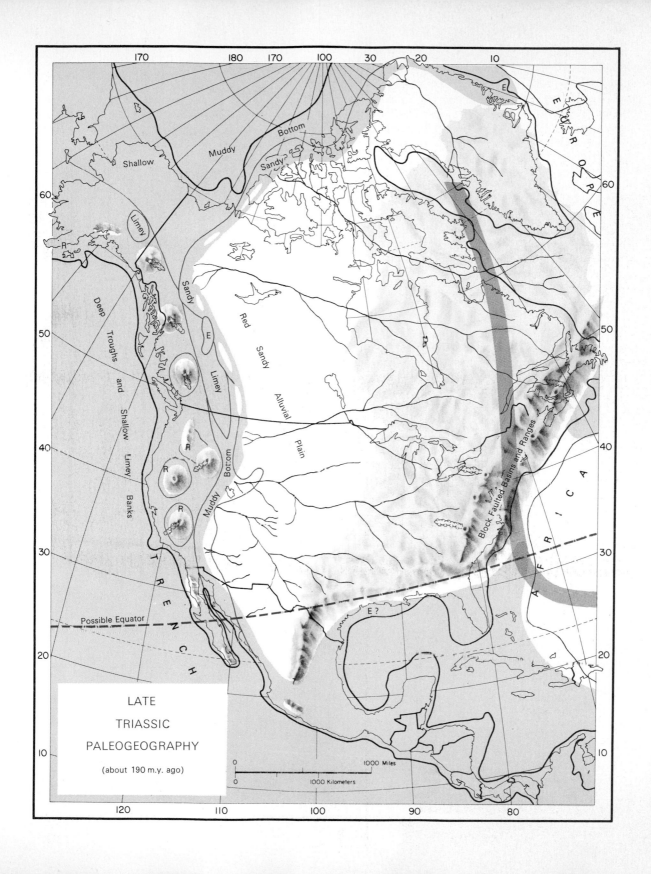

LATE

TRIASSIC

PALEOGEOGRAPHY

(about 190 m.y. ago)

Japan-like island arc separated from the mainland by a small oceanic basin similar to those behind all the western Pacific arcs (see Fig. 15.2). Ultramafic rocks and basalts conspicuous in central British Columbia (Fig. 16.8) and extending at least as far south as Oregon are regarded as remnants of the oceanic crust and mantle of that basin (in other words, oceanic ophiolites). Thick, late Paleozoic carbonate rocks associated with the basalts represent reef and bank deposits that accumulated around volcanic islands and shallow shoals as occur today in tropical volcanic island areas (see Chap. 9). On either side of the central oceanic rocks today are found late Paleozoic andesitic volcanic rocks interpreted as remnants of volcanic arcs formed on either side of the central ocean as a result of symmetrical subduction of oceanic crust (Fig. 16.8). Blueschists reflecting high pressure but relatively low-temperature metamorphic conditions—now considered characteristic of subduction zones—formed at several places in the central British Columbia ophiolite zone.

The Cassiar Orogeny is now interpreted as the result of closing of the central oceanic belt in Late Permian and Early Triassic time by complete subduction of the oceanic lithosphere and the jamming of the western island arc against the eastern arc at the edge of the continent. An alternate hypothesis, which

seems less likely to us, suggests that the western arc, rather than moving directly eastward, moved laterally from far to the south in the United States to its present position.

Cassiar tectonic lands were short-lived, for by Late Triassic time marine sedimentation renewed almost everywhere in the Cordilleran belt. The sediments are varied, with carbonates, in part reefoid, being especially prominent (Fig. 16.6). Volcanism continued in many local centers, and many small volcanic islands existed with fringing reefs growing around them (Figs. 16.7 and 16.9). Thick successions of conglomerates, mudstones, and very heterogenous sandstones (graywacke) containing volcanic detritus also were deposited (Fig. 16.10). Many of the sandstones accumulated in relatively deep water from turbidity currents. The mobile belt as a whole was subsiding profoundly, producing deep-water areas at the same time that rapid extrusion of lavas and buildup of carbonate banks and reefs by organic activity produced islands and shoals.

After the Cassiar Orogeny, simple subduction of a Mesozoic Pacific oceanic lithosphere plate beneath the western edge of North America occurred. A single deep trench and an associated magmatic arc system with almost continuous volcanism occupied the western edge of the continent from Late Triassic onward.

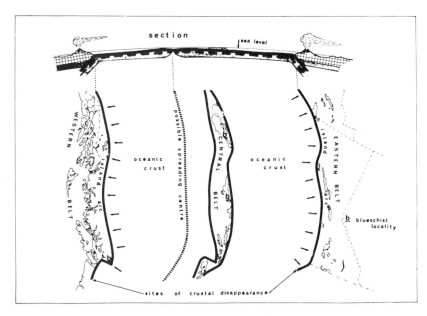

FIGURE 16.8 Restoration of western Canada for late Paleozoic time, showing a plate tectonic interpretation for upper Paleozoic volcanic rocks, blueschists, and ultramafics. Basalts suggest a marginal ocean basin in the center with andesitic arcs on either side. Presumably the basin closed during the Permo-Triassic Cassiar orogeny. Similar arcs and marginal ocean basins are postulated elsewhere in the Cordillera. (From Monger and Ross, 1971, Canadian Journal of Earth Sciences v. 8, p. 259–278; used by permission.)

FIGURE 16.9 Chitistone Valley, Alaska, showing a large syncline with Permian volcanic rocks overlain by Upper Triassic massive limestone and shale with thin limestones; glacier-shrouded peaks in background include Pleistocene volcanoes. *(Courtesy Gene L. LaBerge.)*

The above interpretation of British Columbia illustrates the great significance of different volcanic rock types. The exclusively basaltic rocks with abundant pillow structures generally indicate an oceanic setting, especially where they overlie ultramafic and gabbroic rocks to constitute an ophiolite sequence. Andesitic rocks, with a silica content intermediate between that of granite or rhyolite and that of basalt are characteristic of magmatic arcs (Table 16.1). In the past, andesites were believed to originate through melting of older granitic continental crust, but now it appears that they form instead by partial or selective melting of silica and potassium from the upper mantle in zones of deep earthquakes beneath the arcs (see Fig. 9.16). If this is true, then the andesitic eruptions represent new material of essentially continental

FIGURE 16.10 Microscopic photograph of an immature sandstone composed of andesitic rock fragments such as is typical of Mesozoic and Cenozoic strata deposited along the Pacific margin of the Cordilleran belt in proximity to active volcanic arcs. (Coaledo Formation, Coos Bay, Oregon; average grain size is 0.5 millimeters.)

TABLE 16.1 Comparison of chemical compositions of granitic rocks* with sedimentary and andesitic ones†

Rock type	Composition, %										
	SiO_2	TiO_2	Al_2O_3	Fe_2O_3	FeO	MnO	MgO	CaO	Na_2O	K_2O	H_2O
Granite	70.5	0.4	14.1	0.9	2.4	0.06	0.6	1.6	3.6	5.4	0.5
Quartz diorite	63.2	0.6	17.7	1.8	3.2	0.1	1.9	4.8	4.2	1.9	0.6
Average Middle Pre-paleozoic sediments	65.2	——	14.1	1.7	2.9	——	2.3	3.1	2.8	2.6	——
Average Post-Pre-paleozoic sediments	58.8	——	13.6	3.5	2.1	——	2.7	6.0	1.2	2.9	——
Andesitic lava	60.1	0.5	17.8	2.0	3.4	——	3.5	6.3	4.2	1.3	0.3
Basaltic lava	48.4	2.7	13.2	2.4	9.1	0.1	9.7	10.3	2.4	0.6	——

*True granites are much less common than diorites, which have less quartz and potassium feldspar. Therefore, the broader term granitic rocks is used to include the whole assemblage of coarse, crystalline-textured, light-colored, silica-rich igneous rocks.
†After Engel, 1963; Turner and Verhoogen, 1960.

composition extracted from the mantle to provide potential additions (accretions) to continents.

TRIASSIC OF THE WESTERN CRATON

Red beds continued to form widely over the craton as well as in the Appalachian and East Greenland mobile belts during the Triassic Period. Continental land area remained very large even though average elevation was reduced slightly. Mountains raised earlier in Colorado and New Mexico were lower and became partially buried by nonmarine Triassic sediments (Fig. 16.7). Sediments were transported by rivers flowing westward across an immense alluvial plain to the Cordilleran Sea; red beds accumulated on the plain. Forests covered at least the highlands, as evidenced by petrified logs (Fig. 16.11). Triassic red beds also have yielded important animal fossils, including remains of amphibians, dinosaurs, and other reptiles. Fresh-water molluscs and fish point to some swamp and river environments.

FIGURE 16.11 Late Triassic tree trunks at Petrified Forest National Park, Arizona (Chinle Formation). Large coniferous trees were transported by floods from uplands, then the wood tissue was replaced by silica and, occasionally, by uranium minerals. The same strata also contain cycads and tropical ferns.

Along the western margin of the craton, Triassic red sediments grade into typical marine gray shales and limestones of the eastern part of the Cordilleran mobile belt. Complex facies variations as well as an abrupt westward thickening of Triassic strata characterize the cratonic margin. It was long assumed that the widespread red beds to the east in the entire Rocky Mountain region were nonmarine deposits. But in Wyoming and Utah, thin marine limestone and gypsum layers are interstratified in the red beds, forming tongues of western facies penetrating eastern red strata. Marine conditions existed, at least intermittently, on the western craton where periodic transgressions extended at least 600 kilometers eastward; a considerable amount of the associated red strata probably represents marine lagoonal and tidal flat deposits.

JURASSIC RETURN OF THE EPEIRIC SEA

As the Triassic Period drew to a close, sedimentation changed markedly over western North America. A vast blanketing mass (approximately 40,000 cubic kilometers) of very well-sorted sand was deposited along the entire west edge of the craton in the United States (Fig 16.12). The Late Triassic and Early Jurassic sands consist of 90 percent quartz. They are strikingly similar to the widespread, mature lower Paleozoic quartz sandstones of the craton (see Chaps. 11 and 12.)

Where did so much sand come from? Relative purity and roundness of the grains point to derivation from older sandy sediments through recycling. With much sandstone present in upper Paleozoic and Triassic strata of the western craton, there is no problem of designating potential source rocks. Paleocurrent data indicate general southerly transport of the sand (Fig. 16.12). This points to derivation from the north, chiefly from the craton in Canada where a widespread pre-Middle Jurassic unconformity indicates that upper Paleozoic and Triassic strata indeed were being eroded. Derivation from that region would be consistent with the general pattern of westward tilting of the entire craton accompanied by westward shift of loci of most active sedimentary accumulation and attendant gradual westward stripping of strata that formerly covered most of Canada (see Fig. 14.4).

Large-scale cross stratification is a famous trait of the Jurassic sandstones in southern Utah (Fig. 16.13). This spectacular feature was attributed to preservation of lee faces of dunes thought possible only through wind transport. Thin carbonate rocks with marine fossils are interstratified locally with the upper sandstones, and the formation is overlain by widespread limestone, shale, and evaporite deposits, therefore at least some of it clearly is aqueous. The problem is to determine proportions of wind versus aqueous sand. A wind origin for all the sand was proposed years ago when wind-formed dunes provided the only known modern analogue, but in recent years large underwater dunes have been discovered in shallow seas (see Fig. 11.26). We believe that sand derived from the craton was transported south parallel to an oscillating eastern shoreline as is the rule today on shallow continental shelves. Some of the sand apparently was deposited on the shelf, but onshore winds also produced coastal dunes, upon which dinosaurs left telltale footprints.

Beginning in Middle Jurassic time, the sea advanced widely over the western craton from the present Arctic and northern Pacific regions. Quartz sand deposition was followed by conditions in the northern Rocky Mountain–Canadian Plains region like Paleozoic epeiric seas. Late Jurassic time marked the last significant carbonate and evaporite deposition anywhere on the craton. Middle and Upper Jurassic strata lap across a widespread major unconformity and are younger toward the center of the craton. As transgression proceeded, normal marine conditions with better circulation prevailed, and evaporite deposition practically ceased in the west, though it continued in the Gulf of Mexico region (Fig. 16.14). Some workers believe that Jurassic salt was deposited across the entire present Gulf of Mexico, which was just beginning to form as a new, small-ocean basin through the separation of Gondwanaland from North America.

Nearly every type of epeiric sea deposit discussed before formed in the Late Jurassic sea. Limestones rich in fossil fragments, oölites, and algal material, fossiliferous shales, and cross-stratified glauconitic sandstones all are prominent, but complexity of the facies variations almost defies analysis.

Most of the Appalachian belt was above sea level and being actively eroded at this time, though areas around the present Gulf of Mexico experienced im-

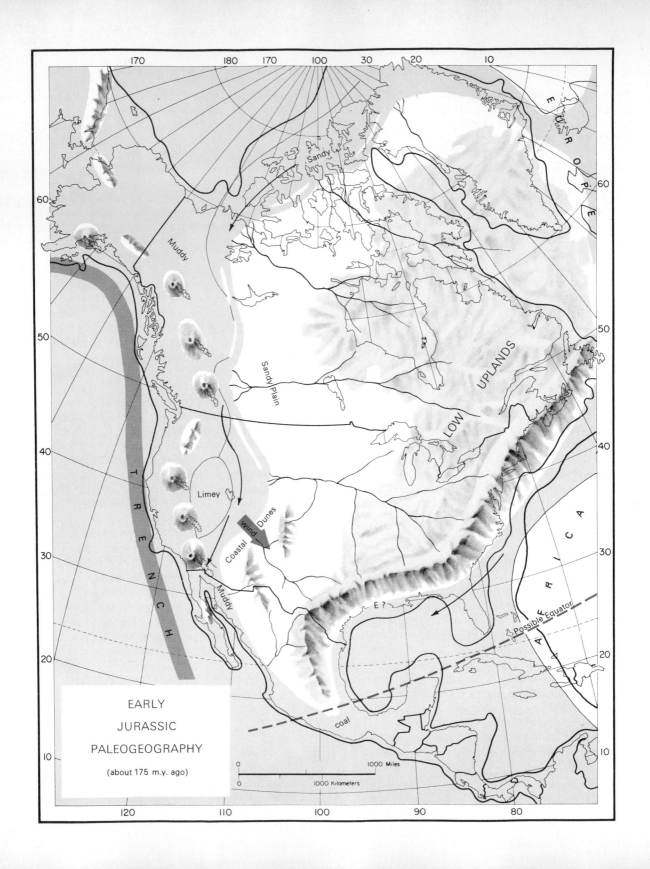

EARLY

JURASSIC

PALEOGEOGRAPHY

(about 175 m.y. ago)

FIGURE 16.13 Large-scale wedge festoon (above) and planar-parallel (below) cross stratification in Navajo Sandstone, East Rim Trail, Zion Park, Utah. The upper type has been interpreted as wind dunes, whereas the lower suggests aqueous deposition. In reality, neither type is diagnostic of environment. *(Courtesy William M. Jordan.)*

mersion. Extensive, thick carbonate rocks as well as evaporites formed in Texas, Mexico, and Cuba. The epeiric sea almost certainly joined with waters encroaching northward from the present Gulf of Mexico region (Fig. 16.15) to produce a more or less continuous seaway stretching to the present Arctic. But there must have been some ecologic barrier to marine animal migration, for Jurassic faunas of the Pacific Coast and Canadian Rocky Mountain regions have affinities with Asian ones, and differ from those of the Gulf region, which were more like those of the Tethys seaway in southern Europe and Africa.

A DINOSAUR GRAVEYARD

Latest Jurassic deposits of the Rocky Mountain region form a famous nonmarine sequence called the Morrison Formation. It is best known for its treasure of dinosaur skeletons. The strata also contain some invertebrate and plant fossils, which, together with the dinosaurs, prove that it was a nonmarine sequence.

FIGURE 16.12 Early Jurassic paleogeography; note slight encroachment of the sea along the western craton with deposition of Navajo quartz sands in and next to the large embayment. Apparently that region lay in the Westerly Wind belt.

The Morrison is made up of a varied assemblage of pastel-colored shales, sand, and rare conglomerate (Fig. 16.16). Volcanic eruptions were still important farther west (Fig. 16.17), and considerable ash blown from that region occurs in the Morrison. Though plant fossils and coal are rare, the very large, herbivorous-dinosaur skeletons indicate that vegetation was abundant; oxidation destroyed the plants almost completely. This reasoning points to a moderately humid climate. Particularly rich concentrations of fossil dinosaur skeletons in a few localities, such as Dinosaur National Monument, Utah, suggest local watering places where beasts gathered and became mired.

Morrison sediments were deposited as the short-lived Jurassic epeiric sea retreated northward and river and swamp deposits pushed forward to form

On the following pages:

FIGURE 16.14 Upper Jurassic sedimentary facies. Patterns in the Rocky Mountain region are simplified here due to many rapid shifts of sediment types. Evaporites in Iowa long were considered Permian, but fossil pollen shows them to be Mesozoic. (See Fig. 11.15 for symbols and sources.)

FIGURE 16.15 Late Jurassic paleogeography. Transgression of the western craton is shown as more extensive than thought previously; note also the importance of volcanism in the western Cordilleran belt.

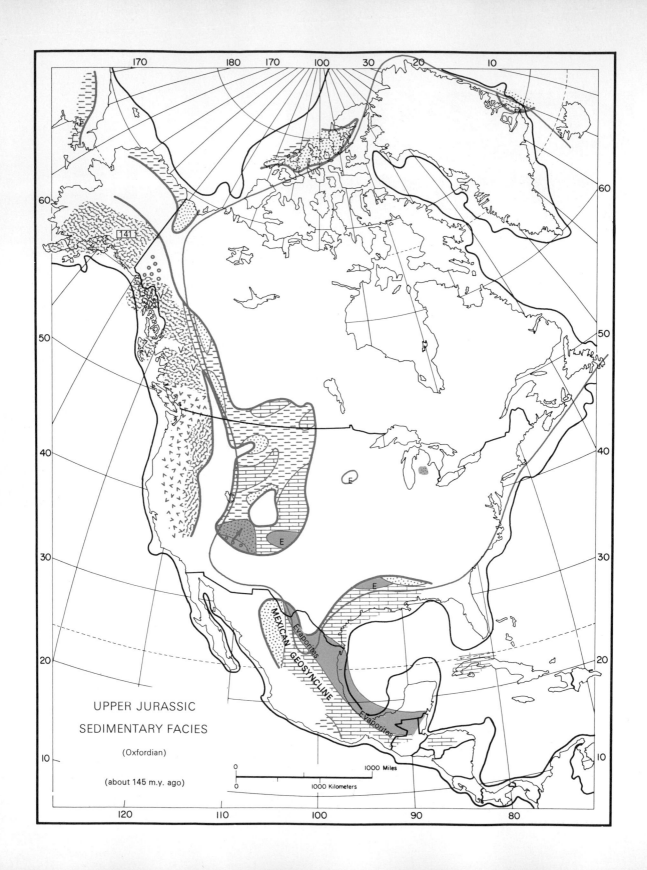

UPPER JURASSIC

SEDIMENTARY FACIES

(Oxfordian)

(about 145 m.y. ago)

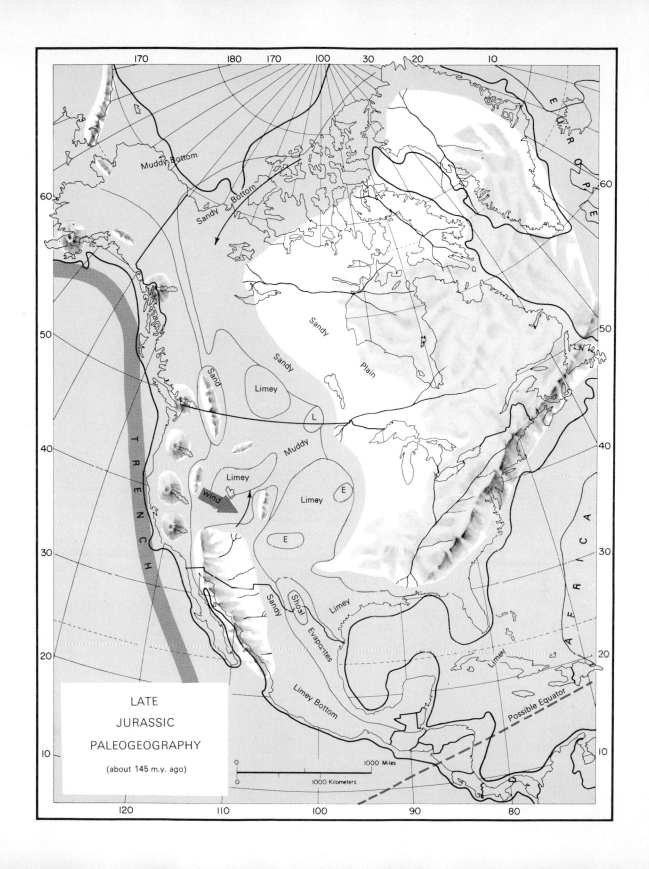

LATE

JURASSIC

PALEOGEOGRAPHY

(about 145 m.y. ago)

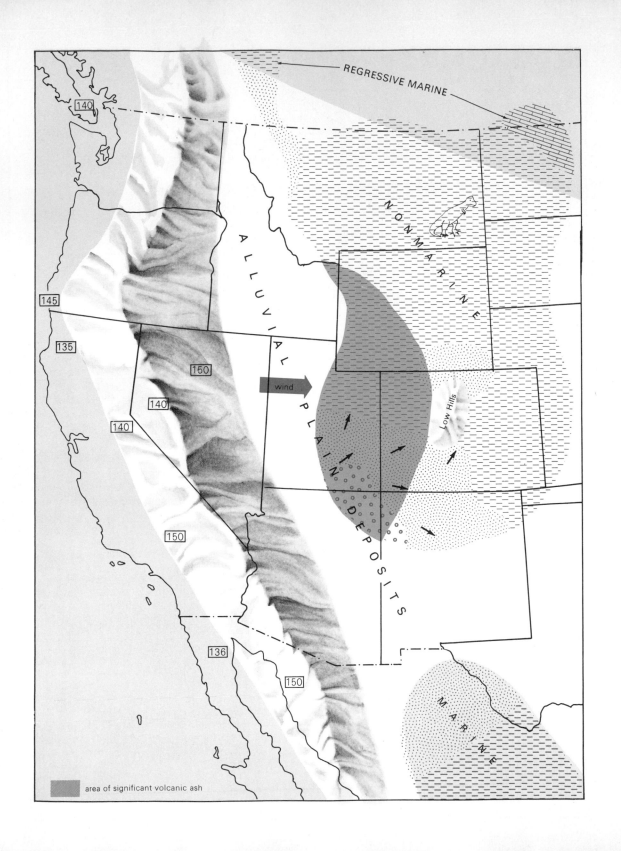

REGRESSIVE MARINE

N O N M A R I N E

A L L U V I A L   P L A I N   D E P O S I T S

wind

Low Hills

M A R I N E

140

145

135

150

140

140

150

136

150

150

area of significant volcanic ash

FIGURE 16.17 Ellipsoidal structures in Late Jurassic porphyritic lavas, Pistol River, southwestern Oregon. Submarine volcanic rocks such as these and thick graywacke sandstones and mudstones typify Mesozoic sequences along the Pacific Coast; volcanic islands also existed there.

an immense alluvial plain extending north and east from Arizona to southern Canada. Lands were beginning to be raised in the Cordillera, especially in western Arizona, so it is probable that the western craton was tilted slightly upward, causing the epeiric sea to retreat into present Canada and toward the present Gulf of Mexico. But this regression was short-lived, for the sea returned to the Rocky Mountain region again during Early Cretaceous time.

THE LATE MESOZOIC AND EARLY CENOZOIC CORDILLERAN OROGENY

Over most of the Rocky Mountain and Gulf Coast regions, Cretaceous strata overlie Jurassic ones with no discontinuity, but farther west in the Cordilleran belt, they overlie a variety of deformed rocks, reflecting onset of major mountain building and deep erosion (Fig. 16.18) The culminating Cordilleran Orogeny spanned Late Jurassic through early Cenozoic time. An entirely different tectonic behavior, distinguished by faulting, has characterized western North America since the Oligocene.

FIGURE 16.16 Combined facies and paleogeographic map for the nonmarine, latest Jurassic Morrison and Kootenay Formations. Both the coarse facies and paleocurrent data indicate highlands in the Arizona area; isotopic dating and unconformities along the present Pacific Coast also indicate beginnings of *Cordilleran orogeny*. Winds blew volcanic ash eastward.

The eastward spread of nonmarine, varicolored Morrison sediments reflects the first major influence in the Rocky Mountain region of the Cordilleran Orogeny, but Lower Cretaceous strata show much greater effects. All Cretaceous sediments in the eastern Cordillera coarsen westward and thicken in the same direction to form a clastic wedge as much as 10,000 meters thick (Figs. 16.19 and 16.20). This wedge resulted from erosion of mountains within the mobile belt (Fig. 16.21). More direct effects of mountain building are evident farther west, where regional metamorphism and igneous activity were widespread. Upheaval began during Jurassic time near the Pacific Coast and then spread eastward. After

FIGURE 16.18 Relations of igneous intrusions and angular unconformities resulting from Late Jurassic and mid-Cretaceous phases of *Cordilleran orogeny* in southwestern Oregon. (F = index fossils.)

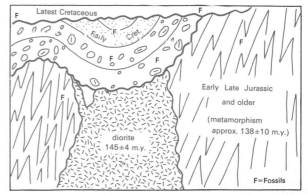

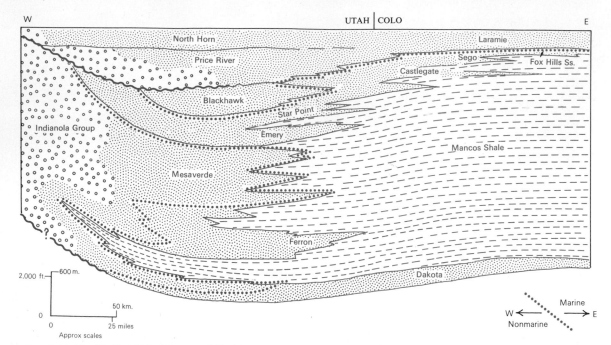

FIGURE 16.19 Stratigraphic diagram of Cretaceous facies across the present Rocky Mountains of the Utah-Colorado region showing coarse clastic tongues extending eastward into marine shale facies, and indicating cyclic transgressive-regressive events reminiscent of Pennsylvanian times. *(After R. H. Dott, Jr., 1964, Kansas Geological Survey Bulletin 169.)*

local Early Jurassic uplifts and volcanism (Fig. 16.12), widespread deformation began in Late Jurassic time. Granitic batholiths began to form (Fig. 16.16), but most of the isotopic dates are Cretaceous (75 to 100 m.y.) (Fig. 16.20). Widespread Late Cretaceous granitic pebbles, feldspar-rich sandstones, and unconformities reflect rapid uplift and exposure of many batholiths by erosion.

In the Rocky Mountain region, there is more evidence of Late Cretaceous and early Cenozoic than of earlier deformation. Thick, coarse conglomerates and sandstones attest to vigorous erosion of Cretaceous highlands within the mobile belt to the west as well as to suddenly accelerated subsidence of the cratonic margin, probably due to westward flow of subcrustal material as the Cordilleran Mountains rose. In central Utah, coarse gravel 3,000 meters thick accumulated close to the mountainous highland (Fig.

16.22). Meanwhile, volcanic activity erupted in western Canada, Montana, and along the Mexican border as well as in the Arctic (Fig. 16.20).

Beginning in latest Cretaceous time, the easternmost edge of the former mobile belt, which had continued to subside profoundly during deposition of thick Cretaceous strata, was crushed eastward against the craton just as strata of other geosynclines had been during earlier orogenies. Cordilleran mountain building had swept eastward like a wave so that in early Cenozoic time, a series of complex, low-angle thrust faults carried immense slabs of rock eastward over one another along a zone extending from Mexico to northwestern Canada (Figs. 16.23 and 16.24). As in older belts, the thrust-fault zone today marks the former cratonic margin. Northern Alaska also suffered folding and thrust faulting (Fig. 16.25). The mobile belt at this stage could be described as having been turned inside out; its former zone of greatest subsidence had become the axis of maximum uplift while subsidence now became concentrated along both margins of the old belt.

FIGURE 16.20 Middle Cretaceous sedimentary facies and isotopic dates for granitic batholiths. (See Fig. 11.15 for symbols and sources.)

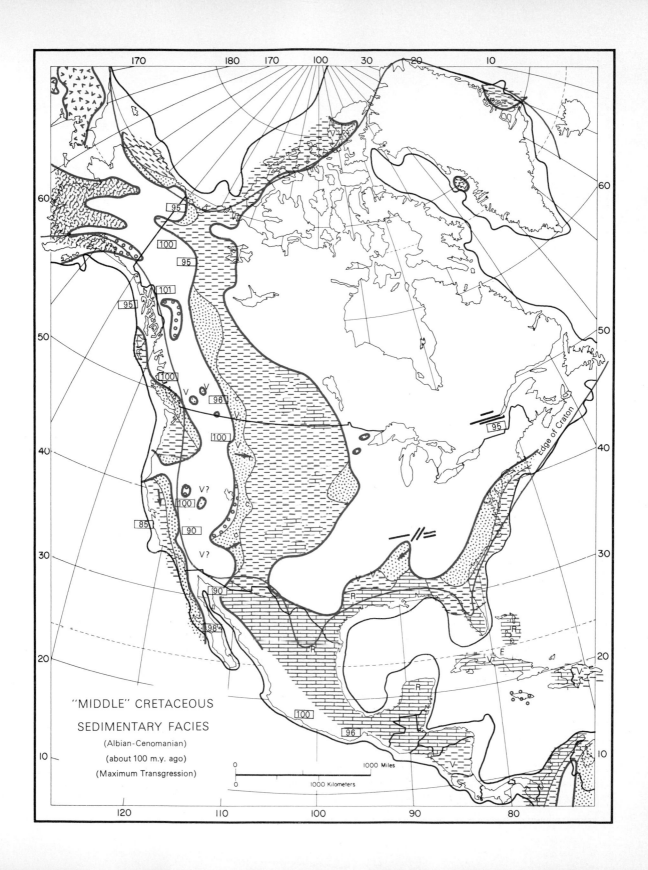

95

100

95

101

95

100

96

100

100

85

90

V?

V?

90

98

90

100

96

95

Edge of Craton

"MIDDLE" CRETACEOUS

SEDIMENTARY FACIES

(Albian-Cenomanian)

(about 100 m.y. ago)

(Maximum Transgression)

0   1000 Miles

0   1000 Kilometers

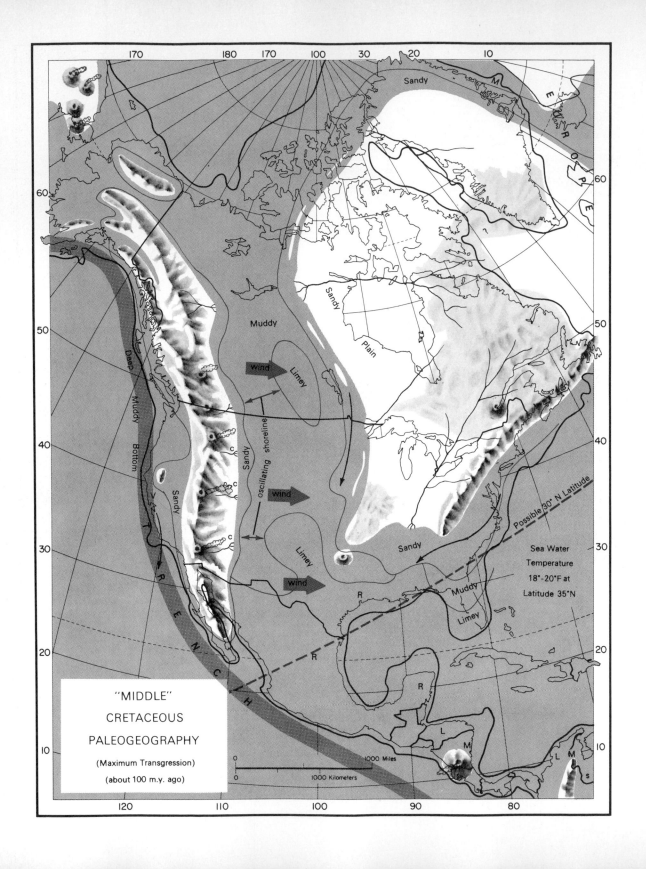

"MIDDLE"

CRETACEOUS

PALEOGEOGRAPHY

(Maximum Transgression)

(about 100 m.y. ago)

FIGURE 16.21   Middle Cretaceous paleogeography at the time of maximum worldwide transgression. Westerly winds blew volcanic ash widely over the epeiric sea. Note sea-water temperatures from oxygen-isotope studies.

FIGURE 16.22   Upper Cretaceous conglomerate (Price River Formation), Maple Canyon State Park, central Utah. Coarse fragments of Lower Paleozoic and Prepaleozoic quartzites were deposited at the foot of mountains raised in western Utah and Nevada during the Cordilleran orogeny. A sandy coastal plain lay east of the gravels (see Fig. 16.21).

FIGURE 16.23   Sentinel Range, northeastern British Columbia, along the Alaskan Highway. The Canadian Rocky Mountains were carved by rivers and glaciers from immense sheets of Paleozoic and Mesozoic strata thrust faulted eastward against the craton during early Cenozoic time (see Fig. 16.24). In this view, Devonian rocks have been folded and repeated by at least two thrusts; to the right of view, Silurian strata have overridden Triassic. *(Courtesy L. R. Laudon.)*

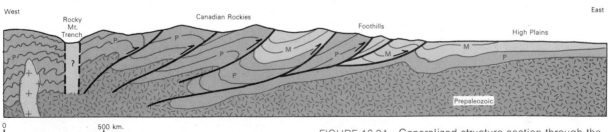

FIGURE 16.24   Generalized structure section through the Canadian Rocky Mountains showing complex thrust faults and the Rocky Mountain Trench, a possible rift or graben feature (P—Paleozoic; M—Mesozoic).

FIGURE 16.25 Faulted anticline in Mississippian (Lisburne) limestone and shale at Cape Thompson, western Alaska; deformation was Late Cretaceous or early Cenozoic. Light streaks at left are reflections from helicopter windshield. *(Courtesy K. O. Stanley.)*

CAUSE OF THE OROGENY

Unlike the other orogenies discussed before, the Cordilleran did not involve the collision of continents. Instead, the deformation and magma generation seem to have resulted only from the subduction of a Mesozoic Pacific lithosphere plate. Figure 16.5 shows the probable Mesozoic circumstances affecting the western edge of both North and South America. As Pacific lithosphere was thrust beneath the continents, earthquakes must have been generated frequently along the inclined zone near the top of the downgoing plate as seems to be the case beneath the modern Andes Mountains of South America. Selective melting of either upper mantle or lower crustal material (or both) by heat generated by underthrusting produced andesitic magma, which rose to the surface to form a volcanic arc on the edge of the continents. A deep trench formed near the present coast of the continent like the Peru-Chile trench today. Subduction of the Pacific plate presumably began in Permian or Triassic time when andesitic volcanism began in both the Cordillera and in the Andes. As it progressed, deformation of the arc rocks and the edges of the two continents occurred intermittently, but gradually intensified until it culminated in Late Cretaceous time with the uplift of the entire mobile belt.

THE GRANITE BATHOLITH QUESTION

The Cordilleran and Andean belts contain immense volumes of granitic rocks, and so it is appropriate at this point to examine further the origin of granitic (or dioritic) batholiths. Through James Hutton's eighteenth century efforts, it became widely accepted that igneous rocks formed by cooling and crystallization of hot, liquid magmas. But close association of large granitic batholiths with high-grade metamorphic rocks also was noted by Hutton and many others, and late nineteenth century French and Scandinavian geologists proposed that granite formed by ultra-metamorphism of sedimentary rocks through slow recrystallization in mobile belts (see Chap. 10). In its extreme form, this metamorphic hypothesis—granitization of geosynclinal sediments—involves little or no liquid magma, but rather the metamorphic transformation in place by recrystallization from sedimentary to igneous-appearing rocks without becoming entirely liquified. Note that, according to the granitization school, granitic rocks may be regarded more properly as metamorphic than igneous. Battle lines were drawn for one of the greatest modern controversies in geology, ironically centered on granite as was the old plutonist quarrel. Even today it is not fully settled. Regardless of whether granites ever have reached their temperatures of fusion, there is no doubt that tempers of men who study them have done so many times.

How could such a war between *magmatists* and *transformationists* have developed to a heated pitch reminiscent of feuds between religious zealots? Hardly anyone quarrels with evidence that small granitic masses, such as dikes, are intrusive *à la* Hutton and must have been more or less fluid when emplaced. But one of the stronger lines of evidence leading to the proposal of granitization is a characteristic banding or layering found in at least parts of many batholiths (Fig. 16.26). Indeed, such layering typically parallels the regional structural grain of, and commonly can be traced into, surrounding strata. Because of this, most geologists long ago acknowledged that some recrystallization of sediments probably occurs near margins of batholiths. But this is a chicken-or-egg argument. Did a liquid magma intrude sediments and cause recrystallization of more susceptible strata near its margin due to heat and gases, *or* did ultrametamorphism cause wholesale recrystallization of the sediments to form granite?

It seems easy to test the plausibility of transformation of geosynclinal sedimentary and volcanic rocks simply by comparing chemical composition of granitic rocks with them. But when we do this (Table 16.1), similarity is not so striking that it proves granitization. Instead it suggests volcanic rocks as more plausible parents than sediments alone. But does this similarity mean that granites are igneous and merely originate from the same magmas as andesites, or do they represent ultrametamorphism of the andesite-rich rock sequences of volcanic arcs? Clearly composition alone does not provide the answer.

Another question involves the space occupied by large batholiths. Of course this causes no embarrassment if the batholiths formed by in-place granitization, and so the "room problem" has haunted the magmatists for years, especially where strata seemingly can be traced into the batholith itself. Were the granite truly and wholly magmatic, one would expect instead clear signs of deformation of marginal strata shouldered aside by intruding magma. Such is commonly lacking, however. Alternatively, it has been suggested that magma engulfed and more or less "digested" large blocks of older rocks to make room for itself, but this is a form of granitization in disguise.

In areas of only slightly metamorphosed sedimentary and volcanic rocks, dikes and small irregular masses of granitic rocks are clearly intrusive. But at the margins of batholiths, sedimentary and volcanic rocks become increasingly metamorphosed and may fade insensibly into banded granitic rock called migmatite ("mixed rock") whose layering conforms with structural trends outside the batholith; the three-dimensional form of such masses may be extremely complex. Finally, in the batholith's interior, banding and inclusions of ingested sedimentary material are less common.

It may be that deep in the bowels of mobile belts where temperatures were high for long periods, in-place granitization of sedimentary and volcanic rocks occurred, with partially melted rock forming some

FIGURE 16.26 Banded granitic rock or migmatite from a late Mesozoic batholithic complex in northern Cascade Range, Washington. Light-colored sills and pods of quartz diorite represent replacement of darker (hornblende-bearing) diorite gneiss. Some sills were intensely contorted by plastic flowage during replacement of gneiss. Some geologists feel that such migmatites have formed by recrystallization (granitization) of sediments and volcanic rocks. *(Courtesy D. F. Crowder, U.S. Geological Survey; from Crowder, 1959 Bulletin of the Geological Society of America, v. 70, plate 7; by permission of the Geological Society of America.)*

magma that was intruded upward to form plutons at shallower, cooler levels (Fig. 16.27). Some geologists today believe that all granitic rocks are igneous and originate by subduction beneath arcs from the same magmas as the andesites; where some magma crystallized slowly beneath the andesitic piles, diorite or granite formed. Others, however, argue that batholith

FIGURE 16.27 Schematic conception of granitization by partial melting deep beneath a mobile belt and upward intrusion of magmas into shallower levels of the belt. Many shallow intrusions may be tear-drop shaped, having risen like oil droplets through water. Conversely, however, some authorities believe that batholiths are merely thin, tabular masses formed beneath immense volcanic sequences.

rocks are partially or completely melted crust with only the heat being derived from subduction below the crust. There seems to be evidence favoring each view.

CRETACEOUS TRANSGRESSION AND SEDIMENTATION

WORLDWIDE TRANSGRESSION

Worldwide transgression occurred during the Cretaceous Period, for marine rocks of this age unconformably overlap older rocks widely on practically every continent. Maximum dousing, which submerged about one-third of the present land area of the earth, occurred near the middle of the period, roughly 100 million years ago. "High water" occurred at slightly different times in different regions, depending upon local tectonic and sedimentary conditions (see Fig. 4.10). In broad terms, an essentially synchronous worldwide relative rise of sea level occurred, followed by general worldwide regression down to the present time. The Cretaceous flood affected North America profoundly, producing the last epeiric sea over the western craton and inaugurating formation of thick prisms of sediments now beneath marginal coastal plains and continental shelves.

Because there is no evidence of Cretaceous glaciation, a structural cause for the worldwide transgression must be sought. Recently it has been suggested that an acceleration of sea-floor spreading caused the enlargement of ocean ridges and wholesale continental breakup sufficient to displace sea water and cause the transgression. This seems the most plausible explanation to us.

EFFECTS IN THE PACIFIC COAST REGION

Along the western margin of the Cordilleran tectonic land, the Cretaceous shoreline oscillated somewhat, but effects were less pronounced than on the cratonic side. The western coastline was steeper, and local structural disturbances had pronounced masking effects by producing local transgressive and regressive complications; while transgression occurred in one area, regression may have prevailed only 160 kilometers away. A narrow continental shelf received marine gravel, sand, and mud deposits, while farther offshore in deeper water, clastic sediments accumulated

through the action of submarine sliding and turbidity currents (Fig. 16.28). Much of the region subsided very rapidly and more than 2 or 3 million cubic kilometers of sediments eroded from the Cordilleran Mountains accumulated there. Submarine lavas were erupted intermittently along the western edge of the continent in what was probably a deep submarine trench. The Pacific Coast must have looked much like the present west coast of South America with high mountains directly adjacent to a trench.

Cretaceous paleogeography of the Pacific Coast is poorly known because of complex Cenozoic structural disturbances that greatly obscured the record. Important local unconformities within the Cretaceous sequence attest to recurring tectonic disturbances; underthrusting of oceanic crust beneath a narrow continental shelf apparently occurred continually during Cretaceous time, creating a chaotic mixture of sedimentary, volcanic, and ultramafic rocks. Upper Cretaceous sediments along the Pacific Coast and interior Alaska contain more coal and other nonmarine deltaic and shoreline deposits than do Lower Cretaceous ones, thus marking a time of considerable filling of former deep-marine basins and embayments during an interval of temporary tectonic quiescence immediately after the Cordilleran Orogeny had culminated.

EFFECTS ON THE CRATON

The craton suffered flooding both from Arctic and Gulf regions, with merging of waters over the present Plains area in mid-Cretaceous time. Marine strata of this age extend from the Gulf Coast to the Arctic and from Minnesota to western Wyoming. On the east side of this seaway, a widespread transgressive, sandy shoreline facies developed. Farther west, the epeiric sea lapped almost at the foot of the Cordilleran Mountains near the middle of the period, but marine strata became less and less widespread through the Late Cretaceous times as regression commenced. By the end of the period, the sea had retreated to the present Plains region; during early Cenozoic time, the waters parted and drained completely from the craton both northward and southward. The great clastic wedge of Cretaceous strata on the western craton is reminiscent of the Pennsylvanian coal-bearing wedge of the western Appalachian region. Conglomeratic facies grade cratonward into thick, massive, cross-stratified tan sandstones containing coal seams (Fig.

FIGURE 16.28 Very evenly stratified Late Cretaceous or early Cenozoic graded graywacke sandstones and mudstones considered to have been deposited by turbidity currents in moderately deep water (flysch); near San Pedro Point, San Francisco peninsula, California (compare Fig. 14.22).

16.19). The total volume of clastic sediments exceeds 3 million cubic kilometers. These in turn pass eastward into widespread black shales with thin limestone layers and zones of altered volcanic ash blown from the west. Gradually sand was deposited farther and farther eastward during Late Cretaceous time, forming a classic regressive facies pattern. The conglomerates and coal-bearing portions of sandstones are largely nonmarine, whereas the shale and some sandstone are marine. Therefore, we see that the Cretaceous coal-bearing strata have another trait in common with Pennsylvanian coal sequences, namely repetitive or cyclic sedimentation. The intertonguing of marine and nonmarine strata represent wide oscillation of the shoreline due to (1) worldwide fluctuations of sea level, (2) spasmodic uplifts in the Cordillera, or (3) possibly cyclic climatic changes that affected weathering, erosion, and deposition of sand in a rhythmic fashion.

As the shoreline oscillated over the region, large deltas and associated swamps shifted with it as during Pennsylvanian time in the eastern craton. Sand-barrier islands paralleled the shores of marine embayments so that the coastline must have looked much like that of Texas and Louisiana today. Abundant vegetation grew in swamps, and gave rise to widespread coal seams, which constitute North America's second largest coal reserve, which, because of a low sulfur content, has now become our most important one. "Thunder lizards" wallowed in the swamps and roamed the alluvial plain that extended back to the Cordilleran Mountains. Besides coal, Cretaceous strata have long been important sources of petroleum trapped by folds and faults formed during the last phase of the Cordilleran Orogeny.

EFFECTS IN GULF AND ATLANTIC COASTAL PLAINS

By Late Jurassic time, the entire Appalachian mobile belt system from Mexico to Newfoundland had been eroded to a low-lying surface, and the present Atlantic and Gulf of Mexico basins had begun to form. Jurassic transgression brought marine deposition over the southwestern end of the belt, and during Cretaceous time, owing to still greater transgression, marine strata unconformably overlapped farther to cover at least half of the old eroded belt. Cretaceous sediments must have extended still farther inland originally than now, but erosion has stripped them from large areas of the Appalachian Mountains and southern plains.

In the southern United States, quartz-bearing sandstones are common in Lower Cretaceous strata, whereas shale and limestone are more important in the Upper Cretaceous. In Mexico, limestone deposition occurred widely throughout most of the period (Fig. 16.20). Beneath the Atlantic coastal plain, Cretaceous strata are of moderate thickness (1,000 to 2,000 meters), but along the Gulf Coast, thicknesses are about 4,000 meters. The Gulf Coast region was subsiding much more rapidly, but sedimentation kept pace, as most known Cretaceous sediments are richly fossiliferous shallow-marine sands, muds, or carbonates. There is evidence of some nonmarine deposition, too. As subsidence occurred beneath the Gulf Coastal region, worldwide transgression also was taking place so that younger marine Cretaceous strata tended to lap farther cratonward until the latest part of the period. Local structural warping and erosional truncations occurred, and there was considerable faulting. Small igneous intrusions (some containing diamonds) were emplaced in widely scattered areas, and a few volcanic vents in Texas, Arkansas, Louisiana, and Mexico erupted considerable ash.

Paleomagnetic evidence suggests that the possible paleoequator was approaching its present relative position and orientation by Late Cretaceous time (Fig. 16.21). This would have placed the Gulf Coast region at a relatively low, nearly tropical latitude. It is significant that marine invertebrates were abundant and very diverse in the shallow sea there, and that carbonate rocks were of great importance; organic reefs also developed widely.

LATE MESOZOIC PALEOCLIMATOLOGY

Emergence and elevation of most continents in late Paleozoic and early Mesozoic times produced a diversified "continental" climate characterized by considerable aridity. With a much larger proportion of the earth's surface covered by water during late Mesozoic time, we would expect that more solar radiation would be absorbed by the water, and heat would be efficiently distributed poleward by currents, producing an overall warm, mild, "oceanic" climate with ice-free poles. Lack of evidence of late Mesozoic glaciation and widespread distribution of cold-blooded reptiles and mild-climate plants all across the continent strengthen this belief. For example, Cretaceous dinosaur footprints recently have been discovered on Svalbard (Spitsbergen), whose present latitude is 77°N (10 degrees north of the Arctic Circle). This frigid land today enjoys an average annual temperature range from −20 to +5°C! While cold-blooded animals can stand occasional frost and mild freezes, it is inconceivable that they could survive extended periods of subfreezing temperatures unless they hibernated in protected places. But hibernation would have been next to impossible for giant dinosaurs. (It has been suggested recently that dinosaurs could, in fact, regulate their body temperatures to some degree). Fossil plants, which had evolved toward a modern aspect by Late Cretaceous time, also indicate mild-temperate to subtropical conditions over most continental areas. For example, conifer and gingko forests grew in Franz Josef Land, now at 80°N. Ancestors of breadfruit trees, laurels, *Magnolia*, and *Sequoia* (redwood), which could not stand freezing temperatures, thrived in western Greenland at 70°N. Although evaporites were important in Jurassic times, suggesting high evaporative potential at many latitudes, their general absence from Cretaceous sediments suggests relatively humid conditions (or more open-marine circulation) over North America in late Mesozoic time. Cretaceous evaporites did form on the edges of Africa, however (see Fig. 15.21).

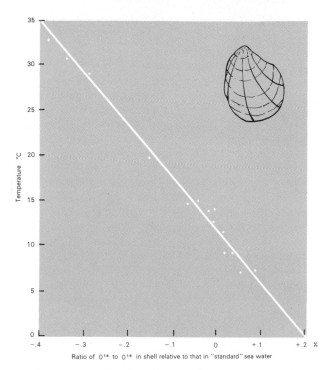

FIGURE 16.29   Relation of oxygen isotopes to sea-water temperature in calcium carbonate shells formed by modern animals; ¹⁸O decreases relative to ¹⁶O as temperature increases. *(After Epstein et al., 1951, Bulletin of The Geological Society of America, v. 62, p. 424; by permission of the Geological Society of America.)*

It has been shown by analysis with the mass spectrometer that relative proportions of the oxygen isotopes, ^{18}O and ^{16}O, in calcium carbonate marine shells vary according to temperature of the sea water in which organisms grew, such that ^{18}O in shells decreases relatively as temperature increases (Fig. 16.29). From experimental studies, chemical relationships controlling this variation are understood sufficiently so that $^{18}O{:}^{16}O$ ratios in shells can be used for paleotemperature determination. Paleotemperature is estimated through analysis of ancient fossil shells under the assumption that identical chemical relationships controlled oxygen-isotope composition of ancient as well as of modern shells. Determinations in 1951 of $^{18}O{:}^{16}O$ ratios in Jurassic belemnoids from 57°N latitude (equivalent to Scotland or southern Alaska) points to an average annual sea-water temperature of 14 to 20°C (54 to 68°F), or roughly 15°C

warmer than is typical at that latitude today. The quantitative oxygen-isotope data, therefore, confirm qualitative fossil and sedimentary evidence pointing to a warm Jurassic climate at midlatitudes. Cretaceous fossils from North America, western Europe, and Russia yield results (Fig. 16.30) that confirm mild ocean temperatures on the order of 20 to 25°C in present middle latitudes from 30 to 70°N (present coordinates). Today, comparable surface ocean temperatures are characteristic of such coasts as Florida, Mexico, Central America, and northwest Africa (i.e., roughly between the low latitudes of 5 and 32°N). Even in east-central Greenland, now at a latitude of 72° north, the sea was a balmy 17°C during mid-Cretaceous time or about 15°C warmer than today. By extrapolation from oxygen-isotope studies of Greenlandic and Russian fossils, it is estimated that Cretaceous north polar water was no colder than 10 to 15°C.

If, in Late Cretaceous time, North America occupied approximately its present position with respect to pole and equator, and *if* there was little complication of ocean-current patterns, then it would be

FIGURE 16.30   Latitudinal paleotemperature gradients for shallow-marine water at two separate times late in the Cretaceous Period, compared with the modern gradient (based upon fossils from North America, Europe, and Russia). Note continuous cooling trend since mid-Cretaceous time. *[Adapted from Lowenstam and Epstein, 1959, El Sistema Cretacico (Primer Tomo), pp. 65–76; 20th International Geological Congress.]*

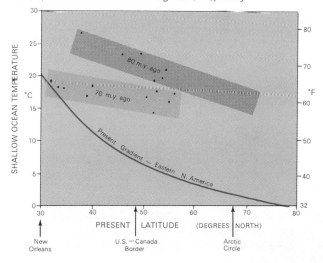

possible to reconstruct a rather detailed paleoclimatic map from the above temperatures and what is known of the paleogeography. Geologic evidence indicates that total earth climate was milder and more uniform than today, and the fossil record suggests that North America had largely subtropical conditions.

Though actual atmospheric and oceanic circulation cannot be accurately restored without detailed knowledge of size and position of the other Cretaceous continents, a few cautious interpretations about patterns over North America seem warranted (Fig. 16.31). Most of the continent would have been within the westerly wind belt, so that moist air would have blown over its western part. The Cordilleran Mountains were well watered by westerly winds, and areas farther east could receive moisture from the epeiric sea east of the mountains and from the Gulf of Mexico. Uniformity of Cretaceous and Eocene plants indicates a lack of sharp climatic zonation of the continent. By analogy with modern continents at similar latitudes, it is assumed that there were distinct seasonal variations of temperature and precipitation, but these cannot be fully assessed.

MESOZOIC ANIMAL LIFE

THE INVERTEBRATES

The great wave of Permian extinctions was not followed immediately by replacement with a new and varied fauna. Early Triassic marine faunas are not well known because epicontinental seas did not spread widely until Late Triassic time. Thus few early shallow water habitats are known. The earliest fossiliferous Triassic strata contain some ammonites (Fig. 16.32) and a few species of clams. Both the latest Permian and earliest Triassic marine-bottom faunas are represented by a few conservative genera, but the species are represented by large numbers of individuals. These highly unusual faunas are similar to those found today in hypersaline lagoons. No single satisfactory cause has been advanced to account for the late Permian extinctions, because both land and marine organisms were affected. Some major groups such as the fishes were not altered at all.

During the Triassic Period, the ammonites diversified and filled a wide range of environments. Some 400 genera and families evolved. For unknown reasons, almost all these became extinct at the close of

the period save a few conservative groups that were evolutionary *cul-de-sacs*. A single family of ammonites remained to give rise to a vast and complex group of 1,200 genera that dominated the Jurassic and Cretaceous Periods. So rapid was the evolution and extinction of subgroups that Mesozoic ammonites serve as the best guide fossils known for worldwide correlation. As a result, the history of life and paleogeography is perhaps better known for the Mesozoic than for any prior segment of time.

Other molluscs show quite a different pattern. Clams and snails gradually took on a modern appearance by addition of new, mostly still-living families. Clams took up residence in the sediments as burrowers, and diversified quickly. Some characteristic Mesozoic families evolved and waned to extinction at the era's end, such as *Inoceramus* and rudistid clams. Another important mollusc found in many Mesozoic faunas is the curious cigar-shaped, internal skeleton belonging to belemnoids, a group related to modern squids.

We saw that during the latter part of the Paleozoic Era there was an appreciable decline both of reef- and nonreef-forming rugose and tabulate corals. Rugose corals became extinct in Middle Permian time and it was not until the Middle Triassic that a new group of corals (scleractinians) became prominent (Fig. 16.33). Since Late Triassic time, the latter have been the principal reef formers; however, it was not until the Cretaceous Period that reefs acquired a fully modern aspect. We are unsure if the scleractinians were derived directly from rugose corals or if they were derived from some form that lacked a skeleton (e.g., the modern anemones).

THE VERTEBRATES
Mammal-like Reptiles
By far the most important preserved record of Early Triassic vertebrates is found in the Karroo nonmarine

FIGURE 16.31 Hypothetical Late Cretaceous paleoclimatic map. Dominance of westerly winds is indicated by volcanic ash fallout, land climate from fossil plants and meteorologic theory, sea-water temperatures from oxygen-isotope studies, marine faunal migrations from paleobiogeography, and probable latitudes from paleomagnetic evidence. Even if the continent occupied its present latitude, climate would not have been greatly different. (For a paleoclimatic interpretation assuming a fixed continent-pole-equator relationship, see the suggested reading by Millison, 1964.)

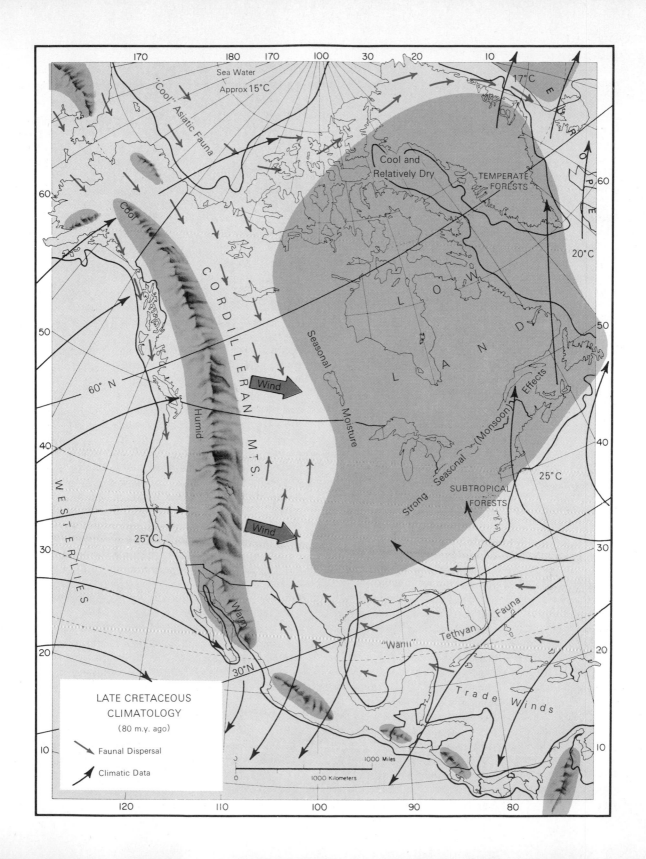

LATE CRETACEOUS
CLIMATOLOGY
(80 m.y. ago)

Faunal Dispersal

Climatic Data

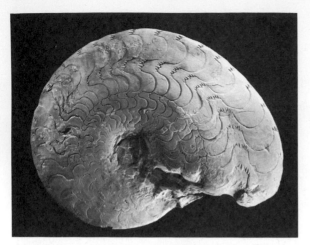

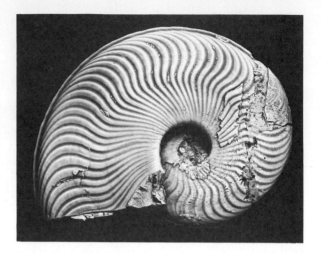

FIGURE 16.32 Mesozoic ammonoids. *Left*: Most Triassic ammonoids are characterized by having subdued ornamentation and ceratitic sutures; however, by Late Triassic time ammonitic-type sutures became more common. *(Photograph by G. R. Adlington.)* *Right*: With the exception of a few Cretaceous pseudoceratitic patterns, all post-Triassic genera have ammonitic sutures. Most evolution and variation in the latter involved development of ribs and nodes. *(Courtesy U.S. National Museum.)* By Cretaceous time, families arose having members that became uncoiled.

red bed sequence in South Africa (Fig. 15.17). In addition to the interesting depositional history of this important Permo-Triassic sequence and its *Glossopteris* flora, it contains the richest and most varied reptile fauna known for this age. One of the most important reptilian groups is the mammal-like forms, the therapsids, represented by large types thought to be the ancestors of the mammals. The mammal-like reptiles had anatomical characteristics that suggest the possession of body-temperature regulating capabilities.

Although the mammal-like reptiles persisted into Jurassic time, we are unsure of the exact origin of the mammals. The first true mammals appear in Late Triassic strata, and were rather small, rodent-like forms known only from isolated teeth and jaws; not one complete skull is known. These early forms belong to a prototype group called the Pantotheria, to the multituberculates, primates, and two other families which became common in the Paleocene (see p. 411). Most genera had three-cusped teeth, but an early development was for a more massive, multi-

cusped tooth apparently more efficient for plant eating. Rarity of Mesozoic mammalian fossils possibly reflects the dominance of reptiles. It was not until the beginning of Cenozoic time that mammals "explosively" radiated to become the most successful land dwellers of all geologic history.

The Dinosaurs

The most important reptiles in the Triassic were the Thecodonta, a varied group with bipedal tendencies. It is from thecodonts that dinosaurs arose sometime during Late Triassic time. Not all thecodonts were bipedal. The phytosaurs, for example were a long-snouted predaceous form, which occupied a niche to be filled by crocodiles in the Jurassic. Bipedal forms, such as *Ornithosuchus*, had long muscular hind legs and a slender, flexible tail used as a balancing organ. Front legs were shortened and it obviously was an excellent runner (Fig. 16.34). The earliest dinosaur order (there is no formal name *dinosaur* in reptilian classification), the Saurischia, is represented in Upper Triassic rocks by forms such as hollow-boned *Coelophysis* (Fig. 16.34), which resembled the thecodonts in being a bipedal runner. Numerous sharp teeth of this form indicate that it was an efficient carnivore. The primitive coelosaurs gave rise to the great carnivorous saurischians of late Mesozoic time. The name Saurischia means lizard hips and indicates the most important characteristic of the group (Fig. 16.35). In many regions, such as the eastern United States, saurischians were very abundant during the

FIGURE 16.33   A colonial scleractinian coral from the Miocene of eastern Virginia. *(Courtesy U.S. National Museum.)*

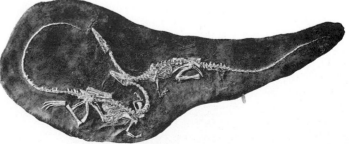

FIGURE 16.34   A Triassic thecodont reptile, *Ornithosuchus* (left). This group is thought to have given rise to the saurischian dinosaurs like *Coelophysis* (right). *(Courtesy American Museum of Natural History.)*

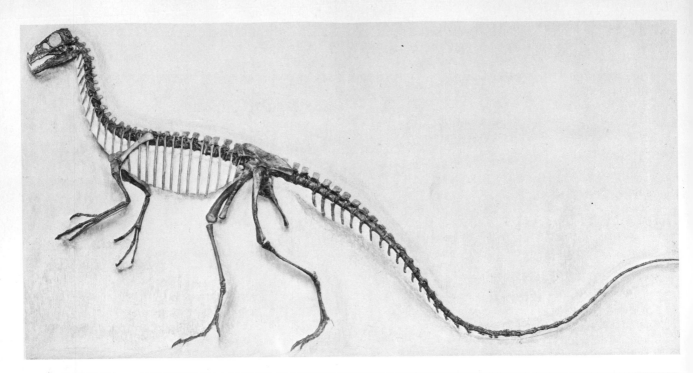

Triassic, judging from extensive tracks they left (Fig. 16.1).

The coelosaurs, which include the smallest (dog-sized) dinosaur, continued into the Cretaceous where they are represented by a small creature, *Struthiomimus*, which strongly resembled an ostrich with a long, slender neck and a small head. The teeth were replaced by a horny, bill-like structure. Only a single dubious fragment from the Upper Triassic South African rocks has been associated with the other important dinosaur order—the Ornithischia.

Relatively small saurischian dinosaurs of the Triassic evolved into dominant and sometimes terrifying monsters in the Jurassic Period. One group, the Carnosauria, retained primitive bipedalism and carnivorous habits. They are represented in the Jurassic by giant *Allosaurus* (Fig. 16.36), which preyed on the gargantuan herbivorous sauropods. By Cretaceous time, *Tyrannosaurus* and *Gorgosaurus* appeared and climaxed the trend toward large size. *Tyrannosaurus* skeletons as large as 15 meters, standing almost 6 meters in a walking position, are known. Front legs were reduced in number of digits and were so tiny that they probably were completely useless. The skull, in contrast, was huge in proportion to body size. It was armed with a vicious set of curved, serrated teeth as much as 15 centimeters in length. Large holes reduced the weight of the huge skull.

When most children hear the word dinosaur, they think of great herbivorous Sauropoda of the Jurassic. These solid-boned quadrupeds retained some ancestral features, such as a small head, long tail, and rather short front legs. By Middle Jurassic time, they reached their zenith in attaining the largest size and weight of any land animal known. *Apatosaurus* (Fig. 16.36) and *Diplodocus* found in the Morrison Formation in Wyoming and Colorado were massive (up to 36 metric tons). Culmination of this trend in size was one

Jurassic genus over 25 meters long. Except in the Southern Hemisphere and India, sauropods are rarely found in Cretaceous strata; and no known later ones achieved the amazing proportions of the Jurassic forms.

Ornithischian dinosaurs never attained the giant size of the saurischians, but they became almost unbelievably bizarre (Fig. 16.37). Even the most conservative suborder, the ornithopods, are represented by such forms as the hadrosaurs, a common Upper Cretaceous group, which was bipedal with a duck-billed skull adapted for feeding on water plants. The hadrosaurs were well adapted for swamp dwelling since they possessed webbed appendages, and some had nasal tubes that swept back up to the skull roof in snorkel fashion, although these tubes have been shown not to function as snorkels.

Though derived from bipedal thecodonts, the majority of ornithischians were quadrupeds primarily with front legs more fully developed than in saurischians. They were exclusively herbivores, and in a number of genera, teeth were reduced or absent. Frequently, jaws were in the form of a beak or were flattened as in the duck-bills.

While bipedal hadrosaurs appeared to be fairly mobile, a prime necessity with carnivores lurking about, the majority of the ornithischian quadrupeds were ponderous. As in most slow-moving organisms, some type of protection was developed and three suborders solved this problem in different ways.

The Ceratopsia developed a very large bony head and neck shield formed of the upper facial bones. The head, with its bony shield and turtle-like beak, occupied almost a third of the entire length of the beast (Fig. 16.37). Almost all had a median horn over the nasal area, and the most famous of the group (*Triceratops*) had an additional pair over the eyes. The ankylosaurs, another exclusively Cretaceous group, were entirely covered by heavy dermal plates accompanied by long, bony spines—not very appetizing fare. Another important suborder, the stegosaurs, are primarily found in Jurassic rocks; the most famous of these is *Stegosaurus* (Fig. 16.37) from Morrison strata. Most conspicuous features of this group were large spines and triangular plates extending down the back and terminating in a vicious-looking spiked tail. The stegosaurs also are noted for having two "brains"—a very small regulation model in its snake-like head, and another, much larger nerve

FIGURE 16.35 These illustrations show the difference in the pelvic girdle of the two dinosaur orders. *Top: Ornitholestes*—note that in the order Saurischia the pelvic girdle is arranged in a triradiate pattern that provided the giant hind-leg muscles with a firm attachment. *Bottom: Camptosaurus*—note that the pelvic girdle in the order Ornithischia is bird-like in that the two lower pelvic bones extend anterior-posterior forming a tetraradiate pattern with the upper bone. (*Courtesy American Museum of Natural History.*)

FIGURE 16.36 *Top*: *Allosaurus*, a carnivorous dinosaur.
Bottom: *Apatosaurus*, a herbivorous dinosaur. Both are
Jurassic saurischians. *(Courtesy American Museum of
Natural History.)*

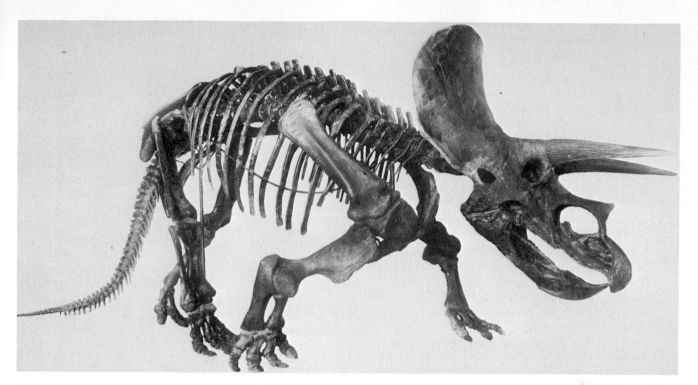

FIGURE 16.37   *Top*: *Triceratops* (Cretaceous). *Bottom*:
Stegosaurus (Late Jurassic). Both are examples of
ornithischian dinosaurs. *(Courtesy American Museum of
Natural History.)*

center at the base of the spine. The latter supposedly controlled the spiked tail.

The dinosaurs underwent gradual adaptive radiation during the Mesozoic Era (over 230 genera are known; Fig. 16.38). At least 70 genera roamed the lands and browsed in near-shore waters during the Late Cretaceous, and then all were gone, utterly and completely, in spite of science fiction novels. Speculations about the cause of dinosaur extinction are legion; we simply do not know the cause.

Finally, the first true birds (*Archaeopteryx*) were found in the Jurassic Solenhofen Limestone of southern Germany in 1861. They represent a classic case of a transitional group in that their skeletons and large teeth strongly resemble small bipedal dinosaurs, but they had many bird features, such as feathers and an

expanded brain case (Fig. 16.39). Specimens fortuitously were preserved in fine-grained limestone in which impressions of feathers were clearly preserved; otherwise *Archaeopteryx* might be classified as a dinosaur. In fact, recent studies have shown that the birds are so close to dinosaurs that a new classification has been proposed.

The Ornithischia, Saurischia, and birds (Aves) would be subclasses of the new class Dinosauria. This has interesting implications because the suggestion has been made that, since the birds are warm-blooded, the dinosaurs also may have been.

Some Other Reptilian Adaptations

Two other important radiations occurred during the Mesozoic—the invasion of the aquatic and aerial environments by reptiles. Ichthyosaurs (Fig. 16.40) appeared in the Triassic Period, flourished during the Jurassic, but became extinct well before the end of the Cretaceous. Their limbs were reduced to fin-like appendages, and the head resembled that of a swordfish, having large eyes and elongated toothed jaws. Another group, the plesiosaurs (Fig. 16.40), are better known because of their impressive size (15 meters long). Apparently all were active predators.

The earliest-known and most primitive reptile to become airborne was a gliding form. It is known from a single Late Triassic specimen found by two high school boys on an afternoon collecting trip to a New Jersey quarry opposite Manhattan. Much more effective animals in the air were pterosaurs, which were able to fly (barely). The most interesting characteristic was the wing structure, which was supported solely by the fourth digit enormously extended and covered by a wing membrane that also attached to the body. Three other digits were short and clawed. Pterosaurs, or "flying reptiles," are found in marine strata of Jurassic and Cretaceous ages, and probably were fish eaters.

SUMMARY

Marine life underwent profound changes during Triassic time after late Paleozoic extinctions, but land life was not so markedly affected. Several ecologic replacements occurred, notably some brachiopods by clams, and swimming trilobites by ammonoids

FIGURE 16.38 A simplified family tree of the dinosaurs. *(Courtesy American Museum of Natural History.)*

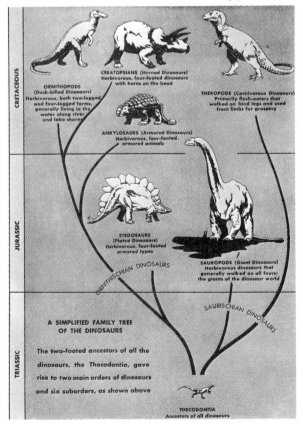

RESTORATION

FIGURE 16.39   *Archaeopteryx*, a toothed bird, the oldest known of all birds, from the Jurassic Solenhofen of Bavaria. *(Courtesy American Museum of Natural History.)*

and fishes. Land plants flourished and diversified; coniferous trees predominated, but flowering plants appeared near the end of the era. Reptiles underwent tremendous diversification with the reign of dinosaurs, swimming and flying reptiles, and mammal-like reptiles. Primitive mammals appeared during the middle of the era.

A distinctive tropical marine Mesozoic fauna developed in the Tethys seaway between Eurasia and Africa and spread westward to southern North America by Jurassic time. The central Atlantic and Gulf of Mexico basins began forming as Africa and South America separated from North America beginning in Early Jurassic time. Europe finally separated from Greenland in Eocene time. As the Atlantic basin widened, later Mesozoic and Cenozoic faunas of America and the Old World differed more and more.

Unlike life, sedimentation in the Mesozoic Era involved fewer innovations. Triassic facies represent a direct holdover of Permian patterns with very widespread red beds on the craton and marine deposition confined to the west. The Permo-Triassic Cassiar Orogeny disturbed most of the western Cordilleran belt, producing a paleogeography resembling the present Japanese region. Red bed deposition nearly

FIGURE 16.40   *Top*: *Ichthyosaur* from the Lower Jurassic of Germany. *Bottom*: *Plesiosaur* from the Lower Jurassic of England. *(Courtesy American Museum of Natural History.)*

ceased in Early Jurassic time, when an immense blanket of pure quartz sand, derived from erosion of Triassic and Paleozoic sandstones in western Canada, spread over the west edge of the craton and eastern Cordilleran belt. There were both shallow-marine sands like those of early Paleozoic time, and wind dunes formed along the southeastern shoreline. In Middle Jurassic time, wider transgression of the western craton commenced. Initially, important evaporites formed in several basins (especially around the Gulf of Mexico), but by Late Jurassic time, normal, shallow-marine deposits prevailed.

Near the end of the Jurassic, due to slight tilting of western United States, the sea retreated both north and south as the nonmarine, dinosaur-rich Morrison Formation was deposited over the Rocky Mountain region. These changes reflected onset of the Cordilleran Orogeny. Near the Pacific Coast, severe deformation accompanied by batholith formation began in Late Jurassic time, caused by the eastward underthrusting of oceanic crust beneath the continent. A great wave of deformation spread eastward, culminating in mid-Cretaceous time, and finishing with eastward overthrust faulting at the edge of the craton in early Cenozoic time. The Cordilleran mobile belt was turned inside out so that by mid-Cretaceous time it was an immense mountainous land from which at least 5 or 6 million cubic kilometers of material were eroded in Cretaceous time. Simultaneously, the west edge of the craton underwent rapid subsidence accompanied by deposition of a vast Cretaceous clastic wedge with great coal-bearing tongues of sandstone grading eastward into marine shale. These same deposits soon were to be involved in Cenozoic thrusting.

A great worldwide transgression affected North America in Cretaceous time, producing the last inundation of the craton. After maximum flooding near the middle of the period, regression occurred and marine conditions in the cratonic interior terminated in the Paleocene Epoch. Short-period tectonic or climatic cycles produced patterns of repetitive sedimentation much like those of Pennsylvanian time.

Paleontologic and oxygen-isotope evidence point to a mild, rather uniform Mesozoic climate over North America. Dinosaurs left immortal footprints from far-north Svalbard (Spitsbergen) to Texas. Late Cretaceous ocean temperatures were no colder than 15°C (58°F) in the Arctic and were about 25°C (78°F) in

California and New Jersey, and 30°C in New Mexico, or from 10 to 20°C warmer than now! Such mild conditions were to continue for another 25 million years, allowing vegetation similar to that of the Cretaceous to persist through early Cenozoic time, yet terrestrial dinosaurs became extinct and marine ammonoids and swimming reptiles followed suit in the seas. Nearly simultaneous extinction on sea and on land is difficult to explain. The sole cause was not climatic, as has been stated frequently, nor is postulated increased cosmic radiation adequate. Most likely, subtle ecologic factors, such as unrecorded changes in the food chain, were responsible.

As is noted in Chap. 14, the Cordilleran region bears striking similarities to eastern Asia as impressive as the eastern North American Paleozoic similarities with Europe. Asiatic Mesozoic similarities include many nearly identical molluscan fossils, dinosaurs, and plants, as well as parallel developments of Permo-Triassic (Cassiar) and late Mesozoic (Cordilleran) orogenies. As in the Cordilleran belt, the greatest granite batholiths of east Asia formed in late Mesozoic time.

Readings

Arkell, W. J., 1956, Jurassic geology of the world: New York, Hafner.

Burk, C. A., 1965, Geology of the Alaska peninsula: Geological Society of America Memoir 99.

Casey, R., and Rawson, P. F., eds., 1973, The boreal Lower Cretaceous; Liverpool, Special Issue No. 5, Geological Journal.

Clark, T. H., and Stearn, C. W., 1960, The geological evolution of North America: New York, Ronald.

Colbert, E. H., 1955, Evolution of the vertebrates: New York, Wiley. (Paperback edition, 1961, Science Editions, No. 099-S)

Dewey, J. F., Pitman, W. C. III, Ryan, W. B. F., and Bonnin, J., 1973, Plate tectonics and the evolution of the Alpine system: Bulletin of the Geological Society of America, v. 84, pp. 3137–3180.

Eardley, A. J., 1962, Structural geology of North America: New York, Harper & Row.

Hallam, A., ed., 1973, Atlas of palaeobiogeography: Oxford, Oxford.

Ladd, H. S., ed., 1957, Treatise on marine ecology and

paleoecology: Geological Society of America Memoir 67, v. 2, Paleoecology.

Logan, A., and Hills, L. V., eds., 1974, The Permian and Triassic Systems and their mutual boundary: Calgary, Alberta Society of Petroleum Geologists, Memoir 2.

Maxwell, J. C., 1969, The Mediterranean, ophiolites and continental drift, in What's new on earth?' Newark, Rutgers.

McKee, E. D., 1954, Stratigraphy and history of the Moenkopi Formation of Triassic age: Geological Society of America Memoir 61.

―――― et al., 1956, Paleotectonic maps of the Jurassic system: U.S. Geological Survey Miscellaneous Publications, Map I-175.

―――― et al., 1959, Paleotectonic maps of the Triassic system: U.S. Geological Survey Miscellaneous Publications, Map I-300.

Middlemiss, F. A., Rawson, P. F., and Newall, G., 1971,

Faunal provinces in space and time: Liverpool, Special Issue No. 4, Geological Journal.

Millison, C., 1964, Paleoclimatology during Mesozoic time in the Rocky Mountain area, Denver: Mountain Geologist, v. 1, p. 79–88.

Phillips, J. D., and Forsyth, D., 1972, Plate tectonics, paleomagnetism, and the opening of the Atlantic: Bulletin of the Geological Society of America, v. 83, pp. 1579–1600.

Price, R. A., and Doeglas, R. J. W., eds., 1972, Variations in tectonic styles in Canada: Geological Association of Canada Special Paper No. 11.

Smith, A. G., 1971, Alpine deformation and the oceanic area of the Tethys, Mediterranean, and Atlantic: Bulletin of the Geological Society of America, v. 82, pp. 2039–2070.

Stokes, W. L., 1973, Essentials of earth history (3d ed.): Englewood Cliffs, Prentice-Hall.

Turner, F. J., and Verhoogen, J., 1960, Igneous and metamorphic petrology (2d ed.): New York, McGraw-Hill.

17

CENOZOIC HISTORY

∘ THRESHOLD OF THE PRESENT

The face of places,
and their forms decay;
And that is solid earth,
that once was sea;
Seas, in their turn,
retreating from the shore,
Make solid land,
what ocean was before.

Ovid,
Metamorphoses, XV

Researchers have already cast
much darkness on the subject,
and if they continue their
investigations,
we shall soon know nothing
at all about it.

Mark Twain

FIGURE 17.1 Rare view of southeast face of Mt. Everest (9,000 meters) showing metamorphic rocks at bottom. *(Courtesy L. N. Ortenburger.)*

Figuratively speaking, the story of early geologic exploration of the Cordillera is as colorful as the scenery. For many years, the geologist was not safe there without a gun. Moreover, in the United States, unlike Canada, his activities were not supported consistently for many years. Several state geological surveys were organized in the East as early as the 1830s, but for years all of the West was under the direct care of the federal government. After the Louisiana Purchase in 1803, one expedition after another was sent out to probe the unknown. Explorations accelerated after the Civil War, and several competitors politicked officials to support "their" survey. Finally the United States Geological Survey was founded in 1879—nearly four decades after its Canadian counterpart.

Important geologic discoveries were made in rapid succession. The Grand Canyon had been explored, and several pioneer geologists contributed new concepts of erosion based upon the first systematic observations of arid landscapes. Structures in the arid West were more clearly exposed than in any regions previously studied. Valuable mineral resources were discovered and studied intensively, ultimately leading to development of a revolutionary theory of metallic ore formation. Untold numbers of rich fossil localities also were discovered, many of which yielded hitherto unknown extinct vertebrate animals. As incomparable scenic beauty was discovered, geologists became instrumental in publicizing and working for preservation of such marvels as Yellowstone, the continent's first national park. Particularly notable in this endeavor were F. V. Haydn and two associates, famous landscape artist Thomas Moran and pioneer photographer W. H. Jackson.

The magnificent mountains and canyons explored by early geologists in western Canada and the United States were formed largely in Cenozoic time. In the last chapter, we examined the Cordilleran Orogeny, which terminated a long geosynclinal subsidence history, and produced structures typical of ancient mobile belts (Fig. 17.2) There was one outstanding peculiarity, namely the almost unique reactivation in the western United States of a large part of the craton. Unusually high uplifts of relatively simple anticlinal and monoclinal structures were formed by vertical fault-movements of blocks of the Prepaleozoic basement.

Completion of our analysis of Cordilleran history as well as that of other parts of the world (especially the Alpine-Himalayan belt) in this chapter will show that structural disturbances of great magnitude occurred with surprising rapidity in late Cenozoic time,

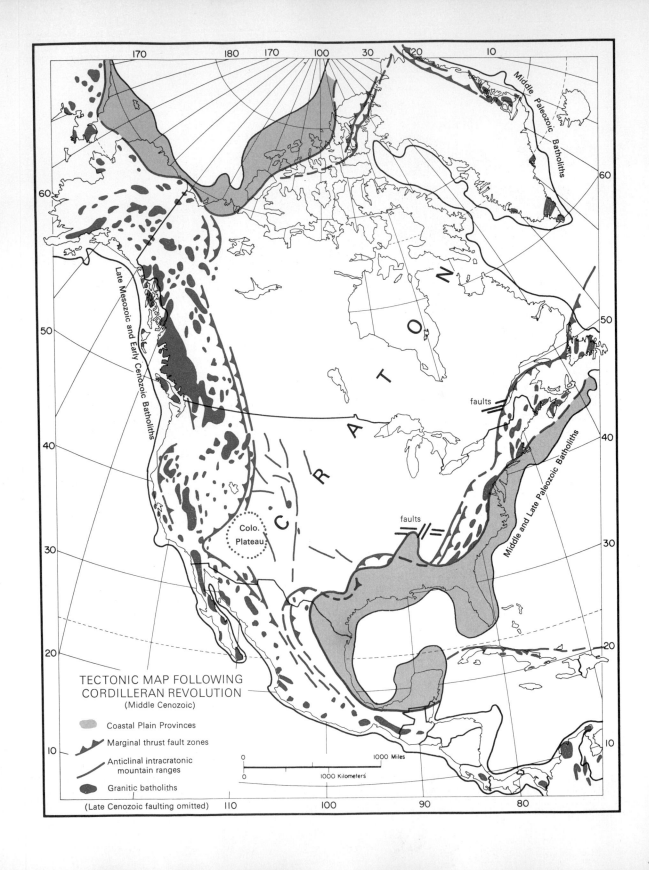

TECTONIC MAP FOLLOWING
CORDILLERAN REVOLUTION
(Middle Cenozoic)

Coastal Plain Provinces

Marginal thrust fault zones

Anticlinal intracratonic
mountain ranges

Granitic batholiths

Late Mesozoic and Early Cenozoic Batholiths

Middle and Late Paleozoic Batholiths

Middle Paleozoic Batholiths

C R A T O N

Colo.
Plateau

faults

faults

0   1000 Miles

0   1000 Kilometers

(Late Cenozoic faulting omitted)

FIGURE 17.3 Teapot Rock near Green River, Wyoming, containing Eocene lake deposits (Green River Formation) composed of laminated "oil shales" under a massive, dark deltaic sandstone. (This photograph was taken by pioneer photographer W. H. Jackson, 1869, who had to coat glass plates with emulsion in a tent before each exposure.) *(Courtesy U.S. Geological Survey.)*

thus pointing to an extremely mobile condition. Intense fragmentation of the crust accompanied by unusually widespread volcanism characterized the late Cenozoic. In marked contrast, the other margins of North America were relatively tranquil tectonically as shown by strata of the coastal plains and broad continental shelves. Finally, we shall review the history of Cenozoic life over the past 65 million years with respect to changing geography and climate as all continents emerged from the seas to assume their present configuration.

CENOZOIC CORDILLERAN HISTORY

THE ROCKY MOUNTAINS
Early Cenozoic

Between the early Rocky Mountain ranges elevated during Paleocene and Eocene time lay low areas,

FIGURE 17.2 Tectonic map of North America for middle Cenozoic time (after the Cordilleran orogeny but prior to late Cenozoic upheavals). Note batholiths and thrust-fault zones along inner margins of all mobile belts.

which became dumping grounds for sediment eroded from the high ranges. These intermontane basins were filled largely with river and lake deposits (Fig. 17.3). At many places the strata contain remarkably well-preserved plant, mammal, fish, reptile, and insect fossils. Volcanic ash is abundant throughout, attesting to great eruptive centers over most of the West (Figs. 17.4 and 17.5).

In the Rocky Mountain and Great Plains regions, Paleocene and Upper Cretaceous strata are conform-

On the following pages.

FIGURE 17.4 Lower Cenozoic lithofacies of North America. The majority of rocks shown in the Cordilleran region are nonmarine, intermontane sedimentary and volcanic deposits. (See Fig. 11.15 for symbols and sources.)

FIGURE 17.5 Early Cenozoic paleogeography of North America. Note that the configuration of the continent for the first time was approaching that of today, but climate still was milder. Europe had just separated, to produce Greenland-Iceland volcanism. Eocene oil shales formed in the large lakes.

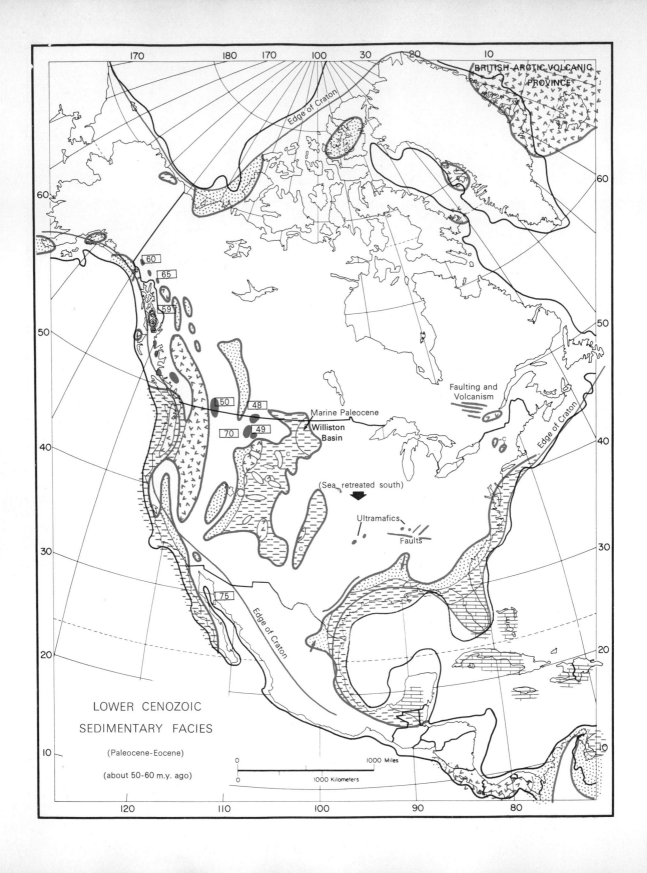

LOWER CENOZOIC

SEDIMENTARY FACIES

(Paleocene-Eocene)

(about 50-60 m.y. ago)

BRITISH ARCTIC VOLCANIC PROVINCE

Edge of Craton

Faulting and Volcanism

Marine Paleocene

Williston Basin

(Sea retreated south)

Ultramafics

Faults

Edge of Craton

0   1000 Miles

0   1000 Kilometers

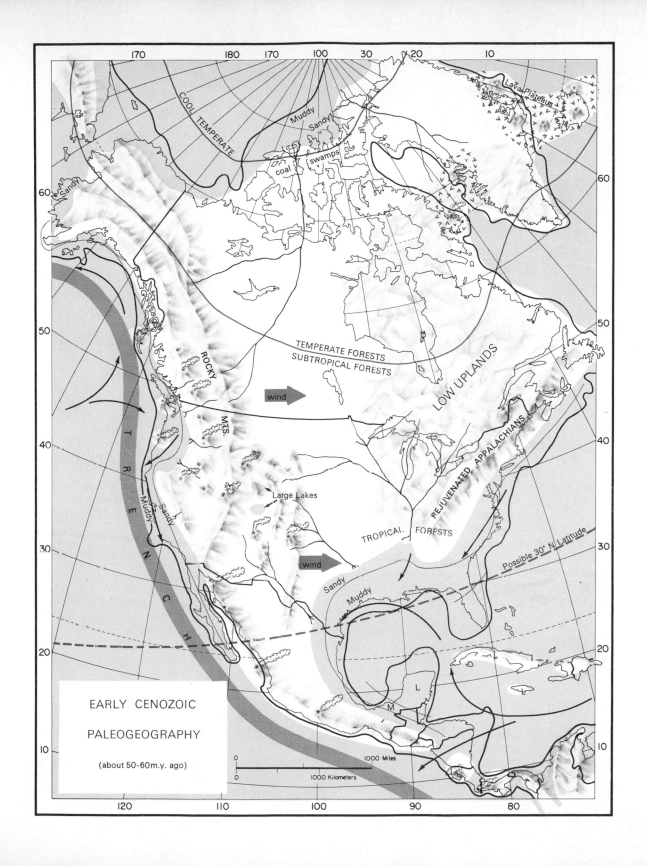

EARLY CENOZOIC

PALEOGEOGRAPHY

(about 50-60 m.y. ago)

FIGURE 17.6 Badlands topography eroded in varicolored, soft Eocene river and lake deposits of the Wasatch Formation, Bryce Canyon National Park, Utah. *(Courtesy R. B. Doremus.)*

able, except near ranges. There was no appreciable change in types of sedimentation except that coal, which formed intermittently during Cretaceous regressions, now was deposited even more widely. The last recorded vestige of marine conditions is found in North Dakota. About the end of Paleocene time, North America's last epeiric sea apparently retreated to the Gulf of Mexico.

During the Eocene Epoch, basin-filling continued, but with gradual diminution of coal. Pink, yellow, and red silts and shales are especially characteristic of the Eocene, and typically are eroded into picturesque badlands topography (Fig. 17.6). The most unusual and economically important deposit is the famous oil shale in Wyoming, Colorado, and Utah (Fig. 17.3). It originated in two huge lakes (each larger than Salt Lake) in which organisms of many kinds flourished. Microscopic plankton were so abundant periodically that their remains enriched the accumulating fine muds with an organic compound called kerogen, which can be distilled to yield petroleum-like compounds. The lake deposits also contain fresh-water fish skeletons and other fossils. On uplands surrounding the Eocene lakes, a lush forest of redwood and other trees thrived. By ecologic analogy with modern counterparts, the fossil flora points to uniformly mild, humid temperate conditions in marked contrast with the present climate (Fig. 17.5).

The last phase of the Cordilleran orogeny oc-

curred during the Eocene Epoch. By this time, the ranges stood a few thousand feet high. At several scattered localities, Paleocene and Eocene strata have been tilted steeply and are overlain unconformably by late Eocene or Oligocene gravels (Fig. 17.7), and at a few localities, Paleocene and Eocene gravels have been overthrust by Paleozoic and Prepaleozoic rocks. Most of the structural movement east of the thrust zone (Fig. 17.2) was vertical, involving large basement blocks and producing monoclinal flexures in overlying strata (Fig. 17.8). Small granitic plutons and volcanism (Fig. 17.9) also attest to structural unrest.

Late Cenozoic

Basin filling was largely completed during late Eocene and early Oligocene time. Oligocene or Miocene strata overlap beyond older ones to rest unconformably upon eroded pre-Cenozoic rocks. The net result of erosion and filling was to smooth the topography, and, as uplift and erosion slackened for perhaps 5 or 10 million years, the landscape became relatively stable. Besides filling intermontane basins, Oligocene and Miocene sediments also were spread in a great apron eastward onto the plains to form a clastic wedge sequence (Fig. 17.10). The Great Plains deposits contain some of the world's finest mammalian fossils. Associated plants show that grasslands were replacing forests at lower elevations, apparently reflecting a drying of climate, which re-

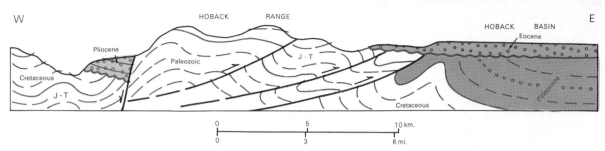

W · HOBACK RANGE · E

Pliocene

Cretaceous

Paleozoic

J-T

J-T

Cretaceous

HOBACK BASIN
Eocene

Paleocene

| 0 | | 5 | | 10 km. |
| 0 | | 3 | | 6 mi. |

FIGURE 17.7 Restored cross section showing structural and stratigraphic relationships that allow dating of tectonic events associated with formation of the Rocky Mountains in western Wyoming. *(Adapted from Dorr, 1958, Bulletin of the Geological Society of America; and Eardley, 1962, Structural geology of North America.)*

sulted at least partly from rise of the mountains that blocked the flow of moist air from the Pacific.

Beginning near the end of the Miocene time and continuing to the present, almost the entire Rocky Mountain and adjacent regions were warped up as a large unit, producing a profound effect upon erosion. In the central Rockies, considerable faulting and volcanism occurred again. Uplift rejuvenated all streams and rivers by steepening their gradients, causing a long period of downcutting. Early Cenozoic basin-fill sediments were eroded and ranges were partially exhumed (Fig. 17.11). Most dramatic has been the cutting of deep gorges superimposed directly across huge structures in hard, pre-Cenozoic

FIGURE 17.8 Monoclinal structure at Flaming Gorge on north flank of the Uinta Mountains, northeastern Utah. Prepaleozoic basement rocks were raised vertically like a plunger, carrying Paleozoic and Mesozoic strata upward to form the nearly right-angle fold. The Green River cut a deep canyon directly across hard rocks of the range during late Cenozoic rejuvenation of the Rocky Mountains. Major John Wesley Powell navigated this canyon en route to explore the Colorado River in 1869, one year before the photo was taken by pioneer photographer W. H. Jackson. *(Courtesy U.S. Geological Survey.)*

FIGURE 17.9 Yellowstone Falls and Canyon, Yellowstone National Park, Wyoming. Rock in the Canyon is altered early Cenozoic volcanic material, which gave the region its name. Volcanic activity occurred throughout Cenozoic time here and is represented today by thermal springs and geysers. Seismic as well as thermal evidence suggests that Yellowstone overlies a hot plume of magma.

rocks (Fig. 17.8). The most famous example, the Grand Canyon, was formed during late Cenozoic time as the Colorado Plateau was raised vertically.

THE CENTRAL CORDILLERA
Early Cenozoic

As the Rocky Mountains and Colorado Plateau were evolving, the axis of the mobile belt farther west was undergoing a somewhat different history. Mountains formed in late Mesozoic time still dominated the region during the early Cenozoic. They sloped westward to the Pacific Ocean, where thick marine Cenozoic strata accumulated (Fig. 17.5). Present Sierra

Nevada, Cascade, and Coast Range Mountains did not yet exist, and so rivers flowed directly west to the ocean.

Early Cenozoic sediments are poorly represented in the central Cordillera because much of the region was high and undergoing active erosion. Volcanic activity is known, however (Fig. 17.4). Famous early Cenozoic fossil floras with palms and associated mammal fossils in Oregon and Washington indicate a warm, humid subtropical climate, as do mid-Cenozoic bauxite deposits there. Meanwhile, warm-temperature hardwood and redwood forests grew in Alaska.

FIGURE 17.10 Upper Cenozoic lithofacies, reflecting great increase of volcanism and widespread deposition in intermontane basins. Note major faults in west and beneath Pacific Ocean. (See Fig. 11.15 for symbols and sources.)

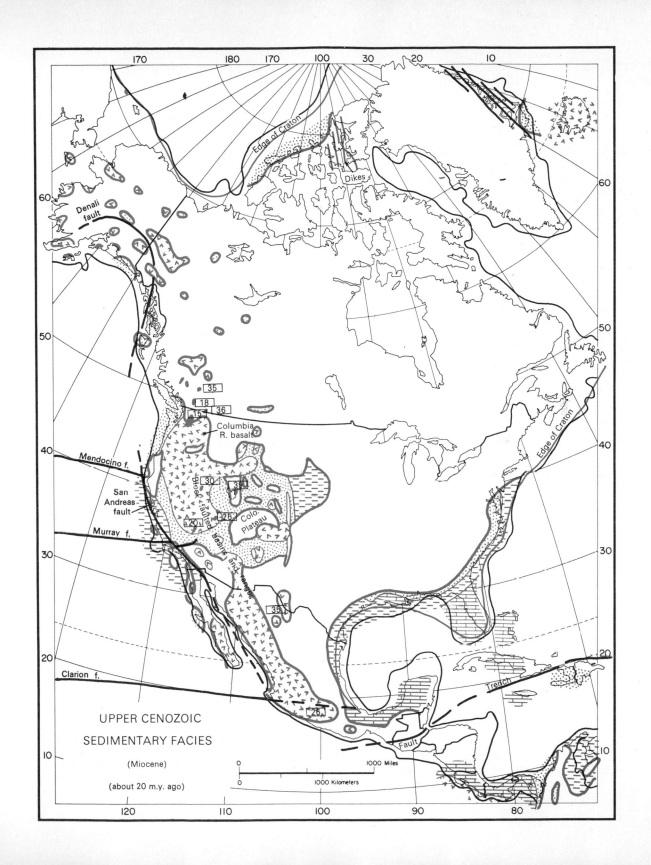

UPPER CENOZOIC

SEDIMENTARY FACIES

(Miocene)

(about 20 m.y. ago)

170   180   170   100   30   20   10

Edge of Craton

Dikes

Denali
fault

Mendocino f.

San
Andreas
fault

Murray f.

Clarion f.

Columbia
R. basalt

35

18

15   36

30

30

25   Colo.
Plateau

20

Block faulted basins and ranges

35

26

Edge of Craton

Trench

Fault

0   1000 Miles

0   1000 Kilometers

120   110   100   90   80

60   50   40   30   20   10

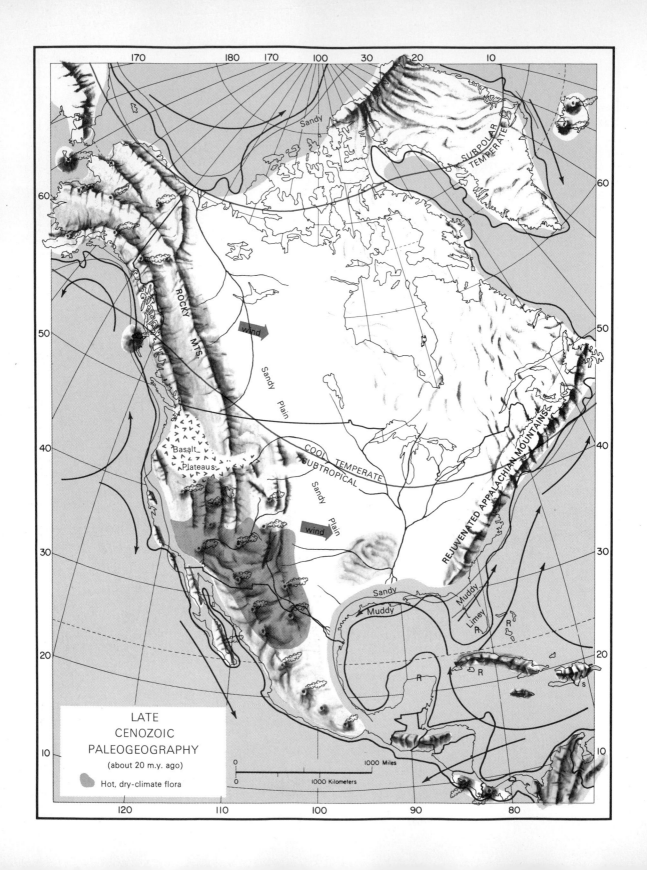

LATE
CENOZOIC
PALEOGEOGRAPHY

(about 20 m.y. ago)

Hot, dry-climate flora

Late Cenozoic Plateau Basalts

In the Oregon-Washington-Idaho region, an unusual accumulation of basalt began during Oligocene and continued locally into Pleistocene times (Fig. 17.10). Swarms of dikes represent fissures related to late Cenozoic faulting. Each lava flow spread rapidly out over very large areas (Fig. 17.12). The total region affected exceeds 300,000 square kilometers in the Columbia and Snake River regions. Flow after flow spilled out from fissures, each successive one spreading over its predecessor. Former valleys were filled with lavas totaling as much as 4,000 meters in thickness until a vast plateau without volcanic peaks was built up.

Late Cenozoic Block Faulting

Renewed structural disturbances began in middle Cenozoic time in the central Cordilleran region, as in the Rocky Mountains, and have continued to the present. But here they were much more severe, involving chiefly block faulting (Figs. 17.13 and 17.14). In Nevada, southeastern California, western Utah, southern Arizona, and adjacent Mexico, parallel, northerly trending faults produced alternating narrow ranges and valleys, giving rise to the name Basin and Range Province. The general structural configuration is that of a series of parallel horsts and grabens (Fig. 17.15).

Fragmentation of the crust of western North America was the hallmark of late Cenozoic time. Faulting provided paths of escape of magma from the bowels of the crust and mantle. Flows spread over downfaulted valleys and lapped against ranges. Erosion of ranges produced sediments that also were dumped into the valleys. Renewed faulting offset the lavas and sediments, and new volcanic outpourings then buried older, faulted rocks. Thus was the late Cenozoic block faulting and accompanying volcanism superimposed upon all of the older complex structures in the mobile belt (Fig. 17.15). Faulting also disrupted earlier drainage. For example, rivers that previously flowed west to the sea across the site of the

FIGURE 17.12   Columbia River basalt near Mitchell, north-central Oregon, showing typical flow-on-flow character of the plateau lavas erupted from fissures to cover thousands of square miles. A small fault cuts the flows.

Sierra Nevada were beheaded when the present range was faulted up in latest Cenozoic time. Nearby Death Valley came into being at the same time by unusual depression of a fault block.

PACIFIC COAST REGION

Early Cenozoic

The present coast ranges were sites of marine deposition during early Cenozoic time. Paleocene and Eocene strata appear to represent a continuation of conditions prevailing in Late Cretaceous time. Thick marine graywacke sandstone and shale sequences occur, representing turbidity current deposition beyond a narrow continental shelf. Along the Eocene shoreline, deltaic coal-bearing strata also accumulated widely. Apparently marine Cenozoic strata along much of the Pacific Coast, like Cretaceous ones before, encroached westward to produce a sort of "continental accretion" by thick sedimentation on oceanic crust followed by uplift. The underlying crust, however, has not acquired continental characteristics—at least not yet.

San Andreas Fault System

Today most people have heard of the famous San Andreas fault system of western California (Figs. 17.16 and 17.17). Many have some notion of its relation to recent earthquakes and of the threats of bisection of a winery near Hollister and of the Berkeley campus football stadium. But the full geologic significance of the San Andreas system of faults has

FIGURE 17.11   Late Cenozoic paleogeography. Regression had progressed almost to the present position of the shoreline, and climate was approaching that of today. Basalt plateaus and other volcanic materials became widespread in the west, as did block faulting, which formed most modern ranges and valleys.

been appreciated only in the past decade. The San Andreas is but one in a family of northwest-trending faults slicing diagonally across the California and southern Oregon coast ranges, and the Baja California Peninsula (Fig. 17.10).

FIGURE 17.14   Small fault scarp produced by the 1915 Dixie Valley earthquake in northwestern Nevada. Such recent movements prove the fault origin of basins and ranges in the central Cordillera of the United States.

FIGURE 17.13   Air view looking west at eastern escarpment of the southern Sierra Nevada, California; Mt. Whitney (named for California's first State Geologist) left center skyline; Alabama Hills, town of Lone Pine, and Owens River in foreground. More than 3,000 meters of topographic relief are shown. The Sierra and Alabama Hills escarpments, westernmost of the Basin and Range faults, have been active from late Miocene to present. *(Courtesy P. C. Bateman and the U.S. Geological Survey.)*

FIGURE 17.15   Diagrammatic sections showing history of the Basin and Range region; Cordilleran thrust faulting and folding (*A*: Late Mesozoic to early Cenozoic) with block faulting superimposed (*B*: Late Cenozoic or today).

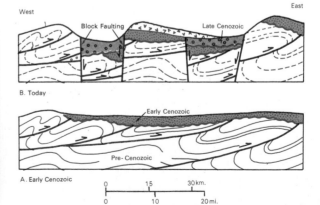

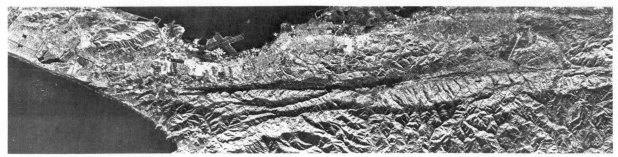

Many years ago it was argued from offsets of fences and roads during historic quakes that the San Andreas had suffered largely lateral or transcurrent rather than vertical movement. Measurable historic movements averaged about 1 centimeter per year. Recently it has been suggested that right lateral movement[1] of the San Andreas system may have been on the order of 500 kilometers and may have spanned more than 100 million years. The evidence for such staggering conclusions comes from appar-

[1]So named because as one looks across the fault from either side, the opposite block has moved to the right relative to the block on which the observer is standing (it *is* the same no matter on which side one stands). Such faults also are termed right-handed or dextral, and strike-slip or transcurrent faults.

FIGURE 17.16   Radar image (not a photo) of the San Francisco Peninsula, showing the San Andreas and related faults delineated by straight valleys. Pacific Ocean to lower left, Bay at top (note International Airport runways bordering Bay). Expansion of greater San Francisco and the construction of huge housing tracts directly athwart the fault zone foretell future earthquake damage far in excess of that of 1906. *(NASA photo No. 67-H-1362; by permission of NASA.)*

FIGURE 17.17   San Andreas fault zone of western California. Intricate folds in ductile Pliocene strata within the fault zone near Palmdale, California. In areas where rocks in the fault zone were more brittle, intense crushing is more characteristic. *(Courtesy K. O. Stanley.)*

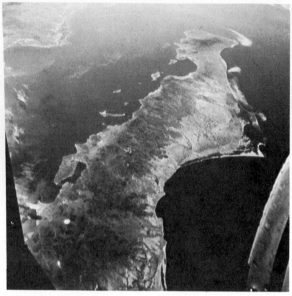

FIGURE 17.18   Baja California Peninsula (right) and the Gulf of California, a *chasm* structure (or hole in the continental crust) formed by late Cenozoic northwestward shearing of the peninsula away from the mainland along faults related to the San Andreas zone in one of the greatest real estate grabs in history (looking south from a Gemini spacecraft). *(Courtesy NASA; photo No. 66.37070.)*

ent offset of peculiar rock types. The most dramatic revelation of all is the suggestion that movements along the system have torn Baja California away from the Mexican mainland, creating a scar or chasm in the crust. Oceanic material now floors the Gulf of California, while continental crust occurs on either side (Fig. 17.18).

What of the antiquity of the fault system? If, indeed, there have been hundreds of kilometers of displacement over a long period of time, then on our maps we should restore the western side of the fault southward to its original position before drawing its paleogeography. But how far back in time would we have to go to find the inception of faulting? Fences, roads, and wineries show only recent movements. Pliocene and Pleistocene strata on either side of the fault suggest a modest movement on the order of a few kilometers. The clearest suggestion of great offset is found in Miocene rocks of the Coast Ranges. In the southern San Joaquin Valley (Fig. 17.19), the thickest Miocene strata are next to the San Andreas·fault. This fact, together with facies data, suggests that they

were chopped off. Across the fault, one has to seek the best apparent match of Miocene facies 200 kilometers farther north!

Other evidences of important disturbances include Miocene and Pliocene volcanic rocks in the California Coast Ranges, an angular unconformity within the Miocene sequence, and initial elevation of the Sierras. In Oregon and Nevada there is also evidence that major faulting was most intense in Miocene and later time. But, although Miocene and younger conglomerates show almost continuous influence of local uplifted fault blocks (Fig. 17.20), early Cenozoic strata of western California appear to be much more homogeneous and to reflect little obvious influence of faulting upon sedimentation. Therefore, it is concluded that the most intensive activity of the San Andreas system began in the Miocene Epoch.

Late Cenozoic Basins

Late Cenozoic sedimentary basins of California are shown in Fig. 17.19. Structure of the San Joaquin basin is relatively simple except along its westernmost, faulted margin. Coast Range basins, on the other hand, suffered remarkable "see-saw" tectonics—that is, areas depressed in one epoch to receive thick sediments were upheaved to form ranges in another epoch of time, thus being converted to sources of new sediments deposited in still other, depressed areas. Local conglomerates and angular unconformities abound.

In the California Coast Ranges, one is most impressed by the extreme youth of the structures, which are reflected clearly by present topography (Figs. 17.16 and 17.19). Most of the anticlines of southern California are so youthful that they still form topographic elevations. Early oil seekers soon recognized this, and immediately began drilling hilly areas within the sedimentary basins with remarkable success. In recent years, oil interest has shifted offshore, for it is known that very similar geology extends seaward (Figs. 17.21). In fact, islands and submerged basins off much of the Pacific Coast represent modern counterparts of the character of the present onshore region a few million years ago.

The Cascade Arc

The last late Cenozoic feature to note is the Cascade Range extending from southern British Columbia southward across western Washington and Oregon into northern California. The Cascade volcanoes (Fig.

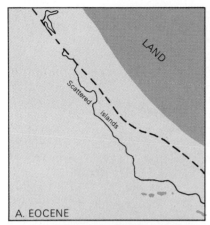

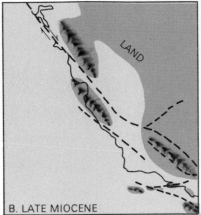

A. EOCENE

B. LATE MIOCENE

C. LATE PLIOCENE

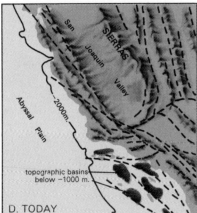

D. TODAY

FIGURE 17.19 Paleogeographic and tectonic maps showing Cenozoic evolution of California Coast Ranges. Intense faulting and rapidly changing uplift and subsidence, and marine to nonmarine conditions, have characterized this region up to the present time. *(Adapted from Hoots et al., 1954, in Guide to the geology of Southern California; Reed, 1933, Geology of California.)*

FIGURE 17.20 Angular, heterogeneous Miocene conglomerate (San Onofre at Dana Point, south of Los Angeles, California). Rubble accumulated rapidly next to an actively faulted island; this conglomerate grades abruptly into marine shales within a few miles.

17.22) represent a new volcanic arc superimposed upon older structures. They were more or less contemporaneous with the young Aleutian and southern Mexico–Central America arcs (Fig. 17.11). Mild folding, local metamorphism, and emplacement of a few small granitic batholiths also occurred along the Cascade trend. Although presently dormant, there is no reason to suppose that the Cascade arc is extinct.

A trench-arc system much like that of present South America (Fig. 17.23) clearly existed along the Pacific margin of North America from Mesozoic through early Cenozoic time, but after the Oligocene no continuous arc-trench can be identified. The Cascade volcanoes represent the only apparent vestige of arc activity between the Aleutians in the far north and the volcanoes of Central America on the south. Some workers have questioned that the Cascade Range is a true magmatic arc, because no deep

FIGURE 17.21 Stationary offshore petroleum drilling platform in Cook Inlet, southwest of Anchorage, Alaska. This is an unusual "monopod" platform designed for stability in very strong tidal currents. Other stationary platforms have three or four legs; floating (ship) platforms also are used. Petroleum occurs here in lower Cenozoic strata. *(Courtesy Marathon Oil Co. and L. C. Pray.)*

trench is associated with it today and no deep-focus earthquakes have been recorded. Certainly the volcanoes and their eruptions were of arc type, and the lack of deep earthquakes is consistent with the fact that the volcanoes are now dormant. The lack of a deep trench is probably due to the great abundance of sediments being supplied by the Columbia and other rivers draining a vast, humid and recently glaciated land; sedimentation simply kept the trench filled. By contrast, the land adjacent to the deep Peru-Chile trench (Fig. 17.23) is still empty because adjacent South America is so arid that little sediment is carried to the coast.

LATE CENOZOIC TECTONIC REVOLUTION IN THE PACIFIC

We now have seen much evidence of a widespread and profound late Cenozoic structural change in the Cordilleran region of North America. This change in structural habit began in middle Cenozoic time, that is, in the Oligocene and Miocene Epochs; it continues today. Late Cenozoic time was dominated in the central and western Cordillera from Central America to Alaska by intense faulting and volcanism, whose patterns were discordant with older structures. Farther east, epeirogenic uplift and resultant canyon cutting occurred.

The very brittle behavior of western North America in late Cenozoic time (as contrasted with its response during the earlier Cordilleran Orogeny) produced unusual heterogeneity. Nevertheless, the seemingly chaotic fragmentation can be shown to be genetically related to disturbances all around the Pacific basin perimeter. The roughly simultaneous onset of such profound changes can hardly be coincidental. We must look to the northern Pacific sea floor for a general explanation of the late Cenozoic tectonic revolution.

The present discontinuous nature of both the East Pacific spreading ridge and its associated magnetic anomalies in the northeastern Pacific (as well as the disrupted trench system) suggests fragmentation of the sea floor as well as of the adjacent continental margin (Fig. 9.13). The cause of all this disruption of formerly continuous tectonic features is now explained as the result of a collision between westward-moving North America and a part of the Pacific ridge system accompanied by the formation of immense transform faults that have since offset large segments of the ridge, the sea floor, and even the edge of the continent itself. Volcanic-arc activity continued only where a spreading ridge segment still lay west of the continent as in Oregon and Washington (Fig. 9.13). From a detailed analysis of sea-floor magnetic anomalies, geophysicist Tanya Atwater concluded in 1970 that what is now northern California first collided with the ridge about 30 million years ago (late Oligocene). As North America continued to move, more and more of the continental margin encountered the ridge, and the San Andreas transform fault grew in length (Fig. 17.24). Not later than 5 million years ago (early Pliocene), Baja California was torn away from the mainland and the Gulf of California began to form. In effect, the sliver of the continent now lying west of the San Andreas transform

FIGURE 17.22 Mount Shasta, northern California, a Cascade volcanic peak. The Cascade volcanoes were active chiefly during Pliocene and Pleistocene times and may only be temporarily dormant at present. *(Courtesy California Division of Highways.)*

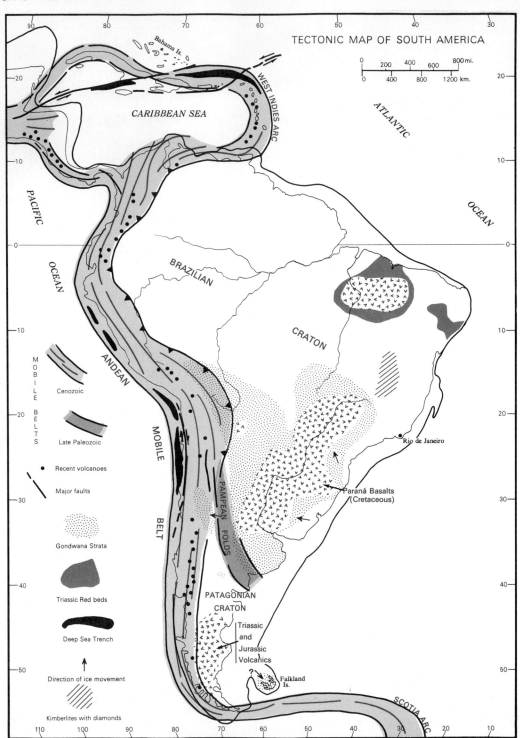

TECTONIC MAP OF SOUTH AMERICA

Bahama Is.

CARIBBEAN SEA

WEST INDIES ARC

ATLANTIC

PACIFIC

OCEAN

OCEAN

BRAZILIAN

CRATON

Rio de Janeiro

Paraná Basalts
(Cretaceous)

ANDEAN

MOBILE
BELTS

Cenozoic

Late Paleozoic

Recent volcanoes

Major faults

Gondwana Strata

Triassic Red beds

Deep Sea Trench

Direction of ice movement

Kimberlites with diamonds

MOBILE

BELT

PAMPEAN
FOLDS

PATAGONIAN
CRATON

Triassic
and
Jurassic
Volcanics

Falkland
Is.

SCOTIA ARC

was detached from the American lithosphere plate and has since moved as a part of the Pacific plate. Eventually western California seems destined either to be completely detached as a long island or to find itself becoming a part of western Canada.

Meanwhile, large volumes of old suboceanic plates were consumed beneath the Aleutian arc in the northern Pacific and beneath the arcs of the western Pacific (as well as beneath the Americas) (Fig. 17.24). Besides the magnetic anomaly patterns, another clue to Pacific plate motions is provided by deep-sea sediments. Because warm equatorial waters support the greatest populations of planktonic organisms, several times more shells of microscopic creatures such as foraminifera are deposited on the equatorial deep-sea floor than at higher latitudes. Deep-sea drilling has shown that a zone of thicker, shell-rich sediments can be recognized back to late Eocene time. The thicker zone, which is presumed to mark a past equator position, is successively farther north in older sediments and suggests a northward movement of the Pacific plate beneath the equator of nearly 10 degrees of latitude in the past 40 million years. Tentative paleomagnetic data from Cretaceous volcanic seamounts near Japan and Cretaceous sediments in the center of the Pacific suggest that the Pacific plate has moved northward about 30 degrees in the past 100 million years.

Lines or chains of volcanic islands and submerged volcanic seamounts, such as the Hawaiian-Emperor chain, give still another clue for Pacific motions. The volcanic centers along this chain become progressively younger toward the southeast (Fig. 17.24). It has been proposed that plumes of hot material rise from the base of the mantle and produce volcanic eruptions through the Pacific lithosphere plate above (see Chap. 9). It is argued that the plumes remain fixed in position for long times, but because the plate was moving continually over the plume, the lines of extinct volcanic centers trace the history of plate motion. Fossils dredged from the northernmost of the Emperor Seamounts are Late Cretaceous. Isotopic dates for volcanic rocks farther

FIGURE 17.23 Tectonic map of South America showing Paleozoic and younger features, notably Gondwana rocks and associated plateau basalts, Andean mobile belt, and modern volcanoes. *(Adapted from Jenks, 1956; Eardley, 1962; Kummel, 1961; Harrington, 1962, Bulletin American Association of Petroleum Geologists.)*

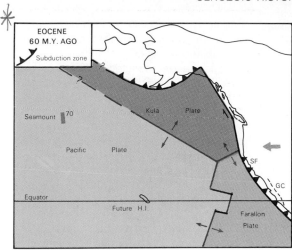

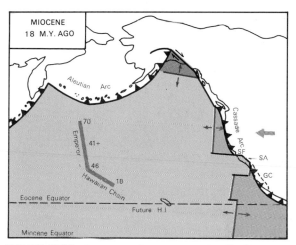

FIGURE 17.24 Evolution of the northern Pacific Ocean basin showing progressive collision of North America with ocean ridges (compare Figs. 16.5 and 17.26). Bend in Emperor-Hawaiian seamount chain may reflect change of motion of Pacific plate about 40 million years ago at the same time as initial collision of North America with the East Pacific ridge and inception of San Andreas fault. (SA—San Andreas fault; GC—future site of Gulf of California; SF—San Francisco; HI—present position of Hawaiian Islands; numbers—dates m.y. ago). *(Adapted from Atwater, 1970, Bulletin of the Geological Society of America; Grow and Atwater, 1970, ibid.; Dalrymple, Silver, and Jackson, 1973, American Scientist; Larson and Chase, 1972, Bulletin of the Geological Society of America; Winterer, 1973, Bulletin American Association of Petroleum Geologists; Pitman, Larson, and Herron, 1974, The age of the ocean basins, special map by Geological Society of America.)*

southeast indicate ages from 46 million years just north of the prominent bend in the chain to 18 million years at Midway Island, and from 6 million to 0 years for the Hawaiian Islands themselves (again with younger dates southeastward). If the Emperor-Hawaiian chain indeed was formed by passage of the Pacific plate over a hot mantle plume, then the sharp bend in the chain must record a major change of direction of plate motion around 35 to 40 million years ago (Oligocene). Each volcanic center was built from the abyssal sea floor 6,000 meters below sea level in only 3 to 5 million years. The older, northern islands subsided beneath sea level to form a submarine ridge.

A mid-Cenozoic date for a major change of Pacific plate motion from south-to-north to southeast-to-northwest is very close to the time of initial collision of the American plate with the East Pacific ridge and with a rejuvenation of activity in the Aleutian and all the western Pacific island arcs (Figs. 17.24 and 17.25). It seems to be one more evidence of major mid-Cenozoic tectonic changes that were of global importance. There is considerable evidence, for example, that Japan and several of the other western Pacific island arcs formed from slivers of continental crust torn from mainland Asia and that New Zealand was torn from Australia, resulting in the formation of new small marginal ocean basins behind them. While the separation of New Zealand is thought to have

occurred in late Mesozoic time, the separation of Japan and formation of the Marianas and other arcs apparently began in mid-Cenozoic time when the other major tectonic changes occurred in the Pacific (see Fig. 9.13). Local sea-floor spreading behind certain arcs seems to produce the marginal basins, but the ultimate cause of that spreading is not yet proven. In the days when continents were seen as fixed and as always growing larger by accretion, the western Pacific arcs were imagined to be embryonic mobile belts that would eventually become additions to Asia or Australia. It is an indication of how profoundly our views of global evolution have changed that now many of those same arcs are seen as slivers *removed* from continents rather than as new additions.

THE LATE CENOZOIC REVOLUTION OUTSIDE THE PACIFIC REGION

EARLY CENOZOIC EVENTS

Major plate reorganizations also occurred in Cenozoic time outside the Pacific region. Greenland began separating from Europe in the Paleocene or Eocene (60 to 65 m.y. ago) as the northern mid-Atlantic ridge formed. Widely scattered Eocene basalt flows and swarms of dikes mark this event in Ireland, Scotland, Iceland, and Greenland (see Fig. 17.4). On the other side of the world, Antarctica and Australia finally separated and went their opposite ways, and India completed its separation from Africa with outpourings of huge volumes of basalts in northwestern India. The southern continents had been very stable as evi-

FIGURE 17.25 Mount Fuji, Japan, one of many late Cenozoic active volcanoes of the western Pacific volcanic belt. Fuji lies along the Fossa Magna rift that cuts central Honshu. *(Courtesy Japan Air Lines.)*

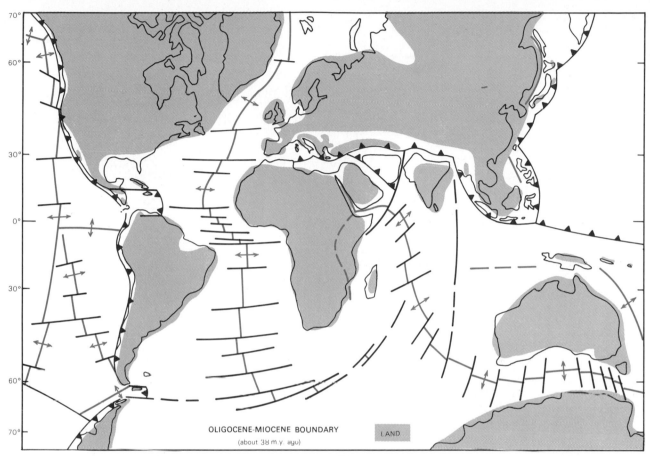

OLIGOCENE-MIOCENE BOUNDARY
(about 38 m.y. ago)

LAND

FIGURE 17.26 Restoration of the continents and ocean basins for mid-Cenozoic time (Oligocene-Miocene), showing impending collisons of Afro-Arabia and India with Eurasia, Australia with Indonesia, and parts of North and South America with the East Pacific ridge. *(Data from same sources as Fig. 16.3, Fig. 17.24.)*

denced by huge erosion surfaces with old lateritic soils. In mid-Cenozoic time, profound tectonic changes occurred on a worldwide basis. New activity in the Circum-Pacific arcs accompanying changes in the Pacific plate was described above. The great Alpine-Himalayan mountain system was formed, and important rifting with associated volcanism occurred on at least four continents—Africa, Europe, Asia, and Antarctica (Fig. 17.26).

THE ALPINE-HIMALAYAN BELT OF EURASIA
While the entire Pacific margin was experiencing intense deformation and volcanism associated with late Mesozoic and Cenozoic mountain building, the Tethyan region between Gondwanaland and Eurasia continued to experience deformation, which culminated in the upheaval of the Alps, Himalayas, and

other, less-famous ranges of Asia Minor in late Cenozoic time (Fig. 17.1). Old continental crust on the southern margin of Eurasia, which was disturbed previously by the late Paleozoic Hercynian Orogeny (see Fig. 14.30), was involved in this Cenozoic upheaval. In addition, however, Mesozoic oceanic (Tethyan) rocks were thrust northward onto the old continental margin and crushed together with the Hercynian basement. As tectonic activity intensified in Cretaceous time, islands were pushed up near the north side of the shrinking Tethys seaway. Rapid erosion of them produced clastic debris that periodi-

LEVELS OF EROSION IN MOBILE BELTS

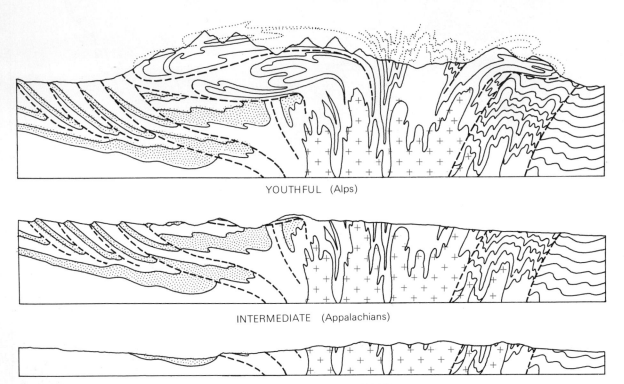

YOUTHFUL (Alps)

INTERMEDIATE (Appalachians)

MATURE (Prepaleozoic Shield)

FIGURE 17.27 Diagrammatic comparison of effects of different levels of erosion on exposed characteristics of mobile belts. Nappe folds shown at top may seem unique to the Alpine belt solely because it is so youthful and shallowly eroded.

cally was swept by turbidity currents into the deep-water basins between the islands much as sediments are now being deposited in basins among the many islands of the Aegean Sea between Greece and Turkey. The result was rhythmic alternations of Cretaceous to Eocene sandstone and shale originally named flysch in Switzerland (see Fig. 16.28). Less frequently, huge, chaotic submarine slides occurred. As compression increased, the area and height of uplifted terrain increased rapidly. The culmination of compressive deformation occurred in Oligocene and Miocene time, when sedimentation changed to very thick nonmarine gravel and sands as well as some deltaic, coal swamp, and shallow-marine deposits,

all collectively named molasse long ago in Switzerland. Some of the world's most complex thrust faulting accompanied by a plastic folding of strata into immense recumbent folds (called nappes in Switzerland) characterize the Alps. Similar structures are inferred for older mountain belts as well, but erosion has cut more deeply into those and removed much of the upper parts of the great folds (Fig. 17.27). Alpine granites formed around 30 million years ago, but metamorphism continued locally until 11 million years ago. The great vertical uplift of the earth's youngest mountains continues today.

In the last chapter we noted that fragmentation of the southern margin of the Tethys seaway accompanied the formation of a series of spreading sites. One result was the formation of a number of microcontinents separated by new, small-ocean basins. It is believed that, as the units moved northward toward Eurasia, there was subduction of considerable oce-

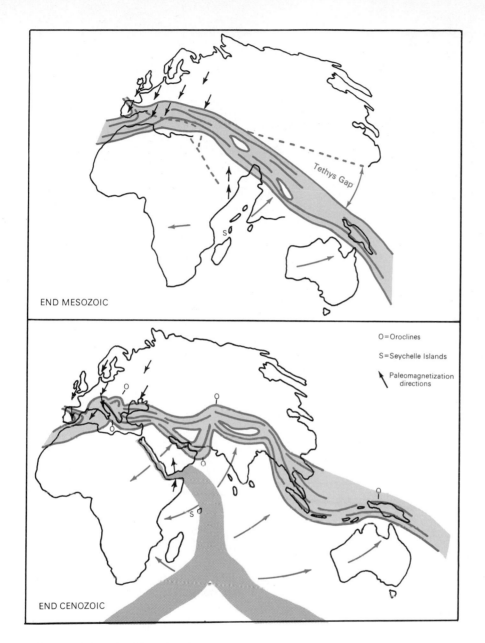

END MESOZOIC

O = Oroclines

S = Seychelle Islands

↗ Paleomagnetization directions

END CENOZOIC

FIGURE 17.28 Possible relation of continental displacements to sea-floor spreading from the Indian Ocean ridges and bending of mobile belts. According to S. W. Carey, the Tethyan and other mobile belts originally were straight; bends developed due to rotations during continental displacements. Divergences of paleomagnetic vectors (lower map) in Europe (Permian) and in eastern Africa and Arabia (early Cenozoic) seem to confirm rotation. (Adapted from Carey, 1959; van Hilten, 1964, Tectonophysics; McElhinny, 1967.)

anic material; andesitic volcanism occurred along the continental margin. Finally, just as F. B. Taylor suggested long ago (see Fig. 8.9), India collided with Asia to produce the Himalayas. Arabia and Iran also collided with Asia to produce the Taurus, Zagros, and other mountains of Asia Minor, and parts of north Africa collided with Europe to give birth to the Alpine mountain systems (Fig. 17.28). Corsica and Sardinia were torn from Spain and France and rotated counterclockwise to their present positions near Italy. Italy itself had previously been torn from northern Africa. Meanwhile, at least part of Indochina may have separated from Gondwanaland and collided with China in late Mesozoic time. It was in turn nudged by

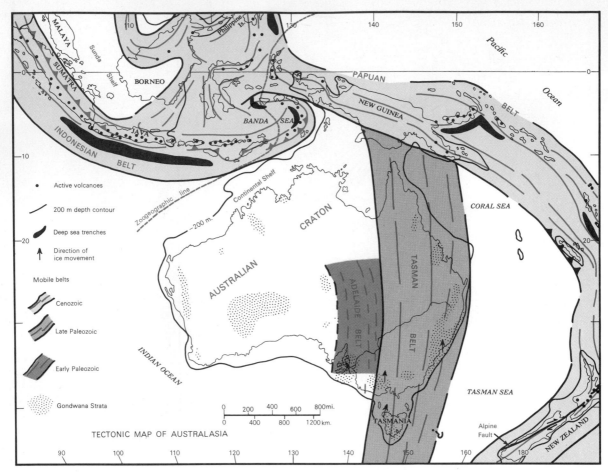

FIGURE 17.29 Tectonic map of Australia and adjacent island arcs with active volcanoes, showing Paleozoic mobile belts and Upper Paleozoic strata containing glacial deposits and *Glossopteris* flora. Note inferred south-to-north movement of ancient glaciers. Collision of Australia-New Guinea with eastern Indonesia occured in mid-Cenozoic time. *Zoogeographic line* marks separation of modern Asiatic and Australian biotas (compare also Fig. 13.34). *(After David, 1950; van Bemmelen, 1954.)*

tary rocks. Gravity measurements in the Himalayan region suggest that the thickest known continental crust in the world probably underlies these highest of all mountains, that is, about 70 kilometers, or twice the normal thickness. It has been suggested that collision of the continents here drove Indian crust beneath Asiatic.

Active subduction is continuing in the northern

north-traveling Australia in the late Cenozoic (Fig. 17.29). Where two continents collided along the old Tethyan zone, the resulting mobile belt is more or less structurally symmetrical. Old oceanic ophiolite rocks commonly mark a suture zone between the two formerly separate continents, and outward thrust faulting generally occurred toward each craton. While most of the former Tethyan oceanic material was consumed by subduction, some was thrust up onto the sedimen-

FIGURE 17.30 Tectonic map of Africa showing late Paleozoic and younger features (see Fig. 15.2 for Prepaleozoic and early Paleozoic "Pan-African" mobile belts). Note especially Gondwana (Karroo) rock distribution, glacial movement directions, Cenozoic rifts, plateau lavas, and recent volcanoes. *(Adapted from Du Toit, 1954, Geology of South Africa; Furon, 1963; Holmes, 1965; and Sougy, 1962, Bulletin of the Geological Society of America.)*

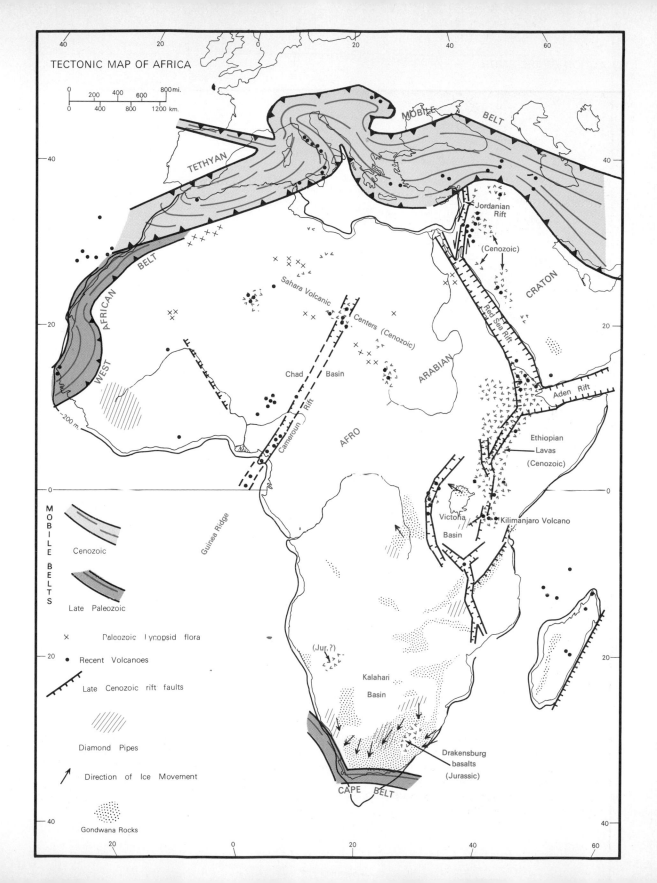

TECTONIC MAP OF AFRICA

Scale:
0 200 400 600 800 mi.
0 400 800 1200 km.

TETHYAN

MOBILE   BELT

Jordanian
Rift

(Cenozoic)

CRATON

WEST

AFRICAN

BELT

Sahara Volcanic

Centers (Cenozoic)

Red Sea Rift

ARABIAN

Aden Rift

Chad   Basin

-200 m.

Cameroun Rift

AFRO

Ethiopian
Lavas
(Cenozoic)

Guinea Ridge

Victoria
Basin

Kilimanjaro Volcano

MOBILE BELTS

Cenozoic

Late Paleozoic

× Paleozoic lycopsid flora

• Recent Volcanoes

Late Cenozoic rift faults

Diamond Pipes

Direction of Ice Movement

Gondwana Rocks

(Jur.?)

Kalahari

Basin

Drakensburg
basalts
(Jurassic)

CAPE BELT

FIGURE 17.31 Mount Erebus (3,794 meters), an "active" volcano on the shore of the Ross Sea, McMurdo Sound, Antarctica; view looking north from Observation Hill with monument to ill-fated Robert F. Scott (British) South Pole Expedition of 1910–1912 in foreground. Erebus formed along a late Cenozoic rift zone. *(Courtesy R. F. Black.)*

Mediterranean region and in Indonesia, where oceanic trenches, deep-focus earthquakes, and andesitic volcanism are still active. In both cases, oceanic material is still being underthrust northward (Figs. 17.29 and 17.30). It seems a safe bet that the Mediterranean will eventually be closed by the ultimate complete collision of northern Africa with Europe. In other words, the Mediterranean seems to represent an intermediate stage in the destruction of the Tethyan realm, which process has already been completed farther east—from Turkey to Burma.

CRATONIC RIFTING AND MANTLE PLUMES

While Pacific arcs were being rejuvenated and the Alpine-Himalayan orogeny was in progress, extensive rift faulting accompanied by outpourings of immense volumes of lava affected large parts of the cratons of Africa, southern Siberia (Baikal rifts), central Europe (Rhine graben; see Fig. 14.30), and Antarctica (Fig. 17.31). This volcanism was unusual in that it occurred within, rather than at the edges of, lithosphere plates. Large-scale upwarping generally preceded rifting. In Africa some plateau areas were raised as much as 2,000 meters, and then their crests collapsed along normal faults to form huge grabens (see Fig. 10.23). The resulting rift valleys in east Africa have become famous for their remains of early humans. Less famous is the west African Cameroun rift and accompanying late Cenozoic volcanism, which extends more than 3,000 kilometers from the central Sahara southwest to the coast and thence into the Atlantic as the Guinea ocean ridge (Fig. 17.30).

Most of the African rifting began in Oligocene or Miocene time and continues today. Associated volcanoes, including Africa's tallest, Kilimanjaro (5,995 meters), produced olivine basalts like those typical of oceanic regions together with peculiar alkaline (calcium- and sodium-rich) lavas. The rifts average 50 kilometers in width, but the Red Sea and Gulf of Aden rifts are twice as wide, and they extend to oceanic depths (2,000 meters below sea level). The latter have oceanic crust beneath their axes and are zones of unusually high rates of heat flow through the crust. Hot solutions beneath the central Red Sea apparently are now creating ore deposits there. Twenty years ago the suggestion by Australian geologist S. W. Carey that these two huge rifts resulted from Africa being literally pulled 200 kilometers away from Arabia, tearing open a long scar in the continental crust was regarded as outrageous. Now it is almost universally accepted, and paleomagnetic evidence has confirmed that Africa was rotated 11 degrees counterclockwise away from Arabia (Fig. 17.28). The Jordan or Dead Sea rift discussed in Chap. 1 is narrower than most, and it represents a dominantly strike-slip offshoot from the Red Sea rift. It is most like the much larger Gulf of California. Offset streams and lavas indicate a total of 100 kilometers of movement since Miocene time. Although displacements have been geologically rapid in this region, one might wish that they could be greatly accelerated in order to solve local political problems by creating natural oceanic moats in drowned rifts between contending nations.

The hypothesis outlined above for the northern Pacific of hot plumes of mantle material more or less fixed in position has also been applied to Africa. Each of the major late Cenozoic volcanic centers (Fig. 17.30) is supposed to mark such a plume, but unlike the Pacific plate, Africa seems to have remained nearly fixed over its inferred plumes for the past 40 million years.

COASTAL PLAINS AND CONTINENTAL SHELVES

EMBRYONIC MOBILE BELTS OR NOT?
The history of the North American craton and mobile belts has been traced from Prepaleozoic through Cenozoic times. This study has shown that mobile belts typically evolved through a long history of mountain building sometimes accompanied by formation of granitic batholiths, finally to become stabilized as parts of the craton. Analysis of continental evolution would be incomplete, however, without a further look at the youngest coastal plain strata and their seaward extensions beneath the continental shelves, which formed partly over the sites of old mobile belts at present continental margins (Fig. 17.2).

The northern margin of the Gulf of Mexico con-

tains beneath it by far the thickest Cenozoic sequence of any of the coastal plains. The strata are almost 12,000 meters thick near the Mississippi delta, while in Mexico and along the Atlantic Coast, they are only one-fifth as thick. Under the Arctic coastal plain, Cenozoic sediments are still thinner. In all cases the strata dip gently seaward and are thickest near present coastlines (Fig. 17.32). Knowledge of submerged extensions of the deposits beneath the shelves comes from offshore deep drilling for petroleum and from geophysical surveys. The staggering thickness of sediments beneath the northern Gulf of Mexico shelf was first revealed by gravity measurements, which give an estimate of the depth to the dense basement. Seismology provides insight into the structure of the young sediments themselves and may clarify the nature of the basement.

Important questions surround the coastal plain–continental shelf strata. Do they represent embryonic geosynclines forming at present continental margins and destined *inevitably* to become mountains? For several years many geologists thought so. As we emphasized in early chapters, however, such strata were deposited on structurally passive margins totally unrelated to mountain building. They may be later

FIGURE 17.32 Cross section of Gulf of Mexico showing general configuration of Cenozoic continental-shelf and coastal-plain deposits (north), Sigsbee abyssal plain (center), and Yucatan carbonate reef platform (right). *(Adapted from G. I. Atwater, 1959, 5th World Petroleum Congress Transactions; and Ewing et al., 1955, Geophysics.)*

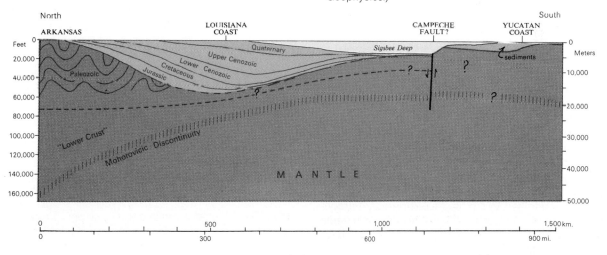

FIGURE 17.33 Miocene, richly fossiliferous, shallow-marine strata in Chesapeake Bay region; these are typical of the Atlantic coastal plain. *(Courtesy Thomas G. Gibson.)*

caught up in mountain-building processes however, if an active (converging) plate margin should encounter them.

THE ATLANTIC COASTAL PROVINCE AND APPALACHIAN REJUVENATION

The Atlantic coastal plain is the simplest of our three examples. Cenozoic strata are only about 1,000 meters thick, and clastic sediments predominate; carbonate facies are confined to southernmost Georgia and Florida (Figs. 17.4 and 17.10). No Cretaceous or Cenozoic volcanic rocks are known. During early Cenozoic time, a tropical marine fauna thrived, implying a clear, warm sea. Texture and composition of sands suggest that the adjacent Appalachian region was low (Fig. 17.5) and undergoing mature weathering. Fossil floras of the southeastern states and the development of bauxite in Arkansas also attest to very warm, humid conditions.

After an Oligocene regression, the sea returned, but brought a rich, cooler-water Miocene fauna (Fig. 17.33). A mid-Cenozoic stratigraphic unconformity also is recognized in the Gulf Coast region, Europe, and many other regions as well. Probably it reflects the same profound cause as the change of spreading direction in the Pacific and the collision of North

America with the East Pacific ridge. Our previous discussions have shown that this has, in fact, happened many times in the past. The point is that coastal plain-type strata are in no way a requirement for mountains; their presence in some ancient mobile belts is accidental.

In the present Appalachian Mountains, there is conspicuous topographic evidence that, after Mesozoic beveling, regional epeirogenic upwarping caused rivers to re-expose the buried old landscape (Fig. 17.34). As the area rose, gradients of major rivers were rejuvenated, and soft, early Cenozoic and Cretaceous strata were stripped off, exposing the harder, deformed rocks of the old mobile belt. Rivers managed to maintain their flow directions originally established on younger deposits, and their drainage patterns became superimposed upon old rocks as the region slowly rose. Entrenched meanders winding through anomalously narrow valleys and water gaps in resistant ridges are the results. Tributaries took advantage of relatively softer strata to carve out long, narrow valleys between parallel mountain ridges to produce ridge and valley topography.

Topographic rejuvenation and the cutting of youthful gorges across resistant, old rocks also typified the Rocky Mountain region during late Cenozoic time, suggesting that epeirogenic upwarping probably affected erosion over the entire continent.

FIGURE 17.34 Diagrammatic evolution of modern Appalachian Mountains by epeirogenic rejuvenation and entrenchment of rivers. *(After Douglas Johnson, 1931, Stream sculpture on the Atlantic slope; by permission of Columbia University Press.)*

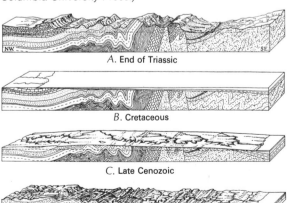

A. End of Triassic

B. Cretaceous

C. Late Cenozoic

D. Today

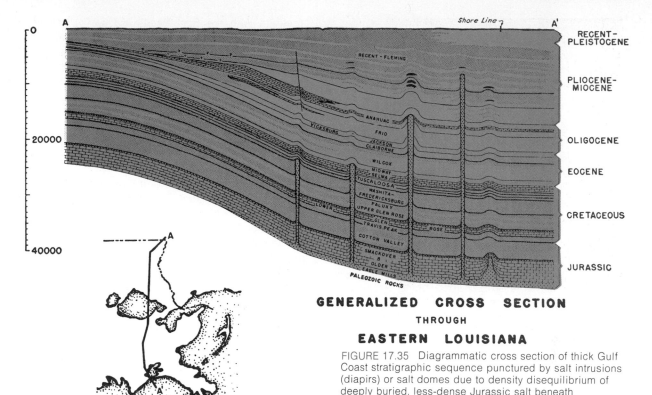

GENERALIZED CROSS SECTION

THROUGH

EASTERN LOUISIANA

FIGURE 17.35 Diagrammatic cross section of thick Gulf Coast stratigraphic sequence punctured by salt intrusions (diapirs) or salt domes due to density disequilibrium of deeply buried, less-dense Jurassic salt beneath more-dense, younger strata. Salt rises in ridges, cylinders, and "teardrops" until isostatic equilibrium is achieved. See also Fig. 15.21. *(After J. B. Carsey, 1950,* Bulletin American Association of Petroleum Geologists, *v. 34, pp. 361–385; used by permission.)*

GULF OF MEXICO COASTAL PROVINCE
General Patterns

After the Appalachian-Ouachita mountain building ceased in Late Triassic time, sedimentation commenced again in the Gulf Coast region during the Jurassic Period (see Fig. 16.14). As noted in the last chapter, it is now thought that the Gulf of Mexico was formed in Jurassic time as Gondwanaland split from North America. As we found in discussing the break-up of Gondwanaland, the first deposits formed were not normal marine ones. Rather they were evaporites and red beds deposited in shallow, restricted lagoons. By the end of Jurassic time, normal marine deposition ensued and continued into the Cretaceous Period. Cretaceous deposits include chiefly limestones and shales with some important reef developments. Cenozoic deposits are composed largely of terrigenous sands and shales with only minor carbonates. Enlargement of the continent, coupled with a gradual change of climate that accelerated erosion, caused this change. The Mississippi and other mod-

ern river systems had formed at least by Miocene time. Having large drainage basins, these streams brought immense volumes of sediment to the northern Gulf, and began building great deltas as today.

Following the great Cretaceous transgression, a Cenozoic regression commenced, and the axis of maximum sedimentation continually shifted southward (Fig. 17.32). The continental shelf prograded seaward, for it is but a sedimentary embankment produced by spreading of material in shallow, agitated water. At the same time, subsidence of the crust has occurred by broad, regional, isostatic downbending accompanied by some faulting under the load of such a large volume of material (see Fig. 12.24).

Evaporite Intrusions

The most interesting features of the Gulf Coast province are intrusions of evaporite material through Cretaceous and Cenozoic strata (Fig. 17.35). Such intrusions occur over an immense area, even well beyond the continental shelf margin. Pollen contained in the salt in Louisiana indicates a Jurassic and Triassic(?) age for the parent evaporites (see Fig. 16.14).

Evaporite intrusions, better known as salt domes or diapirs, take several forms. The simplest are roughly cylindrical, extending downward to the deeply buried parent evaporite zones. Their tops are domal in shape, and they arch overlying sediments as they rise. Complex faulting also is produced over them; graben blocks are common as are radial faults like fractures produced around an impact hole in glass. Internal structure of the domes is fantastically complex due to continuous, slow plastic flow of the evaporite. Sediments are pierced and steeply tilted along flanks of the intrusions. In many cases, the tops are mushroom-shaped, and it is thought that some of the intrusions are teardrop-shaped, being pinched off at depth. Some domes join in elongate evaporite ridges. The rise of the domes has been intermittent as evidenced by unconformities and complex facies patterns around them.

The origin of evaporite intrusions is explicable by simple physical considerations. The density of a buried evaporite stratum is less than that of overlying compacted sediments (2.5 to 2.7 at depths exceeding 3,000 meters). The small density contrast of 0.3 to 1.5 produces an isostatic disequilibrium, and so deeply buried, low-density evaporites will rise slowly in mound-like knobs, first arching, then piercing overlying material. As upward flow continues, the intrusions become more elongated and may pinch off at their bottoms. Upward intrusion ceases when the evaporite reaches a position of density equilibrium. A number of domes that reached the surface were, of course, the first such structures to be discovered. Subsequently, because oil was found to be trapped against and over salt domes, there was a great impetus to discover buried ones. A breakthrough was provided by the torsion balance in the 1920s and gravity meter in the 1930s. The density contrast between evaporites and sediments was easily detectable with these instruments so that a rash of major petroleum discoveries followed their invention. Seismic methods later were used in locating deep domes and associated fault structures (Fig. 17.36).

The Gulf Coast region has the largest North American petroleum reserves, but it also is important for salt and sulfur. Salt domes have a curious cap rock composed of carbonate, evaporite, and pure sulfur. Pressure from the upward rise of the intrusions plays a role in the formation of cap rock through a series of chemical reactions. Bacterial attack of petroleum also may be important. Calcium sulfate ($CaSO_4$) from the evaporites probably reacts with carbon dioxide (CO_2) to form carbonate rock ($CaCO_3$), releasing the sulfate ion, which in turn could react with water trapped in the sediments to produce pure sulfur.

ARCTIC COASTAL PLAIN

Concepts of mobile belts and coastal plain regions already were well advanced when modern geologic studies began in the remote Arctic. So it was with satisfaction that geologists greeted confirmation that the structural symmetry of North America was complete on its north margin. Full documentation of the existence of the Franklin belt proved that almost the entire perimeter of the craton was rimmed by mobile belts of Paleozoic or younger age, and all had similar rocks and structures.

The Arctic coastal plain provides yet another important parallel with the better known margins of the continent. The Arctic example is somewhat unique in having many early Cenozoic basaltic dikes, but it bears a striking resemblance to the Gulf Coast in that it, too, contains evaporite intrusions. Eroded gypsum domes are very evident over the barren northern islands; source of the evaporite is in upper Paleozoic strata. Naturally the prolific petroleum production associated with Gulf Coast evaporite structures has attracted much interest in the petroleum potential of the Arctic.

LIFE OF THE CENOZOIC

EXTINCTIONS

The abrupt change in faunal content at the era boundaries is most noticeable in the extinction of previously dominant forms. This is usually followed by a rapid expansion of previously inconspicuous groups. The Permo-Triassic and Cretaceous-Cenozoic boundaries are both marked by this phenomenon, but different animal groups were involved. Interestingly, marine faunas were most affected in the earlier event, whereas land faunas and floras were most affected at the latter boundary. It should be noted that there was a total of 50 percent extinctions at the close of the Permian versus only 26 percent in Late Cretaceous

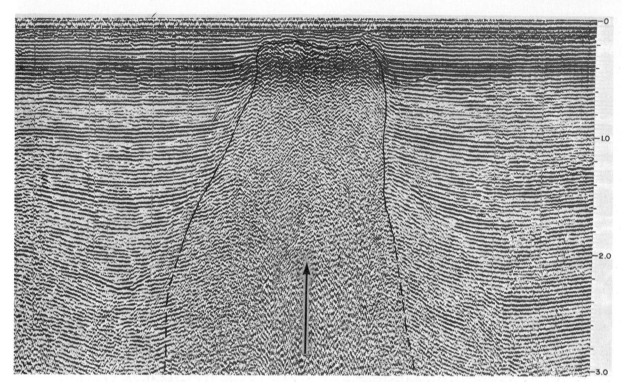

FIGURE 17.36 High-resolution seismic cross section of Freeport salt dome off the Texas coast. Vertical scale is in seconds required for reflection of acoustic energy from within the sediments back to the water surface. Modern geophysical and geological methods have led to an unprecedented 70 percent petroleum discovery record off Louisiana and Texas. *(Courtesy Teledyne Exploration, Marine Sciences Division.)*

time. In the Late Cretaceous there was a plant plankton bloom which sharply decreased CO_2, causing a widespread deposition of chalk (composed of the plant plankton, coccolithids, see Fig. 17.37). By the very end of the Cretaceous this bloom ended, and many marine organisms dependent on plant plankton were directly affected.

In addition the climate at the end of the Cretaceous had become slightly cooler and may have contributed to the extinction of the dinosaurs and the replacement of the flowering plants (angiosperms) by temperate conifers (gymnosperms) in the cooler regions.

INVERTEBRATES

Except for extinction of the ammonoids and extinctions within a few other classes, the marine fauna

FIGURE 17.37 Electron micrograph of plates of a modern coccolith from Atlantic Ocean (scale in micrometers; 1 micrometer = 0.001). Such plates make up most of the famous European Cretaceous chalk deposits. Coccoliths are unicellular plants that float in surface waters. *(Courtesy Andrew McIntyre.)*

FIGURE 17.38 A Jurassic shrimp (crustacea) from the Solenhofen Limestone of Germany. *(Courtesy U.S. National Museum.)*

appears to have been less affected by events of the end of the Mesozoic. Disappearance of conspicuous organisms such as reef-forming clams (rudistids) and some typical Mesozoic snails and sponges made the Cenozoic marine fauna appear different. There was a rapid expansion of the Foraminifera, corals, advanced snails, clams, echinoids, and crustacea (Fig. 17.38). As the climate deteriorated toward the end of the Cenozoic, the broad faunal provinces of today were formed.

FLOWERING PLANTS

The character of the land surfaces changed radically after the Jurassic. Beginning in Early Cretaceous time, the angiosperms (flowering plants) began a rapid period of explosive evolution. Most of today's common plants can be recognized in early Cenozoic floras. This expansion, as mentioned in Chap. 16, was accompanied by a parallel evolution of insects. The final important expansion occurred in mid-Cenozoic time, when prairie grasses appeared, opening many new habitats for grazing mammals. As with the fauna, the flora rapidly expanded and developed provinciality with increasing climatic differentiation. Colder regions were dominated by angiosperms that dropped leaves or otherwise formed hibernation features, while in the tropics most plants were evergreens.

VERTEBRATES

We now must devote our full attention to the adaptive radiation of the mammals. By latest Cretaceous time,

mammals, as seen in deposits of western North America, gained in sheer numbers what they lacked in body size. In addition to the primitive trituberculates, multituberculates, and other groups that became extinct at the close of the Mesozoic Era or in early Cenozoic time, two important groups appeared—the marsupials and the insectivores (Table 17.1). It is the shrew-like insectivores that probably served as the stock from which all other mammals arose.

On the Advantage of Being a Mammal

Except for the curious egg-laying, duck-billed platypus of Australia, and the marsupials (including our opossum and most native Australian mammals), mammals reproduce by placental development so that young are born alive, in contrast to most other vertebrates, which lay eggs. In addition, the young are suckled by the mother. This characteristic has great importance, when combined with the much more highly developed mammalian brain and a very long period of prematurity, for it allows the young time to be trained and provides the groundwork for a structured social order. Social order is observed only rarely in lower vertebrates, which must rely on instinctive behavior. In fact, intelligence, or the use of an ever-evolving brain capacity, to a large extent accounts for the success and evolution of mammals during the Cenozoic. The brain and other features of the skull reflect this evolution.

The mammals were and are far more successful vertebrates in terms of environment because they are warm-blooded. Maintenance of a constant body temperature (by a metabolism utilizing food converted to energy and stored energy, and by hair) allows the animal to maintain sustained activity, unlike the reptiles, fish, and amphibia, whose activity is dependent on air or water temperature. Mammalian teeth are differentiated for biting, chewing, and grinding. This allows them to utilize different types of food and hence to move into a large number of environments.

Early Cenozoic Mammalian Faunas

The earliest Cenozoic mammals were fairly small herbivores not too unlike their Mesozoic ancestors (Fig. 17.39). By middle Paleocene time, however, the picture abruptly changed. The ungulates or hoofed mammals became abundant and are represented by five-toed condylarths.

By Eocene time (Fig. 17.40), the condylarths had

TABLE 17.1 Classification of the Mammalia
  Mammalian groups are based on types of placental or nonplacental development on the higher levels (subclass). Orders and smaller units of fossils are based on skeletal characters, particularly skull and tooth features. The classification below is abbreviated, listing only those groups discussed in the text or of general interest

Subclass Prototheria
    Order Monotremata   Pleistocene-Recent. Includes egg-laying duck-billed platypus
Subclass Allotheria
    Order Multituberculata   Jurassic-Eocene
Subclass Theria
  Infraclass Pantotheria
    Order Pantotheria   Jurassic. Includes the trituberculates
  Infraclass Metatheria
    Order Marsupialia   Cretaceous-Recent. Includes the opossum, kangaroo, etc.
  Infraclass Eutheria
    Order Insectivora   Cretaceous-Recent. Includes the shrews of today
    Order Chiroptera   Eocene-Recent. Bats
    Order Primates   Cretaceous-Recent. Includes apes and man
    Order Carnivora   Paleocene-Recent. Includes cats and dogs
    Order Perissodactyla   Paleocene-Recent. Odd-toed hoofed mammals, such as the horse
    Order Artiodactyla   Eocene-Recent. Even-toed hoofed mammals, such as pigs
    Order Condylarthra   Cretaceous-Miocene
    Order Pantodonta   Paleocene-Oligocene. Includes the coryphodonts
    Order Hyracoidea   Paleocene-Recent
    Order Embrithopoda   Paleocene-Recent
    Order Astrapotheria ⎤ Paleocene to Recent
    Order Pyrotheria ⎟ This group of orders is
    Order Litopterna ⎟ restricted to South America
    Order Notoungulata ⎦ and includes hoofed forms
    Order Proboscidea   Eocene-Recent. Includes the elephant
    Order Cetacea   Eocene-Recent. Includes the whales
    Order Rodentia   Paleocene-Recent. Includes the mice
    Order Lagomorpha   Paleocene-Recent. Includes the rabbits

achieved sizes of our modern large mammals. Just as in all balanced faunas where herbivores exist, the carnivores are sure to be handy to help stabilize populations. The carnivores are first represented by forms which possessed small hoofs rather than the more typical carnivore claw. Their teeth, good carnivore types, were fairly well-developed blade-like teeth that were used to shear flesh like a pair of scissors. The hoofed carnivores became extinct at the end of the Eocene. Other carnivore groups in turn gave rise to cats and dogs by Oligocene time and finally to the hyenas, bears, seals, raccoons, civets, etc., by Miocene time.

In addition to a large number of Paleocene orders that were still extant in the Eocene, many new orders came into existence. Two of these important new orders are the artiodactyls, represented today by pigs, giraffes, deer, and camels; and the perissodactyls, having as modern descendants the rhinos, horses, and the rarely encountered tapir. Two interesting early perissodactyl groups include a curious type that had large claws instead of hoofs and was fairly large sized, and another type which was rhino-like (*Uintatherium*) and well endowed with horns (Fig. 17.40). Both of these groups had molars with fused cusps with the dentine forming a double V pattern well adapted for grinding up plant food.

Horses

The most important perissodactyl of the Cenozoic from our point of view is the horse. There are two reasons for our interest: one is that they have a continuous fossil record and are fairly numerous as vertebrate fossils go, and the other reason is that we can interpret the evolution over the course of the era better than in most other groups. Our friendly Victorian "bishop-eater," T. H. Huxley, first observed an evolutionary trend in the horses, but it was an Ameri-

FIGURE 17.39   An opossum, a primitive mammal (marsupial) from Virginia. Note the differentiated teeth (i.e., front vs. rear). *(Courtesy American Museum of Natural History.)*

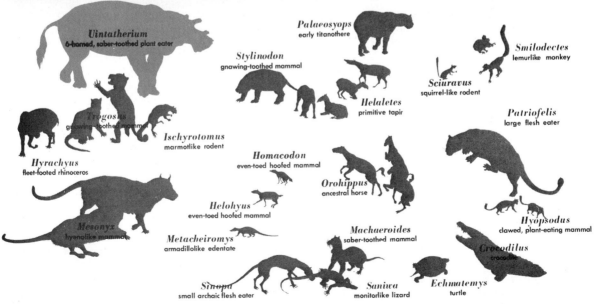

Uintatherium
6-horned, saber-toothed plant eater

Palaeosyops
early titanothere

Smilodectes
lemurlike monkey

Stylinodon
gnawing-toothed mammal

Sciuravus
squirrel-like rodent

Helaletes
primitive tapir

Patriofelis
large flesh eater

Trogosus
gnawing-toothed mammal

Ischyrotomus
marmotlike rodent

Hyrachyus
fleet-footed rhinoceros

Homacodon
even-toed hoofed mammal

Orohippus
ancestral horse

Hyopsodus
clawed, plant-eating mammal

Helohyus
even-toed hoofed mammal

Mesonyx
hyenolike mammal

Machaeroides
saber-toothed mammal

Crocodilus
crocodile

Metacheiromys
armadillolike edentate

Sinopa
small archaic flesh eater

Saniwa
monitorlike lizard

Echmatemys
turtle

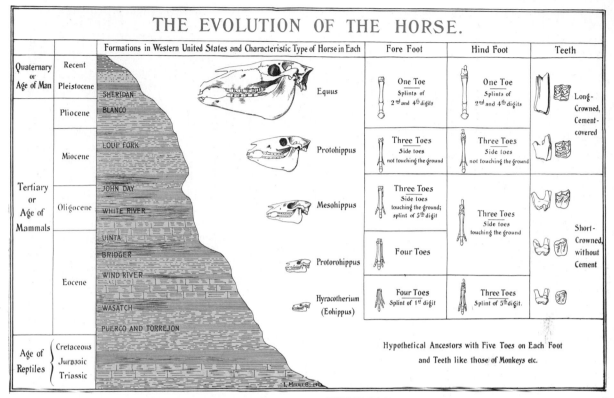

THE EVOLUTION OF THE HORSE.

		Formations in Western United States and Characteristic Type of Horse in Each		Fore Foot	Hind Foot	Teeth
Quaternary or Age of Man	Recent		Equus	One Toe Splints of 2nd and 4th digits	One Toe Splints of 2nd and 4th digits	Long-Crowned, Cement-covered
	Pleistocene	SHERIDAN				
	Pliocene	BLANCO				
Tertiary or Age of Mammals	Miocene	LOUP FORK	Protohippus	Three Toes Side toes not touching the ground	Three Toes Side toes not touching the ground	
	Oligocene	JOHN DAY / WHITE RIVER	Mesohippus	Three Toes Side toes touching the ground; splint of 5th digit	Three Toes Side toes touching the ground	Short-Crowned, without Cement
	Eocene	UINTA / BRIDGER / WIND RIVER	Protorohippus	Four Toes		
		WASATCH	Hyracotherium (Eohippus)	Four Toes Splint of 1st digit	Three Toes Splint of 5th digit.	
		PUERCO AND TORREJON				
Age of Reptiles	Cretaceous / Jurassic / Triassic					

Hypothetical Ancestors with Five Toes on Each Foot
and Teeth like those of Monkeys etc.

FIGURE 17.41   H. F. Osborn's concept of evolution of the horse, considered to be a simplification today (see p. 411). *(Courtesy American Museum of Natural History.)*

can, H. F. Osborn, who first documented the history in a thorough way (Fig. 17.41) and used it as an example of orthogenesis (see Chap. 5). Up until recently, we were under the illusion that his facts were correct and fully developed. Intensive collecting over the past 30 years has shown that the orthogenetic picture drawn by Osborn is a gross oversimplification. Since our treatment must be brief, we will follow his simple picture, recognizing that the trends are not always consistent within groups.

In general, there is a progressive evolutionary pattern involving a whole series of related character complexes in the main phylogenetic line. As in most observed evolution, there is an increase in general body size from the earliest dog-sized horse of the early Cenozoic to the horse-sized horse (*Equus*) of today. It is important to know that this was not a slow persistent increase because the rates of change

FIGURE 17.40   North American Eocene mammals. Note the early horses in right center of the painting by J. H. Matternes. *(Courtesy U.S. National Museum.)*

differed in the various evolutionary lines of descent.

Another relatively persistent evolutionary trend involved the foot and leg complex. The basic mammalian foot consists of five digits. The early Cenozoic horse already had displayed the trend for a reduction of the number of digits because it possessed four toes on the front legs and three on the rear legs. By late Eocene and Oligocene time there were three toes on all legs; the middle toe was elongated with the nail developed into a hoof, and the side toes were reduced. By Miocene time, the middle toe became even more elongated, taking over the principal walking and running function (Figs. 17.41 and 17.42). This trend was accompanied by a change of the molar teeth from a low-crowned cuspate tooth to a high-crowned almost flat-topped tooth which became increasingly elongate and with a more complex enamel pattern through time. With this change there was a deepening of the jaws to accommodate the longer

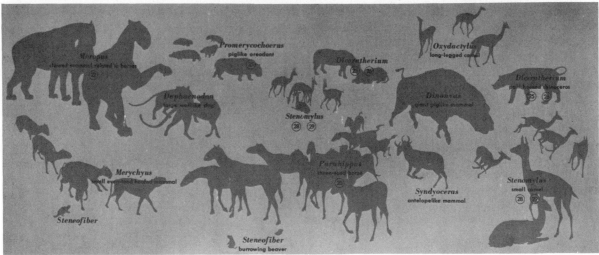

FIGURE 17.42 North American Miocene mammals. Note the large, clawed perissodactyl (*Moropus*) on left; horses in the lower center are three-toed. *(Painting by J. H. Matternes, courtesy U.S. National Museum.)*

teeth and a lengthening of the face. Most of these changes are thought to be in response to a change of diet, and thus a change in habitat, from a leaf-eating and probably forest-dwelling (browsing) animal to an open prairie (grazing) animal. At this time (the early Miocene), prairie grasses had evolved, thus providing a new environment. The deep roots of such grasses fit them unusually well for semiarid climates. Since the horse was becoming a highly efficient runner, it was possible for him to move out into the open country and to escape predators by running rather than hiding. This hypothesis is a simplification, but floral and faunal associations tend to suggest that it has a ring of plausibility.

The enormous number of horse fossils found in the intermontane basins indicate that most of the evolution of the horse took place in North America. The first major invasion of the Eurasian land mass through Siberia probably occurred in Miocene time when a more primitive leaf-eating stock appeared there. It was not until the Pliocene that true grass-eaters became established in the "Old World." By late Pliocene or early Pleistocene time, the modern horse, *Equus*, had evolved, and surviving Pliocene genera became extinct. In Eurasia, *Equus* became highly successful and spread rapidly. However, in the Western Hemisphere the horse became extinct, for no known reason, and was brought back to North America in the 1500s by the Spanish.

Other Mammals

By Miocene time, the diverse perissodactyls (Fig. 17.42) began to decline, mostly by the extinction of primitive groups. The more successful artiodactyls began to expand at their expense. Most of the larger mammals of the mountains, plains, and forests of today, such as the llamas, camels, deer, and the domesticated sheep, goats, and cattle, attest to the success of their ruminant digestive system (the pigs and their relatives lack this unique digestive process).

Together with the artiodactyls, rodents and, to a lesser extent, elephant groups became diverse from middle Cenozoic on, although they declined sharply in the Pliocene-Pleistocene. The elephants have a fairly well-documented phylogeny, and showed some interesting adaptations, particularly during Oligocene to early Pliocene times (Fig. 17.43). The main line, which led to the elephants, mastodonts, and mammoths, rapidly became large-sized and developed tusks of remarkably different shapes and sizes originating from either or both jaws, indicating markedly different habitats.

Cenozoic mammalian faunas of the world are sufficiently well documented to enable us to establish migration routes and patterns for reconstructing their zoogeography. The background and understanding of these patterns are highly important in our analysis of the coming of man.

SOME ASPECTS OF CENOZOIC BIOGEOGRAPHY

PROBLEMS OF DISPERSAL

The biogeography of the recent flora and fauna of the world is the result of a complex history resulting from isolation, evolution, contraction, dispersal, the unique events of Pleistocene glaciation, and, above all, the appearance of the most destructive of all known species through the history of life—man. Contraction of the geographic range of a widespread group is the result of intervening extinctions due to changes of environmental factors, such as the rapid climatic alterations in the Pliocene and Pleistocene. Such contractions resulted in many scattered distributions of living forms.

More than 20 years ago, American biologist and vertebrate paleontologist G. G. Simpson recognized that much of the distribution of Cenozoic land faunas and floras could be explained as the result of migrations from one land mass to another via corridors such as land bridges (like the Isthmus of Panama) or island chains (such as Indonesia), and combinations of subsequent isolation (as on islands) and contraction of geographic ranges due to adverse climatic changes or the appearance of competitors. The most primitive biotas tend to be those that have been most isolated. His explanation still seems to account for the curious disjunctive (isolated) distribution, for example, of the camels in north Africa and southern Asia far from their cameloid relatives (llama, etc.) in South America. The Cenozoic fossil record shows that camels formerly were widespread over Asia and North America in between, and so there must have been a continuous geographic range in Pliocene time. Since then, however, the range contracted due to extinctions in North America and Asia.

Presence of modern oceanic island biotas far from continents, as on the Galápagos Islands (see Chap. 5), proves long-distance dispersal is possible for some organisms. Peculiarities of these biotic communities, however, indicate that dispersal is selective and somewhat accidental. New Zealand, for example, has an "unbalanced" fauna replete with birds, but totally lacking in indigenous mammals, turtles, snakes, and fresh-water fishes. Similarly, Madagascar has a few amphibians and reptiles, one shrew, one primate, one carnivore, and some rodents.

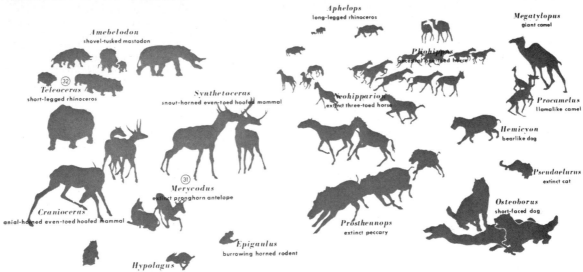

FIGURE 17.43 North American Pliocene mammals. Note the shovel-tusked elephant in the upper left of the painting by J. H. Matternes. *(Courtesy U.S. National Museum.)*

Though that fauna is clearly of African origin, no larger mammals or reptiles are present.

Dispersal across water of many birds, some reptiles, and insects is easily envisioned, as is that of light-weight plant seeds adapted for long-range wind transport; all these are common on islands. But only rare, very improbable introductions of other plant and animal forms have occurred since Jurassic time, probably by bird transport of seeds, or even less-probable rafting of seeds and small land animals on drifting logs, icebergs, or (recently) ships. When G. G. Simpson pointed out the importance of chance or "sweepstakes" dispersals, he noted that if the chances of an event happening were say one in a million in any 1 year's time, it would be raised to a 63 percent chance in 1 million years, and to 99 percent in 10 million years (equivalent to post-Miocene time). Thus a seemingly impossible event (by human standards) may become possible, while the improbable becomes inevitable given enough time! While this is dazzling mathematics, it overlooks the added improbabilities that enough individuals will be dispersed *and* find the proper ecologic niche in which to survive *and* reproduce to establish viable populations. Probability of success in a new location is considerably lower than that of freak arrivals, and probability of successful dispersal of whole communities of many coadapted plants and animals is still smaller.

SOUTHERN TEMPERATE FOREST

Even allowing for chance rare events over long time, some disjunct organisms are difficult to rationalize without island bridges or former closer proximities of lands. Out of some 20 groups of trees and shrubs making up the southern temperate forests, two trees, the southern beech (*Nothofagus*) and a southern conifer (*Araucaria*), together with associated insect communities, are especially interesting (Figs. 17.44 and 17.45). Today they live only south of the equator, and are closely adapted to cool, humid climatic conditions. Yet they are widely separated (i.e., disjunctive) across seemingly insurmountable water barriers. *Nothofagus* seeds typically are windborne only a few hundred meters, even during gales; the tree has failed to populate islands only a few tens of kilometers offshore from New Zealand. On the other hand, *Nothofagus* forests lie in the strongest wind zone on earth. Experiments show that in sea water, which is

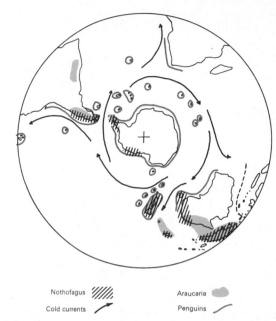

Nothofagus ////   Araucaria ●

Cold currents ⟋   Penguins ⟋

FIGURE 17.44   Distribution of certain distinctive Southern Hemisphere organisms that suggest Antarctica as an important former land bridge. Distribution of penguins is explained easily by patterns of cold currents; early Cenozoic fossil penguins are known in New Zealand, Patagonia, and Antarctica, suggesting origin in the south. Distribution of *Araucaria* and *Nothofagus* trees seems to require at least land-bridge connections among southern continents until Late Cretaceous time (both trees are now extinct in Antarctica; *Araucaria* also in New Zealand and South Africa). Dashed line is Australia–Asiatic biogeographic boundary (see Fig. 17.29). *(Adapted from Murphy, 1928, Problems of polar research, American Geographical Society Special Publication No. 7; Gressitt et al., 1963; Darlington, 1965, Biogeography of the southern end of the world.)*

detrimental to their viability, the seeds soon become water-logged and sink. Yet *Nothofagus* logs are known to drift from Chile to Tasmania. Could seeds have been transported on drifting logs (note currents in Fig. 17.44)? It is judged unlikely that birds dispersed these particular trees because oceanic species rarely frequent the *Nothofagus* forests.

The Cenozoic record in the northern Antarctic Peninsula contains marine and terrestrial fossils, including giant Miocene penguins. Important plant fossils include both *Araucaria* and *Nothofagus*. These and related plants have lived since Cretaceous time

FIGURE 17.45 Straits of Magellan from Bahia Fortesque, southern Chile (Chilean naval vessel in middle ground). This region is characterized by the southern beech (*Nothofagus*) forest community, nearly identical today in Chile, New Zealand, Australia, and New Guinea (see Fig. 17.44). The beech forests were publicized first by Charles Darwin after *H.M.S. Beagle* passed through the Magellan Straits (1832–33).

in widely scattered Southern Hemisphere lands (Fig. 17.44). Their presence in Antarctica leads to two important conclusions. First, some means of biologic dispersal was possible among Australia, New Guinea, New Zealand, Antarctica, and South America in Cretaceous or early Cenozoic time; in all probability, Antarctica served as a land bridge. Since then, however, these lands have become more ecologically isolated. Secondly, the climate, at least in the relatively low-latitude Antarctic Peninsula, still was temperate enough until mid-Cenozoic time (Miocene) for forests to thrive. Continental glaciation then brought to extinction all indigenous land life in Antarctica.

SOME CENOZOIC ANIMALS

When Simpson first proposed his explanation of Cenozoic biogeography, continental drift was not yet

respectable. He attempted to explain all distributions of Cenozoic mammals using the Bering and Panamanian land bridges as well as one Cretaceous or early Cenozoic "sweepstakes" migration of marsupials to Australia from Asia. Now that drift is better accepted and the timing of separations of different land masses is fairly well established, we can improve upon his story considerably. For example, there is no known fossil record of marsupials in southeastern Asia, and we now know that Australia was far from Asia until late Cenozoic time (Fig. 17.46). Therefore, it seems less plausible to derive Australia's fauna from Asia as Simpson suggested.

As noted above, the ancestral southern temperate forest community apparently dispersed over South America, Antarctica, Australia, New Zealand, New Guinea, and New Caledonia in Cretaceous time when these land masses were all grouped in a supercontinent (Fig. 17.46). New Zealand separated from Australia about 80 million years ago (Late Cretaceous), Australia and Antarctica remained together until 45 to 50 million years ago (late Eocene), and it appears that the Antarctic Peninsula maintained a close connection with South America until mid-

Cenozoic time. Apparently early land mammals and birds also dispersed together with the southern forest. Large flightless birds similar to the ostrich reached New Zealand by mid-Cretaceous time. Their relatives also lived in South America, South Africa, and Australia. Of greatest interest, however, are the marsupial mammals, who first appear in mid-Cretaceous strata of North America and in the Upper Cretaceous in South America. Marsupials as well as the placental mammal groups called edentates (such as the armadillos, sloths, and anteaters of today) and ungulates (hoofed forms) presumably reached South America from the north. By Eocene time, the marsupials had begun to diversify, some becoming carnivores, others rat-like. The ungulates and edentates also underwent considerable evolution, which resulted in a rich and unique fauna, including forms that resembled elephants, camels, horses, rabbits, and pigs. None of these groups is known outside South America, indicating isolation there since early Cenozoic time. Monkeys and rodents were added to this fauna from the north by middle Cenozoic time, but the New World monkeys are unrelated to those in Africa. It is thought that these forms might have "island hopped" from Central America. Both primates and rodents were already well established in North America and Europe.

It is in Australia that the marsupials have had their glory, for there they have developed the most fascinating and strangest examples of faunal isolation. Only two native groups of mammals are known, the monotremes (duck-billed platypus; see p. 411) and the marsupials. Unfortunately the fossil record of mammals in Australia is very poor. There is a gap from Triassic to Oligocene, which makes it difficult to establish the origin of the fauna. The monotremes, though very primitive forms, are not known as fossils

FIGURE 17.46 Suggested major dispersal patterns of mammals from Cretaceous to present, showing probable role of continental displacements both in isolating (for example, marsupials in the south in B) and in allowing mixing at different times via land bridges and island stepping stones. Also shown is disjunctive southern temperate (*Nothofagus*) forest, which illustrates contraction of its range. (B—Bering bridge, which lies out of map projection). *(Adapted from Raven and Axelrod, 1972, Science, v. 176, pp. 1379–1386; Fooden, 1972, Science, v. 175, pp. 894–898; Tedford, 1974, in Soc. Econ. Paleo. and Mineral Spec. Publ. No. 21, pp. 109–126; and unpublished sources.)*

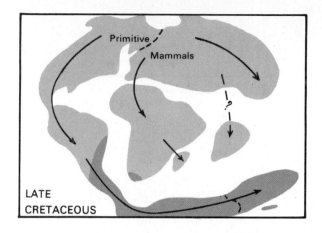

LATE CRETACEOUS

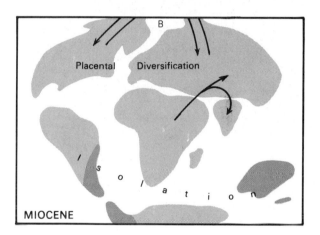

MIOCENE

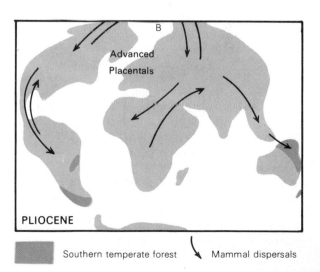

PLIOCENE

Southern temperate forest   Mammal dispersals

prior to the Pleistocene, and marsupial fossils first appear in Oligocene strata. The record does show, however, that the marsupials underwent a great adaptive radiation to fill many unoccupied ecological niches. There are even more different marsupials here than in South America, including forms that resemble placental dogs, cats, wolves, rabbits, squirrels, pigs, rodents, bears, and horses of other continents. Others such as the kangaroos have no counterparts among either placental mammals or marsupials on other continents. New Zealand, on the other hand, which separated from Australia near the end of the Cretaceous Period, had no native mammals until man arrived.

Although we cannot rule out an origin for the marsupials in Australia or Antarctica, the fossil record suggests that they probably originated instead either in North America and migrated into South America in Cretaceous time, or vice versa. From South America, apparently they migrated via Antarctica to Australia (Fig. 17.46). Exactly when they arrived "down under" is not clear, as noted above, but presumably it was before late Eocene time when these two continents separated from one another. The absence of any marsupials from New Zealand may indicate that that island had already separated from Australia before their arrival. Although there are other factors that also could explain such an absence, the lack of snakes in New Zealand may also indicate isolation of the island before snakes evolved in the latest Cretaceous. New Zealand, together with New Caledonia and Norfolk Islands (between New Zealand and Australia), illustrates a commonly observed situation whereby the most (or longest) isolated biotas tend to be the most primitive and least diverse.

Meanwhile the placental mammals had also appeared and were flourishing in the north by the beginning of the Cenozoic Era. Whether they originated in Eurasia or North America is not known, but early placental ungulates and edentates noted above entered South America at the beginning of the Cenozoic, where they became isolated with the marsupial immigrants by separation of the Americas. In Pliocene time, the building of the Panamanian Isthmus by volcanism allowed a new influx of more advanced placentals from the north, including camels, horses, deer, and some carnivores (Fig. 17.46). Meanwhile, ground sloths, armadillos, opossums, and porcupines moved north over the new land bridge (Fig.

17.46). One of the most puzzling questions concerning Cenozoic mammalian distributions is why the early placentals did not reach Australia. They were abundant in North America and Eurasia throughout the Era, and they entered South America together with the marsupials as just noted. Perhaps climate or some other environmental factor permitted the early, probably less-specialized marsupials to filter across the southern access route, but prohibited more specialized placentals from being fellow travelers.

In Africa, unlike South America and Australia, only placental mammals are known. They show no relation to groups found in South America, which probably reflects the relatively early separation of those two continents. The earliest known African mammalian fauna is of late Eocene age. It contains ancestral elephants and other forms that still dominate the African fauna today. Madagascar received an influx of primitive placentals from Africa, but after the Eocene they seem to have become isolated from the mainland fauna. After permanent northern contact was established by the collision of Africa and Eurasia in Miocene time, there were some important interchanges of mammals (Fig. 17.46). Camels, elephants, and lions are well-known examples. The Australia-New Guinea land mass encountered southeastern Asia (Indonesia) in Miocene time. Extensive uplift of the edge of the southern plate created a host of new ecologic niches, and there was a prompt influx of tropical Malaysian plants and animals by island hopping and rafting. This made possible for the first time some mixing of primitive Australian and more advanced Asiatic mammals and plants.

India presents its own peculiar puzzles. The mammal record there is from Eocene to Recent. But the Eocene mammals are most akin to those of Eurasia and North America rather than, say, Africa! Yet, geologic and oceanographic evidence suggests that the continent was far to the south of Asia and had already separated from Gondwanaland. In short, it would seem to have been isolated by the Eocene. So we have an unexplained problem—how could northern mammals have gotten to India so early? In Miocene time, when Afro-Arabia collided with Iran and Turkey, African forms migrated across Asia Minor into India and tropical southeastern Asia (Fig. 17.46). From this Miocene fauna developed the Indian elephant and lion. Still later, other immigrants, including monkeys, also came in.

The puzzles presented by mammalian fossil distributions remind us of the following favorite quotation, which serves as a reminder that we do not have all the answers yet:

If observations agree with our theory, that is nice, but if they do not, that is interesting. (An Anonymous Wise Man)

SUMMARY

In this chapter we have examined the development of western North America since the culminating Cordilleran orogeny. The central Cordillera underwent vigorous erosion during early Cenozoic time, while thick marine sedimentation continued along the Pacific margin. In the present Columbia-Snake River region, immense outpourings of basalt began in mid-Cenozoic time. In late Cenozoic time, a profound change in structural habit affected all of western North America. It was characterized by brittle fragmentation of the crust and extreme volcanic activity, although granitic batholiths formed locally. Block faulting created the Basin and Range Province from Oregon into Mexico. The Sierra Nevada was raised at the west edge of that region, and major transcurrent faulting also commenced (or at least intensified) in the present Coast Ranges. Faulting in western Oregon and Washington led to eruptions that built the majestic Cascade volcanoes, which seem to represent a new volcanic arc. Large-scale fragmentation of western North America is the result of movement of the eastern Pacific basin against the west edge of the continent after collision of the continent with a portion of the East Pacific spreading ridge. Since then, the San Andreas system has acted as a transform fault. Baja California and the California Coast Ranges have been moved north along the San Andreas system up to 200 kilometers in the past 25 million years.

While the Cordilleran region has suffered severe and rapid tectonic changes during late Cenozoic time, the eastern Cordillera, and much of the remainder of the continent, have been gently unwarped. Rejuvenated rivers coupled with Pleistocene glaciation produced the present Rocky Mountain landscape. The modern Appalachian Mountains were topographically (but not orogenically) rejuvenated as

well, causing entrenchment of large, meandering rivers whose courses were established in Cretaceous or early Cenozoic time.

We have seen that, after a last major Cretaceous transgression, regression characterized the Cenozoic Era. But not only has the continent become larger in area, its average elevation also has increased due to mountain building and widespread epeirogenic upwarping. Increase of area and elevation—i.e., greater continentality—naturally has resulted in more erosion and therefore accelerated clastic sedimentation both on and adjacent to the continent. Pleistocene glaciation, a subject of the next chapter, accelerated such sedimentation even more.

The rest of the world also experienced a tectonic revolution. Regression of the sea was worldwide as all continents were elevated and the world sea-floor-spreading regime underwent important changes beginning in mid-Cenozoic time. The change of direction of motion of the Pacific plate, formation of several western Pacific arcs by the tearing away of microcontinent slivers from Asia, the upheaval of the great Alpine-Himalayan mountain systems by continental collisions, and important rifting of at least four continents are the major results. The rifting and spreading changes seem destined to bring about a new phase of continental drift.

Coastal plains and continental shelves represent tranquil accumulations of sediments on relatively passive crust at the edges of continents. Most of the structural deformation in them is due to intrusions of low-density evaporite masses. Relative thickness of strata within the plains and shelves is simply a function of the magnitude of sediment delivered by particular rivers and available for lateral spreading by marine processes. What subsidence of the crust has occurred on the shelves seems due only to isostatic loading and compaction. Rather than being sites of inevitable future accretions of continental crust, the shelves are simply marine accumulations on passive coasts (or trailing edges) following separations of the continents in Mesozoic time.

In the previous chapter, we demonstrated that warm temperature characterized most of the Northern Hemisphere about 90 million years ago (Late Cretaceous). Evidence from early Cenozoic plant fossils shows a persistence of mild conditions, perhaps even with a slight warming. Paleobotanist Ralph Chaney showed in 1940 that humid, subtropical, and mild-temperate climate plants covered large areas of North America (Fig. 17.5). Latest Cretaceous or early Cenozoic fossil redwoods (*Sequoia*) occur at 70°N latitude in Alaska, Greenland, and Siberia, and 78°N in Canada and Svalbard (Spitsbergen); palm trees grew as far north as 50°N latitude. Moreover, climatic zonation suggested by plant distributions was closely parallel with the present equator, conforming with paleomagnetic evidence that the continent had essentially its present relation to the pole and equator. Floras very similar to Asiatic ones spread widely over the continent. Seas were slightly larger, the continent was somewhat lower, and climatic zones were wider than today. By late Cenozoic time, the climatic zones, as suggested both by land and marine fossils, were narrower and were shifting southward (Fig. 17.11). Cooler temperatures are indicated for the first time. A trend toward greater aridity had developed, especially in the southwestern United States and adjacent Mexico as indicated by shifts of humid- and arid-climate plant communities. A contributing factor was rapid elevation of the Cascade Range and Sierra Nevada near the Pacific Coast. Early Cenozoic floras were so uniform as to suggest no sharp east-west climatic differentiation. But by late Cenozoic time, westerly winds were disrupted, creating aridity to the east (rain shadow). In the west, grasslands enlarged and forests shrank to moist river beds and high, cool uplands. A number of plants known today only in Asia (e.g., *Gingko*) vanished from western North America at about this time, and the ranges of others contracted greatly. For example, as recently as Pleistocene time, redwood and Douglas fir trees grew on islands off southern California 200 kilometers farther south than at present. A parallel change from browsing to grazing habits occurred among many mammals as an adaptation to changing environment. Some important mid-Cenozoic extinctions of mammals may have resulted from failures to adapt, but on the whole, mammals proved well suited for late Cenozoic changes of topography, climate, and vegetation.

Evolution of modern land plants and animals was greatly influenced by the breakup of Pangaea and the separation of its component continents. Some groups, such as the marsupials, underwent unique evolution after being isolated early in South America and Australia. The placentals, however, had free run of Eurasia, Africa, and North America, but for the most part reached South America and Australia very late.

Climatic and other environmental changes in late Cenozoic time have reduced former much broader geographic ranges of many organisms, leaving widely separated disjunct outposts (e.g., camels of Africa-Asia and cameloids of South America) due to extinctions over intervening territory. The final character of the modern flora and fauna resulted from events of the Pleistocene ice age, during which evolved the most adaptable and destructive species of all time—man.

Readings

Atwater, T., 1970, Implications of plate tectonics for the Cenozoic tectonic evolution of western North America: Bulletin of the Geological Society of America, v. 81, pp. 3513–3536.

Axelrod, D. I., 1950, Evolution of desert vegetation in western North America: Carnegie Institution of Washington Publication 590.

Brinkmann, R., 1960, Geologic evolution of Europe: New York, Hafner. (English translation by J. E. Sanders)

Burke, K. and Whiteman, A. J., 1973, Uplift, rifting and the break-up of Africa: *in* Continental drift, sea floor spreading and plate tectonics: New York, Academic.

Chaney, R. W., 1940, Tertiary forests and continental history: Bulletin of the Geological Society of America, v. 51, pp. 469–488.

Colbert, E. H., 1955, Evolution of the vertebrates: New York, Wiley. (Paperback edition, 1961, Science Editions, No. 099-S)

Coleman, P. J., (ed.), 1973, The Western Pacific island arcs, marginal seas, geochemistry: report of the 12th Pacific Science Congress, Canberra, 1971: New York, Crane, Russak and Co. and Univ. of Western Australia Press.

Crowell, J. C., 1962, Displacement along the San Andreas Fault, California: Geological Society of America Regional Reviews, Special Paper 71.

Dalrymple, G. B., Silver, E. A., and Jackson, E. D., 1973, Origin of the Hawaiian Islands: American Scientist, v. 61, pp. 294–308.

Eardley, A. J., 1962, Structural geology of North America (2d ed.): New York, Harper & Row.

Fooden, J., 1972, Breakup of Pangaea and isolation of relict mammals in Australia, South America and Madagascar: Science, v. 175, pp. 894–898.

Gansser, A., 1964, Geology of the Himalaya: London, Interscience.

Gilluly, J., 1965, Volcanism, tectonism, and plutonism in the western United States: Geological Society of America Review Articles, Special Paper 80.

Girdler, R. W., ed., 1972, East African rifts: Amsterdam, Elsevier.

Gobbett, D. J., and Hutchison, C. S., 1973, Geology of the Malay Peninsula: New York, Wiley-Interscience.

Hamilton, W., 1969, Mesozoic California and the underflow of Pacific mantle, Bulletin of the Geological Society of America, v. 80, pp. 2409–2430.

Himalayan and alpine orogeny, 1964, Part XI of Report of 22nd International Geological Congress: New Delhi, India.

Jahns, R. H., ed., 1954, Geology of southern California: California Division of Mines, Bulletin 170.

Karig, D. E., 1971, Origin and development of marginal basins in the western Pacific: Journal of Geophysical Research, v. 76, pp. 2542–2561.

Kay, M., and Colbert, E. H., 1965, Stratigraphy and life history: New York, Wiley.

Kurtén, B., 1972, The age of mammals: New York, Columbia.

Larson, R. L., and Chase, C. G., 1972, Late Mesozoic evolution of the western Pacific Ocean: Bulletin of the Geological Society of America, v. 83, pp. 3627–3644.

Minato, M., Gorai, M., and Hunahasi, M., 1965, The geologic development of the Japanese Islands: Tokyo, Tsukiji Shokan.

Murray, G. E., 1959, Geology of Atlantic and Gulf Coast province of North America: New York, Harper & Row.

Raven, P. H., and Axelrod, D. I., 1972, Plate tectonics and Australasian paleobiogeography: Science, v. 176, pp. 1379–1386.

Ross, C. A., ed., 1974, Paleogeographic provinces and provinciality: Society of Economic Paleontologists and Mineralogists Special Publication No. 21.

Rutten, M. G., 1969, The geology of western Europe: Amsterdam, Elsevier.

Scott, W. B., 1937, A history of the land mammals in the Western Hemisphere: New York, Macmillan.

Sharp, R. P., 1972, Geology field guide to southern California: Dubuque, W. C. Brown.

Sugimura, A., and Uyeda, S., 1973, Island arcs: Japan and its environs: Amsterdam, Elsevier.

Williams, H., 1953, The ancient volcanoes of Oregon, The Condon lectures: Eugene, Oregon State System of Higher Education.

Winterer, E. L., 1973, Sedimentary facies and plate tectonics of equatorial Pacific: Bulletin American Association of Petroleum Geologists, v. 57, pp. 265–282.

18

PLEISTOCENE GLACIATION AND THE RISE OF MAN

The more it
SNOWS-tiddely-pom,
The more it
GOES-tiddely-pom
The more it
GOES-tiddely-pom
On
Snowing.

A. A. Milne,
The House at Pooh Corner, 1928

(By permission of E. P. Dutton and Co.,
Inc., and Methuen and Co., Ltd., 1928.)

FIGURE 18.1 Adélie penguins near McMurdo Sound, Antarctica. Penguins eat small marine animals abounding in the cold waters; the birds are delicately adapted to their peculiar niche, but easily could become extinct if land predators swarmed over their rookeries. *(Courtesy National Science Foundation; photo by Cartsinger.)*

James Hutton was among the first to publish (1795) a suggestion that Alpine glaciers in the past had been much more extensive than today. His reasoning was inspired by reading of famous occurrences of erratic boulders on hilltops in the Swiss Plain near Geneva, and on the slopes of the Jura Mountains to the north, more than 50 kilometers from the nearest modern glacier. This was a lucky bit of Huttonian intuition, but it is a curious irony that he himself never recognized the abundant evidence of glaciation in his Scottish homeland.

The glacial interpretation of erratic boulders was overlooked in the tempest over granite and basalt and because of the curious state of thought about features now routinely attributed to continental glaciation. As was pointed out in Chap. 4, the Noachian Flood dogma hung as a pall over early nineteenth century thought. The Diluvialists continued to regard erratic boulders scattered over northern Europe as having been drifted in by the Flood (see Fig. 4.3). In northern Scotland, a famous set of conspicuous parallel benches along a valley, the parallel roads of Glen Roy, also were interpreted as having been formed by the Deluge pouring down a narrow defile; in fact, they are glacial lake beaches.

First full recognition of former glaciation on a continental scale by huge, vanished ice sheets was in Norway in 1824 by J. Esmark, though the great German writer Goethe apparently anticipated Esmark somewhat earlier. This suggestion, too, was largely overlooked, and it was a Swiss engineer (Ignace Venetz-Sitten) who, beginning in 1821 and joined about 1830 by a friend (Jean de Charpentier), confirmed earlier speculation that Alpine glaciers once extended out of the mountains onto the Swiss Plain, leaving moraines and erratic boulders far north of present glaciers. A prominent Swiss biologist, Louis Agassiz, felt compelled to see what manner of heresy these men were promulgating. To his own surprise, he became convinced in 1836 that they were correct. By 1840 Agassiz himself became the leading champion of continental glaciation by gathering evidence of a widespread cover of ice over nearly all of northern Europe, including Britain. Presence of peculiar rock types among the erratic boulders of northern Europe proved that most of the ice had not come from the Alps, but had flowed south from Scandinavia, the only place where such rocks were known to occur. In 1846 Agassiz joined the faculty of Harvard University, and was elated to find evidence of similar continental glaciation in New England.

By mid-century, other geologists were becoming

interested in continental glaciation. Patterns of ice flow were being recognized:

The rock scorings are the trails left by the invader. Their character should reveal the nature of the icy visitant as tracks reveal the track maker. (T. C. Chamberlin, 1886)

And in the midwestern United States, several levels of glacial till superimposed upon one another and separated by obvious buried soils containing plant and animal remains were differentiated. There had been not one but multiple advances and retreats of the ice. It had already been suggested in 1842 that a major, worldwide lowering of sea level must have accompanied the past expansion of glaciers; with recognition of multiple ice advances about 1875, it became apparent that sea level must have fluctuated several times. Finally, only 20 years after Agassiz began his glacial researches, evidence of a much more ancient (late Paleozoic) glacial episode was discovered in India.

FIGURE 18.2 Glacier-enshrouded Antarctic Peninsula near Chilean scientific base Gabriel Gonzalez Videla. This region, still very much under Pleistocene climatic conditions, lies within a Mesozoic and early Cenozoic mobile belt closely related to the Andes Mountains of Chile. Glaciation probably began in interior Antarctica in middle Cenozoic time, but temperate forests existed near this site until the Miocene. *(Courtesy M. Halpern.)*

The Pleistocene Epoch represents a truly unique interval in geologic history. Not only did it encompass the Great Ice Age, but also during its brief 2- to 3-million-year span, man evolved both biologically and culturally to take his pre-eminent position on the ladder of life. Because of its recency, the record of the Pleistocene can be read in fascinating detail. The presence of man makes Pleistocene study the most interdisciplinary facet of geology. Anthropologists, climatologists, botanists, zoologists, and even biblical scholars are very much concerned with latest Cenozoic history.

In most minds, the Pleistocene is synonymous with glaciation, and we have stressed that glaciations have been truly rare events in history. We also have taken pains to show that the typical climate of the past was warmer and more uniform than that of the last 20 to 25 million years. Therefore, neither climatically nor topographically is the present a perfect key to the past! Definition of the limits of the Pleistocene Epoch is by no means simple. It is obvious that, solely on the basis of glaciation, polar and alpine parts of the earth would be classed as still in the Pleistocene today (Fig. 18.2), whereas midlatitudes emerged from the ice age between 10,000 and 11,000 years ago. Similarly, ice caps must have begun forming earlier in polar areas than elsewhere, and so neither beginning nor end of glaciation provides wholly satisfactory defining time limits. Lyell suggested a biologic defi-

nition of the Pliocene-Pleistocene boundary in Italy that remains in general use today; Pleistocene marine strata were defined as containing only marine species that still are living. Problems arise, however, in trying to extrapolate this datum from shallow marine deposits, where it was defined, to deep marine and to nonmarine successions. Isotopic dating places this boundary between 2 and 3 million years ago. The Pleistocene-Holocene boundary is best defined either as the end of the last rapid rise of sea level between 6,000 and 8,000 years ago or as the time of deglaciation of the midlatitudes about 10,000 years ago.

DIVERSE EFFECTS OF GLACIATION

GENERAL

With nearly one-third of the present land area of the globe covered by roughly 43 million cubic kilometers of ice during maximum Pleistocene glaciation (Figs. 18.3, 18.5), it is not difficult to imagine that profound effects resulted. Practically no corner of the globe—not even the deep seas—escaped at least some influence of Pleistocene climate (Fig. 18.4); paleobotanical studies show that tropical land areas were least affected. Direct influences of glacial ice, such as its erosive power, deposition of erratic boulders and moraines, and the like, are well known, but certain other phenomena deserve special mention.

ISOSTASY

Depression of the crust by the load of ice caps up to 3,000 meters thick, and the crustal rise or "rebound" upon their melting provide convincing proof of the general validity of the principle of isostasy and of the ability of mantle material to flow plastically as the crust is warped epeirogenically (see Chap. 8). Actual evidence of postglacial rebound consists primarily of raised ocean and lakeshore features such as beaches. Around the Great Lakes and the northern Baltic Sea, for example, such raised features are not level, but are tilted southward. This reflects thicker ice (thus greater crustal depression followed by greater rebound) at the northern ends of these areas. Rebound continues today, indicating that, by ordinary human standards, it is a slow process spanning 10,000 years or more. If Greenland and Antarctica were to be deglaciated, they too would rise. Figure 1.8 shows differential rise of the Baltic.

LAKES

During much of Pleistocene time, accumulation of water in low depressions formed large lakes far from continental glaciers. Most notable were those of the Great Basin region of western United States. Today, under a more arid climate, only small, saline bodies, such as Great Salt Lake, and playa or alkali flats remain as remnants of once much larger and far more numerous lakes. These were formed in closed basins between the youthful block-faulted ranges of the Utah-Nevada-California region (Fig. 18.6). The largest were glacial Lake Bonneville in Utah and Lake Lahontan in western Nevada, each of which covered up to 50,000 square kilometers and was as much as 300 meters deep. Lake levels fluctuated enormously at least five times as climate oscillated (Fig. 18.7).

Farther east, glaciation itself had a marked effect upon drainage patterns and produced many temporary lakes. Preglacial drainage valleys were gouged by the ice to leave large depressions subsequently filled with water; the Finger Lakes of New York are examples. The Great Lakes basins are of similar origin, and had a complex history during advances and retreats of the ice. Huge postglacial Lake Agassiz—four times as large as Lake Superior—formed temporarily in the southern Canadian Plains when the ice began to retreat; eventually it drained into Hudson Bay (see Fig. 18.5).

Large lakes formed from three to five times in the northern Rocky Mountains. Immense terraces of coarse gravel and sand over 30 meters high line the upper Columbia River valley in Canada, attesting to staggering volumes of meltwater. In western Montana, a large basin filled and drained several times to form Lake Missoula, which was as much as 300 meters deep. About 18,000 years ago its moraine dam in northern Idaho suddenly broke. A wall of water rushed across eastern Washington with incredible velocity. This catastrophic flood scoured channels and deposited immense gravel bars over a large part of the Columbia Plateau (now called channeled scablands). Many river canyons were deepened by great volumes of meltwater (Fig. 18.8), and some such canyons have been abandoned entirely since deglaciation (Fig. 18.9).

WIND EFFECTS

With much meltwater pouring from wasting glaciers near the end of glacial episodes, tremendous quantities of sediment were dumped in front of receding ice

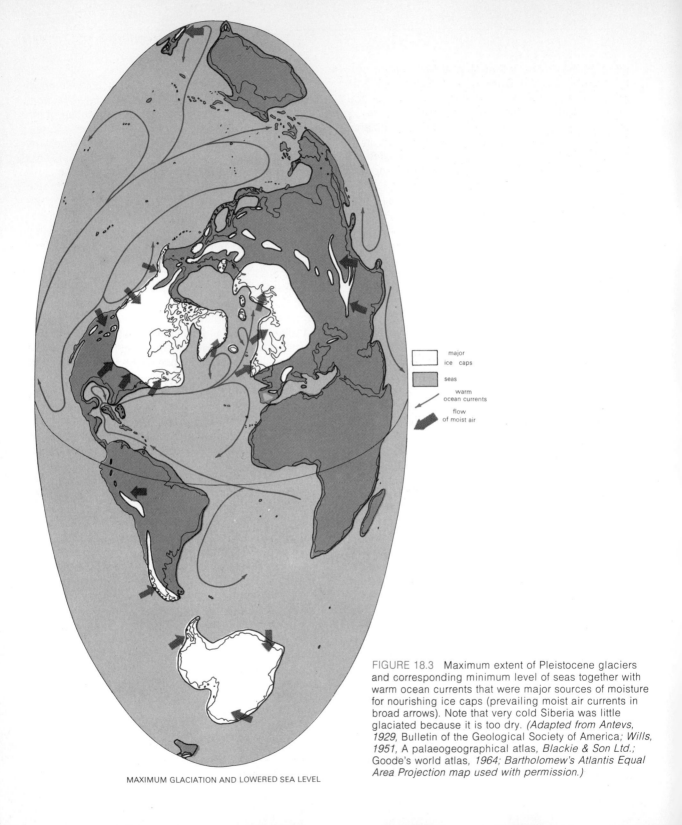

FIGURE 18.3 Maximum extent of Pleistocene glaciers and corresponding minimum level of seas together with warm ocean currents that were major sources of moisture for nourishing ice caps (prevailing moist air currents in broad arrows). Note that very cold Siberia was little glaciated because it is too dry. *(Adapted from Antevs, 1929, Bulletin of the Geological Society of America; Wills, 1951, A palaeogeographical atlas, Blackie & Son Ltd.; Goode's world atlas, 1964; Bartholomew's Atlantis Equal Area Projection map used with permission.)*

major ice caps

seas

warm ocean currents

flow of moist air

MAXIMUM GLACIATION AND LOWERED SEA LEVEL

FIGURE 18.4 Iceberg-rafted cobbles dropped onto the sea floor 4,000 meters below sea level between Antarctica and South America; such cobbles have been carried as far as 3,000 kilometers from Antarctica. Similar ancient rafted pebbles also are known (e.g., Fig. 10.24). *(Official NSF photo, USNS Eltanin, Cruise 10; courtesy Smithsonian Oceanographic Sorting Center.)*

fronts. Rivers became so sediment-choked that they formed mosaics of braided channels. Wind then worked over the sediment bars along such channels, winnowing out fine sand and silt. Huge dune fields were formed and large areas of fine dust (loess) accumulated near major river valleys. As the ice front retreated farther, vegetation became reestablished, and wind effects gradually decreased.

WORLDWIDE CHANGES OF SEA LEVEL

Worldwide fluctuation of sea level in response to glacial oscillations was one of the most profound phenomena of the Pleistocene. Submerged beach ridges far out on continental shelves indicate a maximum reduction of sea level on the order of 100 to 140 meters. Elephant teeth 25,000 years old have been dredged by fishermen from more than 40 sites as far as 130 kilometers off the Atlantic coast and in water as much as 120 meters deep. Melting of all existing glaciers, especially those of Antarctica with 90 percent of present ice, would raise sea level about 70 meters above present level; this would flood nearly 15 percent of existing land area!

Major drops of sea level occurred at least four times, with the greatest about 40,000 years ago. Short-term oscillations were superimposed upon more subtle, long-term worldwide Cenozoic regression of the sea. It is interesting to note that the last rise (Fig. 18.10) operated to produce universal transgression in opposition to the regressive tendency caused by isostatic rebound along many high-latitude coasts.

Hudson Bay, most of which lies less than 600 meters below sea level today, was drowned by rapid rise, but it appears destined to become largely land as crustal rebound continues. Here is another example of the interaction of local versus universal effects operating on different time scales.

Sea-level changes had widespread effects extending from well inland, on the one hand, to the deep seas, on the other. Reduction of sea level lowers effective base level of major rivers. This causes downcutting, which produces river terraces that correlate with former marine beaches and deltas (Fig. 18.11). In coastal regions, vast expanses of former shallow seas were laid bare, in several cases forming important land bridges for migration of organisms between adjacent continents or islands. Finally, lowering of sea level made possible a much greater-than-normal delivery of relatively coarse clastic material to the deep seas. The total rate of sedimentation throughout large portions of the ocean basins must have accelerated. The greatest drops of sea level

On the following page:

FIGURE 18.5 Pleistocene paleogeography of North America showing maximum ice advance and retreat of sea, immediate postglacial Lake Agassiz, Cascade and other volcanoes, and late Pleistocene desert floras in southwest. Note joining of Cordilleran ice caps with main Canadian-Shield cap in western Canadian plains, which blocked migrations of organisms to and from Asia. (Contrast climatic patterns with Fig. 17.11.)

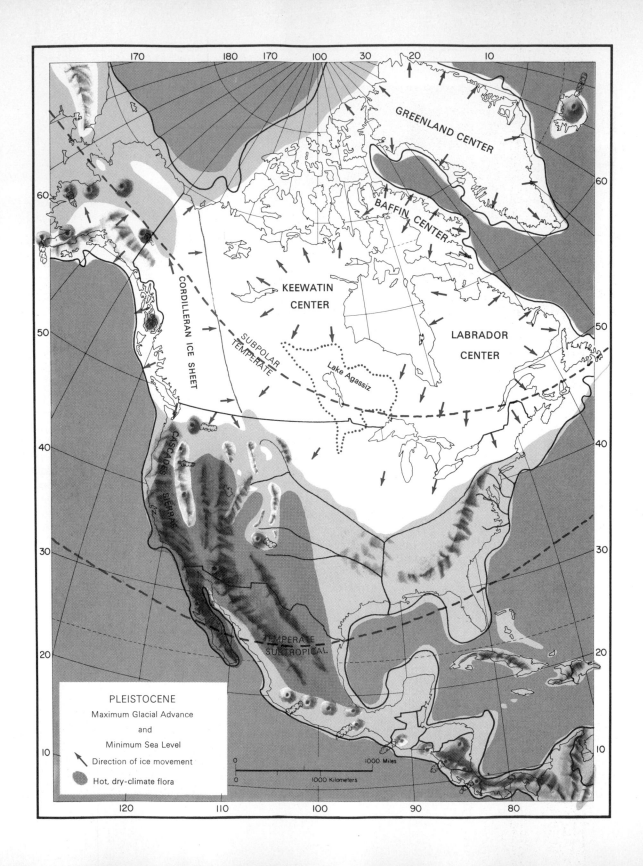

PLEISTOCENE

Maximum Glacial Advance

and

Minimum Sea Level

↖ Direction of ice movement

● Hot, dry-climate flora

GREENLAND CENTER

BAFFIN CENTER

KEEWATIN CENTER

LABRADOR CENTER

CORDILLERAN ICE SHEET

SUBPOLAR TEMPERATE

Lake Agassiz

CASCADES

SIERRA

TEMPERATE SUBTROPICAL

0   1000 Miles

0   1000 Kilometers

FIGURE 18.6   Playa or alkali flat in closed, arid, intermontane basin near Eureka, central Nevada (Diamond Range in background). Rain flows to valleys, evaporates, and alkali is precipitated. Many such playas were large, fresh lakes during glacial advances. Most modern evaporites are forming in this manner (see Fig. 13.21).

carried the shoreline out to continental shelf margins (Figs. 18.3 and 18.5) so that rivers swollen many times during early melting of the ice could carry vast quantities of sand to the very edge of the shelf. Ordinarily, most such sediment would be trapped on the inner shelves, but with little or no submerged shelf, much sand could escape to the deep seas. It is thought by most authorities that gushing, muddy meltwater torrents stimulated the cutting of submarine canyons that nick most continental shelves, though the deeper ends of the canyons extend far below even the lowest glacial sea level. Below that level they must have been cut by submarine currents and mud flows. Once cut, the canyons provided funnels for

continuing export of sediment from the shelves onto abyssal plains.

PLEISTOCENE CHRONOLOGY

As mapping of glacial features progressed in North America and Europe, the different glacial and interglacial deposits were differentiated and named. In

FIGURE 18.7   At least nine former beach levels of glacial Lake Lahontan visible on perimeter of Pyramid Lake, northwestern Nevada. Lake Lahontan, like other Pleistocene lakes in the west, rose four times and then shrank in response to Pleistocene climatic oscillations.

FIGURE 18.8   Snake River Canyon at Twin Falls, Idaho (note oasis of trees and irrigated fields along canyon floor). The river was crowded southward by Pliocene and Pleistocene lava outpourings; this canyon (together with many other western examples) was then deepened by increased late Pleistocene river discharge.

FIGURE 18.9   Ouimet Canyon near Nipigon, Ontario. Nearly 200 meters deep, this gorge was completely abandoned after the deglaciation of southern Canada. (Man in upper left corner provides an idea of scale.)

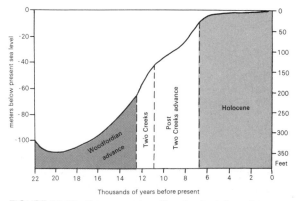

FIGURE 18.10   Curve representing the last rise of sea level to its present position in late Pleistocene time. Note comparison of episodes with Table 18.1.

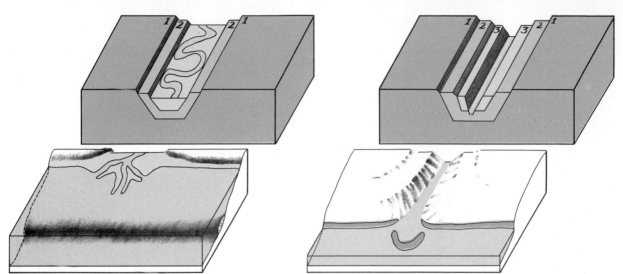

FIGURE 18.11   Effects of sea-level changes on river valleys and coast lines. *Left*: Initial drop of sea level caused cutting of main valley and formation of now-submerged shore features; a rise, then a second fall, and a second rise also are recorded. The final rise produced a branching delta, present beaches, and valley alluvium. *Right*: Recent drop in sea level caused downcutting into all previous valley alluvium, seaward migration of crescentic delta-bar, and left several high terrace levels reflecting earlier events (terraces numbered in order from oldest to youngest). *(Adapted from Scientific American, The Bering Strait land bridge, W. G. Haag, January, 1962; Copyright © 1962 by Scientific American, Inc.; used by permission.)*

central North America, four major glacial advances were recognized, and in Europe, from three to five advances, depending upon location. Apparently each glacial episode was in turn marked by lesser oscillations of the ice. Details are, of course, much clearer for the last, or Wisconsin glaciation, than for earlier episodes. Full understanding of Pleistocene history requires correlations of deposits, first among glaciated areas and then with unglaciated ones. The same principles and problems encountered in correlation and interpretation of more ancient strata apply.

Perfection of carbon 14 isotopic dating about 1950 provided a breakthrough for late Pleistocene chronology. Deposits from widely separated localities containing wood, charcoal, bone, or calcareous shells could be dated and correlation was thus enhanced. Even carbon-bearing deep-sea sediments could be correlated with the glacial standard time scale established on the continents. Now it became possible to compare in detail the effects of glacial conditions on the continents with conditions in the oceans. At the same time, ^{14}C dating became very important to archaeologists studying the last 65,000 years of human history. Attempts continually are under way to extend isotopic dating back farther in time from 65,000 to 100,000 years, at which age the K-Ar method becomes widely applicable (unusually favorable material as young as 5,000 years has been dated). Protactinium 231–thorium 230 dating is being used for deep-sea clays back to 300,000 years ago (see Chap. 6). But even with isotopic dating becom-

ing increasingly available and more reliable, it still is necessary to correlate Pleistocene strata by other, more conventional means as well. Figure 18.12 shows diagrammatically some ways that correlations are accomplished among different types of Pleistocene sequences. In deep-marine sediments, temperature-induced reversal of coiling direction of foraminiferal shells (Fig. 18.13), together with oxygen-isotope analyses, provide bases for constructing temperature fluctuation curves that can be used for correlation together with ^{14}C dating of the same shells. Changes in polarity of the earth's magnetic field leave an imprint on magnetically susceptible minerals in marine sediments. A time scale of polarity reversal events has emerged, and these can now be correlated from place to place (see Chap. 9).

Several persistent and obvious Pleistocene volcanic ash layers occur over western North America,

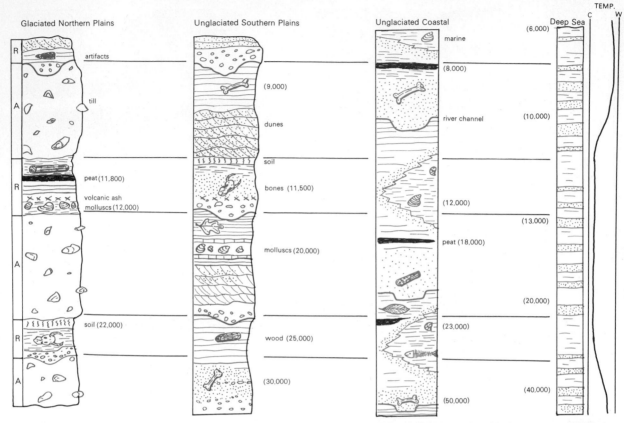

Glaciated Northern Plains

R — artifacts

A — till

R — peat (11,800)
— volcanic ash
— molluscs (12,000)

A

— soil (22,000)

R

A

Unglaciated Southern Plains

(9,000)

dunes

soil

bones (11,500)

molluscs (20,000)

wood (25,000)

(30,000)

Unglaciated Coastal

marine

(6,000)

(8,000)

river channel

(10,000)

(12,000)

(13,000)

peat (18,000)

(20,000)

(23,000)

(40,000)

(50,000)

Deep Sea

TEMP.
C   W

FIGURE 18.12 Idealized representation of typical sequences from four different types of regions showing different records of glacial and interglacial episodes in each as well as means of correlating among them. (A—glacial advance, R—glacial retreat; numbers represent ^{14}C dates.)

and they provide invaluable datums for correlation. One of the best known of these extends through the Great Plains from El Paso, Texas, to South Dakota. Most of the widespread ash layers were derived from a few of the great Cascade volcanoes. Careful geological detective work has shown that the ashes are mineralogically distinctive enough so that their

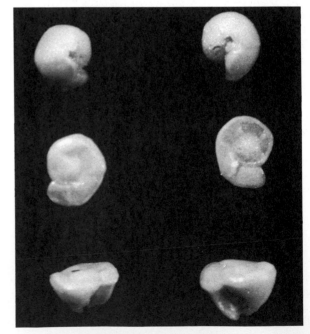

FIGURE 18.13 Contrast of coiling directions of microscopic shells of planktonic foraminifera (*Globorotalia truncatulinodes*) from deep-sea Pleistocene sediments of the North Atlantic; right-hand coiling (left) occurs in waters warmer than 8 to 10°C, whereas left-hand coiling (right) characterizes cooler temperatures. *(After D. B. Ericson et al., 1954, Deep sea research; by permission of Pergamon Press.)*

FIGURE 18.14 North across Crater Lake, Oregon, showing north-south alignment of Cascade volcanoes in the distance. Crater Lake formed when former Mt. Mazama suffered a cataclysmic eruption. Immense volumes of ash were blown hundreds of miles eastward by prevailing westerly winds. The top of the mountain then collapsed, forming a caldera, which filled with water. Ash from Mt. Mazama and other Cascade volcanoes provides important marker layers for correlation of late Pleistocene deposits over many western states.

sources can be pinpointed accurately. Two volcanoes were especially important, Glacier Peak in Washington and Mount Mazama in Oregon. The final eruption of Mt. Mazama (Fig. 18.14) was witnessed by Indians, for its ash buried artifacts that have been dated isotopically as 6,600 years old. Lavas interstratified with glacial tills may provide indirect means for dating glaciations. For example, K-Ar dates on lavas in the Sierra Nevada date an interstratified till as 3 million years old, the oldest known Cenozoic glacial deposit in middle latitudes.

Table 18.1 shows a comparison of the stratigraphic classifications of Pleistocene records for central North America and Europe as evolved through the application of procedures illustrated above. The chronologies, of course, apply strictly only to the glaciated areas where they were first established. Not many years ago, the Pleistocene Epoch was assumed to be encompassed at most within a bare 1 million years. But deep-sea cores and K-Ar dating now suggest that it was at least twice that long. This revelation has required some drastic revision of concepts of subdivision and correlation of the older divisions.

WHAT HAS CAUSED GLACIATIONS?

CLIMATOLOGIC BACKGROUND

By 1850 or so, the outstanding phenomena associated with Pleistocene glaciation had been recognized. Naturally, speculation about the cause of glaciation began, but even a century later, there is still some doubt. Before discussing several alternate working hypotheses, it is important to state some important climatological background.

First, *continental glaciations have been rare events in earth history*. Besides the well-known Pleistocene Ice Age (0 to 3 m.y. ago), there was widespread glaciation in the southern continents in late Paleozoic time (200 to 300 m.y. ago in South America

EVOLUTION OF THE EARTH

TABLE 18.1 Pleistocene stratigraphic classification standards for North America and Europe. (Numbers indicate years before present)

North America (Upper Mississippi Valley)			Europe (Alps and Southern Baltic Regions)	
Holocene Epoch	(0–7,000 or 10,000 years ago)		Holocene Epoch	
Wisconsinan glacial	Post-Two Creekan advance (8,000–11,000)		Younger Dryas advance (10,000–11,000)	Würm glacial
	Two Creekan retreat (11,800–12,500)		*Allerod retreat* (11,000–12,000)	
	Woodfordian advance (12,500–22,000)		Older Dryas advance (13,000–20,000)	
	Farmdalian retreat (22,000–28,000)		*Riss-Würm retreat* (20,000–30,000+)	
	Altonian advance (28,000–70,000+)		Riss advance (30,000–60,000+)	
Sangamonian interglacial			*Mindel-Riss interglacial*	
Illinoian glacial			Mindel glacial	
Yarmouthian interglacial			*Gunz-Mindel interglacial*	
Kansan glacial			Gunz glacial	
Aftonian interglacial (approximately 1 m.y. ago)			*Donau-Gunz interglacial*	
Nebraskan glacial			Donau glacial	
	———2–3 million years ago———			
Pliocene Epoch			Pliocene Epoch	

(PLEISTOCENE EPOCH)

in Siluro-Devonian time (around 400 m.y. ago) and in north Africa during the Ordovician (about 450 m.y. ago), and a very widespread one in Eocambrian time (circa 700 m.y. ago). Prepaleozoic episodes are not as well documented (see Chap. 10). The time period between the recognized great glacial events is irregular, varying from 100 to 1,000 million years, and so it is difficult to claim that glaciation is a regularly cyclic phenomenon of the earth. Because of its rarity, glaciation must have required a very special combination of conditions.

A second important observation is that all available evidence, incomplete as it is, indicates that *the "normal" climate during the past 1 billion years was milder than that of the last 40 million years*. Present lower latitude regions offer closest modern analogies with typical climate of the past. We must recall, however, that relative positions of continents and of poles may have changed more than overall climate (see Fig. 13.24).

The third point, really a corollary to the second, is that *past climatic conditions were not only milder but also more uniform over the earth*. Although the overall mean temperature of the earth cannot have changed

greatly, temperature and moisture gradients were less abrupt, and climatic zones tended to be broader. Nonetheless, because of the geometry of the earth-sun system, polar areas must *always have been relatively cooler* than lower latitudes, although it does not follow that there always was ice at the poles. Polar areas would be coldest when continents lay at or near them because land has a lesser heat capacity than does water (see Chap. 11). Conversely, poles would be relatively milder whenever open, well-mixed seas lay there. This important paleoclimatic principle was appreciated and discussed at length by Lyell over 100 years ago.

The fourth point to note is that *magnitude of overall average yearly temperature differential between a normal and a glacial climate is only on the order of* 5 *or* 10°C. That is to say, as noted in Chap. 7, the earth enjoys a very critical temperature position in the solar system and has a sensitive thermal budget. It is estimated that a drop on the order of only 4 to 5°C from the present mean annual temperature could cause renewal of continental glaciation. One might, therefore, be surprised that glaciations have not been more frequent if the earth's heat budget is so sensitive. The answer must be that a stable thermal equilibrium characterized most of geologic time. This was always Lyell's firm contention because it was demanded by his rigid uniformitarianism. He admitted climatic fluctuations for limited parts of the earth, but held that *overall* temperature of the earth was invariable. Today we can say that the total heat budget of the earth can vary only within narrow limits; glaciation requires not changes in kind, but only in degree!

LATE CENOZOIC
CLIMATIC DETERIORATION

Some evidence for the decline of overall temperature was presented in the last chapter. Long-known land-fossil evidence of cooling is now supplemented with morphologic and oxygen-isotope paleotemperature data from calcareous shells in marine sediments. Figure 18.15 shows the general trend of land and sea temperatures for the past 100 million years. The striking shifts of plant communities in arid southwestern United States and Mexico introduced in the last chapter continued into Pleistocene time (Fig. 18.5). Whatever the cause, it is clear that a steady temperature decline occurred until a critical threshold level was reached when ice caps began to form and grow. Apparently glaciation began at both poles in mid-Cenozoic time. By Miocene time, fossil plants suggest cooling temperatures, and local Miocene tills are known near both poles; some ice-rafted pebbles appear in Miocene deep-sea sediments near Antarctica. It is inescapable that, by Pliocene time, polar and alpine ice caps were sizable, as evidenced by the appearance of ice-rafted pebbles widely in Pliocene deep-sea sediments surrounding Antarctica (Fig. 18.4) and by the 3-million-year-old till in the Sierra Nevada.

The critical threshold necessary for glaciation was reached when annual snow precipitation slightly exceeded summer wastage by melting and evaporation for a number of years. This amounts to saying that if the permanent snow line descends sufficiently in elevation or latitude, ice can begin to form. The difference between present level of descent of glaciers relative to that of past glacial episodes shows effects of fluctuation of overall temperature on snow-line changes (Figs. 18.16 and 18.17). Apparently the snow line was lowered an average of from 600 to 1,000 meters during glacial episodes. The snow-line

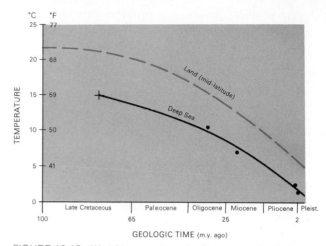

FIGURE 18.15 World temperature-decline curves for Late Cretaceous and Cenozoic times based upon fossil land organisms (upper curve) and oxygen-isotope (^{18}O:^{16}O) studies of shells from deep marine sediments (lower curve). *(Adapted from Chaney, 1940, Bulletin of the Geological Society of America; Emiliani, 1958, Scientific American; Lowenstam and Epstein, 1959, Reports of 20th International Geological Congress.)*

level is critical in the original area of ice cap formation, but once a glacier becomes large, it may flow to lower elevations or latitudes well below the snow line. If rate of flow is sufficient, the ice front will advance in spite of relatively warm temperatures. This accounts for the seeming anomaly of finding mild-climate fossils very near obvious glacial moraines. A notable example exists on the equator in east Africa, where glaciers from the 5,113-meter-high Ruwenzori Mountains left terminal moraines on the edge of eastern Congo jungles.

Fossil crocodiles and giant land turtles occur in

FIGURE 18.16 Elevation versus latitude of terminal moraines during maximum glaciation and at present; dark band shows chief latitudinal span of major continental ice sheets.

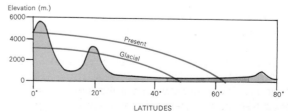

FIGURE 18.17 Grand Teton Range and Snake River viewed across Jackson Hole valley. The Tetons were carved by glaciation from Prepaleozoic metamorphic rocks that had been raised along a late Cenozoic normal fault. Larger glaciers flowed out onto the edge of the valley; only one puny relict glacier remains since the warming of climate ensued several thousand years ago. Several prominent river terraces along the Snake River reflect the Pleistocene climatic oscillations. (This view is 400 kilometers upriver from Fig. 18.8.)

Pliocene and Pleistocene sediments of the plains and southern United States. These were cold-blooded creatures requiring above-freezing temperatures practically all the time. Their distribution suggests relatively warm, moist conditions as far north as South Dakota in early Pliocene time. In late Pliocene, they still ranged as far north as Kansas; the plains region was more humid than now and had scattered forests. Even during the first two glacial episodes, turtles and crocodiles still lived in the southwestern plains region, and during interglacial times they again ranged north to Kansas and Nebraska. It is apparent that a really harsh climate did not grip most of the United States until the last (Wisconsin) glacial advance. A number of significant fluctuations within postglacial times are recorded by fossil and archaeological evidence. The last ice was retreating from mid-latitudes by 10,000 years ago. Soon thereafter early human civilization approached its zenith in Asia Minor. Cooling recurred from about A.D. 1500 to 1880 and from then to 1950 there was a period of amelioration characterized by warmer and drier conditions. Glaciers have retreated since 1890, but cooling again is indicated by recent advances in many mountainous areas. Many fluctuations of climate have occurred, but they have widely differing time periods. At least the shorter ones seem to be controlled by subtle meteorological changes in atmospheric circulation patterns. We know that, historically, alpine glaciers have been extremely sensitive to such shorter term changes; large continental ice sheets must have responded similarly, but on a much larger scale, to longer period changes.

TYPES OF GLACIAL HYPOTHESES

Two principal phenomena associated with Pleistocene glaciation require explanation. First is the general deterioration of climate down to a threshold for ice cap accumulation just discussed, and second is the oscillatory or periodic nature of glacial expansion and retreat. Note that major glacial and interglacial episodes appear to have been rhythmic with a period of roughly 30,000 to 50,000 years. It is natural to assume that both aspects of Pleistocene glaciation were produced by one mechanism, but the great rarity of continental glaciation suggests an *unusual chance combination of several factors.*

Many working hypotheses and speculations have been offered to explain glaciation, but they can be conveniently grouped into a relatively few types as follows:

1 Changes in solar radiation
2 Astronomical effects involving changes of earth-sun geometry
3 Terrestrial changes affecting net heat budget:
   a By changes of atmospheric transparency (e.g., dust, clouds, etc.)
   b By changes of relative land, sea, or snow areas and/or elevations
4 Variations of heat exchange from equatorial to polar regions due to paleogeographic changes

Hypotheses of the first type, while plausible, are difficult to test because no adequate absolute measure of solar radiation has been made. Moreover, such measurements would be needed for a very long time span. Apparent fluctuations of ^{14}C generation suggest solar fluctuations, but appeal to sunspot cycles seems invalid because no corresponding patterns have been verified by analysis of tree rings or glacial lake varves. We do not deny the possibility of solar variations as the major cause of glaciation and other climatic changes, but it is fruitful to see if other factors can be discovered that are more readily testable by geologic means.

ASTRONOMICAL HYPOTHESES

The second category of hypotheses has received a great deal of attention for more than a century. Lyell considered astronomical effects and rejected them as of little climatic importance. But, in 1920, a Yugoslavian meteorologist, M. Milankovitch, made pioneer calculations of the net solar radiation variations due to three known cyclic parameters of the earth's orbit. These parameters are variations of ellipticity of the orbit, tilt of the axis, and a wobble of the axis due to changing gravitational interactions of the moon and sun on the earth's equatorial bulge (also known as precession of equinoxes or longitude of the perihelion). The last determines the time of the year when the earth is nearest the sun. The periods of these perturbations in the earth's movements are as follows:

Parameter	Variation	Approximate periods
Ellipticity of the orbit	0.017–0.053	92,000 years
Inclination of the axis (obliquity)	$21\frac{1}{2}$–$24\frac{1}{2}°$	40,000 years
Precession of equinoxes (wobble)	0–360°	21,000 years

Because the periods of these separate phenomena are different, there can be only a few times when their effects reinforce one another to cause unusually great or small net solar radiation. For example, if the earth simultaneously had its smallest angular tilt, maximum orbital eccentricity, and also was at its greatest annual distance from the sun during the Northern Hemisphere summer, then a small but significant lowering of net radiation in the north would occur. Some meteorologists believe that the tilt of the axis is of greatest importance because of its effect upon the angle of incidence of solar radiation, and therefore upon net radiation, especially at high latitudes (see Fig. 11.33). Wisconsin climatologist R. A. Bryson believes that the north-south temperature gradient, determined by contrasts in net radiation between poles and equator, is the most significant single climatic parameter. It affects not only net radiation, but also large-scale atmospheric circulation patterns.

In 1922 a famous German climatologist (W. Köppen) suggested that Milankovitch's radiation curve for the past 600,000 years might match the glacial advances and retreats. Ever since, there has been a great deal of interest in this possible relationship, and a number of hypotheses invoke it to help explain glaciation. Subsequently the Milankovitch curve was recalculated, and it was found that the combined result, termed the Milankovitch effect, of the three periodicities has itself a periodicity of very roughly 40,000 years at higher latitudes. An American geologist, Cesare Emiliani, who has pioneered oxygen-isotope paleotemperature determinations of deep-sea sediments, has developed a curve of ocean-temperature variations for the Pleistocene, and it shows an impressive correspondence with the recalculated net radiation curve (Fig. 18.18). In marine sediments isotopically dated as late Pleistocene, the period between successive cold peaks on the curve is on the order of 40,000 years. But because the earlier Pleistocene record has not been dated accurately, it is difficult to test fully the comparison for the entire Ice Age. Serious problems have arisen in trying to match the curves with the classic Pleistocene chronology (Table 18.1); these problems are magnified by recent indications that the Pleistocene was nearly twice as long as formerly suspected. In spite of shortcomings, the seeming confirmation of the climatic importance of the Milankovitch effect is impressive The effect presumably has operated over all of geologic time; yet glaciation has not occurred every 40,000 years in earth history. This means that the total effect is relatively small and left no mark in the historical record *except when some other factors also operated* to reduce overall average temperature sufficiently so that the Milankovitch effect might then operate to cause expansion and contraction of glaciers—in other words, when the earth was made susceptible to glaciation by other factors.

HYPOTHESES OF CHANGING HEAT BUDGET

Albedo

Any change that shifts the balance between incoming and outgoing radiation—the heat budget—would affect overall temperature of the earth. If such changes were of sufficient magnitude, they might contribute to forming (or terminating) a glacial climate. The changes envisioned include variations of transparency of the atmosphere either to incoming or outgoing radiation and variations of reflectivity of the earth's surface. Viewed from space, the earth presents a variable face, with dark oceanic and forested areas, and light clouds, snow, and deserts. Photographs of the earth taken from space (Figs. 1.3 and 9.7) prove that much solar radiation is scattered, while a great deal is reflected (more properly, reradiated) back to space from the light-colored surfaces. The magnitude of

radiation reflected from different surfaces expressed as a percentage of the incoming radiation is termed albedo. Albedo can be measured with great accuracy using a radiometer; such measurements show the importance of relative albedo (see Table 18.2). Overall albedo for the present earth is about 30 percent. It is estimated that a 1 percent change of albedo would produce approximately a 1 degree change of temperature. Put another way, only a 2.5 percent increase of total average annual cloud cover could reduce summer temperatures to their probable level during a typical glacial episode.

Transparency of the Atmosphere

Relative area of clouds obviously affects atmospheric transparency as well as earth albedo, hence overall temperature. Dust and other particles suspended in the atmosphere have similar effects. Climatologists tell us that a mere 7 percent increase of overall atmospheric turbidity would lower the average annual temperature on the order of 1°C. Between 1950 and 1965, average temperature dropped about 0.3°C, and rapidly increasing atmospheric pollution probably has contributed to this decline. Long ago it was suggested that unusual volcanic activity might have brought on glaciation by filling the atmosphere with volcanic dust. It was supposed that the Cenozoic was a time of unusual volcanism, but there is no correlation of more ancient volcanism with glaciations. Nonetheless, secondary effects of eruptions do occur. Three months after Krakatoa exploded in 1883 (see Chap. 1), dust from it reached Europe. Instruments in France recorded a sudden drop in solar radiation and

FIGURE 18.18 Comparison of recalculated Milankovitch curve of summer insolation at 65°N latitude (upper) with Caribbean surface sea water temperature variation curve from oxygen-isotope analysis of foraminiferal shells (lower). The time scale is based upon isotopic dating of the marine sediments. *(Adapted from Emiliani,* Science, *v. 154, 18 November 1966, pp. 851–856; copyright 1966 by American Association for the Advancement of Science.)*

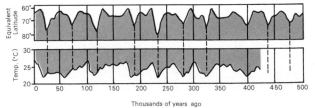

Thousands of years ago

TABLE 18.2 Relative earth albedo
Percentage of total incoming solar radiation reflected

Snow or ice	45–85%
Clouds	40–80%
Grasslands	15–35%
Bare ground; light-colored rocks	15–20%
Forests; dark-colored rocks	5–10%
Water	2– 8%

a level 10 percent below normal was measured for 3 years. Explosion of Mt. Katmai, Alaska, in 1912, was followed by a 20 percent radiation reduction in faraway north Africa. Dust particles play still another important climatic role. They act as potential nuclei for ice crystal formation high in the atmosphere, and so may cause cloud formation, which would further increase earth albedo.

More subtle factors that influence transparency are related to gaseous composition of the atmosphere. As was pointed out in Chap. 7, carbon dioxide, water vapor, and ozone all act as filters. Ozone filters out deadly incoming ultraviolet radiation, but molecules of all three gases inhibit escape of infrared and longer wavelength radiation to space, producing the well-known greenhouse effect.[1] If all long-wave radiation from the earth's surface were to escape to space, the earth would be cold and uninhabitable.

[1]Solar radiation is divisible approximately as follows:

	Wavelength, micrometers*	Proportions	
X-rays and gamma rays	0.0005–0.01	} "Short"	10%
Ultraviolet	0.2–0.4		
Visible light	0.4–0.7	"Medium"	40%
Infrared	0.7–3.0	} "Long"	50%
Heat rays	3.0–3,000		

*1 micrometer = 0.001 millimeter.

Nearly half of incoming solar radiation is medium-wavelength visible light to which the atmosphere is transparent (see Fig. 7.9). But the earth's surface radiates infrared and long-wave energy toward the sky, to which the atmosphere is not transparent. Water vapor and carbon dioxide molecules are of such sizes as to absorb much of this wavelength spectrum, which causes them to reradiate heat that warms the atmosphere. The same principle causes solar heating of greenhouses and closed automobiles. Glass transmits short-wave solar radiation, but surfaces inside reradiate longer-wave energy to which glass is opaque; thus air inside the greenhouse or auto is heated.

Suggestion that changing carbon dioxide content of the atmosphere could be a major factor in climatic change dates from 1861, when it was proposed by British physicist John Tyndall. The carbon dioxide budget is complex. Additions come through igneous and hot spring activity, animal respiration, decay, and combustion. Carbon dioxide is consumed by plant photosynthesis and deposition of carbonate rocks. Unfortunately we cannot estimate accurately changes of past CO_2 content of either atmosphere or oceans, nor is there any firm quantitative basis for estimating the magnitude of drop in carbon dioxide content necessary to trigger glaciation. Moreover the entire concept of an atmospheric greenhouse effect is controversial, for the rate of ocean-atmosphere equalization is uncertain. Until these questions are resolved, the carbon dioxide hypothesis cannot be considered satisfactory for explaining glaciation.

The Emiliani-Geiss Hypothesis

One interesting glacial hypothesis invokes changes of total earth albedo to explain the long-term temperature change that triggered glaciation, but couples with that the Milankovitch effect to explain smaller magnitude cyclic temperature fluctuations indicated by glacial and interglacial episodes. As continents became larger and higher during the Cenozoic Era, authors Emiliani and Geiss argue, total albedo would have increased (Table 18.2). This could produce general temperature decline until the threshold for glaciation was reached at least 3 million years ago. Once a critical heat-budget level was reached, the Milankovitch effect then could become significant in varying average temperature slightly, but critically, for expansion and contraction of ice caps over millennia. As was shown earlier, the Milankovitch effect has roughly the same time period as do at least the last glacial-interglacial episodes. Recently it has been shown that ice caps have an inherent tendency for instability, which has also affected their regimes. As caps expand, the wastage or ablation area increases as the ice flows outward and downward into warmer areas. Meanwhile, the accumulation area has remained unchanged, however. A small ice cap may grow slowly to a critical size and then expand suddenly, but it may then shrink suddenly if it becomes unstable. The most critical factor is the relative elevation of the upper ablation limit versus total size of the ice cap.

HYPOTHESES OF PALEOGEOGRAPHIC CAUSES

It was suggested long ago that the great amount of mountain building and enlargement of continents during late Mesozoic and Cenozoic time may have altered atmospheric circulation so as to bring on glaciation. The average elevation of continents is now 800 meters above sea level, whereas at the end of the Mesozoic Era, average elevation probably was closer to 200 meters. American Pleistocene geologist R. F. Flint has emphasized that an essentially worldwide uplift of continents in Miocene and Pliocene time may have contributed to the onset of glaciation. Interestingly this large-scale uplift also was accompanied by retreat of seas, therefore an increase of worldwide albedo. Result: a double cooling effect. Mountains and plateaus deflect moist air upward to colder levels and induce considerable atmospheric turbulence. The net effect is to cause cloud formation and local precipitation. As already recorded, mountain building dramatically affected local climate in the Cordillera, but, because most large continental ice sheets accumulated on lowland areas far from high mountains, this hypothesis seems of secondary importance. A number of scientists have proposed that geographic factors causing variations of equator-to-pole heat exchange controlled glacial and nonglacial climates. In all probability, such changes also were only of secondary importance. There are several variations of these hypotheses, only one of which will be mentioned.

Geophysicists M. Ewing and W. L. Donn adopted in the 1950s a combination of geographic factors. To explain general Cenozoic decline of temperature, they pointed to the paleomagnetic evidence of relative, apparent polar wandering or continental drift. While the North Pole apparently lay in or near the North Pacific until Mesozoic time, warm, mild climate persisted. But, presumably, as the pole became thermally isolated in the nearly enclosed Arctic Ocean, temperatures declined. Meanwhile, Antarctica had migrated over the South Pole, an even more thermally isolated position.

Continental glaciation is thought by Ewing and Donn to have begun first in Antarctica during Pliocene or earlier time and to have chilled the entire globe. Continental glaciation then began on the northern continents through evaporation of moisture from a presumably ice-free Arctic Ocean. Finally, the Arctic Ocean froze over. Ice caps ceased to advance for lack of nourishment, and melting along their southern margins soon caused retreat. As ice melted, total climate would warm, more ice would melt, and so on. But if the Arctic Ocean were reopened by climatic warming and rise of sea level, then it again might provide moisture for new northern ice caps. Presumably glacial and interglacial fluctuations should continue indefinitely until some major change either of pole positions or of polar ocean exchange occurs.

From archaeological records, we know that between A.D. 800 and 1200, whales could move east along the Arctic coast from Alaska to northern Greenland, where they were hunted by Eskimos. From historic records, we also know that at this same time Scandinavia enjoyed very favorable climate, and the Vikings settled in Iceland, Greenland, and Newfoundland; they met (and were raided by) Eskimos in Greenland. Viking sailors reported nearly ice-free conditions as far north as Svalbard (Spitsbergen). While this suggests a semiopen Arctic coastal condition, it hardly proves a completely ice-free ocean, however.

Many meteorologists doubt that significant evaporation of moisture could occur even *if* the Arctic Ocean were ice free, and studies of oxygen isotopes and foraminifera in Arctic Ocean sediments, as well as land plant fossils, all suggest a continuous major ice cover throughout the Pleistocene. Moreover, centers of the ice caps lay well south of the Arctic Ocean (Figs. 18.3 and 18.5), and practically all moisture today comes to these areas from elsewhere—the Gulf of Mexico for North America and the North Atlantic Gulf Stream for Europe.

PLEISTOCENE CLIMATIC EFFECTS UPON LIFE

RANGES OF ORGANISMS

Continental glaciation in the Northern Hemisphere is reflected in Pleistocene marine faunas as well as among land fossils. For example, species living today along the northern California coast ranged only as far north as San Diego during glacial episodes.

Worldwide changes in sea level during Pleistocene time had a great effect on coral reef growth. For example, in the West Indies, fossil coral reefs occur from 1.5 to 5 meters above present low-tide level. These reefs have been dated as interglacial, when water may have stood slightly higher than at present (alternatively, the islands may have risen). Most nearby living reefs are postglacial, being no older than 4,000 to 5,000 years, and form veneers growing on fossil reefs. Furthermore, the biota is unique to the West Indies. During earlier Cenozoic time, the reefs were closely related to Indo-Pacific and Mediterranean faunas. Alternate cooling, plus many other factors, have caused the recent reef fauna to become more restricted geographically and faunally. Other reefs of the world also show the ravages of sea-level fluctuations, but not as clearly or to the same extent as in the West Indies.

Even in polar regions and in deep-water deposits, glacial and interglacial episodes are seen reflected in the microfossil floras and faunas. Evidence from deep-sea cores shows that new species and extinctions of old ones seem to correlate with alternate periods of normal and reverse polarity of the magnetic field (discussed further in Chap. 19); it is not clear why this should be so.

On land there were even more dramatic effects upon the distribution of plants and animals. For example, large logs, cones, and tree leaves dated as 30,000 years old have been dredged off the west coast of Mexico. Since no such trees grow along this very dry coast today, it is inferred that the climate has become much drier there in the past 30,000 years. In other, widely scattered areas, peculiar relict outposts of cooler-climate floras and faunas are found today, as in many scattered ranges of the southern Rocky Mountains. Cold-adapted species of trees live at high elevations surrounded by vast lower areas with forms characteristic of warmer and drier climate. Many such examples reflect past widespread cool-temperate forests that have shrunk in the past few thousand years leaving these relicts, but some of these "islands" may have resulted from wind transport of seeds to isolated mountain tops.

Faunal and floral changes due to glaciation were most extreme near the southern limit of the ice. Here, alternately glacial and nonglacial conditions prevailed. The Great Lakes region provides a convenient example. The last major ice advance (Woodfordian substage) culminated there about 16,000 to 18,000 years ago. Ice retreated from the southern Great Lakes by 12,000 years ago, and cold-climate spruce

forests, much like those of central Canada today, reoccupied the region. A famous buried forest at Two Creeks, Wisconsin (about 11,850 years old) gave the name to this warmer interval (Table 18.1). Then the ice readvanced south a short distance to cover part of the northern Great Lakes region from about 11,000 to 10,000 years ago. The Two Creeks forest was overridden, trees were toppled and then were buried beneath till (Fig. 18.19). Spruce forests persisted, however, farther south. About 10,000 years ago the glacial climate regime ended and ice began its last retreat. The ice front paused briefly in the western Hudson Bay region about 8,000 years ago and finally completed its retreat from that region by 6,000 years ago (Fig. 18.20). As the ice left the Great Lakes region, boreal spruce forests migrated northward. By about 8,000 years ago, spruce forests were replaced by pine in the Great Lakes region. From 8,000 to about 1,500 years ago, a relatively dry period set in, which is reflected in a change from pine to widespread oak-hemlock forests.

The postglacial climate culminated about 5,000 years ago, but in the past 1,500 years there have been several small fluctuations of climate that are reflected in soils, vegetation, and human and animal activities. Today the northern hardwood forest covers much of the Great Lakes region with boreal spruce forests just touching its northern limits; prairie-oak forest characterizes the southwestern corner of the region. Boundaries between these entities have fluctuated, as is indicated by plant types found in vertical sequences of late Pleistocene and Holocene sediments in lakes and bogs. Wood, leaves, fruits, pollen, animal remains, and soil profiles all are employed in their study (Fig. 18.21). Magnitude of changes suggested by such data is indicated in Fig. 18.22 for Europe where a wealth of evidence is available. As noted earlier, by far the harshest climatic stress occurred during the last (Wisconsin) glacial episode. Cold-blooded reptiles retreated farther south during (and reinvaded less far north after) each advance. Only the Wisconsin episode was able to relegate them completely to the Gulf Coast region where they now reside.

LAND BRIDGES
Formation of land bridges by glacial lowering of sea level was of paramount importance to the present biogeography of the earth. The most important examples include the Bering bridge between Siberia and

FIGURE 18.19 Excavation of Two Creeks Forest on Lake Michigan shore, northeastern Wisconsin. Dozens of spruce logs were buried in lake sediments, which in turn were covered by late Wisconsinan glacial till. Some stumps are *in situ*; prostrate logs were driftwood. Carbon 14 dating indicates that the logs are about 11,850 years old. *(Courtesy R. F. Black.)*

Alaska, the North Sea bridge to Britain, the Sunda bridge between western Indonesia and Asia, and the New Guinea–Australia bridge (Fig. 17.46). The Isthmus of Panama, second only to the Bering bridge in biological importance, was formed by structural rather than sea-level changes.

The Bering bridge deserves special mention. Fossil plant and invertebrate evidence indicates that Alaska and Siberia were connected during most of Cenozoic time. Comparisons of terrestrial biotas of all of North America and Asia indicate that two-way traffic occurred through most of the Cenozoic; west-to-east flow was twice as heavy as the opposite. In late Miocene time, the bridge was flooded, terminating connection for several million years. But during

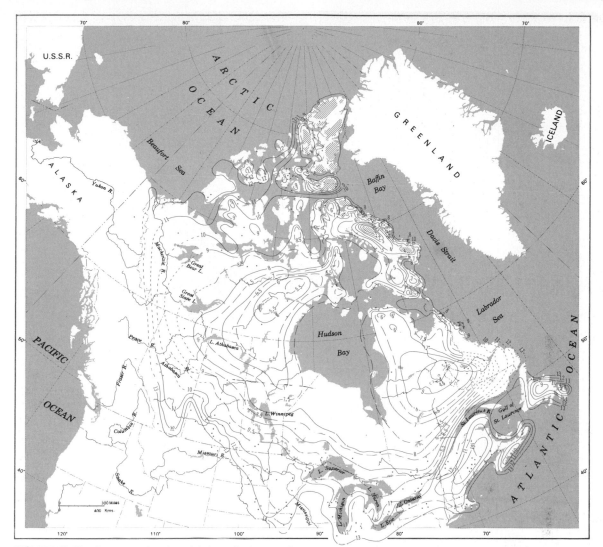

FIGURE 18.20 Contours of carbon 14 dates in thousands of years for late Wisconsin deglaciation of eastern North America. Approximate timing and patterns of shrinkage of the last great ice sheet are shown by successively younger dates. Cross-hatching denotes present ice caps. *(Courtesy Bryson and Wendland, 1967,* University of Wisconsin Department of Meteorology Technical Report No. 35.)

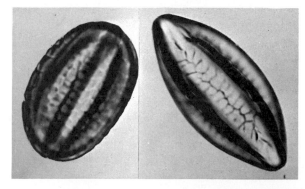

FIGURE 18.21 Microscopic photos of pollen grains from *Ephedra* or "Mormon Tea," a plant found in semiarid regions. Climate must be interpreted carefully from pollen, for some grains are known to be blown hundreds of miles. These grains are 0.05 to 0.07 millimeter long. *(Courtesy L. J. Maher.)*

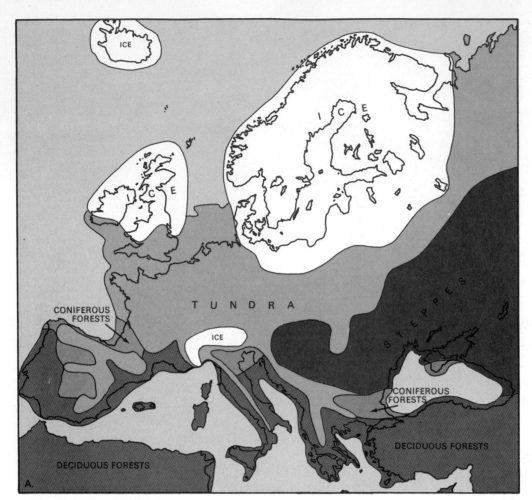

FIGURE 18.22 (*Above and opposite page*) Effects of climatic changes on plant distribution during the past 25,000 years: plant communities of Europe during the last maximum ice advance (*A*, above) compared with native vegetation of modern Europe (*B*, opposite page). *(A adapted from Brinkmann, 1960,* Geologic evolution of Europe, *Ferdinand Enke; Wells, 1951,* A palaeogeographical atlas, *Blackie & Son Ltd.; B after* Goode's atlas, *Rand McNally Corp.).*

glacial advances, Alaska again was biologically more like Siberia because of its isolation from the remainder of North America by a wide ice barrier in Canada (Fig. 18.5). During interglacial high sea levels, however, it showed more affinities with interior North America when an ice-free corridor opened in the western Canadian Plains.

THE END OF AN ERA

As with the close of the other eras, there was (and is) a period of extinction in late Cenozoic time. One might expect that the extreme climatic variations brought about by glaciation would cause widespread extinction. The picture, however, is not that clear. It was thought that many large mammals became extinct because of the cold; yet as we have seen, the climate almost to the ice edge in some areas was relatively mild, especially before the very last glacial episode. An alternate theory has gained some support. It suggests that humans might have been partly responsible for the extinction of the larger mammals. Over 200 genera of mammals became extinct between Late Pliocene and Recent. In some instances these

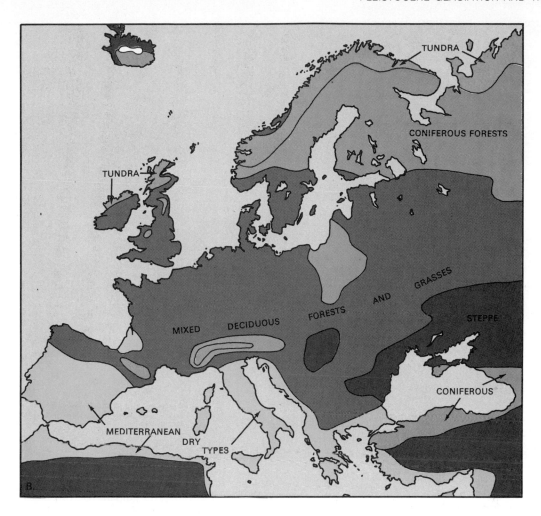

extinctions do not coincide with periods of glaciation. In North America, many extinctions occurred well after the final glacial stage, when fossil pollen indicates that vegetation and climates were similar to prior interglacial times, but after man, the hunter, arrived.

Earliest Pleistocene mammalian extinctions in Africa also seem to be related to the arrival or development of Stone Age hunters. However, in reviewing actual fossil sites, the record is not clear. In Eurasia, artifacts are found along with large mammal remains; yet only four genera of mammals became extinct during man's existence there. In view of the small population of early humans, it would appear unlikely that they alone could have erased so many mammals during their existence, the later near-extinction of the buffalo notwithstanding.

Mammalian extinction began in the Pliocene with loss in North America of some artiodactyl families and the rhinos. It was among the larger types that we find the most striking extinction patterns. While some artiodactyls, such as the buffalo, survived, we lost giant beavers, the mammoth and mastodon, saber-toothed tigers, camels, horses, and many others. A magnificent cross section of a Pleistocene terrestrial fauna was recovered from the classic Rancho La Brea tar pits of Los Angeles, which are preserved in a park (Fig. 18.23).

The interglacial and glacial stages caused rapid migrations of organisms, a process which to this day

FIGURE 18.23 Examples of Pleistocene mammals from La Brea tar pits, Los Angeles, California. The pits apparently were a site of both water and tar springs. Animals came to water, occasionally became mired, and sank in quicksand. The tar preserved their skeletons perfectly in masses of bones, but churning of the springs disarticulated many skeletons. *(Photo courtesy American Museum of Natural History.)*

is still occurring. The range of the armadillo has been steadily expanding northward, and only recently it has crossed the Mississippi River. Similarly the opossum and raccoon have migrated north into the Great Lakes region in recent decades.

THE EVOLUTION OF PRIMATES AND MAN

As we shall see in the final chapter, the most destructive and influential mammal of the Holocene fauna in terms of ecology is man, the culmination of primate evolution. The exact origin of the primates is unknown, but they probably were derived from the insectivores, which early shrew-like forms closely resemble. The earliest forms appeared to be arboreal (tree-climbing) lemurs and tarsiers represented in the Paleocene of Europe (Fig. 18.24). By Eocene time, these groups had become widespread and are found as fragments in Asia, Africa, and North America. The primates possess many primitive mammalian characteristics; for example, they retained the primitive five-digit feet, which are well suited for arboreal life. One important early appearing characteristic was the development of stereoscopic vision, which is an advantage for judging distances and is important for forms living above the ground.

Fossil primates are a very rare part of any Cenozoic fauna, mainly because they probably inhabited forested uplands most subject to erosion. After the prosimian primates (tree shrews, lemurs, and tarsiers) experienced a rapid adaptive radiation in Paleocene and Eocene times, they became somewhat restricted and fairly conservative throughout the later Cenozoic. It is in Africa, where Eocene prosimians became well established, that major events of anthropoid evolution evidently occurred.

The earliest true apes have been found in the Oligocene Fayum beds of Egypt, some 100 kilometers (60 miles) southwest of Cairo, in an area of desert land on the edge of the fertile and historically important Nile Valley. During the Oligocene, this region was the site of a rich, tropical forest with broad rivers and vast swamps inhabited by crocodiles, primitive elephants, rhinos, and ever-present carnivores.

The apes of the Oligocene and Miocene were arboreal fruit eaters. By the end of the Miocene, two lineages of apes spent at least part of their lives in open country. A group of great apes (dryopithecines) appeared during the Miocene in widely scattered areas; they are known mostly from fragments of the skull. These apes had rows of evenly spaced teeth and other features suggesting that they were related to the hominids (the family which includes man); they probably were early forms that might have given rise to later great apes such as the gorilla.

The most important late Miocene genus is *Ramapithecus*, which is the earliest genus included in the hominids. The incisors and other front teeth were small compared to the great apes; the cheek teeth had flat, broad chewing surfaces. *Ramapithecus* fossils are found in forested areas with open spaces near rivers and lakes. *Ramapithecus* was first found in India and is also known from East Africa. The early hominids are distinguished from the apes by head (cranial) features, including dentition, shape of skull, and size of brain, and by the postcranial skeleton (which is rarely preserved).

The similarity between *Ramapithecus* and *Australopithecus* (Fig. 18.25), which appeared in very late Pliocene or early Pleistocene, is close, suggesting to some students that they were on a direct line to man.

By late Pliocene and early Pleistocene, the fossil record of the hominids is much better than the poor Pliocene record. In Africa, two lineages emerged, *Paranthropus robustus* (or *boisei*) and *Australopithecus africanus*, which had smaller teeth but a much larger body. Both show a bipedal tendency, but the

FIGURE 18.24 *Notharctus*, an Eocene prosimian primate (lemur). *(Courtesy American Museum of Natural History.)*

pelvic bones indicate they were not fully or constantly erect and at times were "knuckle-walkers." At the famous East African Olduvai Gorge and other sites, stone artifacts are associated with *Paranthropus* fossils. Associated plants and animals suggest that these genera lived in grassland and open wooded parkland. Both had very small brains—smaller than today's gorilla—but for their small body size their brains were relatively large. Interestingly, *Paranthropus* lived alongside early man (*Homo*) at Olduvai Gorge and outlived *Australopithecus*, in spite of the fact that the latter genus most probably gave rise to man. Most species of *Australopithecus* were apparently hunters and gatherers.

In early Pleistocene deposits just under 1.7 million years old at Olduvai and a few other East African sites are hominid skeletal remains which depart from *Australopithecus* in having larger brain capacities, shorter jaws, and limbs indicating more efficient bipedalism. These fossils show an affinity with *Australopithecus africanus*, and they have been named,

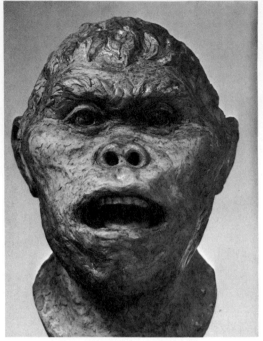
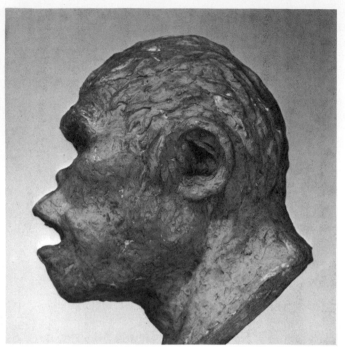

FIGURE 18.25 *Australopithecus*, an advanced ape from South Africa thought to be an immediate precursor of the genus *Homo*. *(Photograph of head model courtesy American Museum of Natural History.)*

by Dr. Louis Leakey, *Homo habilis. H. habilis* had a brain size of 675 cubic centimeters (cm³) compared to *Australopithecus africanus* (600 cm³) or the smaller *Paranthropus* (530 cm³).

In Java and China during the 1920s and 1930s, a number of specimens of a tall, large-brained (over 800 cm³) and small-faced man were found in mid-Pleistocene deposits dated at about 500,000 years. These fossils were given a variety of names but are now placed in the species *Homo erectus* (Fig. 18.26). Legs and associated bones show that *H. erectus* was, in fact, fully erect. Additional fossils found in Africa are somewhat different, but in any event it is likely that *H. erectus* evolved into modern man, *Homo sapiens*. During the 300,000 years of *H. erectus's* existence, the brain size increased to over 1,000 cm³, and this was accompanied by an evolution of complex behavior and the development of social patterns. Evidence for these conclusions is based on presence of tools, use of fire for cooking and warmth, construction of stone and wood shelters, and arrangement of these shelters into communities.

The first members of *H. sapiens* were found in Europe in late Pleistocene sites dated at about 300,000 years. They were called Neanderthal men, and their brain size averaged around 1,200 cm³ (modern man may have brains of about 1,400 cm³). Neanderthal man had heavy eyebrows, as did all earlier hominids, a pronounced chin, and was rather stocky. And despite characterizations as a brutish cave man, Neanderthal man in reality had a well-developed society, as evidenced by complex burial sites and a tool culture. Modern man first appeared over 50,000 years ago, and like Neanderthal man was a hunter and gatherer. Unlike other species, though, modern man had such features as a high flattened brow, a higher shorter skull, and a more rounded face. Both groups coexisted for perhaps 10,000 years, but modern man, with more highly developed technical skills, arts, and language, dominated and rapidly spread throughout the world.

Because of enormous capacities, man has been able to move into more diverse environments than any other organism by the simple device of constructing

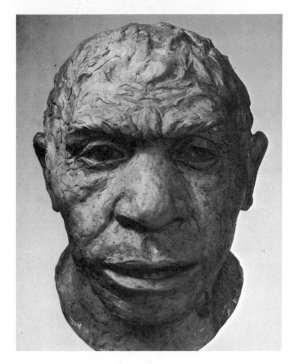

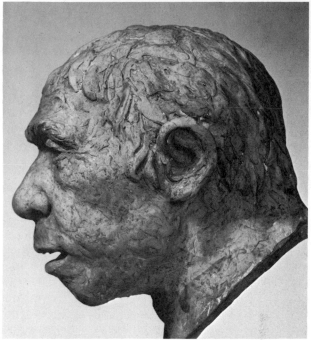

FIGURE 18.26 *Homo erectus (Pithecanthropus)*, a side branch in the phylogeny of man. *(Photograph of head model courtesy American Museum of Natural History.)*

an environment if adverse conditions exist. As a result of this geographic ubiquity, distinct races of man have appeared. Many racial characteristics clearly were adaptive for particular environments. As a single example, we cite skin pigmentation. No one is sure why different skin types evolved, but the strong suggestion is that it was in response to ultraviolet light intensity at different latitudes. Pigmented skin is an effective ultraviolet light filter, which not only protects delicate tissues but prevents overproduction of vitamin D. It is postulated that, as man moved into higher latitudes where ultraviolet radiation is less intense, there was insufficient light to promote the synthesis of vitamin D that occurs in the skin. Selective pressures operated toward light skin, hence the evolution of white skin.

Most anthropologists believe that one ancestral stock spread from Africa or Asia Minor to other, more or less isolated, regions and then became racially diversified in response to local environmental selective pressures. Alternatively, *Homo erectus* may have given rise independently to *H. sapiens* in several separate areas. In any event, by no later than 10,000 years ago, all major races of modern man had ap-

peared and had occupied their primary distribution areas (i.e., their distributions as of the beginning of historic times).

MAN IN NORTH AMERICA

Large mammals such as the woolly mammoth and reindeer apparently crossed over to America about 100,000 years ago. Mammal-hunting man then gained access to the New World by way of the Bering bridge, and dispersed far southward via the Isthmus of Panama. Human occupations as old as 40,000 years have been claimed, but there is doubt as to just when man first crossed. Open-plains paleo-hunters using bifacial, fluted stone weapons became specialized in central and southeastern Europe sometime before 65,000 years ago. The fluted projectile-point culture then spread north and east across Asia, and the people gradually adapted to a cold climate as they dispersed. The culture probably

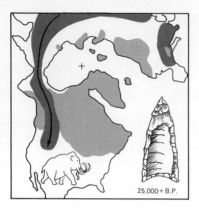

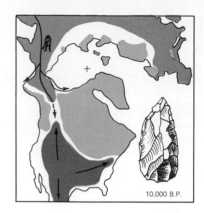

 Ice   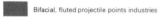 Bifacial, fluted projectile points industries   Unfluted projectile points industries

FIGURE 18.27   Probable history of entry of some early human cultures into the New World, showing importance of glacier histories in western Canada. A primitive hunting culture (represented by fluted Folsom and Clovis projectile points) penetrated into the United States at least 12,500 years ago. New southward immigration was prevented by readvance of glaciers until about 10,000 years ago when the corridor reopened, and a more advanced, unfluted projectile-point culture entered from Asia, soon influencing formerly isolated peoples. *(Adapted from Müller-Beck, Science, v. 152, 27 May 1966, pp. 1191–1210; copyright 1966 by American Association for the Advancement of Science.)*

was carried across the Bering bridge about 25,000 years ago. Artifacts show that it spread south into the western United States and to the southern tip of South America. The immigrants were the ancestors of the Plains Paleo-Indians, who developed in isolation during the Wisconsin glacial advance. This early American culture was distinguished by the use of fluted Clovis and Folsom projectile points, known in the lower United States as early as 12,500 years ago and in South America at least 10,000 years ago.

Additional influxes of humans southward into the New World apparently were impossible until the Wisconsin ice retreated enough to open once again the corridor from Alaska through western Canada (Fig. 18.27). This apparently occurred between 10,000 and 12,000 years ago, when a new immigration brought the ancestors of the Eskimo and Aleut peoples. These mongoloid groups remained in the far north, chiefly along the coasts, but they influenced the older, southern Indian culture. For example, fluted hunting points were replaced by unfluted ones in

North America about 9,000 years ago. This new stone industry had developed in Europe by 27,000 years ago, but the more primitive American Paleo-Indian tradition persisted much later due to its isolation from Asia until 12,000 years ago. Prehistoric Indians lived near the southern ice margin and witnessed the last retreat; climatic changes greatly influenced their habits.

The Eskimos, whose ancestors were early Japanese-Korean peoples, developed the greatest adaptation ever to the harsh Arctic climate. Once in Alaska, rather than moving south as earlier peoples had done, they dispersed eastward several times across Arctic Canada, reaching as far as eastern Greenland. In northern isolation, they evolved both to biologic and cultural uniqueness. Meanwhile, other groups evolved distinctively in at least semi-isolation in other parts of the New World. Human migrations and biological as well as cultural adaptations in the Americas, under the influence of the transitory Bering land bridge, fluctuating Canadian ice barrier, and diverse ecologic niches, provide outstanding illustrations of the basic principles of organic evolution, migration, and adaptation that we have applied before to lower life forms.

SUMMARY

Pleistocene glaciation represents one of the most dramatic episodes in all geologic history. Effects of this profound climatic event affected practically the entire earth from the deepest sea floor to the highest mountain peak and from the poles to the equator, and in a great variety of ways. Practically all kinds of life were affected by it in one way or another. Some became extinct during the refrigeration, while others (such as the penguin) thrived. Animals with body-temperature-regulating capabilities should have fared better than cold-blooded creatures, although many mammals also suffered extinction.

In spite of intensive study for more than a century, the causes of glaciation still are not fully known. Rarity of glaciations suggests that a combination of peculiar conditions must have been required. The appeal of the Emiliani-Geiss hypothesis is its reliance upon earthly mechanisms and its close correspondence to the total record of glaciations. The three best-documented glacial intervals of history—Eocambrian, late Paleozoic, and Pleistocene—all correspond to times when continents were demonstrably much larger than normal. It seems inescapable that the known and measurable effects of total earth albedo must have been of great importance to ancient climates, and may have been instrumental in lowering temperature to the glacial threshold. How important the Milankovitch effect was in controlling oscillations of glacial and interglacial episodes is less clear. Additional factors, such as mountain building and the closing of the Isthmus of Panama to modify interchanges of Arctic and equatorial ocean waters, probably have contributed to glacier changes.

Besides dramatic north-south shifts of ranges of both land and marine organisms as glaciers waxed and waned, reefs were affected by rise and fall of sea level. Similarly, woolly mammoths and other land animals roamed freely over what are now submerged continental shelves during lowered sea level. Marine microorganisms were affected by marked evolutionary changes and temperature-related shell modifications. Most dramatic of all, however, was the evolution of hominids, leading to modern man nearly 1 million years ago. Through migrations, isolation, and natural selection, adaptive improvements occurred, and *Homo sapiens* resulted. Evolutionary principles have applied to man as well as to lower organisms, at least until the past few centuries, when man began to manipulate the environment profoundly. Some believe that man's further evolution will be chiefly cultural and psychological, but it is difficult, based upon geologic history, to believe that all biologic evolution of man has ceased. It is claimed that the capability to control genes is almost at hand. Thus, although man may soon become the first organism to manipulate its own evolution, so too may man become the first organism to so seriously degrade its environment as to invite its own extinction. In the next chapter we explore these possibilities more fully.

Three monkeys sat in a coconut tree
Discussing things as they're said to be.
Said one to others, "Now listen, you two,
There's a certain rumor that can't be true,
That man descended from our noble race.
That very idea is a disgrace.
Yes, man descended, the ornery cuss—
But, brother, he didn't descend from us."

Anonymous

Readings

Bishop, W. W., and Miller, J., eds., 1972, Calibration of hominoid evolution: Edinburgh, Scottish Universities Press.

Black, R. F., Goldthwait, R. P., and Willman, H. B., eds., 1973, The Wisconsin Stage: Geological Society of America Memoir 136.

Campbell, B. G., 1967, Human evolution: an introduction to man's adaptations: London, Henemann.

Charlesworth, J. K., 1957, The Quaternary Era (2 vols.): London, Edward Arnold.

Clark, D. L., 1974, Late Mesozoic and Early Cenozoic sediment cores from the Arctic Ocean: Geology, v. 2, pp. 41–44.

Dort, W., Jr., and Jones, J. R., Jr., 1970, Pleistocene and Recent environments of the central Great Plains: Univ. Kansas Special Publication No. 3, Lawrence, Univ. Kansas Press.

Emiliani, C., and Geiss, J., 1957, On glaciations and their causes: Geologische Rundschau, v. 46, pp. 576–601.

Flint, R. F., 1971, Glacial and Pleistocene geology: New York, Wiley.

Loomis, W. G., 1967, Skin pigment regulation of Vitamin-D biosynthesis in man: Science, v. 157, pp. 501–506.

Match, C. L., 1976, North America and the great ice age: New York, McGraw-Hill. (Paperback)

Müller-Beck, H., 1966, Paleohunters in America: origins and diffusion: Science, v. 152, pp. 1191–1215.

Pearson, R., 1964, Animals and plants of the Cenozoic era: London, Butterworths.

Pilbeam, D., 1972, The ascent of man: Macmillan series of physical anthropology: New York, Macmillan.

Rankama, K., ed., 1964, The Quaternary: New York, Wiley.

Ruddiman, W. F., 1971, Pleistocene sedimentation in the equatorial Atlantic: Bulletin of the Geological Society of America, v. 82, pp. 238–302.

Turekian, K. K., ed., 1971, The late Cenozoic glacial ages: New Haven, Yale.

United States Geological Survey, 1973, The channeled scablands of eastern Washington—The story of the Spokane Flood: INF Pamphlet 72-2: Washington, Government Printing Office.

Washburn, S. L., and Delhinow, P., eds., 1972, Perspectives on human evolution–2: New York, Holt.

Wright, H. E., and Frey, D. G., eds., 1965, The Quaternary of the United States, Review volume for VII Congress of International Association for Quaternary Research: Princeton, Princeton.

19

THE BEST OF ALL POSSIBLE WORLDS?

Life can only be understood backward, but must be lived forward.

Sören Kierkegaard

At the end of our long journey through the maze of geologic history, it is appropriate to ask, What are the greatest implications of earth history? We believe that there is a message of lasting value for all, and in this closing chapter, we enumerate what seem to us the few most important and transcendent implications.

From a scientific standpoint, we have attempted to plot a course through the historical maze based upon an objective relating of evidence and the formulation and testing of working hypotheses in a search for general explanations of the development of the physical earth and its life. But, though we have succeeded in formulating a few comprehensive explanations, it should be obvious that the scientific process is never completed, especially in a young science like geology. Hypotheses are transitory; we should nurture them only until we can construct still better ones to explain phenomena. We are not surprised that some of the speculations presented in our first edition are now outdated. By the time you read this page, perhaps some of the newer ideas incorporated in this edition will already be out of date.

It is common to all intellectual pursuits that each question answered tends to raise new questions. This situation is so universally true that we may question if *anything*, strictly speaking, is fact, for so-called facts have an annoying tendency to be reinterpreted without notice. Skeptical minds question the most basic premise of all, namely that there is some inherent order in nature to be discovered, described, and explained. Voltaire once suggested that "If life has no meaning, man will invent one," and Existentialist writer Albert Camus, in his essay *The Myth of Sisyphus*, wrestled with the dilemma presented by the seeming chaos of constant change and lack of fixed absolutes. Is there a hopeless futility in the ceaseless quest for understanding, or does the search itself give meaning to human activity? Robert Louis Stevenson already had anticipated Camus with the affirmation that "To travel hopefully is a better thing than to arrive."[1]

THREE IMPORTANT MAXIMS

Faced with the reality of uncertainty, it becomes important to ask if there are not at least a few relatively

FIGURE 19.1 "When you have seen one Redwood tree, you have seen them all," said a public official during a 1960s controversy over establishment of Redwoods National Park (Mill Creek Road, Jedediah Smith Redwoods, California). *(Courtesy Clyde Thomas and Save-The-Redwoods-League.)*

[1]Camus took Homer's ancient myth as a point of departure for analyzing the seeming absurdity of life's toil symbolized by Sisyphus's futile labors. Sisyphus was banished to Pluto's underworld, where he was made repeatedly to roll a stone to the top of a mountain from which it rolled back down every time. Camus, like Stevenson, concluded that the relentless struggle for the summit gives life meaning enough.

durable geologic tenets. We believe that the study of earth history carries three maxims of far greater significance and timelessness for humanity as a whole than any alleged facts of history or the hypotheses we have developed to explain them. First, a study of geologic history necessarily develops *wholly new concepts of time and rates of change*. It reveals man's position on the scale of life history. Instead of having appeared together with practically all other life soon after the origin of the earth, as was assumed until the eighteenth century, man has resided on earth for but a brief 2 to 2.5 million years—only 0.04 percent of geologic time! Second, the history of earth and life shows forcefully that *the only certain thing is change*. More importantly, however, is that the change is of a special kind. It is not the perfectly cyclic or "steady-state" change envisioned a century ago, but rather an *irreversible cumulative evolutionary change*. "Nature creates ever new forms; what exists has never existed before; what has existed returns not again," wrote Goethe as early as 1781. The third major implication follows from a study of evolution through geologic time, which is what our book is all about. This most important maxim is the great *ecologic interaction between both the living and nonliving realms of the earth throughout history*. Because an evolutionary sequence represents a series of unique events, history cannot, after all, repeat itself exactly. While organic evolution has been possible because of random changes called mutations, the *results of evolution were not random*. As we have shown, the particular way that life developed on earth is the result of natural selection operating on mutants through a 3.5- or 4-billion-year series of changes. Involved were changes both in the living and nonliving realms in a particular order and at particular rates. Had some of these occurred in a different fashion, the course of evolution would have been different; yet no one can say exactly how it would have differed.

One of the chief goals of science is *prediction of consequences*, and if, as we contend, a study of earth history truly has relevance to the lives of humans, then we feel obliged to make some extrapolations from past history and to suggest some implications for the future of humans and their environment. We shall proceed by reviewing briefly the evolutionary highlights in earth history as they affected life, and conclude by applying our third maxim to the contemporary human world.

THE EVOLUTION OF THE EARTH

THE IDEA OF EVOLUTION

The concept of evolutionary change undoubtedly is one of the greatest of human ideas, and, as we have tried to show repeatedly, it is as important in the nonliving as in the biologic realm. Evolution, we feel, provides a powerful unifying basis for the study of earth history. Indeed, it should be clear that nature cannot be understood fully without special attention to evolution, which is most completely revealed in the historical sciences. It was the eighteenth century before people really began to think of natural change in an evolutionary fashion (see Chap. 5). Previously, practically everything was assumed to be fixed in form and in relation to everything else. All life was assumed to have been created nearly simultaneously soon after 4000 B.C.; thus it seemed necessary for Noah to have taken pairs of every life form into the Ark. No extinctions or new creations of species had yet been demonstrated, though Robert Hooke and G. L. L. de Buffon suspected them.

The seventeenth century cosmogonist Leibniz first preached a principle of grand continuity in nature, which led to a notion called the Great Chain of Being from atoms to God. Allegedly, nature abhors discontinuities; rather, everything is arranged in gradational sequences. Development of calculus by Leibniz and Newton doubtless lent impetus to embryonic evolutionary thought by focusing on the integration of small changes or increments both in space and time. In the nineteenth century, Lord Kelvin, using new principles of thermodynamics, anticipated the modern conception of the physical earth as a dynamic, irreversibly changing sphere. Then in 1896, 37 years after Darwin's momentous *Origin of Species* appeared, discovery of radioactivity provided an important example of inorganic evolution through transformation of one element to another, as described in Chap. 6. It also provided an important clue to an overall chemical evolution of the earth in which radioactive decay has played a major role, as was stressed in Chap. 7.

CHEMICAL EVOLUTION OF THE EARTH

We believe that terrestrial evolution began with the aggregation and densification of the protoearth. Internal heating, especially through radioactivity, facilitated the density differentiation of core and mantle between 4.5 and 4.7 billion years ago, and later the crust. The primitive atmosphere and sea water began accumulating and underwent changes in composition as a result of igneous activity, weathering, and photo-chemical reactions.

As one result of changes of matter on the earth's surface, life developed—apparently as an inevitable consequence of the evolution of complex carbon-bearing molecules (see Chap. 7). Three or four billion years ago, indeed, this *was* the best of all possible worlds for life in our solar system, for it contained abundant carbon, hydrogen, nitrogen, oxygen, and phosphorous; had a critical surface temperature range such that much water was present in the liquid state; and possessed a strong magnetic field to shield the surface from most cosmic radiation. It did not yet have an ozone shield from ultraviolet radiation, however. It has been said by a famous Russian biochemist, A. I. Oparin, that matter always is chang-ing from one form of motion to another, each more complex and harmonious than before. Life, to him, appears as a very complex form of matter in motion that arose at a particular stage in the general evolution of matter on earth. As we have seen, origin of life required very special conditions, especially the lack of abundant free oxygen in the environment, moderately warm temperatures, and high-energy ra-diation.

Subsequent development of life occurred in an irreversibly evolving physical environment. Once photosynthesis became possible, free oxygen (and eventually ozone) began to accumulate slowly in the atmosphere. With the addition of free oxygen, life itself brought about a change in the physical realm that irreversibly precluded its being created anew. Nitrogen was controlled largely by organisms as well, but atmospheric argon evolved continually through decay of radioactive potassium 40. It is the only atmospheric gas wholly independent of life process-es. As noted in Chaps. 7 and 10, an oxygen-rich atmosphere is indicated no later than 1.5 billion years ago; by no later than 0.7 billion years marine animals also had appeared, indicating that oxygen respiration had become possible. But neither plants nor animals had invaded the land as yet, though large land areas long had existed, especially at the end of Prepale-ozoic time. Only after considerable free atmospheric oxygen accumulated could the important ultraviolet-filtering ozone layer develop fully at a level well above the land surface. Such development may have taken until middle Paleozoic time when land organ-isms first appeared (see Chap. 13). Significantly, mountain building at that same time had formed a variety of potential ecologic niches, and these were quickly filled by a sudden invasion of the land both by plants and animals. Here we see a clear example of physical changes influencing the path of organic evolution, but, conversely, once organisms were es-tablished on land, they, in turn, markedly influenced the land by greatly modifying processes of weath-ering, erosion, and sedimentation.

Increasing mountain building on all continents in late Paleozoic time created more new habitats for expanding land life. As land emerged from the sea in Carboniferous time, great coastal coal swamps formed on a colossal scale. Relative land area and elevation are major factors in the earth's heat budget, as are atmospheric and oceanic circulation. As the continents achieved their largest total area and great-est elevation since Prepaleozoic time, climate changed in response. Permo-Triassic red bed depos-its became nearly universal on northern lands, which were then located in low latitudes, while glaciation set in on southern high-latitude ones.

Paleozoic and Mesozoic land animal life was surprisingly homogeneous, indicating connections among most land areas. Plants were somewhat less uniform, apparently due to climatic zonation. By Cre-taceous time, all land organisms began to be more differentiated on present continental units. As the Mesozoic separation of continents occurred, it would have affected distribution and evolution of life, as seems indicated by the fossil record. Presence of several tenuous land bridges has maintained some interchange of land organisms among the different continents up to the present, but Cenozoic land life was far more differentiated than earlier life. In the south, isolation seems to have been more complete and of longer duration than in the north so that more disjunctive distributions of organisms are found there (as discussed in Chaps. 17 and 18).

The late Cenozoic tectonic episode has been more extreme and rapid than geologists appreciated until recently. Since Miocene time, truly profound events have occurred. These included: inception (or rejuvenation?) of all island-arc systems as we see them today, the Alpine-Himalayan mountain building, changes in the ocean-ridge-spreading regime leading to changes of plate motions, and the beginning of large-scale rifting on five continents. As all this occurred, continents have enlarged and risen.

During Cenozoic time, mammals and plants became diversified, and some major extinctions occurred, doubtless influenced by physical changes. Finally, the atmospheric heat budget tipped in favor of Pleistocene glaciation. It was during this geologically recent climatic and structural turmoil that man evolved in Africa or Asia Minor and within the brief span of 2 million years dispersed throughout all land areas. In the past two centuries, man has modified the earth's surface, atmosphere, and water to such a phenomenal degree as to surely represent the most rapid and potentially catastrophic organic event in all geologic history. Anthropologist Loren Eiseley muses that, while evolution of the human body from that of an aggressive ape is essentially completed, man seems to possess an unfinished mind that "could lead humanity down the road to oblivion."

MASS EXTINCTIONS

Mass extinctions of major groups have been used by paleontologists to mark the close of time divisions (Fig. 19.2). The causes of these extinctions are uncertain. No one hypothesis has stood the test of close scrutiny; therefore, probably we must seek more than one cause. Another point is that extinctions were not always instantaneous. The great Permo-Triassic extinction of marine animals began perhaps 10 million years or more before the close of the Permian Period.

FIGURE 19.2 Graph of evolution and extinction of major groups of animals through time showing unusual crises when mass extinctions occurred, especially at the end of the Paleozoic and Mesozoic Eras and in Pleistocene time. *(Adapted from N. D. Newell,* Crises in the history of life; *Copyright © 1963 by Scientific American Inc. All rights reserved.)*

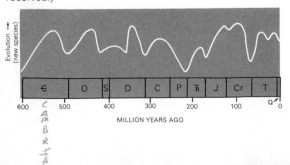

That extinction was so devastating that it took 15 to 20 million years to evolve a normal, balanced fauna in the Late Triassic (see Chaps. 14 and 16). The Permian extinctions involved major groups of marine organisms, and of course, it is tempting to suggest that emergence of land caused the eradication of many shallow-water environments. But why did many nonmarine amphibia and reptiles also become extinct? The Late Cretaceous mass extinction was not as severe, affecting only about 25 percent of the animal families. However, this extinction again affected both land and sea organisms, but on a highly-selective basis (see Chap. 17).

CAUSES OF EXTINCTIONS

Extinctions apparently resulted from failure of species to adapt at a rate equal to or greater than, the rate of environmental change. Extinction hypotheses involve countless ideas ranging from abrupt Cuvieran catastrophism to a kind of evolutionary old age or overspecialization.

Probably there are more theories on the extinction of the dinosaurs than on any other group. Suggestions run the gamut from excess atmospheric oxygen to the supposed chill of impending glaciation and even slipped vertebral discs in the oversized monsters. To a seemingly endless list, we offer our own tongue-in-cheek asthmatic hypothesis. One spring a few years ago when our eyes began to itch and our noses to tingle, we recalled that flowering plants first appeared in late Mesozoic time. Could the appearance of many pollens to which dinosaurs had not previously been exposed produce violent allergic reactions that culminated in wholesale death by hayfever? The new mammals would, of course, also be subjected to the selective stress of pollen, but, seemingly, mutations were producing mammalian strains resistant to the pollens more rapidly than reptilian ones. We might say that mammals were flower children, while blossoms were sprung on the dinosaurs too late in their racial life so that they, poor things, sneezed to extinction.

Some dinosaur groups began dropping out well before the close of the Cretaceous, but up to the end they were still relatively abundant. Both large and very small dinosaurs were present until the end so that ideas that the dinosaurs got too big to survive in the total ecosystem are not satisfactory. Some students have suggested that a pathogenic fungus or some disease wiped them out, but if we take diseases of today as a model, they usually affect only a single species and there is no known species that has been eradicated completely, even by a pandemic. Another theory has considered that mammals might have eaten the dinosaur eggs, but again it is difficult to believe this would have been 100 percent effective. Still another concept involved deadly radiation perhaps from a supernova or from cosmic rays in lethal quantities during periods of zero magnetism when reversals of the earth's field were occurring. If this were true, why did the radiation selectively eliminate some groups and not others? Surely the mammals, other reptiles, and plants would receive the same dosages; yet they were not noticeably affected. Even more puzzling is the simultaneous extinction of marine forms such as the ammonoids, for a moderate depth of water is an effective shield from radiation. Moreover, recent calculations indicate that complete removal of the magnetic field would increase cosmic radiation at the earth's surface only about 15 percent, which is deemed by many authorities inadequate to cause either mass extinctions or accelerated mutations.

Mountain building and its relation to climatic change has been cited traditionally as a cause for extinction, which is based on the fact that many unconformities coincide with faunal breaks. Yet some of the most profound faunal discontinuities occurred during times of relative crustal and apparent climatic stability. The Cretaceous extinction occurred during a time of mild climate; significant cooling did not begin until considerably later (see Fig. 18.15). Climate may have played little role in the Pleistocene extinctions as well, for most occurred during interglacial episodes (but it may be that an extinction effect lagged after some climatic cause).

The most likely cause of extinctions is the most difficult to substantiate. The finest balance in nature involves the ecosystem; that is, the interaction of organisms in the biological and physical structure of the community. The success of any community is largely dependent on a few key species of plants and/or animals in the food chain. If any of these species are removed, the effect will be felt immediately through the whole food pyramid of the community—a sort of domino effect. It is possible that a key species, which is completely unknown as a fossil,

could have become extinct, causing the dinosaur community to collapse.

MAN'S PLACE IN THE ECOSYSTEM

EFFECTS ON OTHER LIFE

Astronomers estimate that there are about 10^{20} planets in the universe capable of supporting intelligent life. But such life has one chance per planet, for the resources of each body are finite and evolution is irreversible. Man, though not nearly as strong as many other animals, is the cleverest and most adaptable organism that has yet evolved on earth. Through the development of tools and the utilization of stored energy (i.e., fuels), man has more than compensated for any lack of brawn. Tools, however, have also given man an enormous capability for manipulating and contaminating the environment, leading to man's belief that a complete control of the environment will soon be possible. Indeed, Western man long ago assumed the self-appointed role of predestined master over all of nature. Thus the natural environment came to be thought of as a *commodity* possessed by man.

Man has brought about the extinction of 450 species of other organisms, for an average of about one species per year over the past century, and now threatens an additional 500 species. Fur coats, for example, take such a merciless toll of the wild cat family (e.g., six leopard skins per coat), that several of the cats are all but extinct because of human vanity. The alligator is suffering a similar fate. Besides overtly killing and inadvertently poisoning organisms, man also threatens many forms by destroying their habitats—often quite unwittingly. In some cases man has introduced new predators, and in other cases competitors for food. It is estimated that, in the past, the average span for isolated island bird species was about 200,000 years. Since aboriginal man populated islands, the bird species have averaged only 30,000 years, and since Europeans arrived, the span has dropped to only 12,000 years.

Many of the results of man's tampering with natural communities have been quite unexpected, for the effects of upsetting ecologic balance are difficult to predict with present limited knowledge. For example, oak forests have increased in area since Anglo-Saxon settlement of the northern Mississippi Valley region, just the reverse of the general pattern of conquest of the frontiers. In this case, cessation of the Indians' habit of regularly burning prairie grass has allowed oak saplings to thrive and mature as never before. Here the settlers had to cut down forests *after* settlement, rather than before, as a consequence of their own activities.

Many changes induced by man have been beneficial to certain organisms. The skunk, opossum, raccoon, deer, rabbit, cow, robin, dog, and sparrow, to name but a few, have thrived in company with man. And no doubt urban slums and dump grounds have been great for rats. Roadsides also have produced a whole new ecologic niche successfully exploited by rabbits, foxes, pheasants, skunks, ground squirrels, toads, and a host of other birds, mammals, insects, and, of course, so-called weeds. After many years of fire control in western American forests, it is becoming apparent that former periodic burning greatly enhanced the propagation of certain desirable forest trees that do not reproduce well in shady, thickly forested areas. Examples include the *Ponderosa* or western yellow pine, one of our major lumber trees. This serves as an example of unforeseen results of well-meaning actions. While no one is advocating wanton burning, foresters are now letting some fires burn themselves out with no interference. Some deliberate manipulations of animal communities were motivated by sentimentality, such as the introduction of the English sparrow into North America and the rabbit into Australia; devastation of rangelands resulted from the latter. All this stresses how little we really can predict of the long-term consequences of ecologic manipulation, and also suggests some parallels with the results of natural competition among species in past ecosystems.

Most of the changes cited seem relatively innocuous, but many others have been more detrimental and have adversely affected man. Current degradation of the Great Lakes provides an outstanding example. Those lakes are said to hold nearly one-third of the world's fresh-water supply, but most are hardly fresh any more; Lake Erie must be described as a gigantic cesspool. Insecticides and herbicides, among other things, find their way into the lakes and then into the food chain—including that of humans (DDT has even been found in Antarctic penguins). The entry of the parasitic sea lamprey into the upper Great Lakes produced catastrophic effects upon fish populations. Early in this century, the Welland Canal provided access for the lampreys from Lake Ontario around the Niagara Falls barrier to the upper Great Lakes. By 1921 they were found in Lake Erie, by 1937 in Lakes Michigan and Huron, and in 1946 in Superior. They multiplied rapidly in the three upper lakes, and soon decimated the fish populations upon which they are parasitic. Of most direct concern to man were the quick inroads in lake trout fisheries. In Huron and Michigan, large trout production dropped to nothing; Superior's annual production shrank from 4.5 to 0.3 million pounds in only 15 years. Since 1953, a joint Canadian–United States control program has been in effect, which has successfully reduced the lamprey population in Lake Superior so that trout are beginning to increase.

It can be argued that man-caused extinctions of other life forms, as well as acceleration of modifications of the natural environment, simply are contemporary examples of selection in action—"survival of the fittest." According to such a view, history is repeating itself as the human species evolves and expands its domain at the expense of others, producing a colossal ecologic replacement analogous to many examples illustrated for past geologic periods. But, although man lives more and more in artificially controlled subenvironments of steel and concrete, man is still the product of billions of years of evolution in the natural environment. Moreover, man always will be dependent upon the mineral and biologic resources of the earth as well as upon its water and atmosphere. Just as plants help to shape their neighbors, man is influenced by other animals, plants, and microbes as surely as individual people are affected by their cultures and intellectual atmospheres. Man is one part of a fantastically complex ecologic community or ecosystem—unable to extract more from the earth than it can produce, and ultimately subject to selection processes acting upon genetic makeup much as are other organisms. It behooves man to attempt to maintain an ecological balance that will ensure the well-being of future generations of people as well as of the rest of the system upon which we are dependent. Man, rather than regarding nature as a commodity to be exploited, must think of nature as a complex community of which man is uniquely both a member and the custodian. It is well, therefore, to remind ourselves in the next section of some of the consequences *to man* of a membership in the community of nature.

MAN'S IMPENDING ECOLOGIC CRISIS

FOULING OF THE ENVIRONMENT

Pollution of the environment is receiving increasing publicity, although it has been with people a long time. For example, it is estimated that New York City had 150,000 horses at the turn of the century, and each produced about 9 kilograms (20 pounds) of manure per day. Apparently in those good old days, horses far outdid the dogs of today in littering urban streets.[2] Broadly defined, pollution includes every-

[2]Still, pets contribute significantly. The 500,000 dogs in New York City deposit on the streets about 70,000 kilograms (150,000 pounds) of feces and 320 thousand liters (90,000 gallons) of urine *each day.*

thing from dumping of raw sewage, DDT, and fertilizers into streams to the uglification of the landscape with billboards, junked cars and other solid wastes (Figs. 19.3 and 19.4) and electric utility poles and wires (which, incidentally, might be missed by birds). Even adverse noise from industries, aircraft, vehicles, and television constitutes a sort of environmental pollution. The sonic boom generated by fast jet aircraft, for example, is not only an annoyance in urban regions, but has triggered massive rock slides in western national parks.

Unfortunately, indifference toward degradation of the environment still is the rule. Seemingly only a city garbage-collectors' strike can arouse very many people. For example, as reported by *The Times* in London

FIGURE 19.3 Roadside art I. A road cut in Cretaceous sandstones on California Highway 128, 35 miles west of Sacramento. Ironically, very old autographs on rocks (as along the Oregon Trail) are prized and protected, but surely wanton autographing by millions of people is intolerable.

FIGURE 19.4   Roadside art II. A "sculpture" made from litter collected by Boy Scouts along four blocks of a typical street in Madison, Wisconsin.

the resort. We have been doing this for ages past and it can safely be done for ages in the future." But, then, New York City does it, too, to the tune of 750 million liters (200 million gallons) per day into the Hudson River.

In recent years, chemical insecticides, fertilizers, and detergents have caused serious damage to many aquatic ecosystems. Lakes rapidly are becoming fertile with nutrients from sewage plant effluent and dissolved agricultural fertilizers carried by runoff. What once were clear fishing and swimming havens are being converted rapidly to smelly slime ponds as algae thrive. Nitrates from fertilizers ultimately can have an ironic deleterious effect upon humans because, even from drinking water, some nitrate inevitably is taken into the body. Intestinal bacteria convert it to nitrite, which hinders hemoglobin's ability to transport oxygen in the blood stream, and can cause suffocation of infants.

No objective conservationist arbitrarily advocates complete outlawing of all fertilizers and pesticides, for clearly they are of great benefit to man, but consequences of their application are being studied intensively, as well as alternative control measures. For example, there are clever biologic ways to cope with insect pests. Release of sterile males into populations impedes propagation, and introduction of natural predators also can control them. Methods such as these are preferable for several reasons. Not only do they pose no threat of poisoning, they also do a more efficient job of extermination *over the long term*. Insects have such short life cycles and propagate so rapidly that chemical insecticides produce unnaturally rapid selection in their populations. Only the fittest survive, and so evolution of hardier strains actually is accelerated by the insecticides, as one can understand from our consideration of mutations and selection. Soon the chemicals become less effective even in dosages that begin to be toxic to man. Unfortunately, chemicals may not be very selective, either: when roadsides are sprayed, wild flowers as well as ragweed are killed—and a brown fire hazard is created as a by-product.

Atmospheric pollution is well known, especially as a result of concern over radioactive fallout and killing smogs (Fig. 19.5). Still there is much ignorance about long-term climatic and biologic effects of the pollutants that man has pumped into the air. Some

recently, the city fathers of an English coastal resort voted to continue pumping raw sewage into their bay for "that is the cheapest, and the proper method for

FIGURE 19.5   Effects of atmospheric pollution in Los Angeles, California. *Upper*: City Hall on a clear day in 1956. *Lower*: same view on a smoggy day in the same year. A temperature inversion—warm air above a cool layer near the ground—prohibits diffusion of contaminated air into the higher atmosphere. The inversion, seen here at an altitude of about 300 feet, is present approximately 320 days per year. Recent control measures have cut Los Angeles smog back to approximately the 1954 level. *(Courtesy Los Angeles County Air Pollution Control District.)*

possible effects have been mentioned before (see Chaps. 7 and 18). Until recent years, the atmosphere, like the oceans, seem an infinite reservoir into which wastes could be dumped forever without harm. In fact, both of these reservoirs are closed systems, and so can take on and disperse only finite amounts of pollutants before life becomes adversely affected.

The atmosphere has acquired untold quantities of pollutants from human activities—both agricultural and industrial—much of it as smoke. In addition, combustion has released carbon dioxide and toxic gases in ever-increasing quantities since the industrial revolution. The most serious culprit is exhaust from internal-combustion engines. Tetraethyl lead released from burning of gasoline affects our nervous systems and can be fatal. In recent decades, the average person's lead content has risen a hundred-fold, bringing it near present estimates of the human tolerance level. Lead also is accumulating in the shallow seas, which are our major fishing grounds; apparently it diffuses into the deeper ocean much more slowly than it is being added.

Besides known toxic effects of gaseous wastes in the atmosphere, gross climatic changes might be triggered by atmospheric pollution. Local meteorological effects are well known around large, industrialized cities, where rainfall is increased by smoke and dust particles; cities receive about 10 percent more rain and fog than do their surroundings. Yet in cities, most rain water runs off because more than half of the ground surface is covered with pavement and buildings, and so infiltration to the ground water table is correspondingly reduced. Also as a result of smoke, dust, and fog, the intensity of visible sunlight in cities is only about 60 percent and ultraviolet radiation about 10 percent as great as in open areas.

Cities alone are not guilty of atmospheric pollution. Rural areas also contribute, especially where agriculture is primitive. Widely practiced burning of brush and overgrazing contribute so much smoke and dust that the ground commonly is invisible from a plane. Worldwide increase of turbidity of the lower atmosphere results in diminished incoming solar radiation, thus a climatic cooling tendency. A 10 percent increase in turbidity causes an estimated reduction of average world temperature of 1°C. From Chap. 18 we recall that reduction of annual temperature of a mere 5 to 8°C over many years would be enough to cause the readvance of ice caps. Possible increase of atmospheric carbon dioxide through combustion also is of concern because of its entrapment of infrared energy radiated from the earth's surface after heating by the sun. Curiously, the latter "greenhouse" effect would tend to warm the climate in opposition to the dust effect. Therefore, it is important to learn the magnitude of both and the rate at which each is operating.

Besides the poisonous and climatic effects of atmospheric pollution, there is growing concern that exhaust from supersonic jet aircraft (the SST) and freon from the convenient but unnecessary little aerosol spray can will disrupt the atmospheric ozone layer. This would subject all life forms to dangerous levels of ultraviolet radiation and bring an abrupt end to sun bathing, as well as to life itself. Are we really in such a hurry that we must be able to cross the continent or the Atlantic in a mere 2 or 3 hours less than is now possible? And is it so exhausting to rub on rather than spray your deodorant?

Changes in atmospheric (or oceanic) composition are sluggish things that have a certain momentum; thus they may be very difficult to change quickly even if methods of cleaning were developed tomorrow. Damage produced over more than a century probably cannot be corrected in less time.

LIMITS OF THE EARTH'S RESOURCES
Renewable Resources

It is not known just how many people the earth can support. Minimum estimates suggest something on the order of 10 billion, which would be achieved in about the year 2015 at the present rate of population growth (Table 19.1). The most optimistic estimates place the figure nearer 100 billion, which easily could be achieved before the end of the twenty-first century. In other words, the Malthusian saturation point discussed in Chap. 5 well may be reached a mere century from now—*either in our children's or grandchildren's lifetime*!

Some of the earth's resources, such as lumber and food, are renewable up to a point. Seemingly, any plant-based resource can be regrown, and even water can be cleaned up (or sea water desalinated) if we are willing to pay the costs. In reality, however, even so-called renewable resources are limited by the earth's ultimate productive capacity, which is the limit of how much and how quickly energy can be supplied to life in usable forms. The total earth energy reservoirs are like a bank savings account; how long the savings will last depends upon how rapidly we make withdrawals. In the face of rapidly growing human population pressure, an *International Biological Program* is under way to try to evaluate the total productivity of the earth and the degree of man's dependence both upon native and agricultural vegetation.

The human food chain or pyramid is fantastically complex. Primary producers of food for the chain are, of course, plants, the only organisms capable of manufacturing food. But in the total energy budget of ecosystems, some of the energy stored by plants must be released for recycling in the system. Perpetu-

TABLE 19.1   The growth of human population*
Note especially the rapidly increasing *rate* of population growth

Year	World population	Doubling time	Year	U.S.A. population
A.D. 1	¼ billion	——		
1650	½ billion	1650 years	1790	4 million
1850	1 billion	200 years		
			1915	100 million
1930	2 billion	80 years		
1960	3 billion	44 years		
1975	4 billion	40 years	1967	200 million
2000	6 billion	35 years	2000	300 million
2025	50 billion (?)	(?)		

*After S. Cain, 1967: Geoscience News, v. 1, and other sources.

al recycling affects every single creature. Your body contains carbon that perhaps was used for awhile by a dinosaur, later loaned to some mammal, and possibly borrowed again by some cockroach or dandelion before reincarnation in you. Recycling is accomplished largely by decomposition to maintain an energy balance. Plants would deplete all carbon dioxide rather quickly from the atmosphere if fire and animal respiration did not recycle it. Certain bacteria could deplete the atmospheric nitrogen in about a million years, but other organisms release it back to the atmosphere as ammonia. Much of the decomposition is accomplished by lowly "pests," which, like it or not, are essential to the whole system. Primary consumers of plant production are herbivorous animals and decomposing microbes. Secondary consumers, in turn, eat them or their products, such as milk and honey, and so on. Organisms are notoriously inefficient, however, and the last link in the chain, such as man, may consume the equivalent of hundreds of millions of pounds of primary plant food annually.

There is no immediate prospect that man can short-circuit photosynthesis artificially. But if the human population grows unchecked, man may be forced to short-circuit the food chain by eliminating all intermediate animal and plant consumers that are not absolutely essential in the human food chain so as to gain exclusive use of earth resources. Shade trees, garden flowers, and pets would not be permissible. Moreover, diet eventually would be entirely vegetarian, for even now, the total existing human population could not be supported on a largely meat diet. But even an intensive, rigidly controlled agricultural system still would be limited in total productivity by soil fertility, which is only extended by a limited amount and for a limited time through chemical treatment. Already conflicts over the use of land for housing and industry versus agriculture are upon us; they surely will intensify in the future. Perhaps we could ease both dilemmas by depending upon algae grown in glass water tanks on the roofs of huge apartment houses or upon bacteria, fungi, and yeast grown under conditions requiring little land and little energy.

Nonrenewable Resources

Ultimate limits of nonrenewable mineral resources, such as ores of the metals, building materials, and the natural hydrocarbon fuels (with all of which geologists have had intimate concern) are even more obvious, especially since the 1973 fuel crisis, than the limits of renewable resources. But for many nonrenewable raw materials, supply exceeds current demand by so wide a margin that there may seem no immediate problem. A recent director of the United States Geological Survey only a few years ago described the earth's crust optimistically as a veritable cornucopia of resources. On the other hand, forecasters of doom repeatedly have predicted impending exhaustion of mineral reserves. We have been "running out of petroleum within a decade" ever since 1910; yet more reserves constantly were discovered by geologists. Nonetheless, all such resources are finite, and most people have no inkling of the magnitude of our conspicuous consumption. If we had to depend only upon presently known mineral reserves, our industrial society might collapse in a mere two or three decades. On the other hand, estimates of this sort are elusive, because definitions of reserves are based so much upon economics; ore that is too low grade to be considered today may become workable ore in the future as demand forces economic redefinitions. New York sewer sludge might even become an "ore deposit" one day, for it contains a greater percentage of silver, chromium, copper, tin, lead, and zinc than do ordinary sedimentary rocks.

When we compare present demands upon nonrenewable resources with population increase, then the picture becomes more sobering. On the average, every member of the human race is increasing usage of energy fuels by 3 percent per year. From 1935 to 1967, world petroleum production grew from 6.5 to 30 million barrels per day; by 1980, requirements will be about 60 million barrels per day (one barrel is 42 gallons). Put another way, a 1-billion-year supply for a constant consumption rate will last only half as long at the presently increasing rate. It is estimated that,

since 1940, North Americans have burned more mineral fuels than did the *entire world in all of previous history*. Today, North America alone requires five or six times as much fuel on a per capita basis as the rest of the world (Fig. 19.6), and by the year 2000 we shall *double* our present consumption! Consumption of metals in the United States is increasing at about double the rate of population growth (Fig. 19.7) so that by 1980 our demand for raw materials will increase 50 percent. From a purely humanitarian point of view, the disparity of exploitation of the earth's alleged cornucopia is shocking. Industrially developed societies with roughly 20 percent of the world's population consume at least half of all the earth's annual total mineral raw material production, and a great deal of this is imported from underdeveloped countries. North America alone, with only about 7 percent of the world's population, consumes nearly 30 percent of the mineral raw materials produced by

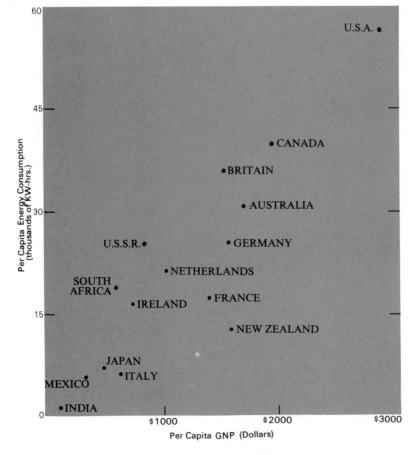

FIGURE 19.6   Relation between per capita wealth expressed as gross national product and per capita energy consumption expressed in thousands of kilowatt-hours. Obviously climate also plays some role in these figures; for example, India's energy consumption is low partly because it has a warm climate. *(Adapted from data in* The Plain Truth, *July–August, 1973.)*

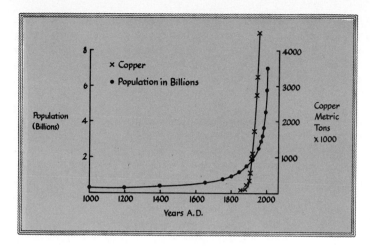

FIGURE 19.7 Graph comparing world population increase and a typical example of the increasing demand for metals. The latter rate of increase is even more rapid than that for population. Such increases of rate are the most alarming facets of man's ecologic dilemma. *(From C. F. Park, Jr., 1968, Affluence in jeopardy; by permission of Freeman, Cooper and Co.)*

the entire world! From such figures, it is clear that every child born in a developed nation puts a much greater stress on world resources than does his counterpart in an underdeveloped country.

The United States has a material wealth one-third greater than the next wealthiest nations (Fig. 19.6). Can such disparities be tolerated? Can we sustain an ever-increasing gross national product, or will we be forced to adjust to a lower material standard of living? Currently the rich grow richer and the poor grow poorer (Table 19.2). But if underdeveloped countries should bring their populations under control, their demands for metals and fuels will increase exponen-

tially as they, too, aspire to higher standards of industrialized living. With such increasing rates of demand, surely we would be much closer to ultimate resource limits, especially of certain rare metals, than we now think.

Salvage and reuse of discarded materials will become increasingly necessary, which is as it should be, for we are frightfully wasteful (Fig. 19.4). The average American, who is the world's worst litterbug, disposes of 6 to 8 pounds (3 to 4 kilograms) of solid wastes per day, or nearly 1 ton (0.9 metric ton) per year (and we already have begun littering the moon, too). As is illustrated by the composition of New York sewer

TABLE 19.2 Comparative population growth rates and approximate economic standards of major nations*

Economic status	Countries	Rate	Doubling time
Developed countries (per capita GNP $1,000–$3,000 per year)	Ireland; East Germany	Losing	——
	Hungary	+ 0.4%	100 years
	Southern Europe	+ 0.8%	88 years
	Northwestern Europe	+ 1.1%	63 years
	United States	+ 1.6%	44 years
	U.S.S.R.	+ 1.7%	40 years
	WORLD MEAN RATE	+ 2.0%	35 years
	Canada; Australia	+ 2.1%	33 years
Underdeveloped countries (per capita GNP mostly less than $200 per year)	India; Indonesia	+ 2.3%	31 years
	West and South Africa; Southeast Asia	+ 2.4%	29 years
	North and East Africa	+ 2.5%	28 years
	South America	+ 2.7%	26 years
	Central America; Pakistan	+ 2.8%	25 years
	Brazil	+ 3.1%	23 years

*After S. Cain, 1967: Geoscience News, v. 1.

sludge noted above, a fundamental redefinition of "waste" as "resources out of place" is being called for by many authorities because we can no longer afford to waste anything. Development of substitutes already has alleviated some resource pressures and no doubt will become increasingly necessary. Substitutes probably will have to supersede hydrocarbon fuels long before supplies are threatened for at least two reasons: first, burning of fossil fuels may become

intolerable because of the atmospheric pollution it produces, and second, hydrocarbons may become too precious for lubricants and as raw materials for the chemical industry (e.g., for plastics). As use of fossil fuels declines, nuclear, solar, geothermal, and perhaps even wind and tidal energy will become more important. Of most importance, ultimately, may be the earth's great internal thermal energy. Besides building mountains and displacing continents, as noted in previous chapters, it very well might be tapped by deep heat wells to drive man's society.

Other serious by-products of the extraction of nonrenewable resources are atmospheric, water, and sound pollution, as well as modification of the landscape through extensive excavations (Fig. 19.8). These ills can be alleviated, however, if society is willing to pay for corrective measures through higher

FIGURE 19.8   Bingham open-pit copper mine. Oquirrh Mountains, Salt Lake City, Utah. This is one of the largest open-pit mines in the world and a major producer of copper from low-grade ore disseminated in a late Mesozoic granitic pluton. Low-grade ore of this sort already is the major world source of copper. Still lower grade ores will be mined in the future for the many metals essential to complex, industrialized societies. *(Courtesy Bureau of Mines, U.S. Department of the Interior.)*

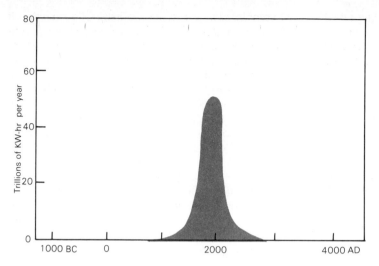

FIGURE 19.9 Exploitation of fossil fuel energy (in kilowatt-hours) from 1000 B.C. to A.D. 5000. According to this estimate, almost the entire earth's reserves will have been consumed in a mere 15 or 16 centuries. *(From Hubbert, 1962, National Academy of Sciences—National Research Council Publ. 1000-D.)*

costs; the producer cannot be expected to carry the entire burden of reclamation. The magnitude of excavations required to mine coal and oil shale in the amounts needed to supplant declining petroleum reserves staggers the imagination, however, and makes one wonder if other energy sources might be preferable.

Energy—The Ultimate Factor

Most of what has been said so far was written before environmental consciousness had reached its present level and was contained in our first edition, which appeared in 1971. Since that time, North America has depleted its huge grain and strategic mineral stockpiles, and there have been threatened food shortages due to weather-caused crop failures. Reduction of grain supplies in turn has adversely affected the livestock industry by increasing the cost of feed. Still more recently a world shortage of sugar also had developed. All these agricultural problems are further aggravated by a fertilizer shortage, which is itself partly the result of the now well-known fuel crisis. As if these examples of shortages were not enough, every geologist knows as surely as you are reading this page that serious shortages of many other mineral resources are just around the corner.

Energy is required for all human activities, including the growing, harvesting, processing, and transportation of the food from which our bodies derive energy. Indeed, today there is 10 times as much energy input to the total food system as is gained by man from his food! More alarming, however, is the fact that our increasingly technological food system now requires four times as much energy for approximately the same level of per capita caloric food energy value as in 1940. Increasing mechanization apparently is a dubious blessing. Because every aspect of life is ultimately dependent upon energy, much attention has been focused upon it in recent years. Every activity must be examined in terms of its energy budget. For example, wise as it may seem to recycle materials, it may be more costly in total energy expended for recycling than for processing the raw material in the first place. Glass bottles seem to be a case in point. There is no shortage of the raw material silica, and when all hidden energy costs (e.g., transportation) are added to the energy required for processing raw silica, the result is discouraging. It simply does not make sense to recycle materials that are not rare when to do so costs more in energy than to start from scratch.

Figure 19.9 shows the situation for the fossil fuel resources of the earth. From it we see that about 80 percent of the total fossil fuel reserves that took 500 million years to accumulate will have been used within a mere 3 to 4 centuries of human history. Considered on the geologic time scale, this represents a catastrophic event. Finding new reserves will help buy time, but it is getting harder (and more expensive) to do so. Best-informed projections of total petroleum reserves indicate that the peak of possible crude oil production was reached in the United States when the first edition of our book was published. For the world as a whole, that same peak will probably be reached about the end of the century. The peak for world coal production probably will occur around 2100 or 2200.

Clearly, alternative energy sources must be developed. Nuclear energy does not now seem as promising as in the flush of post-World War II atomic optimism. Not only were the Atomic Energy Commission and the power industry naïve in grossly overestimating the available nuclear ore reserves, but the waste-disposal problem seems to have been badly underestimated. Even if solar, geothermal, and other new energy sources ultimately can provide our major needs, we shall still be dependent upon fossil fuels for several more decades. And because the demand for these fuels is increasing at such a phenomenal rate, we face serious economic and environmental problems in addition to the fuel pinch itself. Consider these figures. World petroleum consumption is doubling twice as fast as is population. In the past decade, the number of automobiles per capita has increased 25 percent in the United States, 50 percent in Britain, and 90 percent in Japan. Yet, the private automobile, which consumes half the total energy expended for transportation, is notoriously inefficient. A municipal bus is a three-times-more-efficient way to carry people than the ordinary car, and 10 times more so than a large car. Similarly, the train is five times more efficient than a jumbo jet and 10 times more so than an SST. We must conserve our fossil fuels not only to gain time but also to minimize environmental pollution and the possible economic chaos that could result from high crude oil pricing by oil-rich countries. (The United States dollar outlay for imported oil increased sixfold from 1972 to 1974). Canada and Mexico are well fixed with domestic reserves adequate for their modest demands for the foreseeable future, but the energy-hungry United States already must import one-third of its petroleum needs. The famous new 10-billion-barrel (420-billion-gallon) Prudhoe Bay oil field in northern Alaska, largest in North America, by itself could supply the present total United States demand for but 2 years and the total present world demand for only 6 months. A few more large oil fields like this one doubtless are yet to be discovered, especially beneath continental shelves like the North Sea, but with any finite, nonrenewable resource, the race between discovery and consumption cannot be won in the long run.

We have chosen to emphasize here the fossil fuels as an illustration of man's resource plight. Precisely the same story could be written for several mineral resources which are now also in short supply. Examples include nickel, tin, manganese, cobalt, and even aluminum. Perhaps the greatest service anyone interested in geology can perform for society in the next quarter century is to participate in the search for and development of new fuel and other mineral resources. Obviously there will be plenty of work to do.

WHAT QUALITY OF LIFE?

What sort of life will our children and grandchildren lead? Will it be a steak-and-convenience-blessed paradise, or mere existence at a bland algae-and-tenement subsistence level? Is concern for lower life forms and for untrammeled open spaces free of billboards and beer bottles mere sentimentality irrelevant to pre-eminent man of the twenty-first century? We have shown that the only real certainty in nature is change. Among laymen there are two prevalent—and opposing—views of social and technological change. One, the blindly optimistic view, assumes uncritically that *all* change is good and represents "progress." This has been aptly dubbed the *Gospel of Growth*, and part of its creed reads: "That which can be built must be." On the other hand, the pessimistic or "good-old-days" view holds that most change is bad because it undermines some glorified *status quo*. Regardless of whether one is wholly an optimist, a pessimist, or some complex mixture (as are most of us), such attitudes clearly are irrational and nonobjective. It is as absurd to assume that all technological and environmental changes that man produces, even if masquerading as "advances," are beneficial in the long run as it is to weep, ostrich-like, for a sentimentalized past. Therefore, if change *is* so inevitable as we claim, then surely it behooves mankind to study, plan, and control biological and other environmental changes for the best long-run effects on the total ecosystem. This is very much a personal as well as a social problem.[3]

[3]John W. Gardner, in *Self-Renewal—The Individual in the Innovative Society*, discusses in an eloquent and succinct manner the importance on the social level of critically analyzing and planning for change. His inspiring treatment should be required reading for any thinking person concerned about the future of man in the biologic as well as the socioeconomic realm. Surely a creature clever enough to travel to the moon has not exhausted all potential for intelligent management of its own destiny.

Certainly man is the most adaptable creature ever, and perhaps someday he can control the environment absolutely and manipulate it purely for self-convenience. One inescapable truth, however, is that the amount and location of mineral resources and of agriculturally favorable climatic and soil conditions have already been fixed for man by geologic history. Faced with the complex ecological interdependence both among life forms and with their physical environment throughout geologic history, together with our rather pathetic understanding of the intricacies of ecosystems, it is unwise to plot a course for the future on blind faith. History teaches that evolution is irreversible, and once any creature becomes extinct, it can never return no matter how important it may turn out to have been to man's own well-being. This applies even to the lowliest microbes, some of the most beneficial of which probably have not yet even been discovered by man. Examples cited of man's unwitting and commonly unwanted degradation of the environment provide ample evidence of the folly of any policy of blind faith that "everything will work out for the best because science will find a way." Famous conservationist Aldo Leopold once said that the height of ignorance is the man who asks what good is an animal or plant. "If the land mechanism as a whole is good, then every part is good, whether we understand it or not. If the biota, in the course of aeons, has built something we like but do not understand, then who but a fool would discard seemingly useless parts?"

It is a basic principle of biology that diversity is the key both to racial survival and to ecologic stability. The more genetic diversity a species has, the less susceptible it is to complete extinction in the face of environmental change. The highly touted green revolution created very productive hybrid crop plants through careful breeding. But those same plants require more care and are far less resistant to disease, pests, and climatic adversity than are the natural strains. Because of their narrow range of genetic diversity, the hybrids could be completely decimated overnight. Diversity is also a boon to ecologic communities, for it assures a greater degree of stability for the entire community, which suggests that we should take Aldo Leopold's advice very seriously.

Let us suppose that complete ecologic understanding will be achieved and with it the ability to manipulate safely everything in the environment for the benefit of man. Would the quality of life be acceptable in what would likely be some kind of Brave New World or Orwellian 1984 or beyond? Can we presume to judge what is best for future generations?

In North America today, natural water and landscapes are being "lost to development" at the rate of one acre per minute! Because of land filling, the area of San Francisco Bay is only two-thirds of what it was 200 years ago. In 1967, there was an average of about 60 persons per square mile of land on earth, though distribution was very irregular (70 percent of Americans live on 10 percent of the available land). At the present rate of population growth, by about the year 2025 there would be 10,000 persons per square mile. Many, if not most, people require contact with the natural environment for their emotional well-being, and so space may turn out to be one of the most precious commodities as population increases. For example, it took more than 40 years for National Parks in the United States to tally their first billion visitors, but only 12 years for the second billion (see Fig. 19.10). Already reservations are required long in advance for camping in several of the parks.

Laboratory experiments show that rats become psychopathic when too closely crowded for a long time. Unplanned sociological experiments in urban ghettos suggest that crowding under subsistence-level conditions has adverse psychological effects upon *Homo sapiens* as well, even blurring the distinction from brother rat. If population continues to expand unchecked, the entire human species will be subjected to selective pressures imposed by over-urbanization.

SUMMARY

Any way we examine the future of man, even for the geologically minuscule span of the next century, the quality of life will be dependent first and foremost upon population size and energy resources. Problems of food, raw materials, pollution, and space all are secondary to population. Figure 19.7 and Tables 19.1 and 19.2 illustrate the magnitude of this problem. It is the phenomenal *acceleration of the growth rate* that is so alarming together with the resulting *acceleration of demands placed upon the environment*. World population will double in a mere 40 years, achieving 6 or 7 billions by the year 2000. India, with nearly 500 million now, will add 200 million people

FIGURE 19.10 The pressure of a growing population on space. *(Reprinted from: Burdened acres—the people problem, by R. Wendolin with drawing by Monroe Bush, in* The living wilderness, *spring–summer 1967, edited by M. Nadel. By permission of The Wilderness Society, Washington, D.C.)*

before 1980—equivalent to the entire 1968 population of the United States! Unless nutritional standards were dropped, even North America would have difficulty feeding 200 million more mouths if added in such a short period. At the time of publication of this second edition of *Evolution of the Earth*, this fact was just being realized, although the 1974 World Food Conference in Rome seemed unwilling to acknowledge it. In many parts of the world, nearly half of the population is undernourished, and before 1980 a major famine appears unavoidable for some countries.

Studies of rates of change in past geologic times emphasize the truly catastrophic impact of man upon this earth in the past two centuries. Consider the phenomenal acceleration of technological changes alone. For at least 1.99 million years, man was a hunter and gatherer; the agricultural revolution occurred no more than 10,000 years ago; the industrial revolution is but 200 years old; and only within the past century have the electric light, automobile, airplane, television, nuclear reactor, antibiotics, electric toothbrushes, electric matches, and computers appeared. But there is some doubt that man is prepared either emotionally or socially to cope with further change at still faster rates. Eiseley's concern is relevant—has man's mind evolved sufficiently to deal with such rapid change?

The population problem is not just the result of increase in absolute numbers of births, but also improvement of birth-survival expectancy and longevity. A little-appreciated consequence of disease control is the perpetuation in the breeding portion of the population, or gene pool, of all types, including the mentally and physically deficient, which by natural selection would have been reduced. Together with chemicals and radiation, such perpetuation presents a genetic (thus evolutionary) problem for the race, and at the same time complex sociological problems. It cannot be ignored, for man like so many of his predecessors on earth, could become overspecialized in this complex, artificial civilization, and perhaps become genetically weakened. On the optimistic side, however, there are predictions of genetic engineering just around the corner that may correct defects and even provide prenatal controls.

As a product of a 3.5-billion-year history of genetic mutations and natural selection by countless changes of environment, and as a member of a highly complex ecosystem, apparently man must reform if a tolerable ecologic equilibrium is to be achieved. And there will be but one chance. For at least the foreseeable future, man will not be omnipotent, but rather will remain very much dependent upon, and a part of, the total environment. Above all, man needs to achieve a balance of human population in order to conserve that environment and the present way of life, for unlimited growth disrupts stability of orderly ecosystems. As that modern philosopher Pogo put it, "we have met the enemy, and he is us."

Conservation, simply defined, is applied ecology, and it does not mean only to preserve or lock up. About 35 years ago former Wisconsin geologist C. K. Leith aptly described conservation as the "balancing of natural resources against human resources and the rights of the present generation against the rights of future ones." Conservation should ensure maximum present and future benefit from resources. Both public and private machinery are required to make it succeed, but inevitably serious conflicts arise between public and private interests. The conflicts involve long-cherished property rights and freedom to exploit private property as one sees fit. Yet, who is to say what interest is the more important? Should the public interest take precedence over individual self-interest and private initiative? It seems self-evident that as societies increase in size and complexity, some individual freedoms must be restricted, and it is presently popular to cry for governmental control of every socioeconomic problem that arises. But, as Orwell warned, one of the greatest threats to the quality of individual lives is the gradual, insidious restriction of personal freedoms by large organizations of all kinds—governmental, military, business, and even educational. Bigness itself may, after all, be the greatest threat to individualism; therefore the development of responsible environmental and of equitable socioeconomic policies without making a brainwashed slave out of the individual is the greatest social challenge facing all nations.

Certainly greater respect for the natural environment is required on several grounds. Apart from the esthetic and sentimental arguments for conservation, the steepness of the population growth curve is somber warning that already it is later than we think. Our present course may be like Russian roulette with a bullet in all but one chamber. But any really meaningful program for man's maintenance of an acceptable ecologic balance with the environment seemingly will require first a major change in attitudes toward nature that date back several centuries. Aldo Leopold's eloquent plea for a new land ethic characterized by an ecological conscience seems

even more relevant today than when he wrote it in 1949.

In Western industrial societies, both religious and economic philosophies long have been invoked to rationalize the contention that man was the ultimate earthly being (the perfect creation) destined to rule over all else, a tradition that can be traced at least back to medieval Europe (see Chaps. 2 and 3). The Scriptures contained further precedent both for this attitude and for uncontrolled human propagation. In Genesis 1:28, God said to Adam: "Be fruitful and multiply, and replenish the earth, and subdue it, and have dominion over the fish of the sea, and over the fowl of the air, and over every living thing that moveth upon the earth." Nature existed for the benefit of man—a commodity to be arbitrarily exploited. From the practical necessity for frontier societies to conquer nature, a subtle moral justification for exploitation was an easy outgrowth. It became virtuous in the eyes of man to cut down the forest and plow up the ground, that is, to make the land "useful." Wilderness was wasteland, and anything judged useless also was regarded as "bad." Therefore, the world actually was improved by man's efforts to make nature "useful." Such an attitude is still conspicuous today; for example the presumptive value judgments incorporated in elaborate justification for some grandiose reclamation schemes. As Western societies became more industrialized, a self-justifying morality was invented, which preached that those who would develop (exploit?) and thereby improve (plunder?) nature were somehow superior to other, "primitive" peoples. Thus were the subservience of certain peoples as well as of nature itself conveniently rationalized in the nineteenth and early twentieth centuries.

Fundamental overhaul on a social level of some of the most cherished tenets of industrial society is the prerequisite to future welfare of the earth's surface, which is largely in the hands of man. Inasmuch as history reveals, above all else, the inevitability of change, it also suggests that thinking man has the choice of how to influence future changes in his environment. Surely a knowledge of the history of the earth and of its life in concert with wisdom from other disciplines makes clearer the urgency of social and economic overhaul for the future well-being of man and all fellow travelers in space. The ecologic plight is aggravated by the fact that it is a "quiet crisis," as then-Secretary of Interior Stewart Udall described it in

1963, although it has been getting much noisier lately.

As we have seen, the size of any natural population is constantly adjusted through natural selection. But such selection operates with no regard for the individual, and much hardship and deprivation inevitably result. If man wishes to preserve a civilized way of life and to avoid a second Stone Age, the ultimate control of the population must not be left to a deferred process of natural selection such as starvation. Clearly, then, the quiet crisis requires the attention of the best of many diverse talents—nonscientific as well as scientific—at the same time that other crises are being attacked. Ultimately, *all* of the problems converge upon the same great question: *Will this continue to be the best of all possible worlds?*

Readings

American Association for the Advancement of Science, 1974, Energy: Science, v. 184, (special issue), pp. 211–402.

Cain, S., 1967, The problem of people: Geoscience News, v. 1, no. 1, p. 6.

Cameron, E. N., ed., 1973, The mineral position of the United States, 1975–2000: Madison, Univ. of Wisconsin Press.

Cannon, H. L., and Davidson, D. F., eds., 1967, Relation of geology and trace elements to nutrition: Geological Society of America Special Paper 90.

Carson, R., 1962, Silent spring: Boston, Houghton-Mifflin.

Cole, L. C., 1958, The ecosphere: Scientific American. (Reprint 144)

Flawn, P. T., 1966, Geology and the new conservation movement: Science, v. 151, pp. 409–412.

Gardner, J. W., 1963, Self-renewal: the individual and the innovative society: New York, Harper & Row. (Colophon Paperback CN 54)

Hardin, G. 1968, The tragedy of the commons: Science, v. 162, pp. 1243–1248.

Hibbard, W. R., Jr., 1968, Mineral resources: challenge or threat?: Science, v. 160, pp. 143–148.

Hoyle, F., 1964, Of men and galaxies: Seattle, Univ. of Washington Press. (Washington Paperbacks WP-1)

Kesler, S. E., 1975, Our finite mineral resources: New York, McGraw-Hill. (Paperback)

Leopold, A., 1949, A Sand County almanac: Oxford. (Paperback edition GB263, 1968)

Nash, R., 1967, Wilderness and the American mind: New Haven, Yale.

National Academy of Sciences Committee on Resources and Man, 1969, Resources and Man: San Francisco, Freeman.

Newell, N. D., 1963, Crises in the history of life: Scientific American. (Reprint 867)

Osborn, F., 1948, Our plundered planet: Boston, Little Brown.

Park, C. F., Jr., 1968, Affluence in jeopardy: minerals and the public economy: San Francisco, Freeman, Cooper Co.

Steinhart, J. S., and Steinhart, C. E., 1974, Energy: Sources, use, and role in human affairs: North Scituate, Duxbury.

President's Science Advisory Committee, Environmental Pollution Panel, 1965, Restoring the quality of our environment: Washington, U.S. Government Printing Office.

Udall, S., 1963, The quiet crisis: New York, Holt.

White, L., Jr., 1967, The historical roots of our ecological crisis: Science, v. 155, pp. 1203–1207.

APPENDIX I

A SYNOPTIC CLASSIFICATION OF PLANTS AND ANIMALS

Any classification of organisms is a progress report on the status of knowledge of different groups of organisms. Several things must be kept in mind regarding this report. For one thing, it is partly a subjective concoction. Even though there is a general agreement among specialists, synthesizers who scan the whole field of classification may differ with those specializing on individual groups. Thus, some classifications will show the graptolites as a Phylum Graptolithina, while others may place them as orders of either the Phyla Coelenterata or Protochordata. Another point to keep in mind is that some groups have been worked on more intensively than others so that we have more modern concepts of the systematics of the Mammalia than, say, the Porifera (sponges). The classification presented here is abbreviated; categories rarely or never found in the fossil record have been omitted, such as various "worm" phyla. Conversely, some groups are expanded so that important subcategories are outlined.

The following classification is a compromise between new and somewhat older systems in order to present as simplified a classification as possible.

KINGDOM ANIMALIA

Phylum Protozoa: Single-celled or unicelled animals subdivided on the basis of locomotive devices. (Prepaleozoic-Recent)

Class Mastigophora: Flagellated forms which may have calcareous or siliceous skeletons; includes coccoliths. Most students believe this group belongs to the plant kingdom. (See Fig. 17.37)

Class Sarcodina: Forms with pseudopods used for feeding and locomotion.

Order Foraminiferida: Sarcodinids with calcareous or siliceous skeletons, the most important and common fossil Protozoa; includes fusulinids, globigerinids, miliolids, and nodosarids. (Cambrian-Recent) (Fig. 14.39)

Phylum Porifera: A primitive multicellular group generally consisting of vase-shaped body with three wall layers and with ciliated canals that carry out digestive and other functions; there are no organs. The skeleton is composed of calcite, silica, or spongin (a chitinous, flexible substance). The skeleton

FIGURE AI.1 A Silurian sponge (*Astraeospongia*) from Tennessee. (*Courtesy U.S. National Museum.*)

uniquely is composed of discrete particles called spicules which are manufactured in the middle body layer (includes the "bath" and glass sponges). (Cambrian-Recent) (Fig. AI.1)

Order Stromatoporoidea: A group having a fibrous skeleton of pillars and laminae; important reef formers in the Silurian and Devonian. (Ordovician-Recent)

Phylum Archaeocyatha: An extinct, coral-like group. The calcareous skeleton has a double wall with connecting and supporting plates. (Early-Middle Cambrian) (Fig. 11.5)

Phylum Coelenterata: Solitary or colonial animals with a three-layered wall and a body cavity called the coelenteron in which most body functions occur, such as circulation, digestion, and reproduction. Forms are either naked or have a calcareous or chitinophosphatic skeleton. They are polymorphic; that is, during the reproductive cycle there may be an alteration of form. The two principal types are the medusa, a jellyfish stage, and the polyp, a stage attached to a hard substrate. Both forms possess tentacles for food gathering. (Early Ordovician-Recent)

Class Scyphozoa: Exclusively medusoid forms; includes many of the common jellyfish. They are preserved rarely as impressions. (Prepaleozoic-Recent) (See Fig. AI.2.)

Class Anthozoa: The largest and most important class; includes sea anemones and all corals. They are polypoid only and have the coelenteron subdivided by numerous blade-like tissues which, if calcified, are called septa. Greatest majority of genera had calcareous skeletons. (Ordovician-Recent)

Order Rugosa: Solitary and colonial forms with a basic fourfold symmetry of septa; important reef formers during the middle Paleozoic. (Ordovician-Permian) (Figs. AI.2 and 12.7)

Order Scleractinia: Solitary or colonial corals with a basic sixfold symmetry of septa. They formed reefs from Triassic times onward, and are the most important reef formers today. (Triassic-Recent) (Fig. 16.33)

Order Tabulata: Exclusively colonial forms related to the corals; reduced septa but with well-developed horizontal plates (tabulae) subdividing the elongated corallites. The walls have characteristic pores. (Late Cambrian-Jurassic) (Fig. AI.3)

Phylum Bryozoa: Exclusively colonial, very minute forms that build a variety of colonies. They possess a stomach and gastrointestinal tract, which serves several functions; reproduction is accomplished in a separate chamber. The calcareous skeleton, if present, is external, and the animals can completely draw into it for protection. These were important animals during the early Paleozoic

FIGURE AI.2 *Left*: *Heliophylum*, a rugose coral polyp, from the Devonian of Ontario. *(Courtesy U.S. National Museum.) Right*: *Brooksella.* a jellyfish (medusa), from the Cambrian of Arizona. The specimen is a sand cast of the bottom side. *(Photograph by G. R. Adlington.)*

and Tertiary, both in numbers and distribution. They are used for correlation and ecological interpretations. (Ordovician-Recent) (Fig. AI.4)

FIGURE AI.3 *Halysites*, a tabulate coral colony. This is a good index fossil for the Silurian; from Louisville, Kentucky. *(Courtesy U.S. National Museum.)*

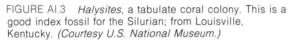

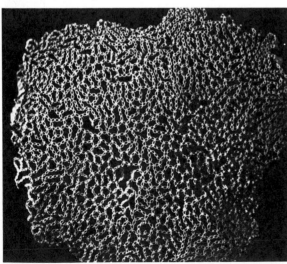

Phylum Brachiopoda: Forms with two valves serving as exoskeletons; they are bilaterally symmetrical. The bryozoans and brachiopods share a respiratory organ or gill which is a tube with a series of hairlike cilia. The brachiopods have a primitive heart and kidney, which is housed with the stomach and reproductive glands in a visceral sac. The anterior cavity houses the circulatory vessels and sinuses. Shell may be attached to bottom by a fleshy stalk (pedicle). (Cambrian-Recent)

Class Inarticulata: A conservative group in evolution having chitinophosphatic or calcareous shells. They lack hinging structures; articulation (opening and closing the valves) was accomplished by complex muscles. (Cambrian-Recent) (Fig. 11.6)

Class Articulata: The dominant, highly evolved, and most common group with calcareous, hinged shells. (Cambrian-Recent)

Order Orthida: With wide hinge line or oval-shaped biconvex shells. Most common during lower Paleozoic. (Cambrian-Permian) (See Fig. 12.5.)

Order Pentamerida: Short hinge line, large internal platform for muscles and other struc-

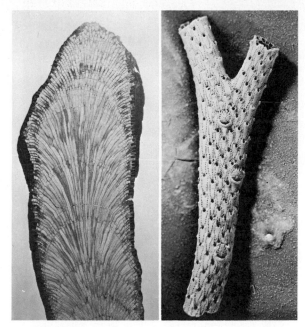

FIGURE AI.4 *Left*: A microscopic photograph of a thin section of a trepostome bryozoan from the Devonian of New York. The trepostomes were the dominant Paleozoic bryozoa. *(Photograph by G. R. Adlington.) Right*: A cheilostome bryozoan from the Eocene of Alabama. The cheilostomes have been the dominant bryozoan group since the Cretaceous. *(Photograph courtesy Alan Cheetham.)*

Order Rhynchonellida: Convex shells with sharp ribbing. The *beaks* may be large; hinge line is short. A relatively conservative group. (Ordovician-Recent) (Fig. AI.6)

Order Spiriferida: Typically with a long hinge line and with the interarea on pedicle valve only. Internally there is a spiral gill. A very important group particularly in the Devonian. (Ordovician-Jurassic) (See Fig. 13.2.)

Order Terebratulida: Biconvex, punctate shells with a short hinge line and an interarea as in the spiriferids. Internally there is a "looped" gill support. (Silurian-Recent) (Fig. AI.6)

Phylum Mollusca: A highly variable group of forms with a two- or three-chambered heart (rarely with more), kidneys, gills or lungs, and with a highly developed nervous system that may include eyes. Single, coiled shells (as in the gastropods and cephalopods), or two-

FIGURE AI.5 *Pentamerus* from the Silurian of New York. *(Photograph by G. R. Adlington.)*

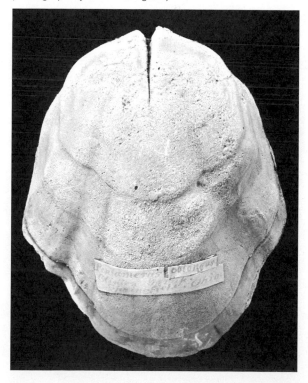

tures. Important index fossils in the Silurian. (Cambrian-Late Devonian) (Fig. AI.5)

Order Strophomenida: One of the very common lower and middle Paleozoic groups. Has a pseudopunctate shell and a wide hinge line. One valve is plane or convex. Pedicle opening very small or absent. (Ordovician-Recent) (See Fig. 12.5.)

Order Productoidea: A large, varied, and sometimes bizarre group most common and important in the upper Paleozoic. Interarea (between valves) small or absent, hinge line is moderately long. The shell is pseudopunctate and generally covered by spines. Attachment to hard substrate with spines, or spines used as stilts to raise shell off a muddy bottom. (Ordovician-Permian) (Fig. AI.6)

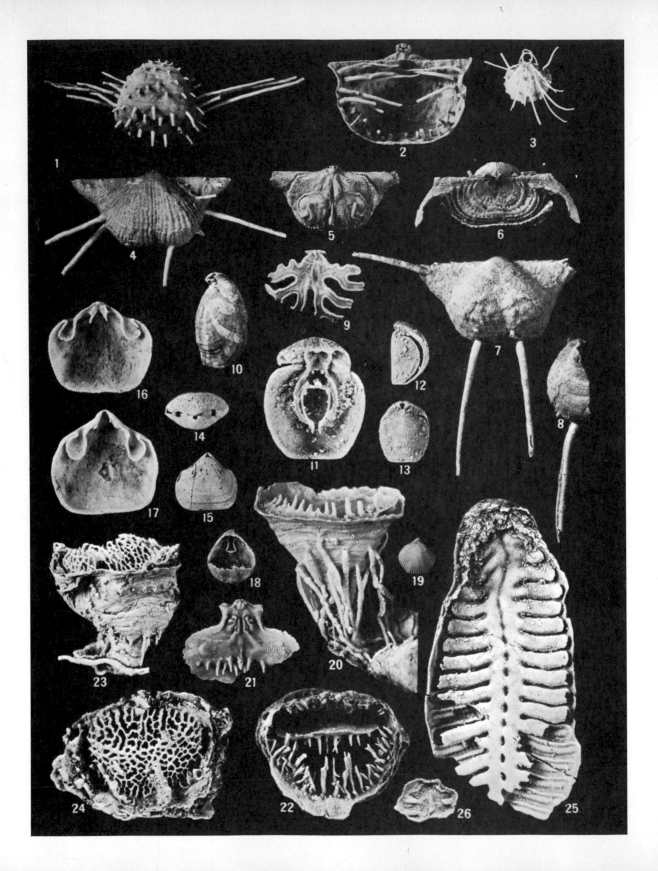

valved shells (as in the pelecypods), or eight-plated shells as in the amphineurans (chitons). One of the most common invertebrate forms living today. (Cambrian-Recent)

Class Monoplacophora: A segmented bilaterally symmetrical form with a single cap-shaped shell. Fairly common in certain lower Paleozoic strata, not known after the Devonian until Recent. One of the best examples of a "living fossil" is *Neopilina*, which was dredged up from deep water near Central America. (Cambrian-Recent) (See Fig. 11.7.)

Class Polyplacophora: This group also is called Amphineura or chitons. Its forms have bilateral symmetry; most possess a shell consisting of eight separate calcareous plates. They are adapted for living on rocks in the surf zone or hard substrate. They are rare in the fossil record, being represented only by isolated plates. (Late Cambrian-Recent)

Class Gastropoda: Single-shelled, generally coiled forms, which may have an operculum for protection. Distinct from all other molluscs by having the gastrointestinal tract contorted into a figure-8 pattern so that the anus and mouth are close together. One of the most successful of all invertebrate classes. (Cambrian-Recent) (See Figs. 5.12 and 14.34.)

Class Cephalopoda: Molluscs that have evolved into highly efficient swimmers. Because of their mobility, they have developed excellent eyes and a jet-propulsion system of locomotion in addition to having tentacles for feeding and locomotion. Includes the living octopus, squid, and chambered nautilus. Fossil cephalopods include the subclass Nautiloidea, dominant in the lower Paleozoic, and the subclass Ammonoidea, dominant in Mesozoic seas. (Late Cambrian-Recent) (See Figs. 14.35 and 16.32.)

Class Bivalvia (Pelecypoda): Molluscs having

FIGURE AI.6 Permian brachiopods from West Texas. Most of the spiny types in the photograph are productids. Specimen 10 is a terebratulid. Specimen 19 is a rhynchonellid. Specimens 20 and 23 (*Prorichthofenia*) are examples of convergence on solitary rugose corals (see Fig. AI.2). *(Courtesy U.S. National Museum.)*

FIGURE AI.7 *Trigonia*, a clam, from the Upper Cretaceous of Tennessee. *(Courtesy U.S. National Museum.)*

two shells that are mirror images in contrast to the brachiopods in which the left and right sides of each shell reflect bilateral symmetry. Clams share with gastropods the honor of being among the most diverse of the invertebrates in today's seas. (Middle Cambrian-Recent) (Fig. AI.7)

Phylum Arthropoda: The largest group of invertebrates includes the insects (probably numbering close to a million species compared to the molluscs with 80,000). The chief characteristic is a chitinous, jointed exoskeleton and specialized, jointed appendages used for locomotion, capturing prey, respiration, and reproduction. The appendages are most important in classification of many groups. (Cambrian-Recent)

Class Trilobita: The earliest appearing class. Most trilobites had *cephalon, thorax,* and *pygidium* segments, and were bilaterally symmetrical. The name comes from the fact that there is an axial lobe which separates two lateral lobes of the body. The exoskeleton frequently was reinforced by granules of calcite. Classification is based on characters of the exoskeleton rather than the appendages, because they are so rarely preserved. Trilobites were very common in the early Paleozoic when they underwent considerable evolution. They became progressively rarer until their extinction in the Permian. (Cambrian-Permian) (See Figs. 11.3 and 11.4.)

Class Crustacea: A varied group which may have a tough calcite-impregnated exoskeleton consisting of a fused cephalothorax and an abdomen. Appendages are highly specialized and are used in classification. The class includes barnacles, crabs, shrimp, copepods (the most important food source for many animals in the sea), and the minute bivalved ostracods, which are very common and important in the fossil record. (Late Cambrian-Recent) (Figs. AI.8 and 14.33)

Class Meristomata: Another large group which has four pairs of appendages and whose exoskeleton is more flexible than that of the Crustacea. They also have appendages modified into antennae as do the trilobites and crustaceans. The respiratory organs are specialized, enabling many of the animals to live on land. The class includes the spiders, scorpions, eurypterids, and the horseshoe crab (*Limulus*). (Cambrian-Recent) (See Fig. 13.5.)

Class Insecta: The most diverse and common of all living invertebrate classes. Insects are six-legged air-breathing forms with flexible exoskeletons and a body divided into head, thorax, and abdomen. Practically all bear paired wings on the second and third thoracic segments. Although there are about

12,000 described fossil species, they are relatively unimportant and rare in the record because of their soft exoskeletons. The class includes mosquitoes, cockroaches, wasps, boll weevils, locusts, and the common housefly. (Silurian-Recent) (Fig. 2.2)

Phylum Echinodermata: A large group of organisms characterized by having a skeleton composed of polygonal plates and having a dominantly radial symmetry with a predominant fivefold structure. Circulation is by a system of tubes and canals associated with food-gathering appendages (isolated in feeding-ambulacral areas having perforated plates). Groups are based on symmetry, ontogeny, and mobility. (Prepaleozoic-Recent)

Subphylum Echinozoa: Globose echinoderms that do not develop arms; most forms are unattached.

Class Helicoplacophora: A unique Lower Cambrian group that is the earliest echinozoa known. They are free living, top-shaped with their plates arranged as helical spirals. (Early Cambrian) (See Fig. 11.9.)

Class Edrioasteroidea: A Paleozoic group, may be attached or free living. From three to five feeding areas are twisted into sigmoid shapes resembling starfish. (Prepaleozoic-Pennsylvanian) (Fig. 11.8)

Class Holothuroidea: The sea cucumbers, sediment feeders having a reduced skeleton consisting of calcareous spicules imbedded in a tough skin. Radial symmetry is observable internally. Known mainly from curiously shaped disarticulated spicules in the fossil record. (Ordovician-Recent)

Class Echinoidea: The largest group of echinoderms. Dominantly with a bun-shaped exoskeleton composed of fused or articulated polygonal plates. The anus is on the upper surface, the mouth on the lower. Typically with protective spines on the intra-ambulacral plates. The echinoids tend to be sessile or move very slowly by moving their spines. One group is more mobile and tends to develop a secondary bilateral symmetry. When this occurs, the anus and

FIGURE AI.8   A Jurassic shrimp (crustacea) from the Solenhofen Limestone of Germany. *(Courtesy U.S. National Museum.)*

mouth migrate to posterior-anterior positions. (Ordovician-Recent)

Subphylum Homalozoa: A small group of lower Paleozoic forms with flattened, asymmetrical bodies. Another name for the group is Carpoidea; representatives are rare. (Cambrian-Devonian)

Subphylum Crinozoa: Forms attached by means of a calcified jointed stem.

Class Cystoidea: Globular to pear-shaped forms attached directly to the substrate by means of a stem. The porous head (calyx) plates are not stabilized in numbers or shape. (Ordovician-Devonian)

Class Blastoidea: A relatively small class of Paleozoic attached echinoderms having rather short stems. The calyx has prominent ambulacra and the 13 interambulacral plates are arranged in three circles. The top of the calyx has five openings leading to complex, calcified internal circulatory structures. (Ordovician-Permian) (Fig. AI.9)

Class Crinoidea: The most important class of Paleozoic echinoderms, which has gradually diminished in numbers and kinds until today. The majority of the crinoids were attached by a well-developed and jointed stem and a root system. The viscera are contained in a calyx, which is variable in structure but with a regular system of plates, and a system of food-gathering arms. (Ordovician-Recent) (See Fig. 14.37.)

Subphylum Asterozoa: A group of stemless echinoderms that are vagrant benthonic. Mobility is achieved by arms and tube feet.

FIGURE AI.10 *Devonaster*, a starfish (asteroid), from the Devonian of Central New York. *(Courtesy U.S. National Museum.)*

Class Somasteroidea: Starfish with flattened, petal-shaped arms, and lacking an anus. The tube feet are nonextendable (Ordovician-Recent; quite rare in the fossil record)

Class Asteroidea: The common starfish with a water vascular system carrying water by radial tubes from near the mouth to the tips of the arms. The mouth is on the underside and anus on the central dorsal surface; water for the circulatory system enters a sieve plate near the anus. (Ordovician-Recent) (Fig. AI. 10)

Class Ophiuroidea: Arms set off from a central disc containing all organs. Arms are jointed and the organism is highly mobile; the common name is the brittle starfish. (Ordovician-Recent)

Phylum Protochordata: A group of sessile forms and burrowers that includes acorn worms. The most important feature is the presence of a notochord (dorsal nerve chord) sometime during the life history.

Class Graptolithina: A group of colonial genera having a chitinous exoskeleton. Found most commonly as carbon films in black shales

FIGURE AI.9 *Nucleocrinus*, a blastoid, from the Devonian of Ontario. *(Courtesy U.S. National Museum.)*

of the lower Paleozoic (primarily in the Ordovician). The graptolites were apparently planktonic forms, either as floaters or attached to some floating or swimming organisms. As a result, genera are found worldwide and are very important for intercontinental correlation. (Cambrian-Mississippian) (See Fig. 12.8.)

Phylum Chordata: A highly organized group of animals with a protected central nerve cord located dorsally; gill slits are usually present, at least in the embryo. (Ordovician-Recent)

Class Agnatha: Fish-like forms without paired appendages or jaws. These were precursors of the fish and the early genera had a heavy bony armor covering the head and anterior part of the body; includes the lamprey eel of today. (Ordovician-Recent) (See Fig. 13.6.)

Class Placodermii: Primitive fish with diamond-shaped scales and, in some, spines. Jaws were either bony or cartilaginous and appendages were paired. They dominated Devonian seas, and gradually were replaced by more efficient and advanced bony fishes. (Devonian) (Fig. 13.7)

Class Chondrichthyes: The sharks long have been successful marine predators. They have cartilaginous skeletons, two pairs of appendages, and lack lungs or air bladders. (Devonian-Recent) (See Fig. 13.6.)

Class Osteichthyes: These are the bony fishes and have well-developed skeletons, complicated skulls, and paired appendages. Most modern fishes belong to this class. (Devonian-Recent) (See Fig. 13.6.)

Subclass Actinopterygii: The ray-finned fish. "Primitive" forms had diamond-shaped scales. Nearly all common fish, such as the salmon, tuna, and sardine, belong to this group. (Devonian-Recent) (Fig. AI.11)

Subclass Choanichthyes: A very important fish group that gave rise to the amphibians. They have internal nostrils and their fins are supported by bony rods rather than rays. The subclass includes the coelacanths represented by the very rare living fossils, *Latimeria* and *Malania*. (Devonian-Recent) (Fig. 13.8)

Class Amphibia: The most primitive tetrapods and the first of the terrestrial vertebrates.

FIGURE AI.11 Ray-finned fish (Actinopterygii) from the Green River Formation, western Wyoming. *(Courtesy U.S. National Museum.)*

They have internal nostrils, lungs, and paired bony limbs. They are dependent on being near water, for the eggs must hatch and early larval stages must live in water. Much "experimentation" in the ossification of the vertebrate skeleton occurred in this group in response to the body-weight problem engendered by life on land. (Devonian-Recent) (See Fig. 14.46.)

Class Reptilia: A large and varied group of vertebrates with an advanced, sturdy, bony skeleton. They evolved from the Amphibia and are independent of water in the sense that their amniote eggs can be laid on land. The eggs are protected by a shell which prevents drying out. Includes the dinosaurs, crocodiles, snakes, and lizards. (Pennsylvanian-Recent) (See Fig. 16.35.)

Subclass Anapsida: Includes the cotylosaurs, which are the earliest reptiles, as well as the turtles. (Pennsylvanian-Recent)

Subclass Synapsida: Includes the pelycosaurs and the Therapsida (mammal-like reptiles). (Permian-Jurassic) (See Figs. 14.47 and 15.17.)

Subclass Ichthyopterygia: The Mesozoic ichthyosaurs. These forms, which superficially resemble sharks or porpoises, are good examples of convergent organisms. (Triassic-Cretaceous) (See Fig. 16.40.)

Subclass Archosauria: Includes the thecodonts, phytosaurs, pterosaurs (Mesozoic flying reptiles), the plesiosaurs, and the dinosaurs. (Permian-Recent (See Figs. 16.34 and 16.40.)

Class Aves: The birds are very efficiently developed flying animals. The breastbone is enlarged for emplacement of large muscles that operate the anterior limbs, which have been modified into wings. Hollow bones reduce weight and feathers retain body heat and aid in flight. Examples include the dove and the hawk. (Jurassic-Recent) (Fig. 16.39)

Class Mammalia: This is a diverse group of vertebrates that underwent the most rapid and spectacular period of adaptive radiation of any animal class. Within a short span during early Cenozoic time mammals came to dominate the majority of vertebrate land habitats. They are warm-blooded, have mammary glands and hair as distinguishing features. Includes the rat, the skunk, and man. (Jurassic-Recent) (See Chap. 17 for mammalian classification and applicable illustrations.)

KINGDOM PLANTAE

Phylum Thallophyta: The most primitive group of plants, which includes such single-celled forms as bacteria and diatoms, and such multicelled types as fungi, lichens, and algae. The earliest-known living organisms, bacteria and calcareous algae, indicate the long and varied history of this important phylum. (Prepaleozoic-Recent) (See Figs. 10.25 and 10.26.)

Phylum Bryophyta: The bryophytes have developed special reproductive cycles and have differentiated tissues, which permit them to live on land. They include the mosses and liverworts, neither of which have left any significant fossil record. (Devonian-Recent)

Phylum Psilopsida: The most primitive and earliest appearing of the vascular phyla, which developed a method of raising water from the substrate to the upper portions of the plant to make possible a nonaquatic habitat. There were no true roots or leaves. (Silurian-Devonian) (See Fig. 14.40.)

Phylum Lycopsida: Scale trees represented today by a few genera such as the club mosses and the ground pine; they had true roots and leaves, and dominated the Paleozoic both in abundance and size, some reaching 100 feet in height. Most commonly found in Carboniferous coal forests; includes *Lepidodendron* and *Sigillaria*. (Devonian-Recent) (See Figs. 14.1, 14.41.)

Phylum Sphenopsida: (Also called arthrophytes.) A moderately common upper Paleozoic group characterized by vertical ribbing and jointed stems; *Calamites* of the late Paleozoic is most typical. The modern *Equisetum* or horsetail found along railroad tracks is the only living member and can be said to be a "living fossil." (Devonian-Recent) (See Fig. 14.41.)

Phylum Filicineae: The true ferns, which have a very well-developed vascular system and a unique spore-bearing reproductive cycle. (Devonian-Recent) (See Fig. 14.41.)

Phylum Gymnospermae: The most primitive seed-bearing plants. Members are quite different-appearing; includes forms that dominated the land during the late Paleozoic and early Mesozoic. The earliest order, the cycadofilicales, had fernlike leaves and are difficult to separate from the true ferns in the fossil record. The cycadofilicales became

FIGURE AI.13   A carbonized film and impression of a sweet gum leaf from the Miocene, Hidden Lake, Oregon. *(Courtesy U.S. National Museum.)*

FIGURE AI.12   A *Ginkgo* leaf from the Paleocene of Wyoming. *(Courtesy U.S. National Museum.)*

extinct in the Mesozoic. The group includes the important Permian Southern Hemisphere form *Glossopteris.* (Devonian-Mesozoic) (See Figs. 14.42 and 14.43.)

Other important orders include the cycadales (Mesozoic-Recent) and the cycadeoidales (Mesozoic). These look like palms in having a long, unbranched trunk with a crown of palm-like leaves.

The conifers are the most common gymnosperm of today. They have blade or needlelike leaves and possess cones with exposed seeds. They began in the Late Carboniferous, but it was not until the Jurassic that they reached their climax. Competition with the angiosperms is responsible for their subsequent decline. (late Paleozoic-Recent)

Another group is the order Ginkgoales (Mesozoic-Recent) which have peculiar fan-shaped leaves. This group is another example of a "living fossil" because they have existed virtually unchanged since the Permian. This resistance to change and tolerance of grossly unfavorable environment is reflected in the observation that the ginkgo happily survives along the streets of New York City. (Fig. AI.12)

Phylum Angiospermae (Magnoliophyta): The largest, most diverse and most successful phylum of plants, characterized by internal, covered seeds, and true flowers. The very rapid adaptive radiation is parallel to that of the Mammalia, which synchroneity is interest-ing. The phylum arose from the cycadeoids in the Jurassic, and by Cretaceous time dominated the world flora. Examples include all hardwood trees, the rose, and the Venus's-flytrap. (Cretaceous-Recent) (Fig. AI.13)

APPENDIX II

ENGLISH EQUIVALENTS OF METRIC MEASURES

Units of Length
1 millimeter (mm) = 0.1 centimeter (cm) = 0.039 inch (in.)
1,000 mm = 100 cm = 1 meter (m) = 39.37 in. = 3.28 feet (ft) = 1.09 yard (yd)
1,000 m = 1 kilometer (km) = 0.62 mile (mi)
1 in. = 2.54 cm
1 mi = 1.61 km

Units of Area
1 m² = 10.76 ft²
1 km² = 0.39 mi²
1 ft² = 0.09 m²
1 mi² = 2.58 km²

Units of Volume
1 m³ = 35.32 ft³
1 km³ = 0.24 mi³
1 ft³ = 0.03 m³
1 mi³ = 4.17 km³

Units of Weight
1 gram (g) = 0.001 kilogram (kg) = 0.0022 pound (lb)
1 kg = 1,000 g = 2.2 lb

Temperature Scales
1°F = 0.56°C
1°C = 1.8°F
0°C = 32°F (freezing point of water)
100°C = 212°F (boiling point of water)

INDEX